WHAT EVERY ENGINEER SHOULD KNOW ABOUT

STARTING A HIGH-TECH BUSINESS VENTURE

WHAT EVERY ENGINEER SHOULD KNOW
A Series

Series Editor*

Phillip A. Laplante

Pennsylvania State University

1. What Every Engineer Should Know About Patents, *William G. Konold, Bruce Tittel, Donald F. Frei, and David S. Stallard*
2. What Every Engineer Should Know About Product Liability, *James F. Thorpe and William H. Middendorf*
3. What Every Engineer Should Know About Microcomputers: Hardware/Software Design, A Step-by-Step Example, *William S. Bennett and Carl F. Evert, Jr.*
4. What Every Engineer Should Know About Economic Decision Analysis, *Dean S. Shupe*
5. What Every Engineer Should Know About Human Resources Management, *Desmond D. Martin and Richard L. Shell*
6. What Every Engineer Should Know About Manufacturing Cost Estimating, *Eric M. Malstrom*
7. What Every Engineer Should Know About Inventing, *William H. Middendorf*
8. What Every Engineer Should Know About Technology Transfer and Innovation, *Louis N. Mogavero and Robert S. Shane*
9. What Every Engineer Should Know About Project Management, *Arnold M. Ruskin and W. Eugene Estes*
10. What Every Engineer Should Know About Computer-Aided Design and Computer-Aided Manufacturing: The CAD/CAM Revolution, *John K. Krouse*
11. What Every Engineer Should Know About Robots, *Maurice I. Zeldman*
12. What Every Engineer Should Know About Microcomputer Systems Design and Debugging, *Bill Wray and Bill Crawford*
13. What Every Engineer Should Know About Engineering Information Resources, *Margaret T. Schenk and James K. Webster*
14. What Every Engineer Should Know About Microcomputer Program Design, *Keith R. Wehmeyer*
15. What Every Engineer Should Know About Computer Modeling and Simulation, *Don M. Ingels*

*Founding Series Editor: **William H. Middendorf**

16. What Every Engineer Should Know About Engineering Workstations, *Justin E. Harlow III*
17. What Every Engineer Should Know About Practical CAD/CAM Applications, *John Stark*
18. What Every Engineer Should Know About Threaded Fasteners: Materials and Design, *Alexander Blake*
19. What Every Engineer Should Know About Data Communications, *Carl Stephen Clifton*
20. What Every Engineer Should Know About Material and Component Failure, Failure Analysis, and Litigation, *Lawrence E. Murr*
21. What Every Engineer Should Know About Corrosion, *Philip Schweitzer*
22. What Every Engineer Should Know About Lasers, *D. C. Winburn*
23. What Every Engineer Should Know About Finite Element Analysis, *John R. Brauer*
24. What Every Engineer Should Know About Patents: Second Edition, *William G. Konold, Bruce Tittel, Donald F. Frei, and David S. Stallard*
25. What Every Engineer Should Know About Electronic Communications Systems, *L. R. McKay*
26. What Every Engineer Should Know About Quality Control, *Thomas Pyzdek*
27. What Every Engineer Should Know About Microcomputers: Hardware/Software Design, A Step-by-Step Example. Second Edition, Revised and Expanded, *William S. Bennett, Carl F. Evert, and Leslie C. Lander*
28. What Every Engineer Should Know About Ceramics, *Solomon Musikant*
29. What Every Engineer Should Know About Developing Plastics Products, *Bruce C. Wendle*
30. What Every Engineer Should Know About Reliability and Risk *Analysis, M. Modarres*
31. What Every Engineer Should Know About Finite Element Analysis: Second Edition, Revised and Expanded, *John R. Brauer*
32. What Every Engineer Should Know About Accounting and Finance, *Jae K. Shim and Norman Henteleff*
33. What Every Engineer Should Know About Project Management: Second Edition, Revised and Expanded, *Arnold M. Ruskin and W. Eugene Estes*
34. What Every Engineer Should Know About Concurrent Engineering, *Thomas A. Salomone*
35. What Every Engineer Should Know About Ethics, *Kenneth K. Humphreys*
36. What Every Engineer Should Know About Risk Engineering and Management, *John X. Wang and Marvin L. Roush*
37. What Every Engineer Should Know About Decision Making Under Uncertainty, *John X. Wang*

38. What Every Engineer Should Know About Computational Techniques of Finite Element Analysis, *Louis Komzsik*
39. What Every Engineer Should Know About Excel, *Jack P. Holman*
40. What Every Engineer Should Know About Software Engineering, *Phillip A. Laplante*
41. What Every Engineer Should Know About Developing Real-Time Embedded Products, *Kim R. Fowler*
42. What Every Engineer Should Know About Business Communication, *John X. Wang*
43. What Every Engineer Should Know About Career Management, *Mike Ficco*
44. What Every Engineer Should Know About Starting a High-Tech Business Venture, *Eric Koester*

WHAT EVERY ENGINEER SHOULD KNOW ABOUT

STARTING A HIGH-TECH BUSINESS VENTURE

Eric Koester

CRC Press is an imprint of the
Taylor & Francis Group, an **informa** business

CRC Press
Taylor & Francis Group
6000 Broken Sound Parkway NW, Suite 300
Boca Raton, FL 33487-2742

Printed in the United States of America on acid-free paper
10 9 8 7 6 5 4 3 2 1

International Standard Book Number-13: 978-1-4200-7697-4 (Softcover)

Library of Congress Cataloging-in-Publication Data

Koester, Eric.
What every engineer should know about starting a high-tech business venture / Eric Koester.
p. cm. -- (What every engineer should know ; 44)
Includes bibliographical references and index.
ISBN 978-1-4200-7697-4 (alk. paper)
1. Engineering firms--Management. 2. New business enterprises. I. Title.

TA190.K637 2009
620.0068--dc22 2008036297

Visit the Taylor & Francis Web site at
http://www.taylorandfrancis.com

and the CRC Press Web site at
http://www.crcpress.com

Contents

Preface

> To think like an engineer is to think of situations in their entire context, including the laws and regulations of society and the actions of all the people necessary for success.
>
> **Dr. John H. Marburger III**
> *Director of the Office of Science and Technology Policy, Executive Office of the President*

Innovation is the centerpiece of the global economy, and the seeds of that innovation are sown by individuals starting new businesses to tackle the world's challenges.

So if you've made it this far—welcome. Picking up this book to learn about starting a high-tech business venture is a good first step. This book has been written to assist both brand new entrepreneurs and those who have been around the block once or twice before. Hopefully, it will be a useful tool and something you turn to at numerous points in your startup life.

Now is a tremendous time for entrepreneurship. More people than ever are starting new businesses. Although there have been periodic slowdowns for new technology businesses, the long-term trend over the past 25 years has been increased availability of human and financial resources available to high-tech entrepreneurs.

This book is part of the Taylor & Francis/CRC Press series "What Every Engineer Should Know About." As a result, it is focused primarily at entrepreneurs and potential entrepreneurs coming from a technical background such as engineering, medicine, science, computers, biotechnology, and numerous other related "high-tech" fields. But don't fret; anyone should be able to use this book, with or without a high-tech background or experience. To aid those entrepreneurs without a strong business background or training, I have included plenty of business, finance, marketing, and related acumen.

Why are engineers, scientists, and technical individuals so successful as entrepreneurs? Some experts believe it is just because engineers approach problems from a different angle and find solutions. The website http://engineeringteacher.com describes how engineers think:

Engineers are problem-solvers.
Engineers use knowledge.
Engineers are creative.

Engineers solve problems by using knowledge and creativity, making them some of the best-positioned people to find a solution to a problem and hence why more and more engineers are becoming entrepreneurs . . . and succeeding at it.

As you read through this book, you will find it is chock-full of statistics, data, anecdotal evidence, tools, spreadsheets, questions and answers, and research to help guide an entrepreneur through the startup process. But I'd be kidding myself if I said it contained everything you'll need. The resources section at the end of the book has more books, articles, and websites, but I've packed more onto the book's website at http://www.myhightechstartup.com. Visit the website for updates, new articles, answers to questions that are submitted, links to download documents, or spreadsheets found in this book and much more. If that isn't enough, just shoot me an e-mail at eric@myhightechstartup.com, and I'll respond directly to you.

Hope you enjoy the book. It is designed to help answer some of the most common questions faced by a startup company, and I hope it makes your journey a bit smoother and slightly more successful. In any case, enjoy the ride and best of luck with the startup life!

Eric Koester

Acknowledgments

Researching and developing this book has been extremely fulfilling and valuable and certainly would not have come together without the involvement of countless people who provided their support, guidance, time, and energy.

Special thanks go to my wife Allison Koester for her direction, insights, and (sometimes) necessary prodding throughout the process of preparing this book. I also took advantage of her research capabilities as a CPA and a PhD candidate in Accountancy at the University of Washington to solicit her assistance with the chapters on accounting and taxation here in the book.

Thanks are due to several important colleagues at Cooley Godward Kronish LLP. Sonya Erickson and John Robertson each provided encouragement and guidance in this process. I also acknowledge the following individuals who assisted with specific sections of the book and thank them for their contributions to the final product, including research, review, and drafting: John Blake, Eric Wood, Pratin Vallabhaneni, Havila Unrein, and Alan Hambelton. Special thanks to each of you.

In addition, thanks are due to my assistant Lauree Lingenbrink who gave the manuscript a review and provided constructive feedback to keep the book valuable and interesting for a first-time entrepreneur. Heather Anichini, Kari Annand, Blake Ilstrup, and Gordon Empey each gave helpful thoughts and comments that have improved this book immeasurably.

The outstanding character drawings at the beginning of each chapter were done by Travis Fox, an artist from the Kansas City area.

I thank the entire team from the Taylor & Francis Group. First to Allison Shatkin, my acquiring editor, who helped me to create what I hope will be a nice addition to the WEESKA portfolio. Next to Jill Jurgensen, the Senior Project Coordinator on the book, who was a valued resource to move this manuscript to completion.

Finally, thank you to my friends and family (including my wife Allison) who put up with me missing numerous social engagements to get this to completion. I greatly value these friendships and each of your support.

How to Use This Book

Any tried-and-true entrepreneur will tell you that starting a new business involves challenges, countless learning experiences, reworked priorities, and, inevitably, much change. The aim of this book is to provide useful information for new entrepreneurs, as well as experienced entrepreneurs and managers of a growing high-technology startup company. To accomplish this goal, I've organized the book into four sections:

I. Becoming a High-Tech Entrepreneur
II. Starting It Up
III. Building a Startup Success
IV. Next Steps and Stages

Each of these sections groups important concepts together. In Part I, the book's focus is on the entrepreneur himself, including what to expect, what makes a successful entrepreneur, how ideas are generated and developed into a business concept, and a summary of some of the key challenges and lessons for any entrepreneur. Part II addresses the early challenges a new entrepreneur faces when forming a new business, including building the founding team, working with your attorney, departing your former employer, forming the business entity, and developing your business plan. Part III focuses on the five key challenge areas new businesses will face: funding, talent, technology, marketing and sales, and operations. Finally, Part IV is all about the next steps and stages after your business is up, running, and growing, including international expansion, mergers and acquisitions, and initial public offerings, as well as thinking about starting your next venture.

Should you just sit down and read this book straight through? Well, that is one approach, and the book is organized in such a way that it is a fairly logical progression from idea generation to business formation through funding, recruiting, product development and sales, and finally into expansion internationally or from sales or acquisition. Ultimately, most readers will probably find this book to be most useful if you are able to read important sections when the question pops up. Think of it as another resource in your corner.

With so much information in one book, the question really becomes, "How do I actually use this book?"

This book focuses on the five key stages in the life cycle of a high-technology startup company. In each stage, there are various priorities, challenges, and opportunities for the company. The stages are as follows:

1. Idea stage ("opportunity recognition")
2. Startup stage ("business formation")
3. Launch stage ("product release")
4. Growth stage ("ramp-up")
5. Expansion stage ("next steps")

Within each of these stages, this book identifies the five sets of challenges faced by a high-technology startup company. These challenges are as follows:

- Funding challenges ("How do we pay for this?")
- Talent challenges ("How should we recruit and retain the right people?")
- Technology challenges ("How can we develop and protect our innovation?")
- Marketing and sales challenges ("How do we get people to buy?")
- Operational challenges ("How to we build the company for success?")

At each different stage of the business, these challenges change and so does your focus. For a new company, operations, talent, and technology challenges may be the most pressing to the business. Yet for a company preparing to roll out its products nationwide, funding, sales, and marketing challenges could become the focus of the management team. Obviously, each company will need to prioritize certain challenges more than others.

What follows are tables for each of the five stages, from the idea stage to the expansion stage. For each stage, the book lays out the five primary challenges and identifies specific challenges, opportunities, and objectives that a business in that stage will be faced with. These changes may not exactly match the issues faced by your business but may be helpful as general guidance. As your business moves through the various stages, you can identify the key challenges and find out where in the book to learn more.

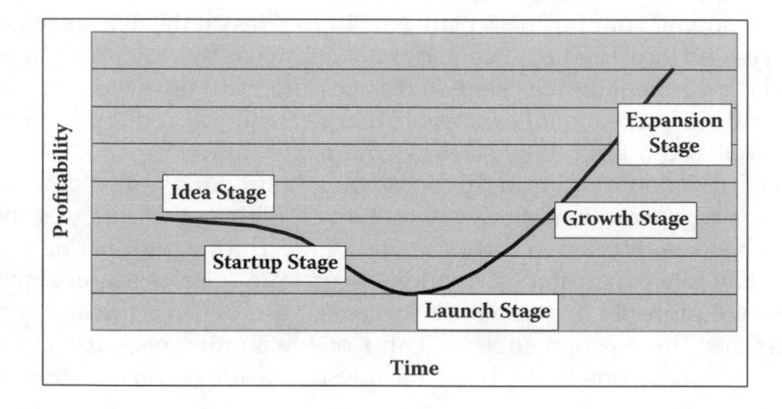

Just as a quick example, let's say your business is in the startup stage. You're deciding how to form your company, hire other founders, design your prototype, and raise some seed money. Obviously, all of these are crucial to your success, but what is the most important? Well, startup stage companies tend to need to focus on operations and technological challenges, because you are starting to build the business itself and lay out the technology, which will be what drives the business going forward. Because of this, your business will likely need to look into things such as how should you best structure your business (as a corporation or a limited liability company, for example) and how to best protect your intellectual property in the most cost-effective manner (using trade secrets and provisional patents, for instance). If those are issues facing your business, the tables throughout this section will try and point you to the place to look. Any entrepreneur will be faced with countless challenges, but if you can anticipate the most likely challenges you'll face and identify tangible solutions in advance, certain obstacles won't be able to stem your progress.

Obviously, an entrepreneur has many things to juggle on a daily basis, so the tables should help you to dive into the book and find information that can help you get results . . . and quickly. That's why it is up front. So read it cover-to-cover or jump to the place where you can find the information you need.

Idea Stage ("Opportunity Recognition")

You've identified a technological idea that appears to have a sizeable target market and fills an important need, but you may not have a great deal of business or startup experience. At this stage, your funding needs are fairly low, the team is either just you or perhaps a core group of founders, and the business is not yet operational.

Key challenges	Focus	Priority	Objectives	Useful book sections
Funding				
• Do we have the initial capital to start the business? • Is it feasible to raise the amount of funds necessary for product development and market penetration?	• Initial capital • Research	♦	• Prepare personal financial summary • Prepare monthly budget • Obtain initial capital • Researching "comparable" companies for information • Operate cheaply	**Chapter 3** • "Initial Capital" (p. 43) **Chapter 9** • "Seed Funding" (p. 212) • "Angel Investors" (p. 223) • "Government Funding and Public Sector Support" (p. 229) • "Bank Loans" (p. 226)
Talent				
• Can we recruit cofounders and management needed? • Does our team have the necessary business and entrepreneurial skill set?	• Networking • Training	♦♦♦♦	• Develop a contact list and solicit input, feedback, and search on the business and to find potential cofounders • Find a mentor • Use training resources (seminars, entrepreneurial organizations, SCORE, etc.)	**Chapter 1** • "What Are the Skills an Entrepreneur Will Need?" (p. 16) **Chapter 4** • "What Should You Be Looking for in Cofounders?" (p. 67)

Key challenges	Focus	Priority	Objectives	Useful book sections
Technology				
• Is our idea really unique?	• Intellectual property • Product development	♦♦♦	• Basic research on intellectual property for United States and key international markets • Develop a list of competitors • Document your progress on product development • Identify potential proprietary aspects of your technology	**Chapter 2** • "The Opportunity Recognition Process" (p. 24) • "Development of the Business Concept" (p. 27)
Marketing and Sales				
• Is the market large enough and growing sufficiently?	• Market competition	♦	• Identify your potential products • Identify your potential target markets • Research buying patterns • Identify trademarks and trade names	**Chapter 2** • "Researching the Market" (p. 30)
Operations				
• Can we build a business around this idea?	• Business planning	♦♦	• Prepare business concept summary • Develop financial projections • Identify staffing, equipment, and product development needs for the first six months of operations	**Chapter 2** • "Getting It in Writing: The Business Concept Summary" (p. 32) **Chapter 3** • "What to Expect in the First Months of Your New Startup" (p. 42) • "Risks and Challenges of the Startup Life" (p. 44) • "What Type of Business Are You Starting?: Startup versus Small Business" (p. 47)

Startup Stage ("Business Formation")

You are going forward with the business, so you have started to recruit cofounders and early employees, but your team is not yet complete. You are putting together a business plan and beginning early stages of product development. Funding needs have increased somewhat, and you are beginning to require founder investments to start the business and looking for additional funding sources from friends and family, angel investors, and government grant programs. During the startup stage, there are likely to be little to no revenues and only an informal organizational structure.

Key challenges	Focus	Priority	Objectives	Useful book sections
Funding				
• Where can we raise our initial and first-round capital?	• Networking with potential investors • Development of financial models	♦♦	• Obtain capital necessary for product development • Build contact list for funding sources • Attend local events with members of the startup community	**Chapter 8** • "Financing and Liquidity Plan" (p. 180) **Chapter 9** • "Seed Funding" (p. 212) • "Angel Funding" (p. 214) • "Government Funding and Public Sector Support" (p. 229) • "Bank Loans" (p. 226)
Talent				
• How do we recruit talent for our organization? • How do we structure our equity ownership and compensation of founders and employees?	• Recruitment and retention	♦♦♦♦	• Enter into agreements among founders • Create a capitalization structure • Create a hiring plan • Retain an attorney • Approve an option or stock plan • Create partnerships with third-party recruitment companies	**Chapter 4** • "Agreements among the Founders" (p. 72) **Chapter 5** • "When Should You Find Yourself a Good Lawyer?" (p. 97) • How to Select Your Attorney?" (p. 100) **Chapter 8** • "Management" (p. 188) **Chapter 11** • "Practical Considerations of Hiring an Employee" (p. 321)

Key challenges	Focus	Priority	Objectives	Useful book sections
Technology				
• How does the technology become our product?	• Intellectual property • Product research	♦	• Create an intellectual property strategy • Hire intellectual property counsel	**Chapter 5** • "Specifics for Intellectual Property and Patent Counsel" (p. 113) **Chapter 13** • "Understanding Intellectual Property" (p. 386)
Marketing and Sales				
• How do we market the business to potential investors and employees?	• Business planning	♦♦	• Develop an internal business plan • Develop an external business plan, including solicitation tools	**Chapter 8** • "Business Planning" (p. 160) • "Building a Business Plan" (p. 168)
Operations				
• How do we structure the new business?	• Business formation • Departing from previous employers	♦♦♦♦	• Select the state and type of entity to form • Obtain employer identification number • Qualify to do business in other states • Create policies for following formalities of the corporation • Implement record-keeping policies	**Chapter 6** • "Best Practices When Departing Your Employer (or When Hiring from a Competitor)" (p. 119) **Chapter 7** • "An Inc., LLC, S-Corporation, or Something Else? Which Entity Do I Choose?" (p. 135) • "Incorporating the Business" (p. 139) • "Formalities of the Corporation" (p. 149)

Launch Stage ("Product Release")

Your team has begun to grow and you are starting to create a simple organizational structure to manage the team and its operations. You have developed a prototype or beta product, and additional refinements and research and development are ongoing for a full-scale launch. Fundraising is an important priority for the business because you need funds to rapidly grow the business. You may have limited product revenues, but expenses have increased as you have begun to recruit sales and marketing staff.

Key challenges	Focus	Priority	Objectives	Useful book sections
Funding				
• What amount of funds do we need to raise to launch our product? • Where can we raise the funds necessary?	• Fundraising to launch the product • Fundraising for future product development	♦♦	• Create fundraising tools • Meet with potential fundraising sources • Identify alternative sources of funds (loans, bootstrapping, grants)	**Chapter 8** • "Building a Business Plan" (p. 168) **Chapter 9** • "Funding Sources for Your Business" (p. 218) **Chapter 10** • "Is Venture Capital Right for Your Company?" (p. 238) • "Attracting Venture Capital" (p. 252)
Talent				
• How do we hire additional business staffing? • How do we recruit members for our advisory boards?	• Recruitment and retention • Board of advisors	♦♦	• Develop documentation and policies for new employees • Monitor hiring goals and retention efforts • Target high value advisors for your board of directors and advisory boards	**Chapter 11** • "The Hiring Process" (p. 316) • "Employee Costs" (p. 327) • "Equity Compensation" (p. 335) **Chapter 12** • "Recruiting Directors" (p. 365) • "Advisory Boards" (p. 378)
Technology				
• How do we retain proprietary rights over our technology?	• Product development • Formal intellectual property protections	♦♦♦♦	• File for applicable intellectual property protections • Implement a trade secret policy • Develop your product or products	**Chapter 13** • "Managing Your Intellectual Property as a High-Tech Startup" (p. 394) **Chapter 14** • "Startup Product Development" (p. 413)

Key challenges	Focus	Priority	Objectives	Useful book sections
Marketing and Sales				
• What is our target market and how do we reach them?	• Target customers	♦♦♦	• Identify the target market • Create a public relations strategy • Allocate sales between internal and external sales resources • Identify competitor responses to marketing strategies	**Chapter 15** • "Marketing Basics" (p. 422)
Operations				
• How do we handle logistical issues? • How do we make the business more efficient? • How do we manage our accounting needs?	• Office space • Record-keeping • Reporting	♦	• Retain proper space for your business • Hire internal and external assistance for accounting, finance, insurance, staffing, and tax needs • Develop regular results reporting system and schedule • Update internal business plan	**Chapter 17** • "Choosing an Office Location" (p. 444) • "Leases" (p. 445) • "Insurance for the Business" (p. 453) **Chapter 19** • "Basics of Bookkeeping" (p. 474)

Growth Stage ("Ramp-Up")

You have likely secured necessary first-round funding to grow the organization. As a result, your team has grown and perhaps even expanded beyond a single location. This growth requires the company to have a more formal management structure. With the addition of outside investors, your board of directors now plays a more vital role within the organization. Your technology is likely to be commercial, and development efforts are underway on new applications or products. Sales and marketing efforts have become very vital as the company begins to target new customers in different geographies, sectors, industries, or positions within the prospects. Although you have secured first-round funding, for the business to expand, you have already begun looking for additional funds for expansion of your sales, marketing, and engineering resources.

Key challenges	Focus	Priority	Objectives	Useful book sections
Funding				
• How to finalize investments from venture capital funds or other institutional investors? • How will you raise funds for expansion of the business?	• Securing new funds • Future fundraising to expand the business • Efficient use of proceeds	♦♦♦♦	• Obtain financing from outside investors • Investigating joint ventures and strategic alliances • Educating shareholders of impacts of additional financings	**Chapter 3** • "Dilution: What Am I Left with at the End of the Day?" (p. 48) **Chapter 9** • "Joint Ventures and Strategic Alliances" (p. 230) **Chapter 10** • "Understanding the VC Term Sheet" (p. 272) • "Deal Documents in a VC Financing" (p. 296)
Talent				
• How do we transition our hiring to a more formalized organization? • How do we recruit members for our board of directors and formalize board operations?	• Recruitment and retention • Management • Board of directors	♦	• Create a formal management structure • Revise and update hiring plans • Implement a retention and reward bonus program • Formalize management and board operations	**Chapter 6** • "Best Practices When Departing Your Employer (or When Hiring from a Competitor)" (p. 119) **Chapter 11** • "Practical Considerations of Hiring an Employee" (p. 321) **Chapter 12** • "Recruiting Directors" (p. 365) • "Working with the Board of Directors" (p. 372)
Technology				
• How do we enhance our product and reach new markets? • How do we retain proprietary rights over additional developments?	• Product enhancement and development • Licensing and intellectual property	♦♦	• Identify technology to license or markets to license to • File for applicable intellectual property protections in foreign markets • Use partner relationships • Use resources such as advisory board to further development efforts	**Chapter 13** • "Licensing of Intellectual Property" (p. 405) • "International Intellectual Property" (p. 408) **Chapter 14** • "Product Development Resources" (p. 415)

Key challenges	Focus	Priority	Objectives	Useful book sections
Marketing and Sales				
• How do we efficiently use our marketing and sales budgets? • What is the sale process for our products?	• Efficient use of marketing spending • Selling Process	♦♦♦	• Identify the target market • Create a public relations strategy • Allocate sales between internal and external sales resources • Identify a sales structure and growth program • Create sales forecasts	**Chapter 15** • "Startup Marketing" (p. 429) **Chapter 16** • "Sales" (p. 434)
Operations				
• How do we manage our key contracts? • How do we manage our tax needs?	• Securities issuances • Contract management • Tax planning	♦♦♦	• Develop contract management system; identify key contracts for review from your attorney • Meet with tax professionals to identify tax minimization strategies • Transition to long-range strategic planning model • Create procedures for securities and option issuances	**Chapter 5** • "Working with Corporate Attorneys" (p. 105) **Chapter 18** • "Basics of Contracts" (p. 458) • "Typical Contract Provisions and Clauses" (p. 462) • "Electronic Contracts & E-Commerce" (p. 465) **Chapter 19** • "External Uses for Financial Statements" (p. 477) **Chapter 20** • "Key Tax Considerations for a Startup Company" (p. 482) **Chapter 21** • "Key Considerations for Issuing Securities" (p. 506)

Expansion Stage ("Next Steps")

The company is focused on expansion into new markets, new products, and new geographies. At this point, your products are likely producing revenues and you are exploring ways to increase efficiency and expand the sales of those products. However, you are also exploring new products and diversifying your revenue stream. The business is likely to have grown to a size where there are now multiple levels of management. You may be exploring possible joint ventures, acquisitions, or opening offices abroad. In addition, to fund these expansion plans, you may begin discussing liquidity strategies with investment banks and larger institutional investors.

Key challenges	Focus	Priority	Objectives	Useful book sections
Funding				
• Should the business consider an initial public offering? • What is our exit strategy?	• Merger and acquisition activity • Initial public offerings	♦♦	• Maintain an updated corporate and business recordkeeping system • Research underwriters and retain appropriate third-party advisors • Identify market trends in fundraising methodologies	**Chapter 23** • "Pursuing an IPO or Mergers and Acquisitions" (p. 548) • "Preparing the Company for a Potential Transaction" (p. 552)
Talent				
• How do we comply with rules and regulations regarding employees? • How do we expand our hiring globally?	• Formalization of human resources function • International hiring	♦♦	• Develop standardized documentation for employment matters • Create protocols for departing employees to manage trade secrets and intellectual property matters • Research and select appropriate markets for expansion • Identify international partners to assist with growth efforts	**Chapter 11** • "Labor and Employment Laws" (p. 349) • "Departing Employees" (p. 354) **Chapter 22** • "Direct Overseas Activity: How Big a Footprint" (p. 528) • "International Issues with Intellectual Property" (p. 531)
Technology				
• How do we develop next generation products and new products? • How do change our products to sell them into international markets?	• New product innovation • Gen II product development • International product modifications	♦♦♦	• Expand product portfolio and target markets for products • Research key partnerships for product innovation • Identify product modifications necessary for international expansion	**Chapter 22** • "Expanding into Global Markets" (p. 522) • "International Issues with Intellectual Property" (p. 531)

Key challenges	Focus	Priority	Objectives	Useful book sections
Marketing and Sales				
• How do we reach customers who are not "early adopters" or customers wary of startups? • How do we sell into international geographies?	• Sales growth • International sales growth	♦♦♦♦	• Modify sales structure to match target markets • Expand and retool your marketing strategy • Research necessary sales size and organization to reach growth targets • Explore alternative structures to sales such as franchising, licensing and direct expansion • Identify the appropriate organizational structure to match international sales efforts	**Chapter 16** • "Sales" (p. 434) **Chapter 22** • "Selling Your Products Abroad" (p. 523)
Operations				
• How do we expand our operations globally? • How do we grow by acquisition or sell aspects of our business to others?	• Global expansion • Merger and acquisitions	♦♦	• Select international counsel for selected markets • International tax planning • Open offices or identify partners in foreign markets • Develop a strategic plan to identify where and how expansion or contraction by merger and acquisition activity is appropriate	**Chapter 22** • "Tax Planning" (p. 527) • "Special Concerns Relating to Overseas Subsidiaries" (p. 530) • "Factors to Consider When Choosing Overseas Counsel" (p. 534) **Chapter 23** • "Mergers and Acquisitions Activity: Mergers, Assets Acquisitions, and Stock Acquisitions" (p. 553)

Author

Eric A. Koester is a business attorney with Cooley Godward Kronish LLP and has a practice focused on emerging technology companies, venture capital firms, and investment banks, with particular emphases on venture capital and bank financings, corporate partnerships, commercial agreements, intellectual property licensing, public offerings, and mergers and acquisitions. He has started two businesses focused on environmental and technology consulting and website design and has a passion for working with entrepreneurs and technology-focused thinkers. Eric is a graduate of the business school at Marquette University, the Environmental Leadership program at the Brexgata University in Brussels, Belgium, and a graduate of The George Washington University School of Law. Eric is also a certified public accountant. Eric's wife Allison is a PhD candidate in the University of Washington's Foster School of Business. They have two pugs and enjoy travel, sports, and outdoor activities in their free time. Visit Eric's LinkedIn page at http://www.linkedin.com/in/erickoester or his startup blog at http://www.myhightechstartup.com for more information.

Part I

Becoming a High-Tech Entrepreneur

So I think relatively early on I probably was on a path to be more of an entrepreneur, and I think everybody in my family kind of sensed that.

Steve Case
CEO and Founder of America Online

1

The Startup Life

The entrepreneur is our visionary, the creator in each of us. We're born with that quality and it defines our lives as we respond to what we see, hear, feel, and experience. It is developed, nurtured, and given space to flourish or is squelched, thwarted, without air or stimulation, and dies.

Michael Gerber
Entrepreneur and author of E-Myth Mastery

What to Watch For

Whether you are a recent (or soon-to-be) graduate from a college, masters, or doctoral degree program considering starting a new business or an established professional pondering a transition into a new startup career, the numbers of individuals looking to form a new business continue to climb annually.

Technology has made it possible to run a virtual business from your home or with cofounders around the globe. Private investments into private companies by "angel investors" and venture capitalists (VCs) have reached nearly $45 billion annually in the United States alone. The federal government is in the act as well, with small firms winning $77 billion, or 22.8%, of a total of $340 billion in federal government contracts eligible for small business competition in 2006, and the value of awards to small firms through the Small Business Innovation Research (SBIR) program reached an estimated $1.9 billion in 2006. With hundreds of universities creating undergraduate and graduate programs in entrepreneurship and private and public agencies offering training programs for startups and small businesses, more trained professionals begin their new ventures with key skill sets.

Yet starting a new business is not without its share of risks. Data from the U.S. Bureau of Labor Statistics shows that approximately 35% of new establishments fail in their first two years, and 56% fail in their first four years. New businesses fail for a variety of reasons, from financing problems to inability to attract and retain talent to general market conditions.

Those businesses that wind up succeeding require hard work and determination from the founders and management team, a unique and compelling vision, and, frankly, quite a bit of luck. However, by learning from previous successes (and failures), you can build a business with the potential for success. You've started in the right place; this book is written for you and designed to be a resource you can use from day one until the day you move on from your startup venture.

Strap yourself in for a wild ride ahead in the startup life.

Why Startups?

As recently as the 1980s, only 1% to 2% of graduating MBA students had the goal to become an entrepreneur at graduation, according to research by the University of California at Berkeley. Today, these numbers have jumped to between 10% and 20% of graduating MBAs, a 10-fold increase over that period. According to the Kauffman Foundation, entrepreneurial activity was at a higher level in the mid- and late-2000s than it was even during the Internet boom of the late-1990s. The Kauffman Foundation estimates that more than 437,000 people are actively creating new businesses every month over the first decade of the 2000s.

So the obvious question to ask is: Why? Why would individuals be willing to take the substantial personal and financial risks of starting their own business?

That's a tough question, but there seems to be a fairly consistent reason given by new entrepreneurs. Most entrepreneurs decide to start a new business to find a greater sense of personal satisfaction in their work. Basically, entrepreneurs start businesses to try to do what they love. Study after study shows that entrepreneurs are happier than their peers who have traditional employment, and that these entrepreneurs wouldn't give up their independence of a company founder for additional pay.

ACCORDING TO ENTREPRENEURS

Would you give up your small business to earn more?

No — 61%
Yes — 39%

Source: Discover Small Business Watch.

So, the entrepreneurial lifestyle continues to grow year after year. Today, forming a business has become much easier than ever before, and a huge customer base is now available that wasn't just 15 years before, with more consumers willing to buy products via the Internet or other new distribution methods, even if it means buying products from a startup business. Want to open a store in a weekend? Just try one of the dozens of "storefront" Internet portals, post your products online, do a bit of virtual advertising or use Google's AdWords, collect your payments via Paypal, and —*voilà*—you've got yourself a brand new business. With the increased availability of information on the Internet and in informational sources around the world, starting a business is something nearly anyone can do (but building a *successful* business, well, that is a bit more of a challenge, as you'll soon find out!).

Professor Elizabeth Gatewood from Indiana University studied individuals considering whether or not to form a new business. In her personal interviews, Gatewood identified the top reasons why these individuals desired to start their new business. The results were varied, from identifying an unmet need for the business in the market to a desire for autonomy or independence to the aim of making additional money:

- Identification of a market need — 29%
- Autonomy and independence — 18%
- Desire to make more money — 18%
- Desire to use knowledge and experience — 16%
- Enjoyment of self-employment — 7%
- Desire to show that it could be done — 5%

Your upbringing may also affect your reasons for starting a new business. Research has concluded that individuals in different countries often had different reasons for starting a business than citizens of another country. For instance, in the United States and Denmark, researchers found that entrepreneurs surveyed listed the "need for independence" as their most important reason for forming a new business; in China and Portugal, entrepreneurs listed the "need for approval" as the most important factor; whereas entrepreneurs in Sweden, Norway, and Denmark actually listed "wealth building" as their least important reason for forming their businesses.

ACCORDING TO ENTREPRENEURS

Why did you start your own business?

The 2006 QuickBooks Small Business Survey by Intuit asked 1,300 small business owners why they decided to start their own businesses.

- Desired to be my own boss — 40%
- To turn my passion into a business — 23%
- Needed a job to supplement my income — 10%
- Desired more flexible hours — 10%
- Saw the need for the business in the marketplace — 9%
- Wanted to spend more time with family — 4%
- Wanted better pay — 3%

No matter what the exact reason for the desire to form a business, more individuals are considering starting their own business and being their own boss, and, in many cases, being your own boss might be the best decision for a budding entrepreneur.

Successes of High-Tech Startups

According to the National Venture Capital Association, U.S. companies that received venture capital between 1970 and 2005 accounted for approximately 10 million jobs and $2.1 trillion in revenue in 2005 and accounted for nearly 17% of the gross domestic product, additional evidence of the key role technology development plays in economic development. One of the reasons more and more individuals have considered forming high-technology startup companies is the success in the past 20 years of startup companies.

WHO'S WHO OF SUCCESSFUL STARTUPS

Successful startups are a verifiable who's who of our daily consumer life.

- We buy our books from **Amazon.com**, our iPods from **Apple Computers**, our computers from **Compaq** with **Intel** chips inside, and drugs developed by **Amgen** and **Genentech**.
- We search online with **Google** or **Yahoo!**.
- We send our packages with **FedEx**.
- We use software from **Microsoft** and **Oracle**.
- We buy office supplies and home building items from **Staples** and **Home Depot**.
- We get our coffee from **Starbucks**.
- We eat at **Outback Steakhouse**.
- We listen to concerts at the **House of Blues**.
- We buy groceries from **Whole Foods**.

What does each of these companies listed above have in common? Well, each was funded by venture capital—a fairly common mechanism for certain fast-growing startups—and each of these companies began as a unique or novel idea that eventually became the well-known company that, in many cases, didn't even exist as recently as 15 years ago.

For that big thinker inside of most entrepreneurs, these success stories serve as an example of what can happen if the right entrepreneur builds the right product and grows that business into a market leader. In these cases discussed above, the businesses' founders were handsomely rewarded for their efforts (just ask Bill Gates and Paul Allen of Microsoft fame, who seem to be doing fairly well for themselves these days . . .). These success stories (as well as the countless others) are just another reason some entrepreneurs decide to build a business. The risks may be high, but the rewards can be handsome.

Startup companies play an important role in U.S. and global business and the development of innovative technologies used in everyday life. According to James Utterback of Massachusetts Institute of Technology (MIT) Sloan School of Management, innovation in today's business environment has been driven by emerging companies and not established firms. In Utterback's research, in the past 100 years, none of the radical innovations that have been developed were pioneered by a market leader. In fact, in many cases, the market leader may have developed or been researching the radical innovation but ultimately decided not to pioneer the invention because of concerns it could cannibalize the sales of its own existing products or services. Look no further than Xerox, which was the initial developer of the software platform that served as the inspiration for and would eventually become the Windows platform of Microsoft. In fact, Xerox gave the technology to Bill Gates and his team when management failed to recognize the market potential of the Xerox team innovations.

In addition, other research suggests that technology startup companies fill the need for radical innovation in economy. University of Melbourne researcher Joshua Gans notes that there may be efficiency gains as a result of the division of labor between the upstream innovation by startup companies and the downstream commercialization by established firms. Today, market-leading companies from Microsoft to Cisco to Amgen to Medtronic include the acquisition of emerging companies as growth and market-share acquisition strategies, taking advantage of market-driven technology development by startups.

LIKELIHOOD OF A STARTUP SUCCESS

The thought of building the next Google or establishing the next Genentech is obviously very attractive to a budding entrepreneur, but the likelihood of success for an entrepreneur is far from certain. In a 1998 article, author Bob Zider laid out the fragile environment for startups. According to Zider, if a startup company had an 80% probability of success in each of the eight key factors set forth in the study, the combined probability of success was only 17%, and, if just one of the factors for a company dropped to 50%, then the combined probability of success was only 10%. These eight factors are set forth below.

Individual event probability of success:

- Management is capable and focused — 80%
- Product development goes as planned — 80%
- Production and components sourcing goes as planned — 80%

- Competitors behave as expected — 80%
- Customers want the product — 80%
- Pricing is forecasted correctly — 80%
- Patents are issued and are enforceable — 80%
- Company has sufficient capital/resources — 80%

Combined probability of success — 17%

Source: "How Venture Capital Works" by Bob Zider, *Harvard Business Review*, November–December 1998.

An Entrepreneurial Lifestyle

Successful entrepreneurs often point to a strong work ethic as one of the most important traits that lead to success in a new business venture. Surveys of entrepreneurs have identified that entrepreneurs tend to work more than peers working in the corporate setting. What are a few of the personal health habits of entrepreneurs that you should watch out for?

ACCORDING TO ENTREPRENEURS

How much do you work?

- 28% of business owners surveyed work at least six days a week.
- 52% took off seven days or fewer for the year (*only* 36% of the overall population took off seven or fewer days).
- 36% of the small-business owners surveyed described a day off as "not working at all."

Source: Discover Small Business Watch.

ACCORDING TO ENTREPRENEURS

Healthcare

Do you provide healthcare to your employees?
No — Approximately 75%
Yes — Approximately 25%

If you offer healthcare, have you considered curtailing coverage for yourself or your employees because of costs?
No — Approximately 33%
Yes — Approximately 66%

Source: Discover Small Business Watch.

ACCORDING TO ENTREPRENEURS

Peter Kooiman asked "Entrepreneur's Hot 100" business owners about their "health" habits. This is what he found:

Smoker

Yes — 21.1%
No, I quit — 31.6%
No, I never have — 47.4%

Work-life balance

Average hours per week you work — 59.39
Average hours per week you relax — 16.93

Body weight

Overweight — 13.9%
Average weight — 83.3%
Underweight — 2.8%

How often do you exercise?

Regularly — 52.6%
Some of the time — 42.1%
I don't — 5.3%

Do you suffer from

High blood pressure — 10.5%
Headaches — 10.5%
Ulcers — 2.67%
Backaches — 13.2%
Stress — 34.2%

Source: http://www.entrepreneur.com, November 2000.

YOUR (OTHER) SIGNIFICANT OTHER

What could happen

You are considering starting a new business, but your spouse is highly skeptical of your plans.

What to expect

Starting a new business is a stressful endeavor; add to that a skeptical or even unsupportive spouse or family and the process could move from difficult to nearly impossible. Although this may seem obvious to some, starting your own business isn't simply like changing jobs. Being your own boss involves greater risk and uncertainty and may take some time before you regain the financial stability you find at a more established business. If possible, you should discuss the expectations of a new business venture with your spouse early in the process and hopefully try to obtain an agreement with respect to the amount of jointly held funds you are both willing to use in the endeavor. In addition, you and your spouse should discuss the increased work hours involved in forming a new business venture, the limits of vacation, and the fluctuations in the ability to draw a salary from the company. Some cities have organizations for spouses of entrepreneurs that can help provide more information and a support structure for the nonparticipating spouse.

TIP: Don't forget to include your spouse in your decision-making process to pursue a new business venture.

Do You Have What It Takes to Be an Entrepreneur?

There are a number of successful entrepreneurs who have become celebrities as a result of their successes and innovations. From Bill Gates and Warren Buffet to Richard Branson and Steve Jobs to Oprah Winfrey and the Google guys to Sean "P. Diddy" Combs and Mark Zuckerberg, the "Facebook" founder, to Martha Stewart and Ray Kroc, the wealth and acclaim that society has bestowed on successes in entrepreneurial endeavors is substantial. For this reason, numerous books and articles have been written to try to identify what separates successful entrepreneurs from the rest. For some readers, what you'll quickly discover is that, for many of these successful entrepreneurs, their success came from a combination of hard work, some business savvy, big dreams, and a large portion of luck (or being in the right place at the right time). The truth is, many of these very successful businessmen and businesswomen will tell you that they never expected to hit it big (or at least *that* big)—they were unsure at the start and were fortunate to work with good people who provided them help and insights along the way.

So why couldn't it be you too? Well, it could. Your idea may be as crazy as a graphical operating system for computers once seemed (unfortunately, Xerox's mistake on that one turned out to be Bill Gates and Microsoft's gain), or as unthinkable as a digital music device that could hold thousands of songs, videos, and more (that's right, the iPod). So a bit of luck sprinkled with the right mixture of team, technology, and market potential and you could have a market leader or a solid business.

But before you decide to start a business venture, you too probably wonder whether you have what it takes to be a successful entrepreneur like Sean Combs or Steve Jobs. Were Bill Gates and Oprah Winfrey destined to be successful entrepreneurs as a result of genetics and their upbringings? Or did the educational and experience backgrounds of Warren Buffett and Richard Branson set them onto their respective paths?

Alan Jacobowitz interviewed a large number of entrepreneurs in his research and concluded that nearly all entrepreneurs share certain personality traits, including the following:

- Restlessness
- Independence
- Certain level of isolation
- Very high levels of self confidence

Researchers such as George Solomon have also found that successful entrepreneurs tend to be focused on action over introspection, were inventive and innovative, and operated best when they were in charge. Norris Kreuger found that entrepreneurs were actually highly influenced by their environment, suggesting that entrepreneurial characteristics can be learned and heightened through education and experience.

A very common question is whether an entrepreneurial personality can be taught or whether an entrepreneur is just born with the right combination of inherent entrepreneurial traits. Ultimately, the experts are mixed, citing evidence that certain personality traits tend to be found in entrepreneurs but that many other traits are learned or taught.

What this information ultimately means is left up to you, but it certainly seems to suggest that there are certain skills found in most entrepreneurs, and, whether you are born with them or learn them, a prospective entrepreneur is bound to be more successful with these skills. So if you are looking to leave school directly into a startup venture or after a successful career working for a large corporation, the government, or in academia, focus on building the set of skills likely to be required from a new business founder.

Obviously, it is completely natural for a new entrepreneur or an individual considering a transition into a startup venture to have certain doubts and concerns. Is this the right move for you? Do you have the entrepreneurial spirit? Is your lifestyle and financial situation adaptable to the startup lifestyle? Do you have the right skill set? Can you work to build a successful organization? The answer to each of these questions is a very personal one and is an analysis that each entrepreneur should take before beginning a new venture.

ACCORDING TO ENTREPRENEURS

What traits are important in successful entrepreneurs?

The 2006 QuickBooks Small Business Survey by Intuit asked 1,300 small business owners what they felt was the most important characteristic to be a successful entrepreneur. The respondents identified the following traits as the top three most important:

Hard work — 37%
Visionary — 19%
Good people skills — 18%

ACCORDING TO ENTREPRENEURS

What were you like as a kid?

A Small Business Survey asked 1,300 small business owners what they were like as a kid. The respondents identified themselves as follows:

- Loner — 43%
- Nerd — 25%
- Class clown — 20%
- Jock — 11%
- Bully — 1%

The takeaway from this survey? Looks like self-described bullies don't grow up to be entrepreneurs!

Source: 2006 QuickBooks Small Business Survey by Intuit.

Ask yourself these questions to see whether you have some of the right personality traits that are typically seen in startup founders.

- Do you like unpredictable situations?
- Can you be impulsive?
- Can you make important decisions without all the information?
- Are you a hard worker?
- Are you comfortable without lots of free time?
- Do you enjoy meeting people and building up a network of contacts and associates?
- Do you like to try new things?
- Are you able to take criticism, learn from it, and then quickly brush it off?
- Are you comfortable being uncomfortable?
- Do you listen to others when you make decisions?
- Do you like to build up a team and lead it?
- Do you enjoy forging your own path when rules, records, and protocol aren't yet written?
- Do you mind unpredictability?
- Is your life and your family flexible when things come up?
- Do you like the challenge of difficult problems with complex solutions?
- Do you enjoy working as a part of a team?
- Do you read books, articles, and websites to learn everything you can about a new challenge?

If you answered yes to some or most of these questions, you certainly will have a number of the requisite skills and personality traits that will be tested during your forays into the startup world. In particular, self-professed entrepreneurs have most often pointed to hard work, vision, and strong people skills as some of the most important traits in successful entrepreneurs.

Also remember, although you may not necessary be comfortable in everything listed above (obviously not everyone is a natural networker or thinks of themselves as impulsive or will be comfortable making important decisions without all the information), but the trick here is to be willing to recognize the areas that are currently in need of improvement and improve on them (through both personal growth and proper building of a team to compensate for each other's weaknesses).

ADDITIONAL RESOURCES

***What Color Is Your Parachute? A Practical Manual for Job-Hunters and Career-Changers* by Richard Nelson Bolles**

If you are still wondering whether starting your own business is for you, consider this book. The book identifies two types of job hunts: a traditional job hunt and a life-changing job hunt. If you are considering starting your own business, check out the book or visit http://www.jobhuntersbible.com for more information.

StrengthsFinder 2.0: A New and Upgraded Edition of the Online Test from Gallup's Now, Discover Your Strengths by Tom Rath

This book is based on the popular personality assessment test from Gallup used by individuals to identify their strengths and weaknesses and find careers that most closely match these talents. Check out the book or take the assessment online at http://www.strengthsfinder.com.

Being in Charge of a Startup

"Everybody makes business mistakes. I mean, I take the responsibility, and I did. I was the captain of the ship and I took that responsibility," said Alan Bond after his company, Bond Corporation, went bankrupt.

Author Henry Sedgwick in his article "Hacked by Darkworm" in which he examined key traits of successful entrepreneurs wrote that his colleague Ronald Merrill had once told him the following:

> Reluctance to take on the tough jobs, or disdain for menial tasks, is not becoming in an entrepreneur. When there's danger, you lead from in front. When there's unpleasantness, you lead from below. When I was running Reaction Design Corporation, the company had five employees—three chemists, an administrative assistant, and me, the exalted president. No janitor. Guess who got down on his knees to scrub the toilet?

Why do these previous quotes matter? Well, running a startup company represents a significant responsibility for the founder and the founding team and with that comes a host of new tasks and jobs that an entrepreneur may not have done previously. As soon as your new company moves from a one-person idea or concept into a business involving multiple people and livelihoods, you will begin to see the importance of accepting responsibility for your organization and all those stakeholders in it. If you read books on leadership, you'll often hear the experts and the subjects refer to the concepts of integrity and responsibility. For almost any leader, be it in governmental, civic, or business settings, these concepts cannot be understated, but also remember that from this responsibility that comes from starting your own business also comes substantial opportunity for you and those involved in your new business venture.

For many entrepreneurs, running a startup company will be a transition from a largely technical role as an engineer, a scientist, a designer, an accountant, a salesman, or similar role. As you move into "management" of your startup, remember that you may not be an expert at the role initially. By using various resources at your fingertips, many of the key skills needed to effectively build a company can be learned.

A FEW COMMON "FOUNDER TYPES"

In their book, James Swanson and Michael Baird paint a picture of several common types of founder "characters" they've seen. Here are a few of those individuals:

- **Smart, passionate executioners:** These founders are very focused on the customer, approach the business with passion and discipline, and are the most likely to build successful businesses.

- **The paranoid:** These individuals are very hesitant to disclose any information and will require their own mother to sign a nondisclosure agreement before talking about their ideas. Oftentimes, once you get inside to hear about the idea, you realize it isn't quite the revolution they've sold you on.
- **Entitled youngsters:** These founders "know people" who have made lots of money and seemingly done it without any effort. Obviously, for them, it won't be that hard. They admire the soon-to-come rewards (and planning how to spend it) when success comes their way.
- **Dreamers:** Also known as "idea people," they continue to spit out idea after idea and are convinced that "one of them will stick." There is no separation of good versus bad ideas, nor is there any talk of execution. They don't get distracted with implementation because it just prevents them from thinking about the "next big idea."
- **The "funding is the only problem" founders:** These founders are entirely focused on receiving venture capital. In fact, venture capital money becomes the endgame in and of itself. These founders are oftentimes more concerned about how they can get paid after raising the investment rather than building the business.
- **The "one pizza slice" founders:** The idea of these founders is too narrow, too limited, or too incremental to be a true business. Hence, they focus on their individual slice of the pizza rather than looking at selling the entire pizza. This could be a great "additional feature" of an existing product but probably can't be a stand-alone business.
- **The "never met a customer" founders:** This is the true technocrat of the entrepreneurial world. This founder believes it is all about the product, in which the product is so "technologically superior" or "revolutionary" that customers will rush to buy it. If I build it, they will come. . . .

Based on information from Swanson, James A. and Baird, Michael L. *Engineering Your Start-Up*. Professional Publications Inc. 2003.

Am I Too Young or Too Old to Start a New Business?

Business schools across the country and around the world now offer a full range of entrepreneurship courses and programs of instructions. Undergraduates from technical backgrounds and business-entrepreneurship backgrounds are now graduating with the goal of starting their own business and leveraging their training. Likewise, lifelong employees of major corporations or researchers from government or academia are leaving their professions at a later stage in life looking to leverage on their experience and expertise. With these young and older entrepreneurs looking to start their own ventures, the question must be asked: is there really room for these new entrants? Could you be too old or too young to build a high-tech startup?

A number of researchers and industry trend watchers have examined the respective ages of the founders of successful startups to find the best age to create the proper mix of entrepreneurial "magic." Nick Denton of Valleywag, a Silicon Valley blog, postured that 26 was the age that had produced the most spectacular recent successes (citing founders at the ages of 25–27 at the time they formed companies such as Yahoo!, Skype, Google, YouTube, and MySpace).

Looking at founders of European companies, VenturePedia data suggest that median and mean ages were 38 and 39.5 years, respectively, for founders of startups among their 600 company sample. That data further suggested that founders of information technology companies were likely to be approximately five years younger than founders of healthcare, biotechnology, services, and retail companies. Fred Wilson of Union Square Ventures postured on his blog that founders in their 30s represented the prime time for entrepreneurship because these "thirtysomethings" were looking for "what's next" in their careers and were armed with the right set of skills and experiences to find it.

THINK YOU ARE TOO YOUNG TO BE A SUCCESSFUL ENTREPRENEUR?

A quick look at a few examples of "young" entrepreneurs and their ages at the time they formed their companies:

- Michael Dell (Dell Computers, age 19)
- Mark Zuckerberg (Facebook, age 19)
- Bill Gates (Microsoft, age 20)
- Steve Jobs (Apple Computers, age 21)

Another famous example, Warren Buffett (Bill Gates' bridge partner and the chairman of Berkshire Hathaway) purchased his first stock at age 11, his first farm at the age of 14, and, at age 26, started the partnership that would go on to become Berkshire Hathaway.

But surely you can be too old to be a founder of a successful company? Wrong. There are a number of examples of founders of very successful companies who would buck the research that says thirtysomethings represent the "sweet spot" of entrepreneurship.

THINK YOU ARE TOO OLD TO BE A SUCCESSFUL ENTREPRENEUR?

A quick look at a few examples of "older" entrepreneurs and their ages at the time they formed their companies:

- Ray Kroc purchased and dramatically expanded (from just a single restaurant) McDonald's at age 53
- Jim Clark formed Shutterfly at the age of 55
- Mike Ramsay was 47 when he cofounded TiVo
- Ely Callaway formed Callaway Golf Company at the "'spry'" age of 63
- Harland Sanders was 63 when he opened the first Kentucky Fried Chicken restaurant in his chain

Is there a time that is just right to become a founder of a company? Sure. Each founder will need to find the right combination for him when he holds the "correct" combination of experience and education, with the right timing for the business venture. Research suggests that the most crucial factor is your network, who you know or who you can get to know. Whether you are 21 or 71, bringing a network (and then building that network further) of associates and contacts is crucial to building a successful startup company.

Even if you are "past your prime" in the world of startups, you can sell your "maturity" as the right mix of experience combined with the youthful exuberance of the others in the company. For Ely Callaway, perhaps his company would not have been a success if it had been started in the 1950s before a golf explosion began. If you are a relative youngster, perhaps you can surround yourself with "greybeards" (individuals heavy on experience) the way that founders such as the Google guys have. Remember that the key to success is not fitting the "right" persona or personality but aligning your skill set with the right opportunity.

IMMIGRANTS AND ENTREPRENEURSHIP

Twenty percent of all venture-backed public companies were founded by one or more immigrants to the United States.

Forty percent of high-technology, venture-backed public companies were founded by one or more immigrants to the United States.

Source: Stuart Anderson of the National Foundation for American Policy and Michaela Platzer of Content First LLC and commissioned by the National Venture Capital Association as part of its Maximizing America's Growth for the Nation's Entrepreneurs and Technologists (MAGNET) USA initiative.

What Are the Skills an Entrepreneur Will Need?

Whether you are already in the process of creating your new business venture or are planning for a time in the future when you'll finally be able to be your own boss, it is a helpful exercise to understand what skills are important for an entrepreneur. Although identifying these skills is valuable, it is also important that an entrepreneur not try to drastically change who they are and how they've achieved success to date. Remember that these skills are simply a list of skills that the entrepreneur may need to turn to when the situation demands them.

WHAT TRAITS ARE OFTEN FOUND IN SUCCESSFUL ENTREPRENEURS OR BUSINESS LEADERS?

Research has identified a number of personality traits of successful entrepreneurs from restlessness and independence to heightened self-confidence, innovative, and action orientations. Below is a list of other skills that have been identified in successful founders and leaders of startup businesses:

- Instills a vision (and reminds others when they forget)
- Goal-oriented to keep the business on track and on target

- Practical, focused on solving problems efficiently rather than being "right"
- Leads by example and isn't afraid to answer the phones, make the coffee, or call customers
- Effective brainstormer who gathers information to develop the best solutions
- A good listener who really hears what others say
- Strategic thinker focused on making decisions for tomorrow
- Decisive when the time for discussion has ended
- Acts with integrity to maintain the reputation of the organization
- Good communicator who knows when a mass message won't do or when information has to be provided
- Genuine curiosity about all aspects of the business and all people involved
- Risk-taker who knows that a startup won't succeed by choosing to always play it safe
- Stubborn when the business needs stability and support
- Resilient when negative news hits or criticism arises
- Responsible for the actions of the company who isn't afraid to admit fault when needed
- Objective view of the company and not afraid to actively examine the business from an outsider's perspective

But let's be quite clear: the skills mentioned in this chapter (or even a subset of these skills) alone won't be enough. As Steven Berlack said, "It isn't your skills that are keeping you from succeeding; it is your attitude."

Although you may not always have the right skills to tackle a problem and can't morph yourself into the ideal entrepreneur/leader when times get tough, you can affect success through the right attitude. For a first-time entrepreneur, much will be learned on the fly, so be prepared for the ride, ask lots of question along the way, and remember to get help where you need it. A willingness to adapt, learn from others, and rely on the startup team serves an entrepreneur much better than attempting to alter your personality to be just like Larry Ellison, Bill Gates, or Warren Buffett (although a little of each never hurts!).

Successful entrepreneurs are often identified as individuals who are capable at a broad range of skills. You'll quickly find that there are times when you'll need to be able to make a customer presentation, choose accounting software, and hire an engineer—all before you have lunch.

In the early stages of a business, the entrepreneur will need to quickly adapt along the three key development stages of the venture: moving your business from a simple concept to a formalized organizational and operational plan to implementation of that plan. For each of these key phases, you'll be responsible for tackling the role of each of the following:

- Dreamer
- Planner
- Implementer

Throughout the life of the startup, the entrepreneur will wear each of these hats concurrently, sometimes spending more time in his role as the dreamer, without much focus on the implementation, whereas at other times, the focus of the entrepreneur will be centered on carrying out the plans that have been put into place. A successful entrepreneur will understand when to assume the proper responsibility and to learn to seamlessly move from each role throughout a series of interactions.

The dreamer helps to identify the initial problem and to understand whether a business concept exists to solve this problem. The dreamer provides a big-picture view and will help refocus the efforts of the team on the purpose of their efforts. The dreamer provides passion for the effort and will oftentimes offer the link that will bind the goals of the startup together. The dreamer serves as a motivator, a coach, and a visionary.

Whereas the overarching vision is set by the dreamer, the planner is responsible for setting out specific, tangible, and time-focused goals for the organization. The specific goals must be owned by the key members of the team and will require input from the eventual owners to ensure that this can occur. The importance of the planner will keep the organization focused on the steps needed to reach the overarching goals and will help limit the amount the company strays during its early stages.

The company will require a concrete focus on attaining the goals set forth in the planning effort. Obviously, these plans represent an ongoing and changing obligation of the company, and, as the implementer, the entrepreneur will need to balance the need for flexibility with the demands of rigidity required to successfully meet customer and outsider needs. The implementation stage involves managing people to a high degree.

FIND YOUR ENTREPRENEURIAL PERSONALITY TYPE

Author Darrell Zahorsky identifies nine different personality types for entrepreneurs in his online column "Your Guide to Small Business Information." Zahorsky identifies these key personality types and things to watch for in each personality type.

Are you . . .

- an "**Advisor**" like John Nordstrom, the founder of Nordstroms?
- an "**Analyst**" like Gordon Moore, the founder of Intel?
- an "**Artist**" like Scott Adams, the creator of Dilbert, or Walt Disney of The Walt Disney Company?
- a "**Fireball**" like Malcolm Forbes, the publisher of *Forbes Magazine*, or Herb Kelleher, the founder of Southwest Airlines?
- a "**Hero**" like Jack Welch of General Electric or Lee Ioccoca of Chrysler?
- a "**Healer**" like Ben Cohen, the cofounder of Ben & Jerry's Ice Cream, or Larry Page and Sergey Brin of Google?
- an "**Improver**" like Ingvar Kamprad, founder of IKEA, or Anita Roddick, the founder of The Body Shop?
- a "**Superstar**" like Donald Trump of Trump Hotels & Casino Resorts or Larry Ellison of Oracle Corporation?
- a "**Visionary**" like Bill Gates of Microsoft or Steve Jobs of Apple Computers and Pixar Animation Studios?

You may find that you are parts of each of these personality types and will inevitably find that each of these personality types will serve you well for particular situations you may face in the days of running your business. For more information or to read more, visit http://sbinformation.about.com/cs/development/a/personality.htm.

Do I Need a Particular Educational Background to Be a Successful Entrepreneur?

Countless researchers have attempted to identify the link between educational backgrounds and successful entrepreneurs. The results of this research only prove that there is no single background that yields successful entrepreneurs. Research by Arnold Cooper of Purdue University suggests that technical entrepreneurs (those involved in high-tech fields) tend to be well educated, having at least a master's degree. Conversely, Robert Ronstadt of Babson College found that formal education was actually an obstacle to creativity, innovation, and other skills necessary for successful entrepreneurial activities.

Additional research by Cooper and Javier Gimeno-Gascon does show that entrepreneurs having business or engineering degrees were more likely to grow firms to greater size. This research also concluded, however, that the number of business courses taken by the company's founder was inversely related to the growth for larger startups. The lesson there is, skip that extra accounting minor and pick up a degree in chemical engineering!

The one fact that seems to be widely held by researchers is that skills needed for entrepreneurs can be taught, and, given the growing number of collegiate and university programs, nonprofit trainings, and private-industry coaching programs offering skills training for entrepreneurs, there are numerous opportunities to hone important skills that will likely be used in new venture creation.

DOES PREVIOUS STARTUP SUCCESS MATTER?

If you were at a venture-funded company that went public, does that increase your chances that your next startup will go public? What about if your previous company didn't go public? Or what if you are a first time entrepreneur? Research suggests that experience does matter, but maybe not as much as you might think and certainly is no guarantee of success.

According to a study of nearly 10,000 entrepreneurs who had started venture-funded companies, repeat entrepreneurs were only moderately more likely to see their company go public the second time around. The chances of success (taking a company public) in their second venture-backed company were as follows:

- Entrepreneurs who had previously started a company that went public — 30%
- Entrepreneurs who failed to take their previous venture-backed company public — 20%
- First-time entrepreneurs — 18%

Source: Paul Gompers, Harvard University, National Bureau of Economic Research.

What If I Don't Have a Particularly Strong Technical Background?

Entering into a high-tech business seems like it may require a high-tech background or training. Not so says numerous studies that have looked into the backgrounds of successful entrepreneurs.

EXECUTING ON YOUR BUSINESS

What could happen

You've got a great technical idea but limited business skills.

Watch out for

Technical excellence won't be enough in a new business venture. Make sure that, as a new business founder, you are prepared to focus on a full gamut of areas, including sales, marketing, accounting, law, human resources, and market forces. If you are weak in any of these areas (and even if you aren't), build a team to offer that support and take advantage of seminars, programming, and other opportunities to improve and learn. You won't need to be an expert in everything but remember that, with a lean team, you may need to handle certain responsibilities such as human resources or sales in the early days.

TIP: A successful business requires a founder familiar with business skills and a team with business experience.

According to a study produced by Stanford University, most successful entrepreneurs are business innovators rather than technical innovators, proving that you don't need to have a basket full of patents or be a scientific genius to identify and develop technologies that impact the marketplace. In new business creation the ability to recognize business opportunities is more important than the development of technical breakthroughs. Many successful businesses in high-tech fields result from a team of technical and business innovators.

The Stanford study goes on to posit that generalists (individuals who took a broader range of courses during their studies and have had a broader range of work experiences) are much more likely to become entrepreneurs compared with individuals focused on a particular technology or a specific field. Furthermore, additional studies have shown that a specific technical training does not make you more likely to succeed in the field. The marketplace continues to prove that the best technical innovation without an appropriate business strategy is unlikely to the reach broadest consumer adoption.

REASONS WHY PEOPLE CHOOSE *NOT* TO START A STARTUP

Individuals starting their own businesses appear to be happier, enjoy their jobs more, and can even generate substantial amounts of wealth. So why don't more people start their own businesses given the weight of that evidence? Well, Paul Graham provides 16 reasons that people give for not starting their own business. Graham offers it as a list of all the reasons for reluctance, not to scare people from doing a startup. This list is a tool to help people examine their own feelings about the startup life.

Which of these apply to you?

- Too young
- Too inexperienced

- Not determined enough
- Not smart enough
- Know nothing about business
- No cofounder
- No idea
- No room for more startups
- Family to support
- Independently wealthy
- No reason for commitment
- Need for structure
- Fear of uncertainty
- Don't realize what you're avoiding
- Parents want you to be a doctor
- A job is the default

According to Graham, a thorough self-examination should be done to decide whether to start a business, but many of these excuses are really without merit or can be overcome with effort (i.e., finding a cofounder or an idea; too inexperienced or knowing nothing about business). Several are merited, such as lacking determination or having a family to support. No matter what your situation, examine it through the eyes of these excuses to determine whether the excuse is one that can be overcome or requires you to rethink the startup life. Ultimately, many people have come before and given the same excuses, but, thankfully, many successful entrepreneurs didn't let these reasons hold them back.

Source: List compiled from Paul Graham's essay "Why to Not Not Start a Startup" (http://paulgraham.com/notnot.html).

2

Your High-Tech Business Concept

In a start-up company, you basically throw out all assumptions every three weeks.

William Lyon Phelps
Professor, Yale University

What to Watch For

An entrepreneur is "someone who perceives an opportunity and creates an organization to pursue it," said Professor William Bygrave of Babson College and Charles Hofer of the University of Louisville. Steve Jobs, Chief Executive Officer (CEO) of Apple and Pixar, is renowned for his hands-on approach to identifying and championing new ideas for Apple such as the line of new Mac computers, the iPod, and the iPhone. Many industry watchers attribute Apple's recent turnaround and Pixar's huge successes in digital animation to Jobs' ability to recognize ideas to meet consumer demands.

Hopefully, like Jobs, your next step in developing your business will be identifying the *right* opportunity and creating a successful startup company to take that idea and launch your products or services to market. However, much like the founders of Netscape, Amgen, FedEx, or Home Depot who came before you, how do you find that "next big idea" and make it into a successful business concept?

Opportunity recognition is an important skill set for a technology entrepreneur and a key part in the creation of a successful startup organization. Substantial research has been done to identify how, where, and why successful entrepreneurs are able to identify their opportunities. Many times these ideas are initially rejected by potential partners or mainstream investors as technologically difficult or impossible, or uninteresting to the potential target markets, yet innovative thinkers, such as Henry Ford, Howard Hughes, Jack Welch, and Michael Dell, each built businesses based on ideas others shunned. So where do you find those ideas and how do you know when to make the idea a centerpiece of your business?

There is more to opportunity recognition that spotting an interesting idea, however. Successful entrepreneurs will need to identify an opportunity that satisfies a need in the market and allows the startup company to make money at it. Moving from that initial idea into a business concept for your startup is oftentimes a challenge, and many successful companies have been forced to reinvent themselves early in their lifecycles when that first idea or concept did not prove viable. And recognition of opportunities is not limited to that first idea galvanized into a startup company: successful entrepreneurs will continue to identify new opportunities within their company as it continues to grow.

Keep your eyes open and start thinking like your potential customer: next step, your high-tech business concept.

The Opportunity Recognition Process

Successful entrepreneurs usually describe the decision to found a business as a process, involving observations, numerous brainstorming sessions, detailed conversations, extensive research, and contemplation. Researchers from the University of Chicago and San Francisco State University (C. M. Gaglio and Richard Taub) explained the typical opportunity recognition process for entrepreneurs as four distinct steps:

1. Prerecognition stew
2. The "Eureka!" experience
3. Further development of the idea
4. The decision to proceed

Each entrepreneur comes to recognize ideas differently, but, in most cases, the idea won't come quite like a bolt of lightning. Usually, entrepreneurs are thinking creatively about a problem or set of problems, the time period that Gaglio and Taub describe as the prerecognition stew. Some entrepreneurs are actively searching for a new business idea, whereas others aren't formally looking but are open to possibilities. In either case, once the entrepreneur finds an idea (or two . . . or six), one of the most important steps is further development of the idea and the decision whether or not to proceed with this idea into a business concept.

What Comes First: The Business Idea or a Decision to Start a Business?

Entrepreneurs will often identify three different processes for coming up with their business ideas. In the first case, an entrepreneur will identify the opportunity and then will later decide to form their business. In the second case, an entrepreneur will decide they would like to start a business and begin to identify opportunities after that point. Finally, some entrepreneurs note that the idea or opportunity and the desire to start the business came simultaneously.

Researchers Gerald E. Hills and Robert P. Singh asked nearly 500 entrepreneurs to classify the process for identifying their business opportunity. The research showed that entrepreneurs are quite mixed in how they identify their ideas:

- Business idea or opportunity came first — 36.9%
- Desire to start business came first — 42.1%
- Idea or opportunity and desire to have a business came at the same time — 21.0%

Does it matter which path is chosen? Research by Teach, Schwartz, and Tarpley into idea generation by software firms suggests that the ideas "accidentally" discovered rather than those ideas identified through a formal screening process tend to achieve break-even sales faster. This result isn't entirely unexpected in the high-technology industry in which rapid market entry is necessary. However, there are numerous examples of successful companies that had been formed following each of the paths discussed above.

For entrepreneurs that undertook a screening process for an appropriate idea (deciding to start the business first and then looking for the best idea), Hills and Singh asked the entrepreneurs the number of ideas they considered before selecting one. More than 65% of the respondents said they considered less than four ideas before settling on one:

- One — 27.8%
- Two — 18.4%
- Three —19.1%
- Four — 10.0%
- Five — 7.1%
- Six to nine — 10.7%
- 10 to 19 — 4.4%
- 20 to 39 — 0.7%
- More than 40 — 1.8%

Where Do Business Ideas Come From?

As we've seen, the process for identifying a new business idea is usually either a formal search or review process or a "Eureka!" moment. But, in either of those scenarios, where will the idea for a new business actually come from? Most surveys note that industry or market experience and discussions with individuals in your social or professional network are the primary places where ideas are generated.

Researchers Hills and Singh asked the entrepreneurs what had led to their business idea. Their research found that social and professional network contacts were one of the most important factors in idea identification, with 62% of respondents noting that their idea had come from discussions with business colleagues, friends, or family. Nearly 56% of respondents pointed to their experience as an important factor in the development of their business ideas.

It developed from another idea I was considering	23.1%
My experience in a particular industry or market	55.9%
Thinking about solving a particular problem	22.9%
Discussions with my friends and family	42.4%
Discussions with potential or existing customers	30.9%
Discussions with existing suppliers or distributors	15.9%
Discussions with potential or existing investors/lenders	8.1%
Knowledge or expertise with technology	28.6%
Other	6.9%

Additional research by Chuck Eesley of the MIT Sloan School of Management looked at the sources of product or service ideas for alumni of MIT. According to Eesley, more than 60% of the ideas for these products and services came from working in the industry.

While in school	14.4%
Discussions with social or professional acquaintances	13.0%
Research conference	0.7%
Working in the industry	61.9%
Working in the military	3.1%
Other	6.9%

The First Idea Might Not Be the Best Idea

Successful entrepreneurs note that the early stages of developing their business concept are a struggle between obstinacy and flexibility. On one hand, the entrepreneur wants to stick to her vision in spite of the criticism and doubts. On the other hand, the entrepreneur recognizes the importance of adapting the initial idea as research, experience, and understanding increases.

Researchers Hills and Singh asked the entrepreneurs in their study about the change in their original business idea since its beginning. More than 50% of the entrepreneurs said that their idea is about the same, whereas the other half said the idea or opportunity has changed a "little" to a "great deal":

- Idea/opportunity has changed a great deal — 13.2%
- Idea/opportunity has changed a little — 36.4%
- Idea/opportunity is about the same — 50.4%

According to the results of a survey by Launch Pad, a marketing consulting company, 44% of the high-technology companies they surveyed in 2003 said the company had made a significant change in their business model in the previous year, changes ranging from a new product line or target market to pricing or sales model. The ultimate lesson for entrepreneurs is to recognize that there is a balance in play between obstinacy and flexibility. Recognizing when the idea, focus, model, or strategy needs to be changed, in any amount, is an important skill. However, the flipside of the discussion is that an entrepreneur also must recognize when to stick to his vision even in the face of outside pressures or scrutiny.

"REFOCUSED" SUCCESS STORIES

- The founders of **PayPal** initially began developing cryptography software and a service for transmitting money via PDAs before developing its leading online payment tool.
- **Hotmail's** founders came up with their idea for a Web-based e-mail service because they were tired of being unable to access their personal e-mail addresses when behind their work firewalls.
- **Ironport**, a leading e-mail and network security company acquired by Cisco in 2007, was initially founded to develop technology for sending e-mails rather than blocking them.
- Mitchell Kapor, the cofounder of **Lotus Development**, had been working at VisiCalc, the leading spreadsheet software company. He left the company and leveraged that experience (as well as a few legal conflicts going on at VisiCalc) to trump VisiCalc and create a leading company in the software space opened up by VisiCalc's problems.
- **TiVo** initially received its funding as a "flamboyant, home server network thing," according to its founder Mike Ramsay. However, after realizing that the idea was a hard sell to consumers, primarily because it was difficult to explain, TiVo modified its business to its current digital video recorder model.
- **Viaweb**, eventually acquired by Yahoo! and renamed Yahoo! Store, was originally formed as a company aiming to get art galleries online. After finding out that they couldn't convince art galleries of the benefits of putting their art online, Viaweb's founders realized they should consider making something people actually want—and used the tools developed for art galleries to help consumers create their own online stores.
- The collaborative bookmarking site **del.icio.us** was originally created as a way for founder Joshua Schachter to organize his personal collection of over 20,000 Web bookmarks. Little did Schachter realize that his idea for fixing a problem he had would grow into a business later acquired for an amount rumored to be around $30 million.

Development of the Business Concept

Once an entrepreneur identifies a business idea, the next stage of opportunity recognition involves further development and research into the business concept. Generally, there are several steps in moving an initial idea into a business concept.

An Idea versus an Invention

You've got a great idea for a product that may revolutionize an industry—everyone you've told about the idea really thinks you've found something with lots of potential. So, should you go out and get it patented?

This is a fundamental question that many new entrepreneurs have when they have their business idea in hand and are ready to start their business. Unfortunately, some entrepreneurs fail to realize that an idea isn't an invention. Ideas can't be patented, but inventions can be. So before you rush off to try and patent your new business idea, remember that you can't patent an idea; you'll have to "reduce it to practice" or build and design an invention based on that idea.

HURRYING TO "REDUCE IT TO PRACTICE"

What could happen

You've decided on your idea (and it seems like a great idea). Now should we hurry up and build the product so we can get a patent?

Watch out for

This is a common mistake some new entrepreneurs will make. Once they've decided on their business idea, they throw themselves wholeheartedly into building a product so they can get it patented. Although this approach may help you develop a product you can patent, it may not help you to develop a product you can sell. Before you throw countless resources into development of a product based on your idea, be sure to spend the time and energy researching the market, talking to potential customers, and analyzing the competition.

TIP: Make sure you've done market research and analysis to build a product you can sell rather than just a product you can patent.

Developing a Product Consumers Want

For entrepreneurs with a strong technical background, a common challenge is tying a technical innovation to a consumer demand. The challenge for a successful startup company is not recognizing whether the technology is interesting but whether customers will see a benefit and buy your product or service. Ask this question about your new technology: "Who would buy this?" If you struggle to find an answer to that question, you may need to slow down your development process and refocus on your customers. Here are a few of the key questions you'll need to answer before deciding to move forward with any product or service:

- Is someone willing to buy this?
- Is someone willing to pay enough for this to be profitable?
- Who is the specific user of the product?
- Is this better than the alternatives?

- Is the market for this product large enough?
- Is someone else already filling the market need?

When deciding whether your idea is a good business concept, examine it from the perspectives of your customers. Some of the best business concepts may not be in "hot" markets or involve "cutting-edge" technologies. Remember, your goal is to make a product that people will buy.

COMMON PROBLEMS WITH NEW BUSINESS IDEAS

Where do ideas go wrong? Oftentimes ideas go wrong when entrepreneurs start with their technology and try to sell it to consumers rather than the other way around. The questions isn't, How can I get people to buy my invention? Instead, the question should be, What can I invent that people will buy?

Here are a few of the most common problems for entrepreneurs deciding what idea is best:

- **Marginal niche product:** You can make a successful small business by tackling a small niche problem. However, if you are looking to build a startup that will require outside investors (probably in the form of venture capital), you can't try to carve out a small niche just to avoid any competitors. In some sense, competition is good. It means that the end customer is actually willing to pay money for something. Outside investors rarely will fund a business that doesn't have a potential market of at least $500 million (this means that, if you had 100% of the sales in the market each year, your company would have sales of $500 million). It is okay to tackle a marginal niche market, but realize that it is more difficult to raise outside funds if that is the market you are set to pursue.
- **No specific user/customer:** This is the holy grail of problems with startup business ideas: not having a customer in mind before you build the product. Think like a sales person and consider to whom you would actually sell the product. Before you go out and build any products, interview your likely customers and see whether they would buy it and how much they would pay. In some cases, you've got a separate user and customer: sure, it's the consumer who buys the product off the shelf, but you also need to sell it to the grocery store purchasing manager who would need to purchase the product and stock the shelves.
- **Derivative idea:** Many "new ideas" are actually just improvements or enhancements to existing products. This approach can work, but remember that being "'just as good"' or "'just a bit better"' is a difficult product to sell.
- **Willingness to pay for your product (or pay enough for your product):** Some ideas may seem to make sense and be an improvement over what is available, but if you can't make people pay money for your product or pay enough money for your product, then you may not have a viable business concept. It doesn't mean it is a bad idea or a bad product, but it may not qualify as a viable business concept to start a business around.

Researching the Market

One of the keys to building a successful business is displaying an understanding of the business environment for the company, including the competition, the market, the industry, the technology, the key companies operating in the sector, the partners, and the history involved. According to professional investors, several of the most common mistakes they see in business plans stem from a failure to understand their competitors, the market, and the opportunity. Therefore, beginning to identify and investigate this information early will identify the key aspects that will drive the business.

ACCORDING TO VENTURE CAPITALISTS

What is the most common mistake entrepreneurs make when completing their company's business plan?

- Stating that the company had no competition or underestimating the strength of competitors — 32% of respondents
- Not clearly explaining the opportunity — 27%
- Disorganized, unfocused, or poor presentation — 12%
- Miscalculation of market share and market size — 9%
- Failing to describe a sustainable competitive advantage — 9%
- Failing to address the risks of a venture and failing to provide a contingency plan for coping with the risks — 9%

Source: Profit Dynamics Inc.

Competitors

Any startup company is likely to face a difficult competitive landscape, with competition from larger companies, more established ventures, or international challengers. Underestimating or failing to identify those competitive forces can spell disaster for any business. As a result, one of the key areas that new businesses will need to focus on is research into the competitive landscape. By understanding how your business stacks up against competitors in its markets, you can identify the opportunities and the primary challenges the business may face.

Public Companies

Research into public company competitors will generally prove easier than uncovering information on private company competitors. Public company disclosure requirements offer an insight into the products, research and development, pricing, sales, gross margins, and other key metrics of the company. Services such as Google Finance, Yahoo! Finance, the U.S. Securities and Exchange Commission (SEC) website, and other online research tools can provide access to publicly disclosed information made available in annual reports (10Ks) and quarterly reports (10Qs). Additionally, many of the research tools identified below for private companies may also provide additional research information for public companies.

Private Companies

Private company research is more difficult to obtain than public company information. However, several sources do offer information on private companies. Many of these resources are available for a fee, but many public or university-sponsored libraries provide access to these resources for free.

ONLINE RESOURCES TO RESEARCH COMPETITORS AND MARKETS

- **D&B's International Million Dollar Database** (http://www.dnbmdd.com) provides information on approximately 1,600,000 U.S. and Canadian leading public and private businesses. Fee for access to the database, but many libraries will have free access to the databases available for library patrons.
- **Hoovers** (http://hoovers.com) provides information on public and private companies from sources such as D&B and others. Subscriptions from $75 per month.
- **Hill Search** (http://www.hillsearch.org) provides detailed data on public and private companies in North America and access to current and archived articles from the national and regional newspapers, industry journals, trade magazines, and newswires. $59.95 per month; $650 annually.
- **Integra Information** (http://integrainfo.com) provides various private company research information. Costs range from $9.95 to $200 per report, with subscription fees available for access to multiple reports.

Other free research can be found in industry trade publications, on company websites or from a standard search engine search.

Competing Intellectual Property

For any new technology company building its strategies around a patent portfolio, the company should research competing intellectual property to identify both companies and technology that will provide challenges to the startup's strategies.

ONLINE RESOURCES TO RESEARCH INTELLECTUAL PROPERTY

- **Delphion** (http://www.delphion.com) provides a fee-based Web service for searching, viewing, and analyzing patent documents. Delphion provides access to the following sources: (1) U.S. patents and patent applications (full text searching capable); (2) European patents and patent applications; (3) Patent Cooperation Treaty application data from the World Intellectual Property Office; (4) patent abstracts of Japan; and (5) business research about patent applications.

- **The U.S. Patent and Trademark Office** website (http://uspto.gov/patft) provides free search functionality of its "Issued Patents" (full image search from 1790 to the present; full text search from 1976 to the present) and its "Published Applications" (2001 to the present) databases.
- **The European Patent Office** (EPO) website http://www.epo.org provides access to documents in any official language of the EPO member states.

Most countries have some Web-based services to do country-specific research on intellectual property in that specific country. Find more information on researching intellectual property in Chapter 14.

Markets and Industry

Early-stage businesses are usually unable to spend large amounts of money on market research and industry analysis reports, and few potential investors will expect detailed market analysis that one would expect from a well-funded public company. Investors will instead expect a startup company to have used available research and information from various sources to identify market estimates used by the company in its decision-making.

You can start with many of the previously discussed resources such as Hoovers, D&B, Hill Search, and Integra, which each offer research reports available on specific markets and industries. Look into industry trade associations that oftentimes will prepare market research on the specific industry for the benefit of members. Finally, examine research in industry trade press that will oftentimes analyze various aspects of an industry or market.

Getting It in Writing: The Business Concept Summary

One of the first steps any entrepreneur takes in starting a new business is laying out the business concept on paper. We'll refer to this as laying out the business concept summary; it isn't meant to be a full business plan or to answer all the potential questions about the business but is meant to be the chance to answer the question, "Could this business concept work?" This has also been called the "kitchen table" step, in which you sit around your "kitchen table," discuss the business concept with a few confidants, and map out the initial idea. For many future entrepreneurs, this represents the first step into creating the startup company.

Usually this stage occurs for an entrepreneur once they have vetted their initial business idea and begun to consider how to build a business organization around this idea. Much like the process of building a full business plan for the business (discussed in more detail in Chapter 8), setting out a business concept summary lays out the early vision for your business.

Mike Lazaridis, the cofounder of Research In Motion of BlackBerry fame, discussed the importance of this first step in an interview found in the book *Founders at Work* by Jessica Livingston. Said Lazaridis:

> I went downstairs and got on the computer, and I put on some music and just started writing. Three hours later, I had just put the finishing touches on what became the plan for what eventually became the BlackBerry. . . . That was a turning point, because we've used that document for years. It's still used by people here because it defines the essence of the BlackBerry experience, and it has allowed us to remain true to that and really bring value to our customers. It helped us to stay away from the fads that really didn't bring any value and just made the product more complicated and more expensive.

Many entrepreneurs will begin the planning/research process shortly after they have identified their potential business concept. According to research by Margaret Owen and Patricia Greene, entrepreneurs in this planning stage (before they take a formal step to fully launch their venture) spend just over two hours every day planning and working on their new startups. This time will be spent further developing the technology or products but may also be used in creating a business plan or researching particular aspects of the business plan.

The Reason for Laying Out the Business Concept Summary

In many cases, the creation of the business concept summary is just a one or two page document and some associated spreadsheets created by the founder. Generally, a business concept summary will be prepared very early as a tool or set of tools to identify whether the opportunity might be worth pursuing. As the founders continue to discuss and further investigate the business opportunity, this initial business concept summary will eventually evolve into a full business plan (discussed in greater detail in Chapter 8).

Nearly all new startups create some form of a business concept summary, but it will usually be much less formal than future business planning. The business concept summary represents an early sketch of the business.

What many entrepreneurs will quickly realize about this process is that you have most likely already been performing these activities required to create your business concept summary even without a formal business plan preparation process. However, this process presented here for developing your business concept summary may hopefully save you time in the long run over piecemeal planning efforts.

In the early stages of laying out your business concept, you will begin preparing certain key tools for use in your business, which we'll refer to as preparing your business concept summary. Many of these documents will be spreadsheet tools used for market analysis, sales forecasts, customer tracking, financial forecasting, cash tracking and burn rates, headcount tracking, and the other nuts and bolts tools to help run your business.

STEPS TO PREPARING A BUSINESS CONCEPT SUMMARY

To prepare the business concept summary, you can follow these steps:

- **Prestage:** When to begin? How long will it take? Who to include? What tools to use?

- **Stage 1 (information gathering):** Develop an understanding of your potential products and services, and establish a tentative development timeline; determine the estimated product sales and costs; and collect data used to define the potential market.
- **Stage 2:** Produce a simple business concept summary and a related set of tools for your business.
- **Stage 3:** Initiate a critique and reworking of your business concept summary.

The end result will most likely be a series of spreadsheets, documents, schematics, and reports that will serve as useful tools for the concept phase of your business and should ultimately serve as the basis to prepare a complete business plan.

So how to begin creating your business concept summary? You'll likely begin this process by gathering key information. This process will probably be something that involves multiple rounds of collecting and recollecting information necessary for the planning process. As the information is received, you'll begin to prepare these planning documents and tools. Then, you'll let the critiquing begin among your inner circle and let these questions, comments, and criticisms morph your work into a useful set of tools to assist in the operations of your business.

Prestage

When to Begin the Planning Process and How Long Will It Take?

The process to create the initial documents that will go on to become your business concept summary and business planning tools can and is usually done very early in the founding of your company. In many cases, the "back of a napkin" analysis will be modified slightly and transferred to a spreadsheet. In the development of your business concept summary, you will likely be able to develop some tools in your business concept summary very quickly based on information you have at hand.

Who to Include in the Process?

If you have a cofounder or a core group of individuals you want to involve at the earliest stages of planning, you can involve them. However, many founders will operate individually at this stage based on the information they can gather.

What Tools to Use?

You will most likely be able to produce the spreadsheets, calculations, and documents with typical business tools such as Microsoft Word and Excel. You may consider using a business planning software tool (discussed further in Chapter 8) but may find the software to be too robust at the early stages.

Stage 1: Information Gathering

The first step in putting together a business concept summary will be gathering information, doing research, and interviewing key people. This stage will involve gathering key information to put forth your first efforts at understanding the business and identifying whether and how the business can become successful.

You will primarily focus on three areas to gather information that will serve as your assumptions for your business concept summary.

Products and Services

Gather information in the following areas:

- **Definition:** What product or service are you delivering to the customer?
- **Development:** How much will it cost to develop the product? How long will it take to develop this product? What types of competencies does the company need for development?
- **Future:** Are there improvements to the original product that we can make in the future? Do we have other products or services outside of our primary product and service platform?

Product Sales and Costs

Gather information in the following areas:

- **Customers:** How many potential customers are there? What will it take for them to buy our product?
- **Sales:** How much would a customer pay? What is the sales process like?
- **Costs:** What will it cost to sell this product? Can we decrease costs over time?
- **Intellectual property:** Can this product or service be protected by patents? Can this product or service be protected by trade secrets?

Market Data

Gather information in the following areas:

- **Target markets:** What is the target market? Are there any other potential markets? How large is that target market? What market would be the first market to pursue? Are there opportunities internationally? Who are the target customers and what are the demographics associated with that customer? How is that customer currently being serviced?
- **Competition:** Who would be your primary competitors? What are the relative strengths and weaknesses of each of the primary competitors?
- **Market trends:** What are the trends affecting your target customers? What are the trends affecting this industry?
- **Market research:** What research have you gathered so far? What research has been done? How do you propose to get additional information on your market and customers?

Stage 2: Creating a Business Concept Summary

Using the information you have gathered in stage 1, the information gathering stage, you will begin to prepare a series of spreadsheets, reports, and documents. This may take the form of a short (one to two pages) business concept summary document that contains certain key information or this may not be aggregated into a single report. Again, at this stage of the process, you will not necessarily have detailed research for many of your assumptions. You are presenting a view of the company and its prospects based on the information currently presented. However, you should be clear where you do not have thoroughly researched numbers and need additional information.

What are the five components that should make up your business concept summary?

Business Definition

Your company should define itself by the benefits it provides to its target customers. For example, Amazon.com would not define itself as an e-commerce website. Instead, Amazon is in the primary business of providing books to consumers efficiently and inexpensively. Be specific when you describe your business and the value proposition for your consumers. Too often, even at early stages, a high-tech company will be too focused on the technology rather than the end customer the technology is to reach.

Market Potential

You should prepare a summary of your target market (with research data gathered to date) and identify key points about the target market. Highlight the size of the marketplace, current competitors and their respective strengths and weaknesses, trends in the marketplace, and what additional information that should be gathered to round out the research. Your report may offer pie charts that identify the total size of the market (if available) and the current market share by various competitors and products. Highlight current growth patterns and expectations for future growth in the market.

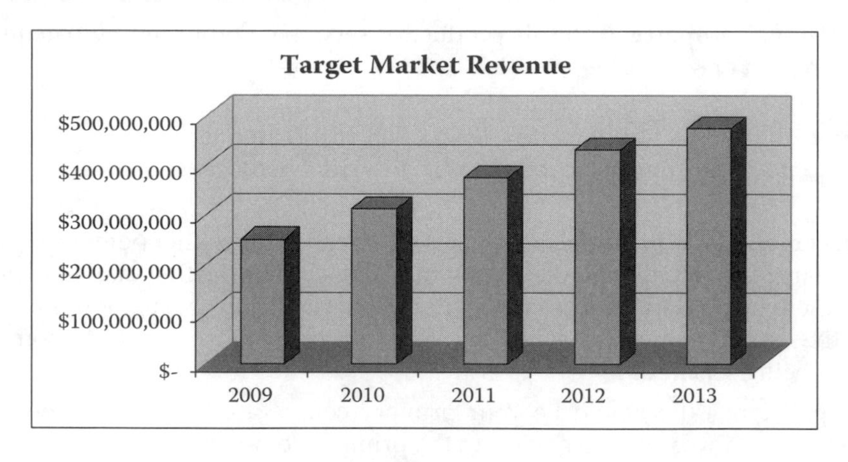

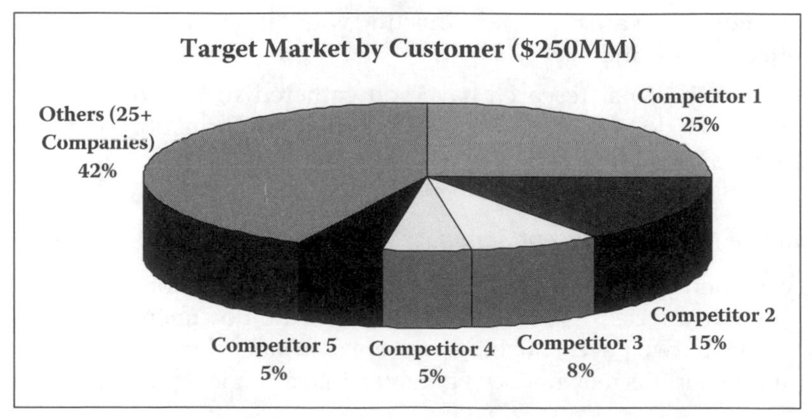

Success Drivers

After looking at your company, the potential products and services, and the development timetable, you should be able to identify what are the top three to five drivers that will separate you from a similarly situated company. For you company to be a success, do you need to be first to market or do you need to attract teenagers? Do you require a substantial investment to develop a successful product? Is it important to partner with certain strategic companies to become linked with another set of products? Identify what are the key drivers that will affect the success or failure of your business. Not every driver is a key, so focus on just those top drivers that affects your success.

Sales and Costs

What are your competitors selling their products for? Will you be able to sell below those products or are you considering a product that will offer a premium product for a premium price? What are the costs associated with the production of each unit (do not include the costs of development or operations)?

You should prepare a simple chart showing the per-unit price and the per-unit cost. From these numbers, you can calculate the gross margin and the gross margin percentage.

	Year 1	Change	Year 2	Change	Year 3
Per-unit price	$100.00	10%	$110.00	2%	$112.20
Per-unit cost	$25.00	–5%	$23.75	1%	$23.99
Gross margin	$75.00		$86.25		$88.21
Gross margin percentage	75.0%		78.4%		78.6%

If possible, provide a comparison of price per unit and gross margins to the primary competitors.

	My company	Competitor 1	Competitor 2	Competitor 3	Average
Per-unit price	$100.00	$125.00	$133.00	$127.50	$128.50
Gross margin percentage	75.0%	67.5%	77.6%	73.2%	

Break-even Point

In the early stages of your company, you should consider including a break-even analysis. The purpose of this analysis is to determine whether, under your assumptions, the company could break even. You'll need to identify your company's per-unit price and cost, as well as estimate the monthly overhead for the business. This will identify the number of sales necessary each month for the business to reach a break-even point.

In the example chart below, the company has a per-unit price of $100.00, per-unit costs of $25.00, and monthly fixed costs of $55,000. Based on these assumptions, your company would need to sell 733 units each month simply to break even. As you can see, if you sell zero units each month, you'll have a loss of $55,000, whereas monthly sales of 2,000 units yields a net profit of just under $100,000 per month. Charts like this are helpful to hone your message and focus your efforts on the key business drivers.

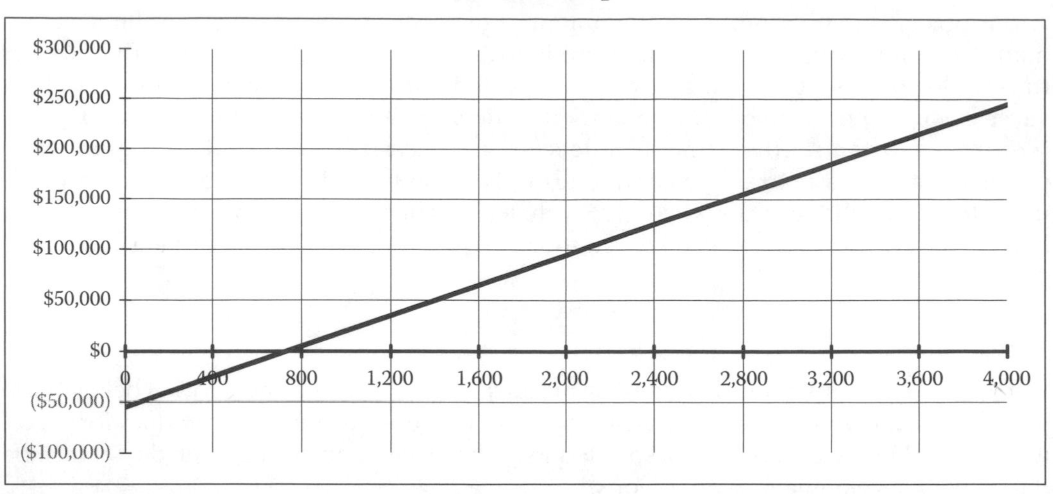

Monthly Fixed Costs	=	$55,000	Break-Even Point (units)	=	733
Per Unit Cost	=	$25.00	Break-Even Point ($'s)	=	$73,333
Per Unit Price	=	$100.00			

Stage 3: Critiques and Reworking

Once you have set forth some basic information on your company regarding the business itself, the product details including sales, costs, and development timelines, as well as the potential product markets, you will have a fairly good set of data points to begin sharing with others outside of the team. Even at this stage, you should continue to insist that outside parties sign nondisclosure agreements to prevent release of your early reporting information. Additionally, you will need to advise any parties who are reviewing your business concept summary that the assumptions are based on limited research.

You should use these early conversations as a chance to determine the relative merits of the business itself and whether additional endeavors are worth the efforts. Be prepared for many comments and thoughts. Does there seem to be a consensus on the market size? Does this appear to be a reasonable need for the target customers? Is there a differentiator in your business and product? Are the time horizons reasonable?

Based on this early feedback, you should continue to revise and rework certain parts of this early-stage planning tool.

3

The Startup Road Ahead

The secret of getting ahead is getting started. The secret of getting started is breaking your complex overwhelming tasks into small manageable tasks, and then starting on the first one.

Mark Twain

What to Watch For

In the simplest world, a successful startup involves coming up with an idea, forming your founders team, creating the business, finding investors, developing your winning product, raising more money to get your product to the consumers who will pay for it, and finally being successful enough to take your company public or to sell your business to a buyer (who will help your company take this next leap). And then, you can kick back on the beach house you purchased with all that money you made and relax . . .

Is it really that simple? Well, not really—otherwise, everyone would be doing it. However, you can get a pretty good sense as to what will lie ahead and the important milestones for the business.

This chapter is designed to lay out some of the big-picture items ahead for a new startup and answer some of the common questions a new entrepreneur is probably pondering. For instance, what are the common stages a business goes through and what challenges are associated with each new stage? What are some of the biggest challenges and risks faced by a startup? How should you budget for those early days to pay for startup expenses? What kind of business am I going to startup? And what kind of payoffs can I look forward to if we hit it big?

The book delves into much more detail on many of these topics, but, even before you make the decision whether to form your new business, you should think about some of these questions. Next step . . . the road ahead.

SOME STATISTICS ON SUCCESS AND FAILURE

In 2000, venture capitalists funded 2,639 companies. At the end of 2005 (approximately five years later), 1,044 of those companies were still private (40%), 35 had gone public (1.3%), and the remainder had either been acquired or gone out of business.

Source: VentureOne.

Stages and Steps Ahead for a New Startup

You've most likely taken your first step on the startup path: considering whether to go into business for yourself and create a new startup. From that point forward, the road in a high-tech startup will move through certain steps and stages as you move your idea to a business concept and then grow the business from there. For the purpose of this book, that path is broken down into five stages (which are detailed in "How to Use This Book" on page xi):

1. Idea stage
2. Startup stage
3. Launch stage
4. Growth stage
5. Expansion stage

The graphic below details a startup's progression through these stages and identifies the profitability throughout. In the early stages, there will likely be expenditures, but those will quickly ramp up as the company prepares to launch its products. After a successful launch, the startup should hope to grow revenues in such a manner that it provides for additional funding for expanding into new markets, geographies, and product lines.

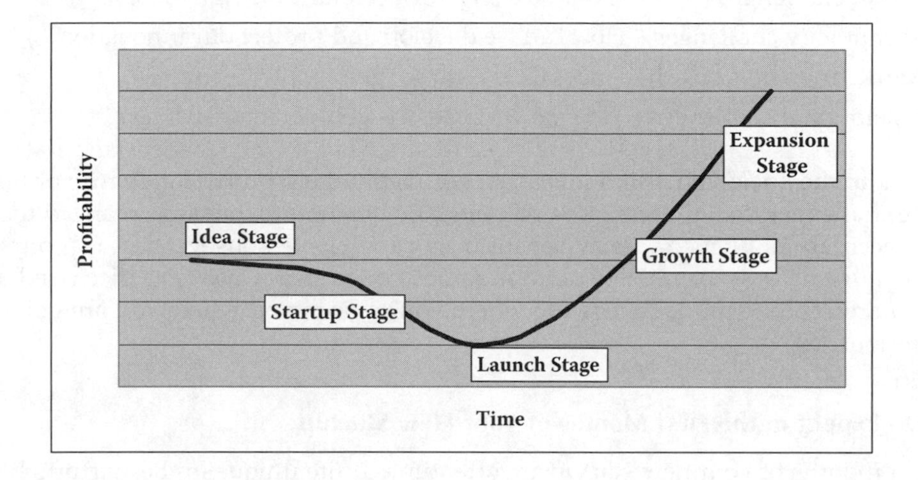

The idea stage is the initial foray into a new endeavor by the entrepreneur. Typically this involves deciding two key things:

1. Is this an idea that we can make money at?
2. Am I interested in forming my own business to do it?

In each case, the answer needs to be yes, and hopefully a resounding yes if you plan to move forward with these efforts. Part 1 is written to help answer some of the questions you might have in deciding whether to begin a startup life and whether your idea can fill an unmet need in the marketplace.

Following a decision to form the business, the entrepreneur and the new business begin the startup stage in which the process of business formation begins. It is in the startup stage that the company begins to recruit its initial hires, finds an attorney, formalizes the business itself, and begins work on its business plan. Part 2 of this book focuses on the key challenges within the startup stage of the business.

The next stage is the launch stage and the resulting growth stage. These two stages oftentimes blur together as the startup develops its initial product for launch and then attempts to grow as its products penetrate the market. These stages represent key times in the startup's life cycle and involve a mixture of fundraising, high-level hiring, product development, sales and marketing, and building the business. Part 3 of the book emphasizes key aspects that a business will face in these two stages of your startup.

Finally, the business will enter into the expansion stage, which is sometimes called "maturity." Don't let this fool you: many challenges continue to face the business. However, with the steady growth of the business, you should begin to discuss new challenges such as expanding into international markets, making strategic acquisitions and sales, and raising larger sources of funds for new product development and expansion. Part 4 of the book addresses some of these next steps and stages.

Within each of these stages, this book identifies the five sets of challenges in a high-technology startup company. These challenges include the following:

- Funding challenges ("How do we pay for this?")
- Talent challenges ("How should we recruit and retain the right people?")
- Technology challenges ("How can we develop and protect our innovation?")
- Marketing and sales challenges ("How do we get people to buy?")
- Operational challenges ("How do we build the company for success?")

Throughout the book, you'll find information aimed to help provide guidance or insights into the challenges the business faces. Although no business is the same, many of the challenges faced are not new and may benefit from previous lessons learned. Of course, the path of your business may not be quite as smooth or as steady as we've discussed above, and you aren't alone in that. Knowing where you've been and where you are going may help the journey.

What to Expect in the First Months of Your New Startup

The first months of your new startup are oftentimes quite unique in the startup's life. For many companies, the first six months to a year are run by a single entrepreneur or a small founder team. In his book *Entrepreneurs in High Technology*, Edward Roberts interviewed entrepreneurs to help identify issues that face a startup in its first days.

Allocation of Efforts

What is the priority in those first six months? According to Roberts, the initial effort by a startup in its first six months is divided among four primary areas: engineering, sales/marketing, manufacturing, and finance/administration. Not surprisingly for a high-tech startup, engineering-related activities required the most effort, with finance and administrative as the smallest percentage of effort.

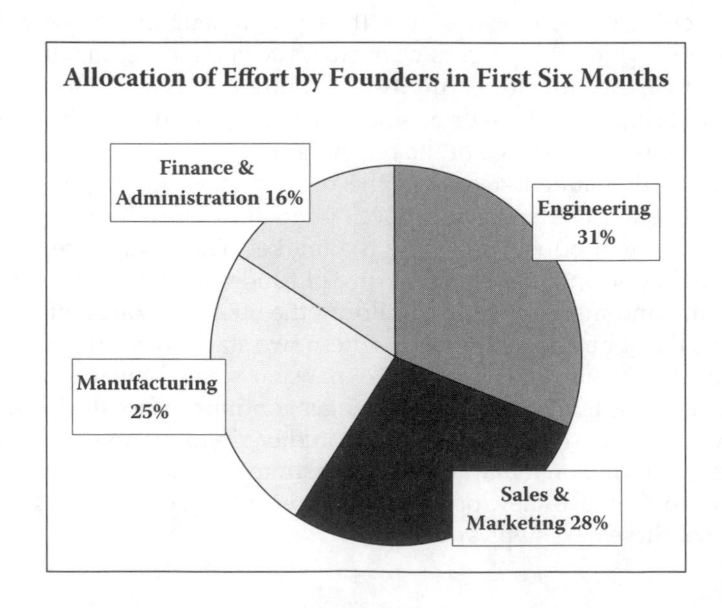

Initial Capital

Roberts's research also examined the amount of funds needed for the initial capital of the business and where startups had obtained those funds. Nearly 80% of those companies surveyed said they had needed less than $50,000 in initial capitalization, with nearly 48% needing less than $10,000. (Note that these figures are as of 1991. Using the consumer price index to adjust these figures to 2007, the numbers would be approximately $76,000 and $15,000, respectively.) Where do these funds come from? The initial funds come primarily from personal savings of the founders, more than 74%.

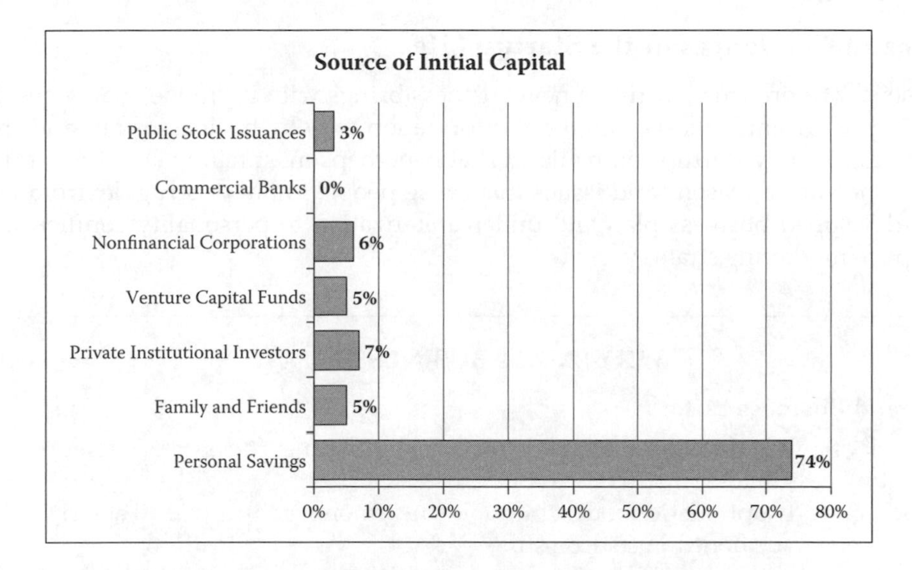

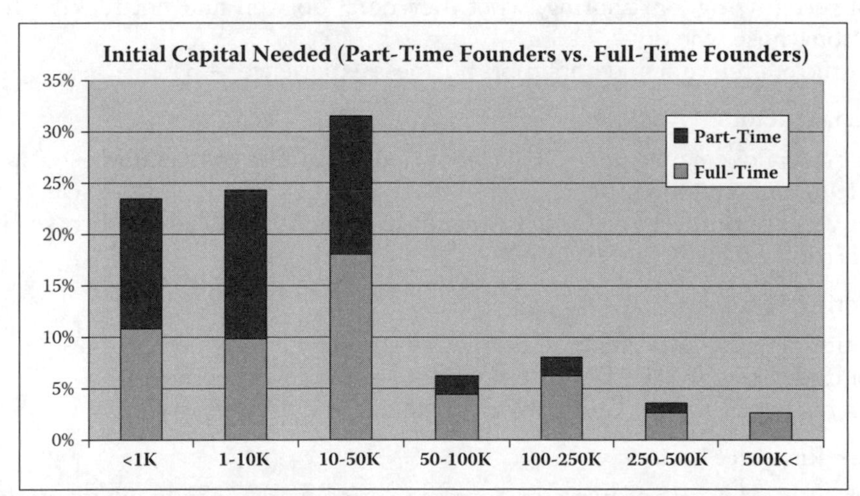

What are these initial funds used for? It somewhat depends on the industry. According to Roberts, hardware companies stated that the primary purpose for their initial capital was for product development, followed by production facilities and working capital. Conversely, software companies needed funds for working capital because software development requires programmers and technical individuals more than expensive equipment. Companies performing contract research and development listed the priority of their needs as laboratory equipment, product development, and working capital.

BUDGETING, PLANNING, AND MODELING

Tables are included at the end of this chapter that can be used to budget your initial startup expenses and determine your personal financial summary. To download copies to use in your startup, visit the book's website at http://www.myhightechstartup.com.

Risks and Challenges of the Startup Life

For a new entrepreneur, starting a new business brings with it unique challenges. In the following figures and charts, you'll see information on why businesses fail and the key challenges faced by startup companies. What is perhaps most telling about this information are the varied reasons and issues that cause people to fail or struggle: from lacking a well-developed business plan and undercapitalization to personality conflicts and an incomplete marketing strategy.

WHY SMALL BUSINESSES FAIL

General Business Factors

Lack of a well-developed business plan, including insufficient research on the business before starting it — 78%
Being overly optimistic about achievable sales, money required, and about what needs to be done to be successful — 73%
Not recognizing, or ignoring, what they don't do well and not seeking help from those who do — 70%
Insufficient relevant and applicable business experience — 63%

Financial Factors

Poor cash flow management skills/poor understanding of cash flow — 82%
Starting out with too little money — 79%
Not pricing properly or failure to include all necessary items when setting prices — 77%

Marketing Factors

Minimizing the importance of promoting the business properly — 64%
Not understanding who your competition is or ignoring competition — 55%
Too much focus and reliance on one customer/client — 47%

Human Resource Factors

Inability to delegate properly or micromanaging work given to others or over-delegating and abdicating important management responsibilities — 58%
Hiring the wrong people or clones of themselves and not people with complementary skills or hiring friends and relatives — 56%

Source: Jessie Hagen of U.S. Bank cited on the SCORE/Counselors to America's Small Business website (http://www.score.org).

WHY SMALL BUSINESSES FAIL

According to a report by the U.S. Bureau of Labor Statistics,[1] approximately one-third of new businesses fail within two years and more than half go out of business within four years. But why do firms fail? In 1999, the U.S. Small Business Administration released a report studying the reasons small firms had been forced to declare bankruptcy.

The reasons for filing were broken down into the following categories: (1) outside business conditions (mentioned by 39% of filers), (2) financing problems (28%), (3) inside business conditions (27%), (4) tax-related reasons (20%), (5) dispute with a particular creditor (19%), (6) personal problems (17%), and (7) calamities (10%). Outside business conditions included such factors as new competition, increases in rent, insurance costs, or declining real estate values. Inside business conditions included a bad location, inability to manage people, the loss of major clients, or inability to collect accounts receivable. Personal problems often included divorce and health problems. One-third of the bankrupt businesses had less than $100,000 in debts, and 79% had less than $500,000 in debts. Mean assets were $841,000, with median assets of $94,700.

[1] Knaup, Amy E. "Survival and Longevity in the Business Employment Dynamics Data," *Monthly Labor Review*. May 2005.

Even businesses that don't go out of business have their share of challenges. In the research done by Roberts, they asked entrepreneurs to identify the primary problem for their business. More than 50% of respondents listed sales as the primary or secondary problem. The next two highest categories include personnel issues and personality conflicts. Research into high-tech companies found similar results, with the top challenge identified as the retention of key individuals.

It goes without saying that, in the high-tech sector and all startup ventures, it is the people that drive the success of the business.

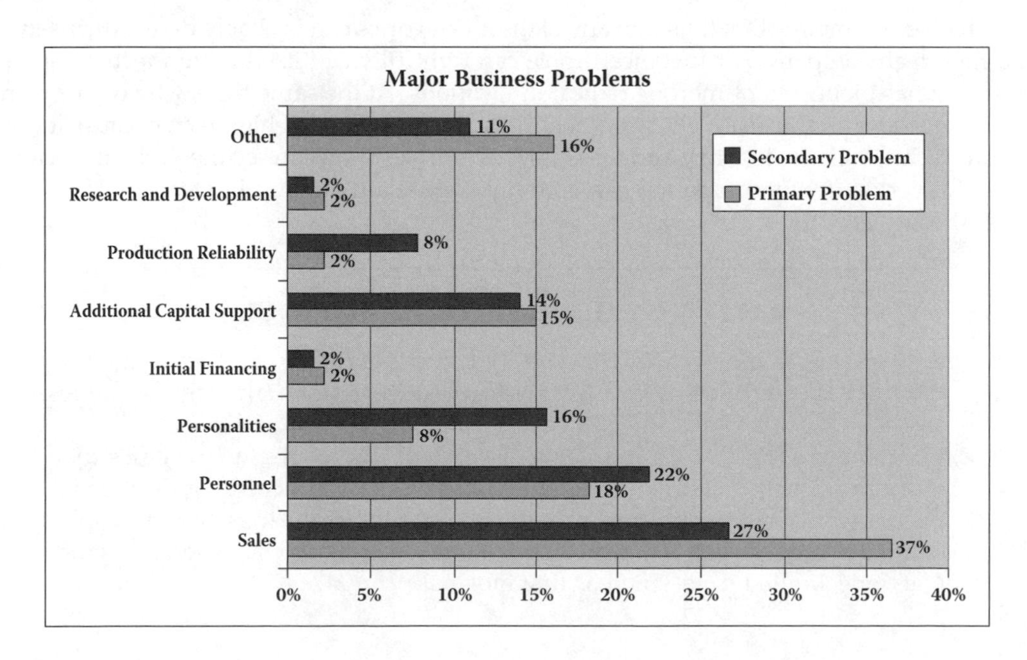

ACCORDING TO ENTREPRENEURS

What are the key challenges you are focusing on for your business?

Entrepreneur magazine and PricewaterhouseCoopers, in their "2006 Entrepreneurial Challenges Survey," asked CEOs of privately held, high-technology businesses about their most pressing challenges they are focusing on for their businesses:

- Retaining key employees — 74%
- Developing new products/services — 47%
- Creating business alliances — 27%
- Expansion to markets outside of the United States — 21%
- Finding new financing — 15%

Going High Tech

This book is aimed at individuals starting high-tech businesses. Lots of businesses claim to be high tech, from software and biotechnology to "old economy" companies that use technology. So what is a "high-tech" company, really? Researchers Kathleen Allen and Timothy Stearns have found that, historically, the term "high tech" has been used to describe three very different types of companies:

- Firms that undertake radical innovation
- Firms that rely on incremental improvements in existing innovations
- Firms that use technology to facilitate their business processes

In each case, a company that falls into any of these categories can probably define themselves as a high-tech company. For instance, Apple can rightfully call itself a high-tech company because it has pioneered numerous radical innovations. At the same time, why can't a company like Exxon, in the relatively "low-tech" oil industry, call itself high tech because it uses various technologies to identify and reach new oil sources? You can see the problem because these two very different companies can both rightfully claim to be "high tech."

WHAT TYPE OF "HIGH TECH" ARE YOU?

Because there are so many different types of companies, Allen and Stearns decided to break down the high-tech sector by the three categories of high-tech companies:

- **First mover:** These companies are in search of disruptive technologies that make previously existing technologies obsolete. The difficulty for these companies yields substantial rewards to success but also brings with it a substantially higher risk of failure. Companies such as Apple, Microsoft, and Netscape would be classified as first movers.

- **Innovator strategy:** A company that relies on an innovator strategy is not structured around creating disruptive technologies. Rather, in many cases, the underlying technology will already exist and the company will combine, merge, or retool existing technologies for use in a new application, market, or area. Oftentimes these entrepreneurs will leverage experience gained at another company in the field to launch a new business, taking an innovative approach to problem solving. Some of these include companies such as Dell Computers, Lotus, and even Google, who initially tackled Internet search using an innovative algorithm.
- **Practitioner strategy:** Many companies that consider themselves to be "high tech" are not actively inventing and innovating. These companies use existing technology to gain competitive advantages, oftentimes using the disruptive technologies of first movers. A company such as Exxon, Wal-Mart, and Home Depot would adopt a practitioner strategy.

As you consider starting your business in a high-tech endeavor, be it developing a radical innovation, incremental innovation, or applying technology to a nontechnical industry, understand how your business fits into the broad high-tech arena. Each strategy has different challenges, and no company can assume all of these strategies at the same time. Business partners, customers, investors, and employees will all be interested in the strategy the business takes within the high-tech sector.

What Type of Business Are You Starting?: Startup versus Small Business

As you begin to consider how to build and structure your business, it is appropriate to ask yourself what size business you are looking to build. In some cases, the goal may be to build a small organization of one to four people that you hope would generate $1 million in sales annually. On the other extreme, you may desire to create an organization with a national brand and operations across the country. In each case, the decision may be driven more by the type of business you want to build rather than your desire to operate a business of a particular size.

The terms "startup" and "small business" are sometimes used interchangeably, but, in practice, there is a difference between the two. According to the Career Action Center, "a startup is a small company, most often with a high-tech focus, that is in the early stages of development, creating a product or service, or having a product or service needing manufacturing and/or marketing. They are looking to grow through possible venture capital funding, initial public offerings (IPOs) or acquisition by larger companies." Conversely, a small business tends to have a more narrow focus and is not limited to high-tech applications. A small business may not have growth goals similar to a startup and may have no plans to grow larger than its small business size.

Startup	Small business
Wealth generation	Income substitution
Importance of technology and proprietary intellectual property	Broader range of businesses
Broad markets	Narrow markets (geographic or target audience)
Goals of $10 million to $100 million in annual sales	Goals of $500,000 to $10 million in annual sales
Seeks venture capital funding	Relies on bootstrapping and bank loans
Staff of 50 or more	Staff of 20 or less

Does it matter what type of business you want to build? Yes, it does in fact. If you intend to build a small businesses with less than 20 people, most likely your concerns will be different from a technology company intending to build its product or service into the market leader for a market or submarket. Small businesses tend to have different financing needs and usually will have a more narrow market, either geographically or in scope. Much of this book will focus on the subset of small businesses that intend to grow to a size in which annual revenues exceed $20 million, which we will refer to as a startup.

In the book *Engineering Your Startup*, James Swanson and Michael Baird separate the businesses into two categories: (1) income substitution businesses and (2) wealth-building businesses. Generally, a small business will fall in the category of income substitution, whereas wealth-building entities are more likely to be startups. According to Swanson and Baird, "People who simply do not want to work for someone else can easily startup a small income-substitution business such as a one-man computer repair service shop." This would qualify as an income substitution business. Conversely, if you wanted to build a business that would be a national franchise or desired to create software allowing for repairs to be made remotely, then this model would be a wealth-building business.

For example, consulting organizations tend to fall into the category of an income substitution business, although the organization can grow quite large depending on the hiring and growth plans.

Dilution: What Am I Left with at the End of the Day?

One of the reasons high-tech entrepreneurs start their own businesses is to generate long-term wealth, oftentimes in the form of sales of your stock in your startup down the road. What are you left with at the end of the day? The question is difficult (if not impossible) to answer. However, what you should do is begin to model your business during the planning stages to give yourself a sense of how dilution will influence what you are left with.

Although it may seem obvious to some, each time you take in additional capital in exchange for equity, your ownership will be diluted (the same is not true of most debt instruments). The chart below details the dilutive nature of financing rounds on the equity ownership of the founder CEO, decreasing from approximately 33% in information technology companies with one round of financing to 14% after four or more rounds, for example.

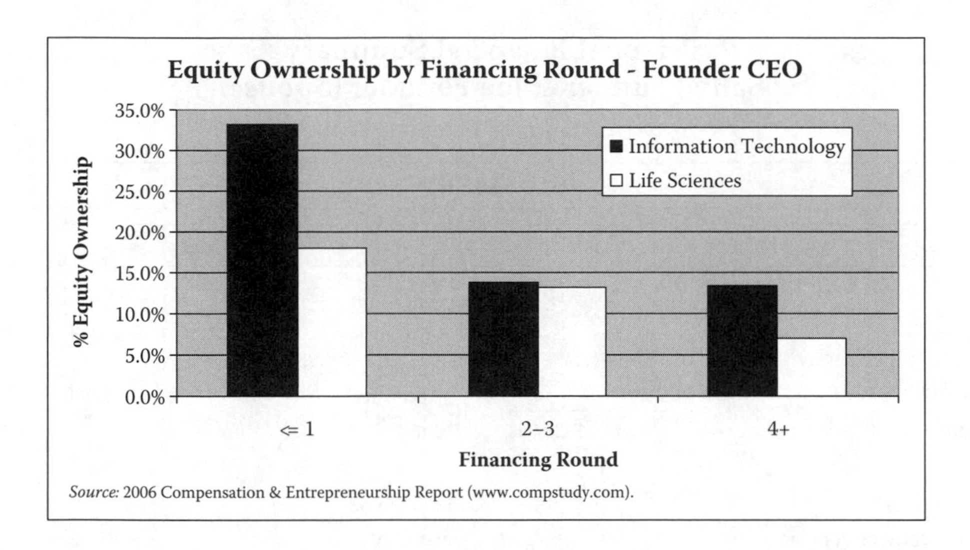

Source: 2006 Compensation & Entrepreneurship Report (www.compstudy.com).

WHAT AM I LEFT WITH?

Seven models are included at the end of this chapter that detail the amount of dilution depending on the capital needs of the company. For example, in the model for a company with very low capital needs, the founder who initially held 25% would be left with 12.4% of the company after raising $3.8 million dollars. Conversely, for a company with medium-to-high capital needs that raises $58.8 million in outside investments, the founder who initially held 25% of the company would be left with 4.9% of the company.

To download the startup documents that follow (such as the personal financial summary), to retrieve the dilution model and use it to model your business, to find updates, or to post other questions, visit the book's website at http://www.myhightechstartup.com.

Personal Financial Summary
Sally Founder & Joe Founder (Spouse)

January 1, 2009

SUMMARY

Assets			**Liabilities**		
Cash			Current Debt		
Checking	$ 8,000		Credit cards	$ 10,000	
Savings	15,000		Other current debt Accounts	2,500	
Total		**$ 23,000**	*Total*		**$ 12,500**
Investments			Other Debts		
Certificates of deposit			Taxes payable	$ 10,000	
Stocks, mutual funds, bonds	$ 10,000		Auto loan accounts	25,000	
Loans/notes due	2,500		Other	5,500	
Life insurance (surrender value)	7,500		*Total*		**$ 40,500**
Education savings	15,000				
Total		**$ 35,000**	Real Estate		
			Mortgages on real estate	$250,000	
Personal Property			Leases due for next 12 months	15,000	
Automobiles	$ 35,000		*Total*		**$265,000**
Real estate (market value)	300,000				
Other	15,000		Other installment accounts	$	
Total		**$350,000**	Other liabilities (specify)	$ 8,500	
			Other liabilities (specify)	$	
Retirement Funds					
IRA	$ 55,000				
401k	75,000				
Other	22,500				
Total		**$152,500**			
Other assets (specify)	$ 4,500				
Other assets (specify)	$				
TOTAL ASSETS		**$565,000**	**TOTAL ASSETS**		**$326,500**

NET WORTH

Assets	**$ 565,000**	Notes:
Liabilities	**$(326,500)**	
Net Worth	**$ 238,500**	
Current Assets	**$ 23,000**	
Current Liabilities	**$ (12,500)**	
Net Current Funds	**$ 10,500**	

Personal Financial Summary
Sally Founder & Joe Founder (Spouse)

January 1, 2009

SUMMARY

You			Spouse		
Employer		Technology Firm Ltd.	Employer		Accounting Firm
Position or profession		Chief Engineer	Position or profession		Accountant
Partner, officer or owner in any other venture?	Yes	**No**	Partner, officer or owner in any other venture?	**Yes**	No
If Yes, please name			If Yes, please name		Joe's Accounting
Average Monthly Sources of Income			**Average Monthly Sources of Income**		
Salary (Post-tax)	$ 5,500		Salary (Post-tax)	$ 4,500	
Bonus and commissions	2,000		Bonus and commissions	500	
Dividends			Dividends		
Other investment income			Other investment income		
Real estate income			Real estate income	800	
Other income (list)			Other income (list)		
			Part-time accounting work	350	
Total		$7,500	***Total***		$ 6,150

Combined

Average Monthly Expenditures				
Mortgage/rent	$ 3,000			
Education	750			
Food	1,025			
Clothing	600			
Travel	500			
Work-related	200			
Discretionary	800			
Other expenses (list)				
Healthcare	600			
Total		$7,475	**Liquid Savings**	$10,500

MONTHLY FINANCIAL PICTURE

Current		**Prospective**	
Income (You + Spouse)	$13,650	Income (Spouse)	$ 6,150
Expenditures	$ (7,475)	Expenditures	$(7,475)
Remaining/(Shortfall)	$ 6,175	**Remaining/(Shortfall)**	$(1,325)
		Months of Savings for Shortfall	7.92

Startup Expenses Planning High-Tech Startup Inc.

January 1, 2009

SUMMARY: Year 1

Startup Expenses		
Real Estate, Office Location, Buildings		
Purchase	$	
Construction		
Remodeling		
Other		
Total		$
Building Improvements (Leased Property)		
(specify)	$	
(specify)		
Total		$
Capital Equipment List		
Equipment	$30,000	
Furniture	10,000	
Machinery		
Other		
Total		$ 40,000
Operating Expenses		
Rent (½ year at $10,000/month)	$60,000	
Deposits	10,000	
Outside legal/accounting	15,000	
Insurance	4,000	
Consulting	8,000	
Salaries	15,000	
Other		
Total		$112,000
Beginning Inventory		
(specify)	$	
(specify)		
Total		$

Sources of Capital		
Owners' Investment		
Sally Founder – 50%	$75,000	
Mike Techie – 30%	45,000	
Jane Designer – 10%	15,000	
Mark Angel – 10%	15,000	
Total		$150,000
Loans		
Banks	$25,000	
SBA		
Other		
Total		$ 25,000
Other Capital		
(specify)	$	
(specify)		
Total		$
TOTAL SOURCES OF FUNDS		$175,000

Marketing, Sales, Advertising		
Marketing	$ 2,500	
Advertising	5,500	
Selling expenses		
Travel/entertainment	10,000	
Other		
Total		$ 18,000

Other Expenses		
(specify)	$	
(specify)		
Total		$
Reserves	$15,000	
Working Capital	$	
TOTAL STARTUP EXPENSES		$185,000

Summary Statement	
Sources of Capital	
Owners' Investments	$150,000
Loans	25,000
Other Capital	
Total Source of Funds	$175,000
Startup Expenses	
Real Estate, Office Location, Buildings	$
Building Improvements (Leased Property)	
Capital Equipment List	40,000
Operating Expenses	112,000
Beginning Inventory	
Marketing, Sales, Advertising	18,000
Other Expenses	
Reserves	15,000
Working Capital	
Total Startup Expenses	$185,000
Funds Available / (Shortfall)	$ (10,000)
Estimated Time to Break-even	24 Months
Estimated Time to Additional Funding	12 Months

Hypothetical: Company with Seed Capital Needs Only

		Amount of Funds Raised		Dilution*
Your Initial Ownership %	25.00%	Angels	$300,000	Series A – 42.9%
		Series A	$0	Series B – 37.1%
Expected Future Sale Price or IPO Value of Company	$10,000,000	Series B	$0	Series C – 25.2%
		Series C	$0	Later – 20.9%
		Total	**$300,000**	*Estimated averages

<u>COMMENTS</u>: This hypothetical represents a company that will require only angel funding ($300,000) in outside investments. If the company is able to grow to a company valued at $10 million, the founder who initially owned 25.0% would continue to own 25.0% at sale time (excluding additional stock awards as the company grew) worth approximately $2.5 million. The angel investors would own approximately 15% of the company.

	Year 0		Year 1		Year 2		Year 3		Year 4		Year 5	
	2010		2011		2012		2013		2014		2015	
Pre-Money Valuation	$1,700,000											
Outside Investment	$300,000											
Post-Money Valuation	$2,000,000	$0.40	$2,000,000	$0.40	$2,000,000	$0.40	$2,000,000	$0.40	$2,000,000	$0.40	$2,000,000	$0.40
Total Shares	5,000,000		5,000,000		5,000,000		5,000,000		5,000,000		5,000,000	
FUNDRAISING ROUND												
	Shares	%	**Shares**	%	**Shares**	%	**Shares**	%	**Shares**	%	**Shares**	%
Your Ownership	1,250,000	25.0%	1,250,000	25.0%	1,250,000	25.0%	1,250,000	25.0%	1,250,000	25.0%	1,250,000	25.0%
Other Founders	2,250,000	45.0%	2,250,000	45.0%	2,250,000	45.0%	2,250,000	45.0%	2,250,000	45.0%	2,250,000	45.0%
Stock Options	750,000	15.0%	750,000	15.0%	750,000	15.0%	750,000	15.0%	750,000	15.0%	750,000	15.0%
Additional Stock Options				0.0%		0.0%		0.0%		0.0%		0.0%
Angel Investors	750,000	15.0%	750,000	15.0%	750,000	15.0%	750,000	15.0%	750,000	15.0%	750,000	15.0%
Investors, Series A				0.0%		0.0%		0.0%		0.0%		0.0%
Investors, Series B						0.0%		0.0%		0.0%		0.0%
Investors, Series C						0.0%		0.0%		0.0%		0.0%
TOTAL	**5,000,000**	**100.0%**	**5,000,000**	**100.0%**	**5,000,000**	**100.0%**	**5,000,000**	**100.0%**	**5,000,000**	**100.0%**	**5,000,000**	**100.0%**
Dilution Effect			0.00%		0.00%		0.00%		0.00%		0.00%	
Owned by Founders/ Employees	85.00%		85.00%		85.00%		85.00%		85.00%		85.00%	
Owned by Investors	15.00%		15.00%		15.00%		15.00%		15.00%		15.00%	
Your Payout (if sold shares today @ <u>Sale Price/IPO Value</u>)	$2,500,000		$2,500,000		$2,500,000		$2,500,000		$2,500,000		$2,500,000	
Your Payout (if sold shares today @ <u>Valuation Price</u>)	$500,000		$500,000		$500,000		$500,000		$500,000		$500,000	

Hypothetical: Company with Very Low Capital Needs

Your Initial Ownership %	25.00%	**Amount of Funds Raised**		**Dilution***	
		Angels	$300,000	Series A – 42.9%	
Expected Future Sale Price or IPO Value of Company	$30,000,000	Series A	$3,500,000	Series B – 37.1%	
		Series B	$0	Series C – 25.2%	
		Series C	$0	Later – 20.9%	
		Total	**$3,800,000**	*Estimated averages	

COMMENTS: This hypothetical represents a company that will require $3.8 million in outside investments. If the company is able to grow to a company valued at $30 million, the founder who initially owned 25.0% would own 12.4% at sale time (excluding additional stock awards as the company grew) worth approximately $3.7 million. Investors would own approximately 50% of the company; founders and employees 50%.

	Year 0		Year 1		Year 2		Year 3		Year 4		Year 5	
	2010		2011		2012		2013		2014		2015	
Pre-Money Valuation	$1,700,000		$4,658,508									
Outside Investment	$300,000		$3,500,000									
Post-Money Valuation	$2,000,000	$0.40	$8,158,508	$0.81	$8,158,508	$0.81	$8,158,508	$0.81	$8,158,508	$0.81	$8,158,508	$0.81
Total Shares	5,000,000		10,080,645		10,080,645		10,080,645		10,080,645		10,080,645	
FUNDRAISING ROUND			A									
	Shares	**%**	**Shares**	**%**	**Shares**	**%**	**Shares**	**%**	**Shares**	**%**	**Shares**	**%**
Your Ownership	1,250,000	25.0%	1,250,000	12.4%	1,250,000	12.4%	1,250,000	12.4%	1,250,000	12.4%	1,250,000	12.4%
Other Founders	2,250,000	45.0%	2,250,000	22.3%	2,250,000	22.3%	2,250,000	22.3%	2,250,000	22.3%	2,250,000	22.3%
Stock Options	750,000	15.0%	750,000	7.4%	1,506,048	14.9%	1,506,048	14.9%	1,506,048	14.9%	1,506,048	14.9%
Additional Stock Options			756,048	7.5%		0.0%		0.0%		0.0%		0.0%
Angel Investors	750,000	15.0%	750,000	7.4%	750,000	7.4%	750,000	7.4%	750,000	7.4%	750,000	7.4%
Investors, Series A			4,324,597	42.9%	4,324,597	42.9%	4,324,597	42.9%	4,324,597	42.9%	4,324,597	42.9%
Investors, Series B						0.0%		0.0%		0.0%		0.0%
Investors, Series C						0.0%		0.0%		0.0%		0.0%
TOTAL	**5,000,000**	**100.0%**	**10,080,645**	**100.0%**	**10,080,645**	**100.0%**	**10,080,645**	**100.0%**	**10,080,645**	**100.0%**	**10,080,645**	**100.0%**
Dilution Effect			42.90%		0.00%		0.00%		0.00%		0.00%	
Owned by Founders/ Employees	85.00%		49.66%		49.66%		49.66%		49.66%		49.66%	
Owned by Investors	15.00%		50.34%		50.34%		50.34%		50.34%		50.34%	
Your Payout (if sold shares today @ Sale Price/IPO Value)	$7,500,000		$3,720,000		$3,720,000		$3,720,000		$3,720,000		$3,720,000	
Your Payout (if sold shares today @ Valuation Price)	$500,000		$1,011,655		$1,011,655		$1,011,655		$1,011,655		$1,011,655	

Hypothetical: Company with Low Capital Needs

		Amount of Funds Raised		Dilution*
Your Initial Ownership %	25.00%	Angels	$300,000	Series A – 42.9%
Expected Future Sale Price or IPO Value of Company	$30,000,000	Series A	$3,000,000	Series B – 37.1%
		Series B	$5,000,000	Series C – 25.2%
		Series C	$0	Later – 20.9%
		Total	**$8,300,000**	*Estimated averages

<u>COMMENTS</u>: This hypothetical represents a company that will require $8.3 million in outside investments. If the company is able to grow to a company valued at $30 million, the founder who initially owned 25.0% would own 7.8% at sale time (excluding additional stock awards as the company grew) worth approximately $2.3 million. Investors would own approximately 69% of the company; founders and employees 31%.

	Year 0		Year 1		Year 2		Year 3		Year 4		Year 5	
	2010		2011		2012		2013		2014		2015	
Pre-Money Valuation	$1,700,000		$3,993,007		$8,477,089							
Outside Investment	$300,000		$3,000,000		$5,000,000							
Post-Money Valuation	$2,000,000	$0.40	$6,993,007	$0.69	$13,477,089	$0.84	$13,477,089	$0.84	$13,477,089	$0.84	$13,477,089	$0.84
Total Shares	5,000,000		10,080,645		16,026,463		16,026,463		16,026,463		16,026,463	
FUNDRAISING ROUND			A		B							
	Shares	**%**	**Shares**	**%**	**Shares**	**%**	**Shares**	**%**	**Shares**	**%**	**Shares**	**%**
Your Ownership	1,250,000	25.0%	1,250,000	12.4%	1,250,000	7.8%	1,250,000	7.8%	1,250,000	7.8%	1,250,000	7.8%
Other Founders	2,250,000	45.0%	2,250,000	22.3%	2,250,000	14.0%	2,250,000	14.0%	2,250,000	14.0%	2,250,000	14.0%
Stock Options	750,000	15.0%	750,000	7.4%	1,506,048	9.4%	1,506,048	9.4%	1,506,048	9.4%	1,506,048	9.4%
Additional Stock Options			756,048	7.5%		0.0%		0.0%		0.0%		0.0%
Angel Investors	750,000	15.0%	750,000	7.4%	750,000	4.7%	750,000	4.7%	750,000	4.7%	750,000	4.7%
Investors, Series A			4,324,597	42.9%	4,324,597	27.0%	4,324,597	27.0%	4,324,597	27.0%	4,324,597	27.0%
Investors, Series B					5,945,818	37.1%	5,945,818	37.1%	5,945,818	37.1%	5,945,818	37.1%
Investors, Series C						0.0%		0.0%		0.0%		0.0%
TOTAL	**5,000,000**	**100.0%**	**10,080,645**	**100.0%**	**16,026,463**	**100.0%**	**16,026,463**	**100.0%**	**16,026,463**	**100.0%**	**16,026,463**	**100.0%**
Dilution Effect			42.90%		37.10%		0.00%		0.00%		0.00%	

	Year 0	Year 1	Year 2	Year 3	Year 4	Year 5
Owned by Founders/ Employees	85.00%	49.66%	31.24%	31.24%	31.24%	31.24%
Owned by Investors	15.00%	50.34%	68.76%	68.76%	68.76%	68.76%
Your Payout (if sold shares today @ <u>Sale Price/IPO Value</u>)	$7,500,000	$3,720,000	$2,339,880	$2,339,880	$2,339,880	$2,339,880
Your Payout (if sold shares today @ <u>Valuation Price</u>)	$500,000	$867,133	$1,051,159	$1,051,159	$1,051,159	$1,051,159

Hypothetical: Company with Medium-to-Low Capital Needs

		Amount of Funds Raised		Dilution*
Your Initial Ownership %	25.00%	Angels	$300,000	Series A – 42.9%
Expected Future Sale Price or IPO Value of Company	$50,000,000	Series A	$3,000,000	Series B – 37.1%
		Series B	$8,000,000	Series C – 25.2%
		Series C	$0	Later – 20.9%
		Total	$11,300,000	*Estimated averages

COMMENTS: This hypothetical represents a company that will require $11.3 million in outside investments. If the company is able to grow to a company valued at $50 million, the founder who initially owned 25.0% would own 7.8% at sale time (excluding additional stock awards as the company grew) worth approximately $3.9 million. Investors would own approximately 69% of the company; founders and employees 31%.

	Year 0 2010		Year 1 2011		Year 2 2012		Year 3 2013		Year 4 2014		Year 5 2015	
Pre-Money Valuation	$1,700,000		$3,993,007		$13,563,342							
Outside Investment	$300,000		$3,000,000		$8,000,000							
Post-Money Valuation	$2,000,000	$0.40	$6,993,007	$0.69	$21,563,342	$1.35	$21,563,342	$1.35	$21,563,342	$1.35	$21,563,342	$1.35
Total Shares	5,000,000		10,080,645		16,026,463		16,026,463		16,026,463		16,026,463	
FUNDRAISING ROUND			A		B							
	Shares	**%**	**Shares**	**%**	**Shares**	**%**	**Shares**	**%**	**Shares**	**%**	**Shares**	**%**
Your Ownership	1,250,000	25.0%	1,250,000	12.4%	1,250,000	7.8%	1,250,000	7.8%	1,250,000	7.8%	1,250,000	7.8%
Other Founders	2,250,000	45.0%	2,250,000	22.3%	2,250,000	14.0%	2,250,000	14.0%	2,250,000	14.0%	2,250,000	14.0%
Stock Options	750,000	15.0%	750,000	7.4%	1,506,048	9.4%	1,506,048	9.4%	1,506,048	9.4%	1,506,048	9.4%
Additional Stock Options			756,048	7.5%		0.0%		0.0%		0.0%		0.0%
Angel Investors	750,000	15.0%	750,000	7.4%	750,000	4.7%	750,000	4.7%	750,000	4.7%	750,000	4.7%
Investors, Series A			4,324,597	42.9%	4,324,597	27.0%	4,324,597	27.0%	4,324,597	27.0%	4,324,597	27.0%
Investors, Series B					5,945,818	37.1%	5,945,818	37.1%	5,945,818	37.1%	5,945,818	37.1%
Investors, Series C						0.0%		0.0%		0.0%		0.0%
TOTAL	**5,000,000**	**100.0%**	**10,080,645**	**100.0%**	**16,026,463**	**100.0%**	**16,026,463**	**100.0%**	**16,026,463**	**100.0%**	**16,026,463**	**100.0%**
Dilution Effect			42.90%		37.10%		0.00%		0.00%		0.00%	
Owned by Founders/ Employees	85.00%		49.66%		31.24%		31.24%		31.24%		31.24%	
Owned by Investors	15.00%		50.34%		68.76%		68.76%		68.76%		68.76%	
Your Payout (if sold shares today @ Sale Price/IPO Value)	$12,500,000		$6,200,000		$3,899,800		$3,899,800		$3,899,800		$3,899,800	
Your Payout (if sold shares today @ Valuation Price)	$500,000		$867,133		$1,681,854		$1,681,854		$1,681,854		$1,681,854	

Hypothetical: Company with Medium Capital Needs

		Amount of Funds Raised		Dilution*
Your Initial Ownership %	25.00%	Angels	$300,000	Series A – 42.9%
Expected Future Sale Price or IPO Value of Company	$100,000,000	Series A	$5,000,000	Series B – 37.1%
		Series B	$10,000,000	Series C – 25.2%
		Series C	$20,000,000	Later – 20.9%
		Total	$35,300,000	*Estimated averages

COMMENTS: This hypothetical represents a company that will require $35.3 million in outside investments. If the company is able to grow to a company valued at $100 million, the founder who initially owned 25.0% would own 5.8% at sale time (excluding additional stock awards as the company grew) worth approximately $5.8 million. Investors would own approximately 77% of the company; founders and employees 23%.

	Year 0 2010		Year 1 2011		Year 2 2012		Year 3 2013		Year 4 2014		Year 5 2015	
Pre-Money Valuation	$1,700,000		$6,655,012		$16,954,178				$59,365,079			
Outside Investment	$300,000		$5,000,000		$10,000,000				$20,000,000			
Post-Money Valuation	$2,000,000	$0.40	$11,655,012	$1.16	$26,954,178	$1.68	$26,954,178	$1.68	$79,365,079	$3.70	$79,365,079	$3.70
Total Shares	5,000,000		10,080,645		16,026,463		16,026,463		21,425,753		21,425,753	
FUNDRAISING ROUND			A		B				C			
	Shares	**%**	**Shares**	**%**	**Shares**	**%**	**Shares**	**%**	**Shares**	**%**	**Shares**	**%**
Your Ownership	1,250,000	25.0%	1,250,000	12.4%	1,250,000	7.8%	1,250,000	7.8%	1,250,000	5.8%	1,250,000	5.8%
Other Founders	2,250,000	45.0%	2,250,000	22.3%	2,250,000	14.0%	2,250,000	14.0%	2,250,000	10.5%	2,250,000	10.5%
Stock Options	750,000	15.0%	750,000	7.4%	1,506,048	9.4%	1,506,048	9.4%	1,506,048	7.0%	1,506,048	7.0%
Additional Stock Options			756,048	7.5%		0.0%		0.0%		0.0%		0.0%
Angel Investors	750,000	15.0%	750,000	7.4%	750,000	4.7%	750,000	4.7%	750,000	3.5%	750,000	3.5%
Investors, Series A			4,324,597	42.9%	4,324,597	27.0%	4,324,597	27.0%	4,324,597	20.2%	4,324,597	20.2%
Investors, Series B					5,945,818	37.1%	5,945,818	37.1%	5,945,818	27.8%	5,945,818	27.8%
Investors, Series C						0.0%		0.0%	5,399,290	25.2%	5,399,290	25.2%
TOTAL	**5,000,000**	**100.0%**	**10,080,645**	**100.0%**	**16,026,463**	**100.0%**	**16,026,463**	**100.0%**	**21,425,753**	**100.0%**	**21,425,753**	**100.0%**
Dilution Effect			42.90%		37.10%		0.00%		25.20%		0.00%	
Owned by Founders/ Employees	85.00%		49.66%		31.24%		31.24%		23.36%		23.36%	
Owned by Investors	15.00%		50.34%		68.76%		68.76%		76.64%		76.64%	
Your Payout (if sold shares today @ Sale Price/IPO Value)	$25,000,000		$12,400,000		$7,799,600		$7,799,600		$5,834,101		$5,834,101	
Your Payout (if sold shares today @ Valuation Price)	$500,000		$1,445,221		$2,102,318		$2,102,318		$4,630,239		$4,630,239	

Hypothetical: Company with Medium-High Capital Needs

		Amount of Funds Raised		Dilution*
Your Initial Ownership %	25.00%			
		Angels	$300,000	Series A – 42.9%
Expected Future Sale Price or IPO Value of Company	$150,000,000	Series A	$8,500,000	Series B – 37.1%
		Series B	$20,000,000	Series C – 25.2%
		Series C	$30,000,000	Later – 20.9%
		Total	$58,800,000	*Estimated averages

COMMENTS: This hypothetical represents a company that will require $58.8 million in outside investments. If the company is able to grow to a company valued at $150 million, the founder who initially owned 25.0% would own 4.9% at sale time (excluding additional stock awards as the company grew) worth approximately $7.3 million. Investors would own approximately 80% of the company; founders and employees 20%.

	Year 0 2010		Year 1 2011		Year 2 2012		Year 3 2013		Year 4 2014		Year 5 2015	
Pre-Money Valuation	$1,700,000		$8,199,411		$33,908,356		$89,047,619					
Outside Investment	$300,000		$8,500,000		$20,000,000		$30,000,000					
Post-Money Valuation	$2,000,000	$0.40	$16,699,411	$1.39	$53,908,356	$2.82	$119,047,619	$4.66	$119,047,619	$4.66	$119,047,619	$4.66
Total Shares	5,000,000		12,019,231		19,108,475		25,546,090		25,546,090		25,546,090	
FUNDRAISING ROUND			**A**		**B**		**C**					

	Shares	**%**	**Shares**	**%**	**Shares**	**%**	**Shares**	**%**	**Shares**	**%**	**Shares**	**%**
Your Ownership	1,250,000	25.0%	1,250,000	10.4%	1,250,000	6.5%	1,250,000	4.9%	1,250,000	4.9%	1,250,000	4.9%
Other Founders	2,250,000	45.0%	2,250,000	18.7%	2,250,000	11.8%	2,250,000	8.8%	2,250,000	8.8%	2,250,000	8.8%
Stock Options	750,000	15.0%	750,000	6.2%	1,651,442	8.6%	1,651,442	6.5%	1,651,442	6.5%	1,651,442	6.5%
Additional Stock Options			901,442	7.5%		0.0%		0.0%		0.0%		0.0%
Angel Investors	750,000	15.0%	750,000	6.2%	750,000	3.9%	750,000	2.9%	750,000	2.9%	750,000	2.9%
Investors, Series A			6,117,788	50.9%	6,117,788	32.0%	6,117,788	23.9%	6,117,788	23.9%	6,117,788	23.9%
Investors, Series B					7,089,244	37.1%	7,089,244	27.8%	7,089,244	27.8%	7,089,244	27.8%
Investors, Series C						0.0%	6,437,615	25.2%	6,437,615	25.2%	6,437,615	25.2%
TOTAL	**5,000,000**	**100.0%**	**12,019,231**	**100.0%**	**19,108,475**	**100.0%**	**25,546,090**	**100.0%**	**25,546,090**	**100.0%**	**25,546,090**	**100.0%**
Dilution Effect			50.90%		37.10%		25.20%		0.00%		0.00%	

	Year 0	Year 1	Year 2	Year 3	Year 4	Year 5
Owned by Founders/ Employees	85.00%	42.86%	26.96%	20.17%	20.17%	20.17%
Owned by Investors	15.00%	57.14%	73.04%	79.83%	79.83%	79.83%
Your Payout (if sold shares today @ Sale Price/IPO Value)	$37,500,000	$15,600,000	$9,812,400	$7,339,675	$7,339,675	$7,339,675
Your Payout (if sold shares today @ Valuation Price)	$500,000	$1,736,739	$3,526,469	$5,825,139	$5,825,139	$5,825,139

Hypothetical: Company with High Capital Needs

		Amount of Funds Raised		Dilution*
Your Initial Ownership %	25.00%			
		Angels	$300,000	Series A – 42.9%
Expected Future Sale Price or IPO Value of Company	$250,000,000	Series A	$10,000,000	Series B – 37.1%
		Series B	$30,000,000	Series C – 25.2%
		Series C	$50,000,000	Later – 20.9%
		Total	$90,300,000	*Estimated averages

COMMENTS: This hypothetical represents a company that will require $90.3 million in outside investments. If the company is able to grow to a company valued at $250 million, the founder who initially owned 25.0% would own 4.9% at sale time (excluding additional stock awards as the company grew) worth approximately $12.2 million. Investors would own approximately 80% of the company; founders and employees 20%.

	Year 0 2010		Year 1 2011		Year 2 2012		Year 3 2013		Year 4 2014		Year 5 2015	
Pre-Money Valuation	$1,700,000		$9,646,365		$50,862,534		$148,412,698					
Outside Investment	$300,000		$10,000,000		$30,000,000		$50,000,000					
Post-Money Valuation	$2,000,000	$0.40	$19,646,365	$1.63	$80,862,534	$4.23	$198,412,698	$7.77	$198,412,698	$7.77	$198,412,698	$7.77
Total Shares	$5,000,000		$12,019,231		$19,108,475		$25,546,090		$25,546,090		$25,546,090	
FUNDRAISING ROUND			A		B		C					
	Shares	**%**	**Shares**	**%**	**Shares**	**%**	**Shares**	**%**	**Shares**	**%**	**Shares**	**%**
Your Ownership	1,250,000	25.0%	1,250,000	10.4%	1,250,000	6.5%	1,250,000	4.9%	1,250,000	4.9%	1,250,000	4.9%
Other Founders	2,250,000	45.0%	2,250,000	18.7%	2,250,000	11.8%	2,250,000	8.8%	2,250,000	8.8%	2,250,000	8.8%
Stock Options	750,000	15.0%	750,000	6.2%	1,651,442	8.6%	1,651,442	6.5%	1,651,442	6.5%	1,651,442	6.5%
Additional Stock Options			901,442	7.5%		0.0%		0.0%		0.0%		0.0%
Angel Investors	750,000	15.0%	750,000	6.2%	750,000	3.9%	750,000	2.9%	750,000	2.9%	750,000	2.9%
Investors, Series A			6,117,788	50.9%	6,117,788	32.0%	6,117,788	23.9%	6,117,788	23.9%	6,117,788	23.9%
Investors, Series B					7,089,244	37.1%	7,089,244	27.8%	7,089,244	27.8%	7,089,244	27.8%
Investors, Series C						0.0%	6,437,615	25.2%	6,437,615	25.2%	6,437,615	25.2%
TOTAL	**5,000,000**	**100.0%**	**12,019,231**	**100.0%**	**19,108,475**	**100.0%**	**25,546,090**	**100.0%**	**25,546,090**	**100.0%**	**25,546,090**	**100.0%**
Dilution Effect			50.90%		37.10%		25.20%		0.00%		0.00%	

	Year 0	Year 1	Year 2	Year 3	Year 4	Year 5
Owned by Founders/ Employees	85.00%	42.86%	26.96%	20.17%	20.17%	20.17%
Owned by Investors	15.00%	57.14%	73.04%	79.83%	79.83%	79.83%
Your Payout (if sold shares today @ Sale Price/IPO Value)	$62,500,000	$26,000,000	$16,354,000	$12,232,792	$12,232,792	$12,232,792
Your Payout (if sold shares today @ Valuation Price)	$500,000	$2,043,222	$5,289,704	$9,708,565	$9,708,565	$9,708,565

Part II

Starting It Up

Going into business for yourself, becoming an entrepreneur, is the modern-day equivalent of pioneering on the old frontier.

Paula Nelson
Cofounder of three successful electronics equipment companies and author of The Joy of Money

4

Forming a Founding Team

There are four of us: me and Sami Shaio, Jon Payne and Arthur van Hoff. All four of us are part of the original core Java team. It was a combination of a personal decision for each one of us, and a realization that this was the right time, the opportune time in the market. Each one of us, I think, had always wanted to be an entrepreneur.

Kim Polese
CEO of Marimba Inc., CEO of SpikeSource

People buy into the leader before they buy into the vision.

John C. Maxwell
Author of The 21 Irrefutable Laws of Leadership *and* The 21 Indispensable Qualities of a Leader: Becoming the Person Others Will Want to Follow

What to Watch For

"Great teams can turn a mediocre idea into a profitable and successful business," says Diane Fox, founder of PartnerUp, a resource for entrepreneurs. "A great team can help your company expand and grow beyond your wildest dreams. On the flip side, a great idea, executed by a shoddy team, is nothing more than a recipe for failure."

And Fox is right: "It's about the people." Without top-tier people throughout the startup organization, it is impossible to compete with competition that can probably throw more money, resources, and time into the fight. Small and nimble organizations stocked with smart, motivated team members have quickly grown into those large organizations, from Apple Computers to ZymoGenetics and others in between.

For a brand new startup, the team starts with its founders, and setting the tone at the top permeates the rest of the organization.

As you continue to read on, you'll quickly realize that a fairly significant portion of this book is spent discussing, arguably, the most valuable asset of you new business: its people. In particular, this chapter will be focused on your core team that sets out to establish the business. There are certain stages of a business that are nearly universal. One of those stages is establishment of a founding team. The process of establishing a founding team could take place in a single event (for instance, you and two other individuals together leave your current employers to start your new business venture), or it could take place over a more extended time period (for instance, you leave your employer and recruit a founding team over the course of a six to nine month period). Either way, the process of recruiting the founding team and forming the overarching tenants of this early partnership with the other founders is vitally important.

In an interesting posting on his blog "How to Change the World" (http://blog.guykawasaki.com), Guy Kawasaki, CEO of Garage Technology Ventures, tells how his former Macintosh Division coworkers had a saying, "A players hire A players; B players hire C players." Said Kawasaki, "Great people hire great people. On the other hand, mediocre people hire candidates who are not as good as they are, so they can feel superior to them."

Kawasaki is right. Hiring excellent talent is a mentality that starts with the founding team. Creating the best founding team will push each of the founders to hire top talent (and will help top talent be attracted to your organization). If you are unable to create a team of highest quality equals, then don't settle for whatever you can find. As will be the theme throughout this book (and hopefully throughout your entire business venture), one of your most important assets (if not the most important asset) is and will continue to be your people.

WHAT ARE INVESTORS LOOKING FOR FROM YOUR FOUNDING TEAM?

Here are some of the potential questions that an outside investor will want to know about a new business:

- Has the team worked together previously? How do we know that they can?
- Does the team seem capable of adapting as the venture and the needs change?
- Does the team have the required technical experience?
- Does the team have the necessary business expertise?
- What is the team dynamic? Is there a natural leader in the group and do the others listen to him or her?

- Does the team accept criticism and challenge each other in a positive and helpful way?
- Can the team recruit and retain top talent?
- Does the team seem willing to adapt to changes in the team dynamic as the need for outside talent arises?

Your Founding Team

Generally, a venture will begin with a small, core group of individuals who will serve as the backbone of the company until it receives outside funding. Additional talent will be added after this point and will supplement in the key areas necessary for the businesses development. This cannot supplant the need for bringing the most outstanding collection of talent together for your founding team.

Most of the success stories of the past three decades in entrepreneurial ventures have come from a team of dedicated founders. Certain stories are obvious in the mass media: Bill Gates and Paul Allen founded Microsoft together; Larry Page and Sergey Brin cofounded Google; YouTube has its famous cofounders Chad Hurley and Steve Chen (the third founder, Jawed Karim, left the company to attend graduate school); and David Filo and Jerry Yang founded Yahoo! together in 1994. In other cases, history may no longer tell the story of the full founding team: Steve Jobs and Steve Wozniak were actually cofounders of Apple (many today have forgotten the other Steve), Oracle Corporation was founded in 1977 by Larry Ellison, Bob Miner, and Ed Oats (but Larry remains in the public consciousness); and although Marc Andreessen and Jim Clark cofounded Netscape, Marc is probably more well known to the general public despite the fact that Jim Clark has led several successful ventures in his career and initially brought Marc into Netscape.

There are certain stories when a single founder was the central figure in the company's success. Stories such as Jeff Bezos from Amazon.com, Michael Dell of Dell Computers, Mark Zuckerberg from Facebook, and Pierre Omidyar, eBay's founder, certainly do exist. However, even in these cases, there were other important early-stage hires at or near the time of founding that were crucial to each of these company's long-term success.

What is the lesson that these stories teach us? Running a startup is an adventure with a series of highs and lows. Most times, facing those lows with a strong team around you will be much easier than facing it alone. Simply put, it is without question one of the most important responsibilities of a founder to build a strong founding team to complement his talents. Although you may want to "go it alone" in the earliest days of a new venture, eventually any successful venture will require building a strong core of early partners and team members to become a sustainable success. To make your venture more than simply an idea or a concept, you must build a competent founding team. If you are unable to build a core group or team of individuals when you begin presenting your story to investors, your story will be judged through a cautious eye.

In research done by David Hsu from the Wharton School, Hsu found that the ability of the founder or founders to recruit talented executives (other founders and management-level employees) for their organization from *within* their own social circles (as to opposed to recruiting talent from within the venture capital firm's social circles) was positively associated with the resulting venture valuation. The result is that a team that is built with outstanding talent before searching for outside investment is likely to see a higher valuation assigned to it by the outside investor. Making these top-notch hires early on in your founding will most likely add immediate value to the organization and increase the amount of equity you (and your cofounders) are able to maintain in your new venture as a result of an increased valuation.

How Big Should the Team Be?

The answer really depends, so much so that you shouldn't spend time worrying whether your team isn't "just right." A team will naturally expand as you begin to enter into future phases of product development and marketing or sales. This means that you don't need to build a complete team from the beginning (and if you did, how would you pay for them anyway?). A team of founders most typically consists of two to four people at its outset. It can be larger or smaller depending on the way this fits into the business plan itself.

Research by Chuck Eesley from the MIT suggests that technology startups with larger founding teams are more likely to be successful than those with smaller teams. According to Eesley, as the number of cofounders increases, so does the likelihood that the company will do an IPO. This may be attributable to a number of factors, such as the increase in contacts a larger founding team will collectively bring to the entity or the ability to tackle broader challenges earlier in the lifecycle of the entity. In either case, Eesley's research also shows that a founding team is more likely to be successful if the team has any previous experience with a company that had done an IPO or has an increased number of connections to venture capital firms. Therefore, this research suggests that building a strong, well-connected team is crucial to future successes of the startup company.

Assuming you begin with a three person founding team, many startups will begin with a president/chief executive officer, a vice president of marketing/sales, and a vice president of engineering/chief technical officer. You will continue to build the organization as you continue to raise funds, develop your product, and begin sales efforts, but ultimately, the core team will likely consist of some combination of these key business aspects and skill sets.

WE'VE FORMED OUR TEAM. ARE WE READY TO INCORPORATE NOW?

What could happen

We've got a founding team in place for our new startup. When is the right time to incorporate the business?

Watch out for

There are different schools of thought on when the "right" time is to incorporate. Some founders have decided to wait until just before they are to receive funding, whereas others have formed their corporation after a meeting of the minds of the founders. Even so, keep in mind that waiting to incorporate until immediately before your Series A investment round may raise a number of tax issues for the founders because of the increased valuation at the time of this investment.

Incorporation is necessary for certain aspects of the business, and the expense is relatively minimal. If you are still in the "research and discussion" phase, deciding whether the idea and concept have legs, there is likely little need to formally incorporate. Instead, the founders should plan the timing of your incorporation based on factors including the timing for extracting themselves from their current employers, the need to transfer existing intellectual property to the company, the need to enter into contracts for the business (with customers or vendors), the formalization of agreed on founders terms (ownership levels, intellectual property matters, etc.), the need to grant options to employees, and receipt of investment funding.

The lesson here is you can incorporate too quickly (creating potential problems for a founder still employed elsewhere) and too slowly (creating potential tax problems for the founders).

TIP: Don't wait too long to incorporate, but also avoid rushing to incorporate and making a rash decision on your entity structure.

What Should You Be Looking For in Cofounders?

The research of Chuck Eesley at MIT focused on founding teams to identify where founders had met one another. Eesley's research looked into founding teams from recent graduates (less than five years after graduation) and established alumni (more than five years after graduation). Among founding teams of recent graduates that had formed companies since 2000, approximately 30% of the founding teams grew out of their MIT research, 20% from work relationships, 20% from extracurricular activities, and 30% from social activities. For founding teams of established alumni that had formed companies since 2000, a larger number of founding teams grew out of work relationships. For these teams, less than 15% grew out of MIT research, 40% from work relationships, 40% from social activities, and less than 5% from extracurricular activities. As graduates advance in their careers, a greater number of founding teams will grow out of work and social relationships, whereas for recent graduates more startup teams will be formed based on previous research and extracurricular activity relationships.

In many cases, a founding team will grow out of personal relationships or a working relationship (for example, Paul Allen and Bill Gates of Microsoft, who became friends in high school in Seattle, and Larry Page and Sergey Brin of Google, who met as Stanford University graduate students). In those cases, you may have a self-contained team in place ready to begin efforts to develop the organization. You may desire to add additional talent to your founding team to round out the relevant talents and skills of the founders; you will need to closely evaluate the current needs you'll have during the formation and early-stage funding phases of the organization.

In other situations in which you do not have a readily identified cofounder (for example, Steve Jobs convinced an initially skeptical Steve Wozniak to join him after Jobs had proposed selling a computer as a fully assembled personal computer board), you may need to begin searching your social network to find additional key members to join you.

As you begin to consider how to form a team of founders for this new business venture, you should put yourself into the seat of a potential investor. What are they going to want from a two to five person team to tackle the problem you've signed up for? Obviously, they are not expecting you to address every area of this challenge, but what is the key talent that needs to be at the table?

Sometimes, founders have likened the early experience with their cofounders to a marriage. After all, remember that, when forming a team, you will need to be able to work well with these individuals in an intense environment.

There are a number of things that can be identified as key traits to look for in any member of your founding team. However, identified here are six key features that an ideal candidate would have (but remember, you may not find anyone who has all of these characteristics, so look for someone who combines the best mixture for your team):

- Compatibility
- Mixture of new and old
- Entrepreneurial experience
- Top-tier technical experience
- High integrity and strong ethics
- A "true" equal

Compatibility

Fancy titles, long resumes, multiple certifications, extensive publication lists, large patent portfolios, and a history of successful startups among your founder team count for very little if the team is unable to work well together. Be sure to have a good sense before you agree to cofound a business with anyone.

The first place to find cofounders may be right in front of you. Oftentimes, entrepreneurs will look among people they know personally (and have worked with or for) or individuals who are one to two degrees of separation away (so you can get a direct reference from someone you know). A cofounder represents more than an employee or member of management: this is an individual who you are comfortable having in your inner circle during the formation and early-stage financing periods of the business.

Mixture of New and Old

When you have decided to start a new business, you should try to avoid forming a team made up of only individuals you had worked with at your last job. Although it is certainly helpful to have some members of your founding team that you've worked with previously or recently (and investors are oftentimes looking for preexisting working dynamic), it is important to build a diverse team to execute your vision. Investors will often be wary of a team made up entirely of individuals who worked together in their most recent job. They are looking for a group that brings together diverse perspectives needed to tackle complex problems and that can recruit additional talents from varied backgrounds.

Entrepreneurial Experience

Many top-tier investors look at previous experience in a successful startup as a key trait needed in a founding team, and the empirical data appear to back this up. In research done by Paul Gompers from Harvard University, Gompers and his team found that repeat entrepreneurs have a significantly higher chance of success (success being defined as an

IPO or a successful merger or acquisition event). In fact, their data suggest that a company founded by an experienced entrepreneur is nearly 15% more likely to be a "success" than a first-time entrepreneur's company. Likewise, these serial entrepreneurs are more likely to receive funding earlier in their company's life cycle and receive funding from more experienced venture capital firms.

Having seasoned entrepreneurs on your team has also been shown to increase a company's valuation at funding and the likelihood of attracting venture capital dollars. In research into the backgrounds of entrepreneurs who had received venture financing, David Hsu found that previous founding experience (especially financially successful experience) increased the likelihood of receiving venture capital funding and was also directly tied to the valuation that the company received. Higher valuations were directly tied to previous founding experience. According to Hsu, in his research on the Internet industry, ventures that had at least one member with a doctoral decree were also more likely to receive venture capital funding and also saw higher valuations, on average.

It may sound obvious to reiterate, but a startup company is a business. Running a successful business requires certain core competencies and business expertise, including management, marketing, sales, finance, and operations. Because founding a new business venture represents a technical and operational challenge, investors look for teams that can execute technically and operationally. Therefore, carefully consider the amount and level of startup experience within your founding team. Individuals with previous startup experience can provide relevant insights into the process and may also have certain contacts (potential investors, startup lawyers or accountants, or other entrepreneurial talent) that first-time entrepreneurs may not have.

Top-Tier Technical Background

In a high-tech venture, you will be responsible for recruiting high-technology talent to solve key problems. One of the most important recruiting tools in the development of your organization will be your technical founders. Therefore, you should look for individuals with an outstanding technical background as well as a wealth of contacts in their field who will serve as your first source of talent for your organization.

High Integrity and Strong Ethics

The nature of a startup (and the "lean" way most are typically formed and organized) is that you are without procedures and protocols to police and prevent certain acts of fraud and deception. It is easier for one individual to act in a way that would be counter to the interests of the founding team. For this reason, you should surround yourself with people who have high integrity and strong ethics (whether you have personal experience or obtain personal referrals to the character of your cofounders).

PIXELON.COM: LESSONS ABOUT (A LACK OF) INTEGRITY

A look at the rise and fall of Pixelon.com highlights the importance of working with high-integrity individuals. Pixelon was founded to revolutionize Internet video. The company claimed to have technology to offer better distribution of high-quality video over the Internet using technologies that never actually existed. The founder of Pixelon, known in this venture by the name of "Michael Fenne," was actually

David Kim Stanley. Stanley was able to burn through $16 million of cash by spending money on an extravagant product launch in Las Vegas highlighted by celebrities such as Kiss and The Who.

Soon after a product launch without a product to show for it, the house of cards built by Stanley started to fall. By 2000, the company had let go of all its employees and filed for bankruptcy. After the company folded when it couldn't deliver on its grand promises, authorities found that Stanley was wanted in connection with other fraud cases, including a previous scheme netting himself $1.5 million from his contacts.

Although Pixelon may be an example of the extreme, recognize that private companies such as startups tend to have less regulation and diligence associated with the companies. A new organization will need to be built on trust and integrity for the sake of other employees, investors, and customers. Be certain that any prospective cofounder of your organization has that high level of integrity.

A "True" Equal

Finally, and perhaps most importantly, your should hire those whom you consider to be your "true" equals. This may seem obvious: of course you should hire a peer and an equal, right? Unfortunately, it isn't quite as obvious as it may seem. When hiring the inner circle of a startup organization (and this should carry over to nearly every hire you will make within your organization), some teams fail when structured in a true hierarchical manner. This approach, although more typical in larger organizations, tends to be less successful in an early-staged startup venture.

Consider building a team that can collaborate as equals and peers. The simple truth is, this strategy builds an exceptional team of equals rather than an organization built around a single leader. In a startup, a one person team with a support staff is rarely the answer. So when considering how to build your team, look to complement the founders with others that can each play a vital role in the development and growth of the business. Surrounding a lead founder with "yes people" is rarely an approach that builds a long-term sustainable venture. As Kawasaki states, "A people hire A people." Create that team of As rather than an A surrounded by Bs and Cs.

Be patient in building your team. Although it may seem that time is of the essence in building the business to take advantage of an unmet need, the long-term health of the business depends on the people involved in the organization.

How Should You Go about Finding and Recruiting These People?

One of the first lessons here should be that looking for a cofounder is very different from recruiting an information technology manager or a vice president of sales. Although hiring members of your management team is crucial, most will not be founders. Founders are partners, and the partnership will set the venture on its initial trajectory.

Founders will each typically join from beginning (or usually very near to the beginning stages, although in some cases an additional founder will be brought on six to nine months after the initial founding group). The founding team will play a unique role and will be responsible for forming the company, setting the initial plans, and raising startup capital. Remember that this process is, in most cases, different from hires made after an initial funding event. (One good way that many companies make this distinction is to issue common stock to the founders and issue stock options to the other management team members.)

How should you find your cofounders? Begin first by looking at your personal network: those with whom you have worked at previous and current employers (but you should be careful with coworkers in your current employment situation if you have a nonsolicitation clause). Who have you worked with in volunteer, civic, or other organizations? Look to a service such as LinkedIn.com or another business networking tool, if you use it. Once you have identified people in your personal network, make a list of these people. This is your starting point of key people to meet with and talk to about the new venture you would like to build. Now that you have a list of useful, helpful, or interesting people, meet with them on an informal basis.

Even founders who believe they have a founding team will likely benefit from this approach. You may not realize what you are missing (or what you already have in your social circle) until you begin these initial discussions. Founder teams have found important members of their advisory boards, future chief financial officers (CFOs), customers, landlords, lawyers, accountants, and countless key contacts they've used in their business from these initial meetings. Each founder should begin, even at this early stage, having these initial dialogues with people in your network.

At this stage of your "meeting circuits," you should not make finding a cofounder the focus of any of your meetings; instead, you should be looking for conversations, meetings, and lunches. As a new business owner, you'll need to begin developing a personal network to help in a variety of areas (for example, finding a good lawyer, recruiting various types of talent, finding customers, and soliciting investors). Ultimately, you may be surprised at what you are able to find from these initial conversations, including the amount of resources you had at your fingertips!

In initial phases of discussions and fact finding, many entrepreneurs will not make an express mention of the fact they are looking for cofounders. Instead, these meetings will just serve as casual meetings with interesting people. Your goal should be to learn as much as you can that may help you in any phase of your business development. Be cautious of revealing potentially confidential or proprietary information. Keep in mind, too, that you should be primarily looking to gather information from these meetings rather than offer detailed information about your business idea or concept. From certain of your contacts, you'll find little of interest or information. However, from other connections, you'll find they are personally excited and interested or may have personal contacts to introduce to you. Again, remember that in many of your initial meetings you are primarily looking for insights and information, not specifically a cofounder.

As you continue to have meetings with your initial contacts as well as with these direct introductions, hopefully you'll begin to get a sense of people with whom you may like to enjoin in this business endeavor. In some cases, a perfect candidate may simply raise her hand to join you as a cofounder. In others, you'll scratch your head after a series of meetings wondering whether you are any closer, but remember, this is just the first phase: developing a list of resources and contacts who you'd be interested in having play a role in your business. You should search broadly at this stage: don't be afraid to meet with a CEO of a company who you are certain won't leave his company. Oftentimes, this person will be a great source of referrals and may be a perfect person to serve on your advisory board. You may not need a biologist or a financial controller, but you may be surprised at who they know and can connect to you.

I Think I've Found Someone Perfect, So What's Next?

Once you've identified someone to be your cofounder, the job has just begun. The next step is convincing this person to join you and enter the "marriage" of a startup. According to Guy Kawasaki, "The art of recruiting is the purest form of evangelism because you're not simply asking people to try your product, buy your product, or partner with you. Instead, you are asking them to bet their lives on your organization."

Remember rule number one: the people you probably want most to join you in your business will most likely not be looking nor will they let on that they are truly interested (at first). Selling a prospective cofounder on your business idea may involve more than just your efforts. If you are family friends, perhaps a discussion may involve the contact and his spouse and family. You may also involve another founder, if you have already brought someone on board. Consider this "sale" to be your first of many selling efforts for your business. This sale will involve a well-thought out description of the vision for the business, as well as a display of your personal passion for the organization, but be sure to avoid overselling the business and the opportunity to "cash in" on success. Starting a business is not without its share of risks, and, for every success story, there are numerous stories in which the business did not achieve on its potential. The startup life is a risk and a challenge.

Agreements among the Founders

Once you've established the founding team, too often founders will forget to have the tough discussions a founding team needs to have. Don't leapfrog these discussions and jump headfirst into the challenges of starting a business.

That first meeting among the founders of a company is most often rosy and focused on the good times ahead. Founding a business represents the first step on this new business journey, and, at the time of formation, few founders are considering the fact that nearly every startup business will face the fact that one or more of the original business founders will leave the venture within the first three years. In some cases, these departures can be completely amicable if a founder realizes that she is unable to stay on the team while waiting for funding. Conversely, departures can stem from conflict and result in hard feelings, frustrations, and, in certain cases, firing of a cofounder.

Recognizing the likelihood of these events, founder teams should discuss and address key issues that will inevitably face the business. During the "honeymoon" period, things always seem terrific, but once the challenges of starting a business set in, be sure to have discussed the what-ifs. What are these issues that the founders will face? Below is a list of the key discussion points a founders team should address, and a more detailed discussion follows this list:

- Personal goals of each founder
- Where will you locate the business
- Leaving your current employer
- Employment
- Company titles and job responsibilities
- Salary and benefits

- Contractual terms of employment (for example, vacation time, severance, noncompetition, and nonsolicitation)
- Termination of a founder
- Cash investments into the business
- Ownership of stock
- Vesting of stock (and acceleration terms)
- Departing the company
- Transfer restrictions
- Future stock issuances
- Day-to-day decision-making
- Major decision-making
- Death or disability
- Proprietary information
- Confidentiality
- Nonnegotiable changes to the business

How Should You Handle These Discussions?

Some founders will address certain matters soon after the team is formed (i.e., when will everyone leave their current employers and who will be employed in what roles?) and other matters when the company is beginning to raise funds or has incorporated the business (i.e., how will our stock vest or do we have any rights in future stock issuances?). Some founding teams choose to have these discussions with their attorney, who may be able to provide some context as to what is typical or "common" among other clients or in certain industries. Still other founding teams will prepare a checklist or term sheet detailing some of these key terms. A sample checklist for a founding team is provided at the end of this chapter that you can use or customize to fit your needs. In either case, the founders should discuss many of these items fairly early in the process to avoid issues that may arise when a founder departs or the business structure needs to be changed.

How Do You Memorialize These Understandings?

There is not a single "founders' agreement" that will lay out all the key terms of your business in a single document. Rather, the list of topics of discussion will tend to be handled through a variety of separate documents and agreements. For instance, you would traditionally handle stock ownership issues in your founders' stock purchase agreements, a stockholders' agreement, or a limited liability corporation (LLC) operating agreement, including vesting and transfer restrictions. Employment issues such as salaries, termination, titles, vacation, severance, noncompetition, and nonsolicitation would all be a part of your employment agreements or offer letters. Confidentiality, invention assignment, and assistance with patent or other intellectual property filings usually addressed in either the employment agreement or a specific confidentiality or invention assignment agreement.

In most cases, your attorney will be able to assist you with drafting certain documents for each of your founders, including employment agreements or offer letters, confidentiality and invention assignment agreements, stock purchase agreements, stockholders' agreements, and other operational documents such as your bylaws and your formation document (articles or certificate of incorporation). These are the key documents (among others) in which you'll handle each of the topics above.

Personal Goals

Each person joins a startup in the earliest stages for differing reasons. Some join for more freedom and flexibility, others for the hope of a sizeable payout at the end, and still others for the challenge. Although the reasons themselves may not be crucial to your business, the personal goals of each founder are important to the continued satisfaction of each team member. If one founder has joined in your business venture because they believe your technology will really help people, be certain that this founder is able to play an integral role in customer development and feedback. Remember, your business is a collection of individuals and each individual may be motivated differently. Your team should ensure that each member's personal motivations are being satisfied as much as possible.

Founders should also consider discussing the risks associated with the venture and the financial implications of joining a startup business. In many cases, startup teams will be unable to draw salaries or will draw salaries at significantly reduced levels. Additionally, startups may require personal financial investments from the founders. Consider asking each founder to prepare a personal financial summary (located at the conclusion of Chapter 3) to allow the founding team to discuss appropriate funding considerations.

Where Will You Locate the Business?

A great deal of research has been conducted to determine whether a startup's chances of success change depending on location. The results are mixed. One study concluded that startup firms in rural areas have an increased chance of survival, whereas firms in urban areas have a decreased chance of survival. Yet other research has shown that product innovation clusters spatially in regions that provide concentrations of the knowledge needed for the commercialization process. That is, the presence of universities, related industries, and specialized business services can create a technological infrastructure that promotes information transfers, thus lowering the risk and costs of undertaking innovative activity. We can casually call this the Silicon Valley effect.

Still other anecdotal evidence suggests that venture capital firms are more likely to fund companies located near their office. The reason? Venture firm partners prefer to avoid the need to travel extensively just to visit their portfolio companies.

Although numerous companies headquartered outside of the traditional startup clusters have been successful, some startup companies will consider relocating their business to a location that may be more favorable, perhaps to be near potential investors, employees with relevant skills and experience, and support industries. Perhaps the technology for your company will be licensed from a university in a rural location, but it may be best to locate your business in a city that has an infrastructure known to support startup technology companies.

Either way, it is important that the founding team discuss and agree on the location of the business. Not all founders will consider themselves sufficiently mobile to relocate themselves and their family, and therefore the founding team should discuss any relocation early in the discussion. In some cases, the founding team may only have some members relocate before receiving initial funding or a founder may choose to telecommute.

DOES IT MATTER WHERE WE "START UP"?

What could happen

You are starting a high-tech business. You live in Nebraska (or name any other city/state of your choice). Does it really matter where you start up your business?

Watch out for

In the same way that New York City is the fashion capital and Los Angeles is the film capital, so is Silicon Valley the startup capital in the United States. Does that mean you shouldn't consider starting up a business if you don't live in that area of California? Absolutely not. However, you should recognize that certain cities are known as places where startups prosper and be sure your business is located in an area that can provide tools for success.

Your startup location matters because you want to be in a place that has the right type of talent for your organization, the partners and joint ventures for your business, the support services and resources, the capital and investment options, and the attitude that promotes and enhances startups. Cities such as Silicon Valley, Boston, Seattle, Austin, San Diego, and Denver have those qualities, and each have produced their own share of successful startup companies. A Kauffman Foundation report listed Massachusetts, New Jersey, Maryland, Washington, and California as the top states in its rankings of locations transitioning to new economy in February 2007.

So if you live in Detroit or Boise or Houston or Tucson, should you scrap the whole idea or move to another city? No. However, be sure that the city you choose will have the right talent and support you'll need. For example, Tucson is known as a leader in the optics space, so choosing it for your optics (or related) business would make perfect sense.

TIP: Locate your startup in a city that has the right combination of resources (talent, competitors, collaborators, and investors).

Employment Matters

Leaving Your Current Employer

The issue of leaving your current employer raises two potential questions: (1) Do you plan to leave your current employer, and (2) if so, when?

Founding teams may quickly realize that the team has differing expectations for leaving their current employers. Sometimes, a founder may want to wait six months or a year before departing. In other cases, a founder may not want to depart the company until funding is received and simply work part-time until that point. As such, it is important for the founders to determine whether each founder will depart his current employer during the prefunding stage.

In these discussions, it is important to remember that a founding team should be clear that any founder who will be required to leave her current job has the financial wherewithal to forgo a paycheck during the fundraising time period (which may last for up to a year). Sometimes, the founders may need to contribute capital to the organization to employ one or several of the founders, and it may be determined that some or all of the founders will continue in their employment during the prefunding stages. Founders who decide to continue to remain employed should be certain that these actions do not raise issues with their current employer and should review their employment paperwork to ensure they are not violating any part of these agreements.

Be clear to establish timetables for any current employer departures by founding team members. You should also determine whether the timetables raise any issues under current employment contracts with the founder's current employer. Be aware of any potential limitations that may be placed on any of the founders when they leave their current employers. Chapter 6 contains helpful information on leaving your current employer, and consult with your attorney if you are unclear on any matters.

MOONLIGHTING IN YOUR NEW VENTURE

What could happen

You and several coworkers have decided that you could run your current employer's business better than they're doing now if given the chance. So rather than wait to become the next CEO of that company, you are going to open up your own business in the same or similar markets as your current employer.

Watch out for

Running (or starting up) a competing business as your current employer raises a number of issues. Numerous court cases have ruled in favor of the employer in these scenarios, especially if the employee was managing those operations or was a key employee. You could even be at risk if you incorporate the business while still employed.

Review your company's policies and your employment documentation before acting formally on behalf of your new business venture. In some circumstances, employees have found that disclosing your business plans before beginning this venture has been met with greater support than expected (but use your judgment on whether or not you should disclose!). Perhaps your employer may be able to reassign you so as not to raise competition issues. No matter how you plan to handle the scenario, you should not misstate or misrepresent the activities of the new business.

Don't risk litigation for your new venture. When in doubt, check with the appropriate parties. Your attorney may be able to provide some guidance for this transitional period.

TIP: Beware of problems that arise from starting and operating your new venture while still employed in your current job.

Employment

Founders should also determine the role each founder will play in the new entity and at what point founders will be employees of the new company. In some circumstances, a founder may serve as simply an initial investor, a board member, an advisory board member, or a part-time employee. Founders should be certain to discuss the best course of action for each founder with respect to their role with the new entity and the timetable with which they will assume that role.

HOW DO WE MAKE OUR (DIFFERENT) CONTRIBUTIONS EQUAL?

What could happen

The four cofounders have all agreed that Joe will work full-time on their new startup and the others will only work part-time on the startup and continue drawing a salary from their current jobs. How do we make this fair?

Watch out for

First, be sure to discuss this arrangement from the outset. Contributions will differ, but if one party feels his or her contributions are not equal, then the relationship could quickly become strained. Discuss it up front rather than wait for tensions to boil over.

Misunderstandings in the role each founder will play, the salary each individual will expect, and the time each founder is expected to contribute are very common. To the extent that the founders will each make part-time contributions to the business, set forth the expectations in terms of number of hours per week or per month each founder is expected to contribute. If one founder will continue to work full-time and the rest will work exclusively for the business, discuss how different contributions shall be treated within the company.

For example, if Joe works full-time on the business, perhaps the other founders should each agree to contribute 40 hours per month (10 hours per week) to the startup. In addition, these founders could agree to each contribute $500 per month to help Joe pay his living expenses or could agree to take less equity than Joe will receive.

TIP: Discuss founding arrangements such as hours expectations, equity ownership percentages, and ongoing capital contributions.

With many of the issues related to employment, you should consult with other entrepreneurs in your area (oftentimes, you will be able to find organizations in your hometown aimed at encouraging interaction between area entrepreneurs, investors, service providers, and others in your community). You may find helpful hints on "market" terms for items such as salaries, noncompetition time periods, and severance. Your attorney may also be able to provide you with insight into the trends his firm has seen with respect to other startups in the area.

For an additional discussion of salaries, benefits, employee costs, and noncompetition and nonsolicitation terms, more information can be found in Chapter 11.

WHOSE INVENTION IS IT ANYWAY?

What could happen

You and your cofounder have invented a better mousetrap, which you are convinced should revolutionize the vermin trap industry. Your coinventor isn't quite sold on the idea so decides not to found the business with you. The rights to the invention are still owned by the business since you are part inventor, right?

Watch out for

No. Under most state laws, an invention is the property of the inventor or inventors. In this case, you and your coinventor have equal rights to the invention. However, when you start the business, this doesn't mean that your half of the invention is automatically assigned to the business. Instead, you will need to assign your rights to your startup company, which is oftentimes done in exchange for your founders stock.

You also have another problem: your cofounder. You both have rights in the invention even if you decide to be the one to invest the time to start the business. So, be sure to have your coinventor assign his or her rights to the invention to the business. In some cases, a coinventor may assign them for minimal consideration (perhaps a couple hundred bucks or perhaps some nominal royalty rate down the road) or you may give the coinventor a small part of the equity of the company in exchange for assigning his rights.

Don't forget that forming a business isn't enough to assign the rights to an invention produced before you started the business. Furthermore, if you coinvented the business with others, be sure to find out how the coinventors would feel about assigning their rights to the business. In many cases, because the technology might be very early stage, you won't have to give up too much for those rights.

TIP: Inventions are initially the property of the inventor, not the business. Don't forget to assign all inventions to the company early on.

Company Titles and Job Responsibilities

The founders should discuss what specific role individuals will play and what title they will hold. In the early stages, each founder will most likely be charged with accomplishing everything from technical development efforts and fundraising to recruitment and office administration. As the entity continues to move along its developmental path, roles will need to be better defined. As such, the founders should discuss both the current scope of activities for each founder as well as the plan, once the company receives funding and begins to employ additional employees. Founders should be aware that job titles assigned initially to members of the founding team are likely to change, and, oftentimes, the company and its investors decide that adding experienced chief executives are necessary down the road. You should be certain that the founding team recognizes that the initial job titles and responsibilities will likely continue to evolve as the business grows.

Many significant disputes will arise related to what role each founder will play with the company. You should spend time early on discussing and deciding which founders will be joining the company full-time and which, if any, will continue to work for their current employers and work part-time on the business. Be aware that individuals may have different expectations of their role and responsibilities to the company in the early stages of founding a business. Avoid problems by discussing these matters before it becomes a problem.

Salary and Benefits

The discussion of salaries is oftentimes a challenging one for many early-stage companies, especially when there isn't any money. In many new ventures, one or more of the founders will have taken a sizeable pay cut for a chance at fulfilling a dream of starting a business. To determine appropriate salary levels, you should check with outside resources in your area, including websites and recruiting services that track industry salary numbers. Additionally, many industry trade associations and entrepreneurial support organizations will have helpful information to determine salaries.

For example, you can draw helpful insights into average annual compensation ranges from the 2006 Compensation and Entrepreneurship Report in Information Technology and Life Sciences. The attached graphs (separated into life sciences companies and information technology companies) provide average salary and bonus figures of founders holding the positions, including CEOs, presidents, CFOs, and high-level technical positions. Remember that these figures are only a helpful guideline—most early-stage, prefunding companies are unable to offer founders much more than some equity and eighteen-hour days. For updated information on salaries and equity awards, visit the book's website at http://www.myhightechstartup.com.

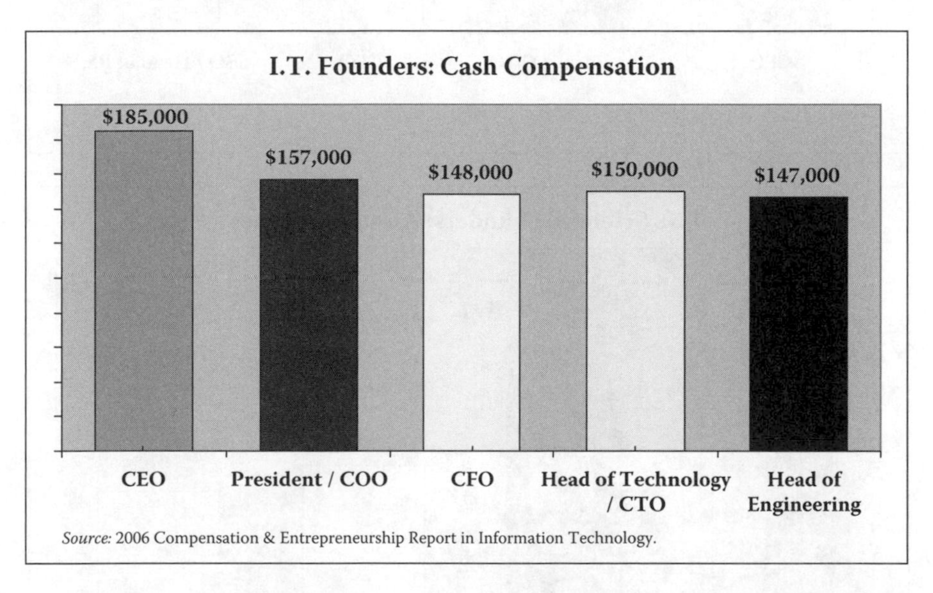

Source: 2006 Compensation & Entrepreneurship Report in Information Technology.

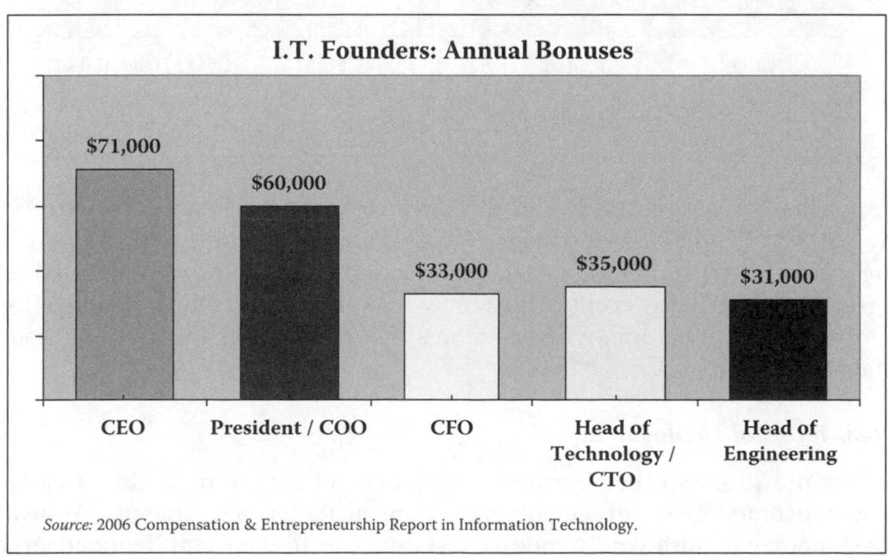

Source: 2006 Compensation & Entrepreneurship Report in Information Technology.

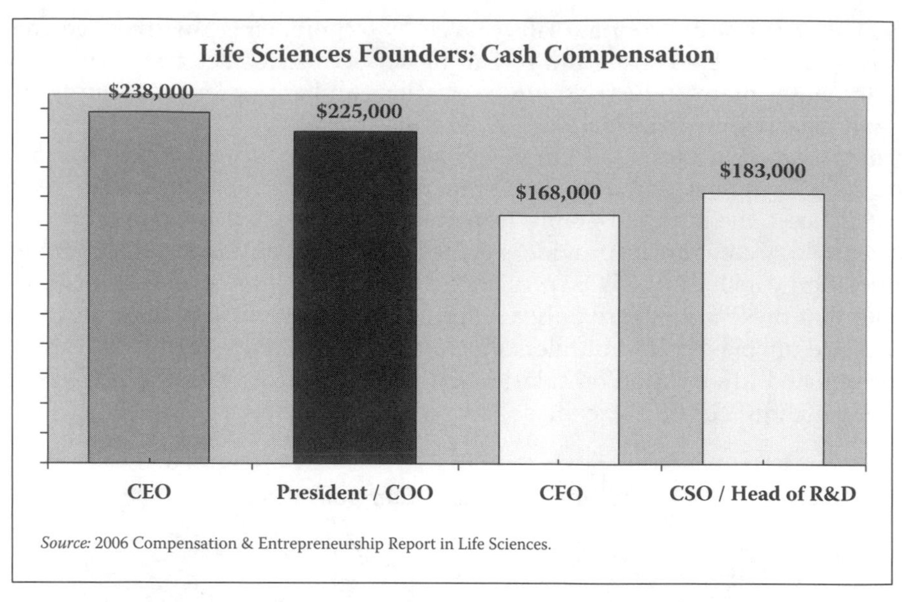

Source: 2006 Compensation & Entrepreneurship Report in Life Sciences.

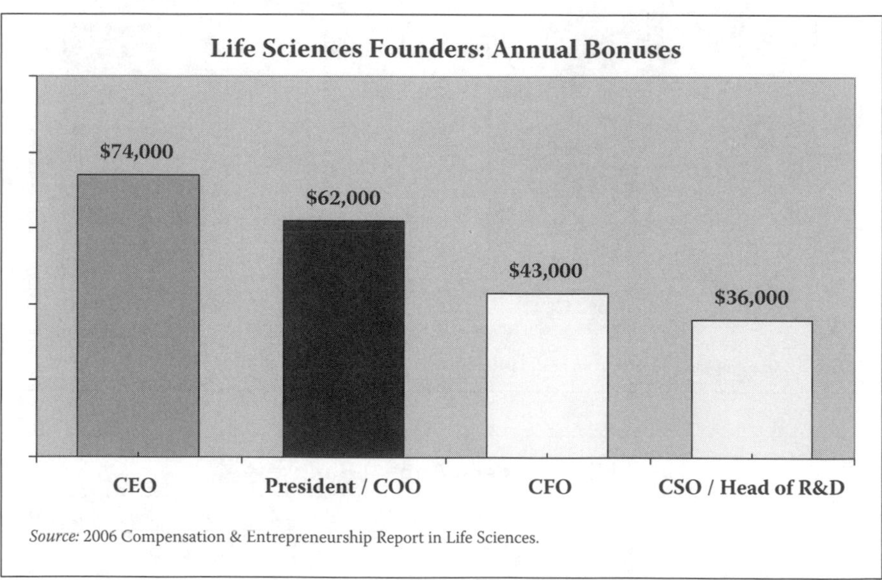

Source: 2006 Compensation & Entrepreneurship Report in Life Sciences.

Be aware when discussing salary and benefits that you will also need to discuss how the company plans to handle deferred salary issues. Some companies will choose to handle this issue through payment of a bonus on completion of financing or a deferred compensation agreement (although recent rulings now make deferred compensation agreements more complicated than before), whereas others will handle this through the issuance of additional stock or options.

Contractual Terms of Employment

Founders should discuss other traditional employment issues, including vacation time, severance, noncompetition agreements, and nonsolicitation agreements. Be aware that your investors may require you to modify certain terms in your employment agreements or offer letters as part of their commitment to provide funding.

To prevent certain issues that could arise with a founder leaving within the first few months of joining the venture, some companies will structure these initial agreements with a three to nine month trial period that would not include any severance on termination and would not have any vesting of stock or options.

Termination of a Founder

Termination issues are a difficult one for founders to consider and discuss because no team ever thinks they will have to fire one of their cofounders. Sadly, these issues do happen and are made to be much more difficult by failing to address the issue before it becomes a problem. The founders should discuss up front how they plan to handle termination of a founder, including what the process will be, what is required to initiate termination, and what type of compensation will be provided on termination.

Company Ownership and Stock Restrictions

Cash Investments into the Business

The founders should discuss how they plan to handle cash investments into the business. Even if the company does not plan to pay salaries to the founders initially, there will still be startup costs that cannot be avoided. The founders should budget for these costs and address plans for handling these amounts. Some companies have successfully financed this prefinancing startup phase on credit cards, others have used savings, whereas still other companies may look for angel funding to bridge the company's needs until it can successfully raise sufficient startup capital.

If the company does need cash to be contributed by the founders, these investments may come as a result of a direct infusion of cash into the company's bank account or with the payment of certain startup costs by the founders. The company should determine how it will handle these investments into the business, when the individuals will be able to be compensated, and how they plan to compensate for these payments.

Some companies will leave these amounts outstanding on their books and reimburse the founder on a successful financing, others may issue promissory notes for this initial capital or these expenses that can then be convertible into preferred stock upon a future financing event, and others will treat these amounts as a loan to be repaid when appropriate.

CREATING A "CHEAP STOCK" PROBLEM

What could happen

The company has waited to incorporate until funding is imminent. Now they'd like to issue stock to the founders immediately before the infusion of funding.

Watch out for

This could create a "cheap stock" tax problem for the company. By waiting to issue this founders stock and then issuing inexpensive stock to the founders while selling more expensive stock to your investors, you may raise tax problems for the company and the founders. The IRS may consider this difference in stock price (between the price paid by the founder and the price paid by the investor) as income for the founder.

TIP: Don't wait too long before issuing founders stock.

Ownership of Stock

Each member of the founding team will contribute something to the venture, from contributions of a business plan, intellectual property, cash, or commitments to provide key services for the company during the startup phase of the company. In exchange for these contributions, the company will typically issue each founder common stock in the company. Once you issue this stock, unless you agree differently, the stock will be owned in its entirety. For this reason, to protect the company and the other founders from the scenario below, many new companies will include vesting on the founders' shares.

EVEN FOUNDERS' STOCK STILL MUST COMPLY WITH SECURITIES LAW

What could happen

You've decided to give away a single share of stock in your new company to all the members of your alumni association to "spread the word." Could this cause a problem? Perhaps. Before undertaking any sale of stock, you should ensure that sale (or gift in the case above) will not require you to register the securities or will be in violation of securities laws. For most private company startups, your stock sales can be exempted from federal and state securities laws, but this requires you to be cautious in your issuance and sale of stock. Simply because your company is small, new, and you aren't selling for large dollar amounts, you cannot forget about the applicable securities laws and your compliance.

Watch out for

For each issuance of stock, or any other security, be sure to check that this sale or issuance does not violate any federal or state securities laws. Each issuance should be made under an applicable federal and state exemption; usually these are not difficult for startups but require continued diligence.

TIP: Private companies must comply with securities laws, so make sure each issuance is reviewed for compliance with applicable laws.

In these early discussions, you may not need to decide the number of stocks to issue to each founder; instead, you can decide the percentage of the initial company to allocate to each person (for example, you may choose to allocate 25% of the company to each of four founders). At a later point when you are prepared to sign founders' stock purchase agreements and issue stock, you can determine how many shares you want to authorize in the company, how many shares you want to allocate to your initial stock option plan for future hires, and the number of shares to be issued to the original founders.

Vesting of Stock

As previously discussed, many founders will choose to add repurchase provisions to the founders stock that will be released over time. This methodology is called "vesting." Companies may choose to apply vesting on all or part of the stock issued to its founders.

DISAPPEARING FOUNDER

What could happen

One of the founders received 50% of the founders stock and then left the company after six months.

Watch out for

In many cases, it is advisable for founders to include vesting provisions on the stock issued to all of the founders. Then, in the event one of the founders leaves before the stock is vested, the company can repurchase the stock at the initial purchase price. As time passes, more of the stock is "released" to the founder and cannot be repurchased. This helps ensure that the individuals who remain as productive members of the founding team will have their shares vest and are rewarded with a higher amount of stock owned outright. This way, if one of the founders leaves, the company can recover the unvested stock (which would be repurchased at par value, which will likely be priced at $0.01 to $0.0001 per share).

What is an appropriate time period for vesting? It depends. However, many venture capital firms or other investors will require a minimum of three to four years of monthly vesting on founders stock. You may also consider a one year cliff on vesting. In this case, no stock will vest until after one year. At the one year mark, the founder will have a portion of their stock become vested (say 25% or one-quarter if the stock vests over four years).

TIP: Founders should consider having vesting terms on the initial stock issued to founders.

TYPES OF VESTING

There are numerous ways to structure vesting restrictions on founders stock, including the use of time-based vesting, milestone-based vesting, and a combination approach.

- **Time-based** (straight line)
  - Stock is released from vesting in equal amounts each month over a particular time period (say monthly, quarterly, or annually over a number of years, usually between two and five years).
  - As an example, a founder has straight monthly vesting over a three year period. After being employed for 12 months, the founder leaves the company. At this point, the departed founder would only own one-third of the original stock issued to him or her. The rest of his or her shares (two-thirds of the original amount issued) would then be repurchased by the company at the par value.

- **Time-based** (cliff)
  - No vesting for a particular time period (for instance, the first six months or first year or until financing occurs). Then, once that initial period is completed, the company will then release a certain portion from vesting. Once the cliff period has passed, the rest will typically vest on a straight-line basis afterward (monthly, quarterly, or annually).
  - As an example, a founder's stock is vested over three years. The vesting will be a one year cliff, followed by straight-line monthly vesting over the remaining two years. If the founder departs after six months, he or she will have no shares vested and the company will repurchase the entire amount of original shares issued. However, if the founder departs after 18 months, then one-third will have vested after the cliff and six more months of vesting would have occurred. So the founder would have 50% of his or her stock vested at that time.
- **Milestone-based**
  - Stock will be released on the achievement of particular milestones rather than based on time periods.
  - As an example, a certain portion of unvested stock will be released from vesting when (1) the company receives at least $1 million in funding (to incentivize fundraising efforts), (2) the company reaches $250,000 in annual revenue (to incentivize sales), and (3) when the company releases its second-generation product (to incentivize product development).
- **A combination-based approach**
  - Stock will be released on a combination of milestone-based and time-based vesting.
  - As an example, half of the stock will vest monthly over a three year period and the other half will vest based on achievement of certain milestones.

In addition, companies may also decide to add acceleration provisions to the terms of the vesting. In these cases, the vesting will automatically accelerate on the occurrence of a certain event or events. The company may provide that, if the company is acquired, then the unvested stock will automatically accelerate. This is oftentimes referred to as a "single-trigger acceleration" provision. In other cases, the company may provide that, if the company is acquired *and* the employee is terminated within 12 months of the acquisition, then the unvested stock will accelerate. This is oftentimes referred to as a "double-trigger acceleration" provision.

Sometimes, the company will accelerate 100% of the vested stock, other times the company will accelerate a particular portion (say 25% of the unvested stock or 25% of the total amount), and still other times a company will only give acceleration to employees that have a certain tenure with the company or a certain level (vice president and above) with the company.

One of the key mistakes that many startups make when they choose to add vesting to the shares issued to their founders is to miss the § 83(b) election (discussed below). Be aware that you only have 30 days to make the election on issuance of the shares, and the Internal Revenue Service (IRS) isn't forgiving to you if you miss it.

THE MISSED § 83(B) ELECTION

What could happen

You issued stock to the founders subject to vesting terms, but you forgot to file § 83(b) election. If a founder is issued stock and the stock is subject to a substantial risk of forfeiture (per the IRS rules), then the stock purchase isn't complete until this risk of forfeiture is gone. Once your founders stock has vested, the risk is deemed to be gone and the IRS judges the stock purchase to be complete. At this point, according to the IRS, this is the difference between the original price you paid (let's say $0.01 per share) and today's fair market value after the vesting has run (let's say $10.00 per share). This difference ($9.99 per share) would be taxed as ordinary income. By filing a timely 83(b) election, you are able to avoid this problem.

When you sell this stock at a later date (after it has appreciated greatly), the appreciation would be taxed as ordinary income at almost twice the rate than if you'd filed the 83(b) election and the gain was taxed as long-term capital gains! In our example, you'd only pay long-term capital gains rates on the $9.99 per share gain.

Watch out for

In the company's early stages, it is easy to miss 83(b) election filings with the IRS, so make sure that this responsibility is delegated to someone. A § 83(b) election must be filed no later than 30 days after the transfer of property [Income Tax Regulations § 1.83-2(b)]. When you issue stock to founders (subject to vesting), make sure to make this filing or have your attorney do the filing for you.

NOTE: If your company allows early exercise of options (which is oftentimes done for the advantageous tax treatment), you also will need to file timely 83(b) elections in this case.

TIP: File timely 83(b) elections for vested stock.

Departing the Company

When a founder departs the company, to the extent the founder has unvested stock, those shares will be automatically repurchased by the company at par value. However, the remaining founders will also need to determine how to treat vested stock held by the departing founder.

In most cases, the company can decide to either (1) repurchase the vested stock at fair market value on the date of departure or (2) allow the founder to hold the vested stock going forward. If the founders choose option 1, the founders will provide that the company or the other founders will be obligated to purchase the vested stock on the founder departing the company. However, if the founders choose option 2, then the departing founder will continue to see his shares increase in value if the venture continues to appreciate in total valuation. Although it is most typical to use option 2, some founders would prefer to provide for repurchase to limit additional dilution (because replacing the departed founder will likely require the issuance of additional stock or options).

THE DISGRUNTLED FORMER FOUNDER

What could happen

One member of your original founding team left before the company got any traction. A year later, the company received funding (or years later is undertaking an IPO) and that former founder is back and is requesting equity in the company.

Watch out for

This founder isn't likely to be entitled to anything unless a binding contract was formed. However, this situation could have been avoided if the company had incorporated and issued stock to the founders (subject to certain vesting terms) earlier. In this case, the former founder would have had stock in the company and knew he'd given it up all or part of it by leaving.

TIP: Don't wait too long before incorporating your startup and consider applying vesting provisions to founders stock.

The founders may also agree to impose additional restrictions requiring a departed founder to follow the vote of their cofounders on departure. These restrictions are called "drag-along" or "tag-along" provisions. Such provisions will force a departed founder to vote her shares with their cofounders and can be helpful in the event of an acquisition, merger, or certain other transactions.

Transfer Restrictions

After you've issued stock to the founders, without restricting the right to transfer the stock, it can be freely transferred. This means that your stock can be split up and transferred to 20 people or it can be transferred to a competitor. In particular, for startup companies that typically rely on certain exemptions to issue their securities (so they do not need to register with the SEC), this can represent a potentially serious problem. Because of these potential problems, most attorneys will advise you to place restrictions on issued stock.

Founders should discuss what type of limitations to place on the ability to transfer their stock. Generally, the founders will agree on a right of first refusal provision, which will allow the company to match the price any third-party agrees to purchase the stock from the transferred stock. Therefore, if a founder wants to sell the stock for $100 for all his shares, the company would have the right to match that price and keep the shares held by the company. Other companies will agree to have the company, the board of directors, or an independent valuation specialist calculate a repurchase price in the event a founder wants to sell the stock and the company agrees to purchase it.

Additionally, the founders may agree to allow certain transfers to be made by the other founders. In some cases, a founder may want to transfer his stock into a trust fund for his family or directly to a family member for tax purposes. In these situations, if the shares are transferred to a permitted third party, the third party would be bound by the transfer restrictions to prevent additional transfers.

With all of these matters, you should also consider the time you would like to impose such restrictions.

Future Stock Issuances

If it is likely that the company will need to raise additional capital and issue additional stock, founders may request to have the opportunity to maintain their current ownership percentage and limit dilution attributable to future issuances of stock. These are called "preemptive rights." In the event the company wants to issue additional stock, the holder of preemptive rights can purchase additional shares on the same terms as the outside investor to maintain her ownership percentage in the company. These rights are somewhat unusual for stock issued to founders; you should discuss with your advisors and attorney before deciding to grant preemptive rights to the founders.

Decision-Making

Day-to-Day Decision-Making

Early in the life of a startup, the founders may all be heavily involved in all aspects of decision-making in the business. As the business begins to grow, the founders and the company will find that it is most efficient to determine that one of the founders will serve as president and/or CEO. The company may also determine that certain founders will have decision-making authority over certain decisions in their respective department or area of expertise.

Major Decision-Making

For events that are outside of the day-to-day realm, such as incurring debt, adding additional investors, approving an acquisition, merger, or sale, or entering into certain significant agreements, the founders may agree that such decisions require approval of a majority or a supermajority (for example, 67%, 75%, or 100%). If the founders hold stock, they will also be responsible for electing the board of directors.

Other Matters

Death or Disability

Many times, founders do not consider the fact that one of their founding team could die or become permanently disabled during the business creation. You should discuss this possibility and determine how the company will handle these events.

For example, should the company purchase life insurance policies for each of the founders? If one founder dies, will you permit stock held by a founder to be transferred to his family? Is this all the stock or just the vested portion? Should the company be obligated to repurchase the stock from the estate of the deceased founder if the estate does not have sufficient cash to pay estate taxes?

Proprietary Information

In the event that a founder has discussed assigning certain intellectual property or inventions to the company, the founders should clearly define what is being assigned, when it will be assigned, and what rights the company will have. In addition, if certain proprietary information held by an employee will not be transferred to the company or will only be licensed for the company to use, then the founders should be clear about the specifics of these items.

For example, you should be clear whether the company have full rights to use this assigned or licensed invention or will only have limited rights in a specific application or market. When a company hires talented individuals who are bringing key skills and intellectual property to the company, it is crucial that these matters are handled with the utmost care.

The company should have the founders execute invention assignment agreements to assign to the company any new inventions or intellectual property created by the founders. These agreements should provide that, even if the founder leaves the company, the founder agrees to assist the company in securing protections for any inventions.

Confidentiality

In high-technology companies, it is recommended that the founders sign confidentiality agreements.

"Nonnegotiable" Changes to the Business

The only certainty in your business will be that it is certain to change. Although this may well be the case, you should be aware that founders may have very strong beliefs with respect to certain items. For example, perhaps two of your founders believe strongly that the business must remain headquartered in your hometown, or perhaps one founder has very strongly held beliefs that the business should not work with "sin" industries such as cigarettes, liquor, or pornography. If these beliefs are held by a member of your team, you should be aware of them and discuss them upfront to prevent wasted efforts or growing frustrations in the future.

MORE INFORMATION

To download a copy of the Sample Key Term Discussion Guideline or the Initial Capitalization Schedule, visit the book's website at http://www.myhightechstartup.com.

Changes in Your Founding Team

The founding team of a significant number of new business ventures will change over the first three years. This may mean that you bring on a new CEO to replace the founder CEO (who may leave the company or take a new role), or you may find members of your founding team depart for a variety of reasons. You should be aware of the fact that changes will occur and take the necessary steps to protect the business from such departures.

DISCUSSIONS, DISAGREEMENTS, AND DEPARTURES—OH, MY!

What could happen

You are worried that this venture might not work and concerned that the team may not work together successfully. However, you are hesitant to broach the subject with the other founders because it may imply that they (or you) aren't serious and may leave the new company.

Watch out for

Remember, one of the major problems for new business ventures centers on relationship issues among the founders. Successful companies may have a number of changes during its formative years, including changes in the founding and executive team. For this reason, it is important to define early on the terms of the founders' relationship, such as vesting on founders stock, assignment of technologies developed by the founders, employment, procedures for disagreements, and other sources of future conflicts.

Include addressing terms of the founder relationship with the incorporation and formation process. For most companies, incorporation is one of the rosiest times and a good time to address potential problems before they become actual problems.

TIP: Plan for future problems that can (and likely will) occur with respect to changes in the relationships of the founders.

SAMPLE KEY TERM DISCUSSION GUIDELINE FOR FOUNDERS

Name: ______________________

Personal goals with company: ______________________

Expectations: ______________________

When do you plan to "start up"? ______________________

Business

Name of the company: ______________________

Business form (Corporation/LLC/Partnership; State): ______________________

Purpose and aims of the company: ______________________

____

Initial location of the business: ______________________

- Would you be open to considering relocating the business? Y / N

What milestones do you have for the business? ______________________

What is the fundraising strategy for the business? ______________________

Roles in the Company

Business lead (i.e., CEO/President): ______________________

Financial lead (i.e., CFO/Controller): ______________________

Technical lead (i.e., Chief Technology Officer): ______________________

Marketing/sales lead: ______________________

Initial directors: ______________________

Will any positions be full-time initially? Y / N

- When do you expect to have or be full-time employees? ______________________

What are the salaries? ______________________

- Do you expect to defer payment? Y / N
- Vacation, severance, health insurance? ______________________

Decision-making

- What decisions require unanimity among the founders? ______________________
- Who is responsible for day-to-day decisions? ______________________
- What are typical day-to-day decisions? ______________________

Current Status

Current job: ______________________

Have you discussed the new business with your current employer? Y / N

Have you shared all the employment agreements with your cofounders? Y / N

Is there any overlap between the activities of the new company and your employer? Y / N

Average hours worked: ______________________

Current employment salary and benefits: ______________________

How many hours per week do you expect to contribute to the business initially?

____

Ownership Interest

Do you expect to contribute any cash initially? Y / N

- If yes, how much? ____________________

Do you expect other founders to contribute any cash initially? Y / N

- If yes, how much? ____________________

How will future cash contributions be handled? ____________________

____

How do you expect the equity and ownership of the company to be structured?

____

Do we have vesting on the ownership? ____________________

- If yes, what are the terms? ____________________

Founder Financial Status

Assets: ____________________

Liquid assets: ____________________

Liabilities: ____________________

Credit history: Excellent / Good / Fair / Poor

Expected money to be contributed to the business in year one: ____________________

Any other personal tax planning issues to consider for company structure? ________

Exit Strategy

How do you value the shares/units on exit of a founder? ____________________

Transfer:

- Can you transfer part of your ownership or only the entire stake? Y / N
- Can you transfer to your spouse/children? Y / N

Departure of a founder:

- "Shoot out" provision (party receiving notice must elect to either purchase shares of other party or sell its shares to that party)? Y / N
- Staggered exit provision (party leaving may be "bought out" over a period of time for cash flow and tax purposes)? Y / N
- "Bring-along" provision (if either party is transferring shares, they must require third party purchaser to offer to buy also the other party's interest at the same price per share)? Y / N
- "Drag-along" provision: (if selling party has at least a certain percentage being purchased [80%+], they can obligate the other party to also transfer its shares to the same purchaser)? Y / N

What happens in the case of:

Death/serious illness: ______________________________

Divorce: ______________________________

Sale of the business: ______________________________

What happens in the case of a deadlock or unresolved dispute (last resort: the right of either party after a minimum deadlock period for either party to call for liquidation of the company)? ______________________________

Other Matters

Intellectual property

Contributing intellectual property to the company? ____________

Entering invention assignment agreements? Y / N

Will the founders execute noncompetition agreements? Y / N

What terms? ______________________________

Will the founders execute nonsolicitation agreements? Y / N

What terms? ______________________________

Will the founders execute confidentiality agreements? Y / N

What terms? ______________________________

Life insurance policies for the founders? Y / N

NOTE: Remember, these are just a sample of some of the issues founders should discuss. Some early-stage discussions might not involve some of these questions, whereas other companies might have other issues not contemplated here.

High-Tech Startup Inc.

Initial Capitalization
Capitalization Table as of => 1/1/2009

Assumptions:	
Authorized Shares	10,000,000
Founders Stock – To Be Issued	8,000,000
Option Pool – To Be Reserved	2,000,000

Remaining shares available for issuance

CLASS	RECIPIENT	SHARES	% of Series	Fully Diluted
FOUNDER SHARES				
	Sally Founder	4,000,000	50.0%	40.0%
	Mike Techie	2,400,000	30.0%	24.0%
	Jane Designer	800,000	10.0%	8.0%
	Mark Angel	800,000	10.0%	8.0%
	[TBD]			
	Issued Founders /Executive	**8,000,000**	**100.0%**	**80.0%**
2009 STOCK PLAN				
	Total Options Reserved	**2,000,000**	**100.0%**	**20.0%**
	Total Options Granted			
	Options Exercised			
	Cancelled/Expired Options			
	Options Outstanding			
	Options/Shares Available for Future Issuance			
TOTAL (fully diluted)		**10,000,000**	**100.0%**	**100.0%**

Total Common Stock (Including Shares Issued under Plan):	8,000,000
Total Options:	2,000,000
Shares Issued on Exercise of Options and Purchase Rights:	0
Reserved but Unissued under Plan:	2,000,000

5

Working with an Attorney

Capital, talent and intellectual property are the steam, steel and barbed wire of the modern age. The most important job of lawyers for early-stage technology companies is preparing their clients so that they can attract, retain and protect those three essential assets of the new economy.

James J. Greenberger
Attorney

The first thing we do, let's kill all the lawyers.

William Shakespeare – *Henry VI, Part 2*

What to Watch For

When many people think of attorneys, they oftentimes think of the trial attorneys popularized in television shows, movies, and fiction (thanks John Grisham!). However, experienced entrepreneurs know that attorneys also play a crucial role in business. Legal counsel provides crucial input and advice at nearly every stage of your company's growth, from the initial formation of your company, to negotiating a lease for office space and hiring employees, to selling or licensing your product or services, to taking on financing or listing on a publicly traded market.

So, despite all the lawyer jokes you may hear, you are about to need to find yourself a good lawyer. (It is even okay to mutter some of those jokes under your breath . . .)

Attorneys in the Business Setting

The attorneys that will assist you in these and other business matters are variously termed "corporate attorneys," "business attorneys," or "transactional attorneys." These terms are essentially interchangeable; they simply signify that the attorney works for clients in a business setting and is not a trial specialist. Depending on the attorney's specialty and the client's needs, corporate attorneys and their clients interact in a variety of ways. Generally speaking, however, there are two ways corporate attorneys work with their clients on business matters: on discrete transactions or through ongoing relationships.

Clients sometimes hire corporate counsel to assist with a single, discrete task, such as drafting a contract or a lease, registering a trademark, or reviewing an employment agreement. This type of interaction allows the client to get legal services as needed.

Other times, the attorney-client relationship will continue even after the completion of the assignment that first brought the attorney and client together. This type of ongoing relationship can be particularly beneficial for the client because, as the relationship develops, the attorney will acquire a deep knowledge of the client's business. With such knowledge, the attorney will be better able to anticipate legal problems likely to affect the client's activities, and the attorney can work with the client to avoid or mitigate these problems before they become serious. Additionally, the attorney will frequently become more than just an ongoing legal advisor. The attorney's understanding of the client's company allows him to also become a business counselor who can share knowledge of the industry and local community and whose legal and business advice can help the client avoid liabilities that could hurt the client's bottom line.

You may find corporate attorneys that specialize further in areas such as patent law, trademark law, employment law, or in very specific fields, such as the Employee Retirement Income Security Act, the Food and Drug Administration, or import-export regulations. For most startups, you will usually require the assistance of general outside counsel (often referred as a startup and emerging companies lawyer) and intellectual property counsel (usually in the areas of patents and trademarks). As you consider entering into transactions abroad, you'll also need to consider retaining local international counsel in the country of interest. More information on selecting international counsel can be found in Chapter 22.

Startup Counsel

For high-tech startups, your attorney will likely serve as your general counsel during the early stages of the business (until it makes sense to have someone provide legal support on a full-time basis). Your lawyer can help in a variety of areas, including the following: preparation of key employment documents, such as offer letters, confidentiality agreements, and invention assignments; management of the stock option process, including issuance of proper paperwork and tracking of share exercises; offering strategy assistance with respect to your intellectual property; and coordination of financing events, including negotiating both debt and equity deals for you. Many of the organizational milestones during your first year will require some level of legal guidance, and a well-trained lawyer can help you sidestep potential missteps.

In short, although corporate attorneys are necessary for many tasks associated with running your business, they can be much more than mere legal technicians helping with isolated transactions. A strong, continuing relationship with your business counsel can be a major asset for your company.

Intellectual Property and Patent Counsel

Companies that intend to file for intellectual property protections should consider retaining appropriate intellectual property or patent counsel, particularly if the company is inexperienced in the patent or trademark filing process. Patent attorneys will typically have a scientific or technology background through education or experience that will allow them to draft patent applications and otherwise assist with issues before the Patent Office. Trademark attorneys are generally specialists in assisting with and managing the process of obtaining trademarks. In each case, these intellectual property attorneys can assist with a single filing or can manage your initial and ongoing filings, which may include working with local counsel in various countries for international filings.

When Should You Find Yourself a Good Lawyer?

One of the questions that face a new founder is when to look for startup or intellectual property counsel. Perhaps the founders are in an exploratory phase in which they are fairly certain about the business but not willing to leave their day jobs quite yet, or perhaps the founders are still uncertain as to whether the business should be structured as a small consulting organization or should look to build a company that can grow into a $50 million a year business. At these times of uncertainty in the early days, when should you begin looking for an attorney?

Perhaps the simplest solution is when an entrepreneur is fully committed to the startup concept (perhaps even leaving a previous job), has a business idea that is well thought out, has or will soon begin product development, and perhaps even has a business plan. With these factors all in hand and a good sense of what the entrepreneur will need from legal counsel, there is little doubt that retaining counsel will add value. Entrepreneurs should evaluate the following factors regarding their readiness and need for counsel:

- **Commitment to the business:** Are you still evaluating whether or not to undertake a startup? Have you quit your job or begun making plans to do so? Do you have other parties engaged in the business as cofounders or early employees?
- **Identification of the business strategy:** Have you decided whether your business will be a smaller-sized consulting organization or a business that could grow to $50 million in annual revenues? Do you plan to follow a more traditional technology funding strategy that will include venture capital?
- **Development of the product and intellectual property:** Do you have intellectual property that has or is being developed? Have you begun product development? Are you looking to hire or involve others in product development?

- **Preparation of a business plan:** Have you identified some basic parameters of the business? Do you know the potential market size for the business's target markets? Do you have a tentative fundraising plan?

Each of these factors shows a certain level of commitment to the startup idea and development of a business. As an entrepreneur moves from the idea stage of the formation of a business into the establishment stage of the business, retaining business counsel will become increasingly necessary.

Even if your business remains as more of an early-stage idea than a business ready to be established and formed, you still want to begin meeting and talking with potential startup counsel. Most entrepreneurs will benefit from early-stage discussions with an attorney or attorneys. This doesn't mean you need to engage a lawyer from day one, but you could begin meetings with a friend who is a lawyer or begin some early discussions with firms specializing in working with high-technology startup and emerging businesses. At the early stages, you can use these meetings to get some insights on when it is best to engage counsel. Depending on the level of certainty you have in the business, this will probably change the answers you get. Meeting with potential counsel may help you make the determination as to when to retain counsel and the type of counsel best suited to assist your firm.

So if an entrepreneur is still uncertain about when makes sense to retain counsel in the early stages of the business, what can you do? Well, some entrepreneurs will attempt to identify certain formation activities and use limited or no legal counsel. Obviously, legal services are expensive, and identifying ways to minimize costs is important for any business.

Listed here are a few of the key transactions and events of any new business and the importance of legal counsel for each. Note: It is usually in your best interest to involve an experienced lawyer in each of these areas, at least as an advisor. However, for a very early-stage business, you may be able to effectively handle some key initial steps and decisions.

Event/transaction	Importance of counsel	Make sure to get legal advice for	What can the entrepreneur do on his own?
Formation of the company	Medium-to-low	Structural questions and implications on taxes	The state formation of the corporation or LLC
Founders stock and agreements	High	Issuance of stock since securities laws are implicated	Determine terms for transfer, vesting, and allocation
LLC operating agreement	Medium-to-high	LLC operating agreements that implicate numerous founders and implicate decision-making authority	Single-member LLC operating agreements
Employee agreements	Medium-to-low	Decision-making on certain employment issues such as noncompetition, invention assignment, trade secrets, severance, and option issuance	Preparing standard agreements
Patents	High	Preparation of a patent or provisional patent application	Research of potential competing patent claims
Trademarks	Medium	Preparation of a trademark application	Doing a basic trademark search and domain name search
Third-party investments	High	Involvement of counsel is highly encouraged (particularly for investments by sophisticated parties)	Preparation of due diligence required by the transaction

Remember, however, that there are risks associated by acting without counsel. Although you can find some good sample and example forms from books or the Internet, no business is fully "standard," and many attorneys can assist with preparation of many form agreements and the entrepreneur can use this reviewed form as a template going forward. Although an entrepreneur may be able to handle certain responsibilities without counsel when the organization only involves two or three founders, as the number of parties involves grows, the need for a qualified outside counsel will rapidly increase.

Although some companies do save money (at least early on) in incorporating their business, preparing employment agreements, or using trademarks without the advice of counsel, other companies find the initial investment in legal services for startup matters to be well worth the costs. Additional costs to "clean up" any problems or issues arising from the failure to involve legal counsel early can be somewhat high.

Smart use of legal services may be more prudent and cost effective than trying to refrain from the use of any legal services. Many experienced startup counsel are familiar with these restrictions and limitations and can help create a cost-effective solution.

How Much Does Your Attorney Really Matter for Your Startup?

Anecdotal evidence certainly suggests that it really does matter: hiring competent counsel can address many of the most common problems that new companies face, can provide business and industry insights that come from representing multiple startups, and can help negotiate market terms in your employment contracts, financings, and other corporate transactions. A good lawyer will also have key contacts with potential investors, potential customers, potential employees, and various industry associations.

Most entrepreneurs will tell you that a trusted business advisor is an important, if not crucial, part of your team (and most can probably refer you to someone they like or trust). The problem is that most will also tell you that they are always surprised by the legal bills.

That's where the real heart of the discussion will begin: should you spend the money and hire the "super lawyer" or does the less expensive reputation by a sole proprietor or local firm represent a better long-term deal for your company?

Unfortunately, this is a difficult, if not impossible, question to answer. However, one thing is clear: hiring an attorney with experience working with startup and emerging companies is crucial and will add value. Hiring the best corporate attorney specializing in work with startups and emerging companies will most likely entail working with someone at a large- or medium-sized law firm and/or working with an attorney that has a relatively high billing rate. Of course, there are always exceptions to these rules. Certain smaller firms or local attorneys will have a great reputation for working with startups (and lower billing rate), but odds are you are looking at hiring a team of lawyers will billable rates that will cause you to raise an eyebrow when you first see them.

So, is there any evidence to suggest that it is worth spending the money for a high-priced lawyer? Well, in a recent study by Shiva Rajgopal, a University of Washington Business School professor, Professor Rajgopal suggests that using a super lawyer does add real value in real-world business settings. In this study, the super lawyers had exceptional reputations and, correspondingly, had exceptionally high billing rates. Despite this, these super lawyers were able to consistently negotiate the highest average dollar value contracts in this study. Was it worth it to use a super lawyer? According to Professor Rajgopal, the super lawyers were able to negotiate better contracts on behalf of their clients than those who did not use super lawyers or negotiated on their own.

The lesson here is this: hire a lawyer who has a good reputation and solid experience working with startups, preferably with specific experience dealing with companies in your broad market. You may find that this will quickly narrow the list to just a handful of attorneys that can fit these requirements. From this list, you will most likely discover that there is a range of billing rates. Select the firm and the attorney to manage your relationship that you feel comfortable with and believe will provide you the highest value for their fee.

WAITING TO "CLEAN UP" LEGAL ISSUES UNTIL AFTER FINANCING

What could happen

Rather than incur the expense of legal counsel, you'd rather wait until you get financing.

Watch out for

Unfortunately, many of the key choices and decisions that a new venture will be making at the start should be made with the counsel of competent startup counsel. Miss 83(b) elections? Out of luck. Forget to impose vesting? That founder who left the company probably won't just hand back those shares. Sold product in an international market? You may have lost your right to patent it. As a client once asked me, "So basically a startup that doesn't get good legal counsel up front is just out of luck?"

When meeting with and interviewing counsel, find out how they can provide payment terms that allow you to use their services but still give you time to raise startup capital. It is in your (and your attorney's) best interest to help form the company properly rather than have to try to clean up the company at a later point.

TIP: Trying to save costs by waiting to hire an attorney may cost you more in the long run.

How to Select Your Attorney

Selecting legal counsel for your business can be difficult, particularly if you do not have much previous experience working with corporate attorneys. The following section presents a few of the main issues to consider when choosing business attorneys to work with your business.

WHO DO I NEED TO SIGN A NONDISCLOSURE AGREEMENT?

Your Attorney (or potential attorney)? Usually not. Attorneys (even potential attorneys) are bound by ethical standards limiting their ability to share confidential information. However, if you are not talking with the lawyer in her role as your attorney or potential attorney and have concerns about the lawyer's intentions, you may wish to have them sign a nondisclosure agreement.

Potential business partner? Most definitely. You should have mutual nondisclosure agreements in place before discussions begin.

Potential hire? Yes.

New employee? You should have a policy in new-hire documentation whereby new employees will sign confidentiality restrictions (among other restrictions).

Venture capital firm where I am sending my business plan? Ideally, you would want the firm to sign a nondisclosure agreement. However, most will not because they see so many similar business plans and presentations that could create problems for the firm. You can try, but know that industry practice is that venture capital firms don't and won't.

Your accountant or auditor? Probably yes, particularly in the case in which they will come into contact with proprietary information or data. However, many accountants (and lawyers) will include confidentiality provisions in their contracts with you.

Firm Size, Locations, and Competencies

Attorneys and law firms come in many sizes, from sole practitioners, to small firms, to large firms with offices across the country and the world. What type of attorney or firm is best to hire depends on a number of factors. Most entrepreneurs would identify ability to respond to immediate legal needs and cost as the two most important factors in choosing corporate counsel, but several other factors are equally important. You should also consider the firm's size, its locations, and its competencies, all factors that combine to define a firm's potential value to clients.

Sole practitioners and small firms can be an effective way to get legal help on general matters. These legal service providers may offer less expensive fee structures and can usually assist with various routine corporate issues.

Other small firms focus on one or a few specific areas of law. Called "boutique" or "specialty" firms, these small firms offer a high level of expertise in the areas of law they practice. They provide excellent representation in their areas of expertise, but, because of their high level of specialization, their costs may not be any less than large national or international firms. Often, however, legal issues do not present themselves in a vacuum. Most boutique firms tend to focus on those few areas of law that occur in episodes (such as litigation) or that can be conveniently separated from other work (such as patent prosecution). General law firms are usually more suitable for an ongoing relationship that will involve a variety of corporate legal matters.

Most large national and international firms have attorneys in a number of practice groups. One advantage of these firms is that they offer services in all of these areas. In contrast, smaller firms may not have many specialists, and so they may need to turn outside of the firm to help with matters beyond their general competency. A related point is that attorneys in large firms interact with their colleagues in other practice areas on a daily basis. This interaction oftentimes helps corporate attorneys in large firms identify potential legal issues that involve multiple legal disciplines. For example, an attorney at a large firm may recognize that a certain intellectual property transaction has important tax implications because of the regular contact between the intellectual property and tax departments of the office.

Another advantage of large firms is the expanded presence of their offices. As you grow your business, you will likely expand into different states and perhaps even different countries. Large firms with a network of offices will have experience with national and international legal issues that smaller local firms will not. These firms will be able to help you expand as your business grows nationally and abroad.

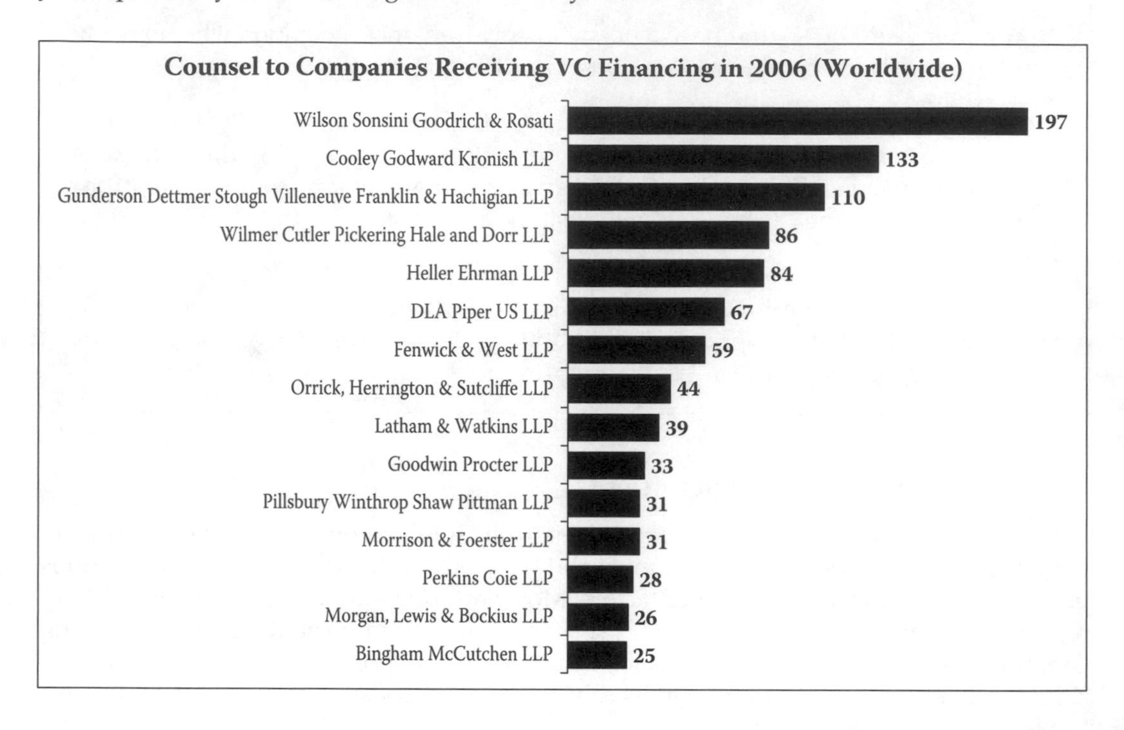

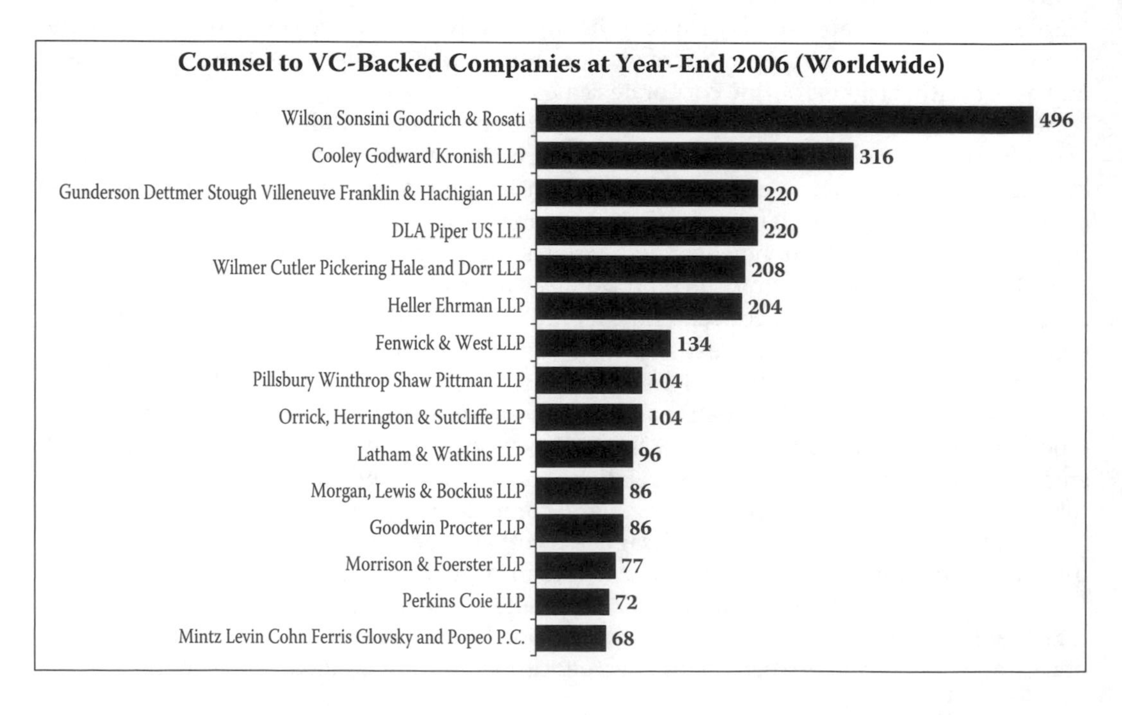

Costs and Fee Structures

Attorneys and firms vary in how much they charge and how they expect payment, and you will want to discuss rates and fee structure before having your corporate attorneys begin work for you. This subsection briefly discusses some of the usual ways that business attorneys charge for their services.

HOW TO PAY WHEN YOU CAN'T PAY?

What could happen

You are considering hiring a startup lawyer to assist with formation matters and provide guidance in the early stages of your company. However, because you aren't planning to be profitable or have any startup capital (other than be bare minimums to start with), how can you afford to pay for legal services?

Watch out for

Most attorneys that specialize in startup work will be able to provide a deferral program that will allow you to use legal service for a set period and up to a set amount without being obligated to pay those fees. You agree, in turn, to pay when the company receives sufficient startup capital.

In many cases, a law firm will defer up to a set amount (i.c., $15,000) of legal fees for up to year while the company is seeking funding. In the event that the company is unable to obtain funding and is forced to terminate the business, you should ensure that the arrangement provides that the firm cannot go after the founders (a typical provision of the agreement).

The startup attorney will usually want to understand the goals of the business and plans for the future (i.e., Are you planning to seek venture capital funding? Who is on your founding team? What is your experience?). Because the startup lawyer understands that an entrepreneur will likely want to leverage the startup lawyer's contacts with potential investors, a prospective attorney will give a critical eye to the business to ensure this is or will be the type of startup he or she can truly refer to potential investors before taking on the engagement.

TIP: Most reputable startup lawyers will be willing to negotiate a deferral program for a startup seeking funding.

Most people are familiar with the legal world's hourly billing. This fee structure is often perceived to be quite high at large firms, and it is true that legal costs can sometimes be considerable. However, working with a large firm on an hourly basis is not always expensive. Large firms have the ability to assign work to attorneys of different seniority depending on the work's complexity. Junior attorneys may be able to handle many general matters and will bill at lower rates, and so use of junior attorneys' time can allow you to limit your exposure to the more expensive time of senior attorneys. Furthermore, large firms also have excellent professional staff, such as paralegals and legal secretaries. Their assistance expedites attorneys' work and can further reduce your bills.

Some law firms will offer a flat fee for routine work. Examples include trademark registration, corporate formation, or certain basic contracts. Working with attorneys on a flat fee arrangement can be an effective method for keeping your legal costs down in a certain area. However, flat-fee arrangements are only advisable when you can be certain that there are few unforeseen issues involved. Situations that require customized, nuanced legal advice will not likely be amenable to flat fee arrangements.

Other fee structures offer additional variations on how attorneys receive their payment. Contingency fee arrangements, in which attorneys receive a negotiated fee only after the achievement of a certain outcome, are more common in trial rather than corporate settings. Deferred fees are somewhat similar to contingency fees and can be useful for entrepreneurs who are seeking financing but have not yet received it. Depending on the arrangement, you might be asked to pay interest on the amount of fees deferred. This type of arrangement might be worthwhile if you have a strong need to defer payment, but it will increase your total legal costs.

Researching, Interviewing, and Making a Decision

With all of these issues in mind, choosing potential counsel can seem daunting. You might not know where to start looking, and you might not feel that you have the ability or experience to select from among the attorneys your search reveals.

There are many ways of coming up with an initial list of attorneys and firms to consider. As you start to plan your business, take note of attorneys that represent companies in your industry, that work with local chambers of commerce and business organizations, and that give presentations or lead business-related workshops. Also, begin to solicit references from your friends and business acquaintances for corporate attorneys that they have worked with and liked. It will not take you long to create a short list of potential attorneys and firms. Your next task will be to get additional information on these candidates, and interviews are a good way to learn more about a potential attorney or firm.

INFORMATION ABOUT ATTORNEYS ONLINE

A recent startup company from Seattle, Washington, **Avvo.com**, is now in the business of providing an online service that provides detailed information on attorneys in your area, including information about education, experience levels, disciplinary actions, industry recognition, and comments from peers and clients. Visit http://www.avvo.com to view attorney profiles or get more information.

You should spend time with all qualified potential counsel. In these meetings, you will want to see whether the attorney has the breadth and depth of legal experience that you will need to help you reach your entrepreneurial goals. Ask the attorney whether she has experience with companies in your line of business and what type of work the attorney has done for those clients. Also important is the attorney's personality and the working environment of her firm. Is this a person or group of people you can imagine yourself coming to with legal problems? Would you be able to work easily with them? Are they proactive, responsive, and approachable?

Many attorneys who work with entrepreneurs have business experience themselves. You will also want to ask about this experience, too, because, as noted above, corporate attorneys can also become trusted business advisors. A few other considerations for your interview include the attorney's communication skills and use of technology in their delivery of legal services and the attorney's connections in the community, because a well-connected attorney may be a source of valuable introductions for your business.

NOT ALL LAWYERS ARE CREATED EQUAL

What could happen

You hire a good lawyer who isn't familiar with startup businesses. The lawyer is a general practitioner but isn't familiar with key issues affecting your company, such as stock option plans, 83(b) elections, or founders stock provisions.

Watch out for

Working with startup companies and small businesses is a specialized practice for lawyers. Hiring the right lawyer who has experience with high-tech startup companies can save you money in the long-run and protect you from countless issues.

Be sure to ask any potential lawyer what experience they have with specific business and legal issues that affect high-tech startups. You may want to ask for client references to other high-tech startup clients. Additionally, if you are looking to find startup capital for your business, you may want to inquire into the strength of a lawyer's contacts in the investment communities.

TIP: Find the right lawyer with the right experience for your startup.

In summary, treat your attorney interviews as you would an interview for any professional business partner. A thorough interview will tell you about the attorney's legal and business experience, personality, communication habits, and local connections. You can use all of these factors to make an informed decision when choosing your attorney.

Working with Corporate Attorneys

Your relationship with your legal counsel will develop naturally through the course of the work that you do together. As mentioned previously, many business attorneys form lasting relationships with their clients that involve general business advice as well as specific legal advice, and these organic relationships can be very valuable for companies.

However, although you should generally treat your relationship with your business counsel as you would any other valuable professional relationship, a few aspects of your attorney-client relationship bear special mention. There are tips, techniques, and strategies for keeping legal costs in line with expectations, and you'll find many of them here. This section also discusses some of the issues that stem from the fact that corporate attorneys represent companies and not necessarily the principals of that company. Next, this section discusses corporate attorneys' obligations to keep your information confidential.

Keeping Fees and Expectations Reasonable

It pays to create an efficient and effective relationship with your attorney. Why? Because your lawyer will typically charge by the hour. As a result, you should consider some of these steps to help manage the relationships with your outside counsel.

Discuss Fees before a Major Transaction and on a Regular Basis

No one likes arguing over legal bills; it isn't fun for you and it isn't fun for the attorney. Many of the situations over billing arise because the client had an expectation as to how much something should cost, but the actual cost was higher. Why does this happen? In many cases, a client asks the attorney to undertake a project without ever asking for an estimate on fees. Research into a question or issue may seem to be a simple process; issues can be more complicated and necessitate a specialist. Therefore, you should ask for estimates up front for any sizeable transaction and review the bill afterward. If the bill looks out of whack with the original estimate, call the billing attorney to discuss the differences.

Most good attorneys view this relationship with a startup client as a long-term relationship. It is in everyone's interest to meet the client's expectations on quality and, to the extent possible, on price. If you have a better sense as to the cost of a particular service or transaction, then the likelihood of miscommunications decreases substantially. Your attorney will appreciate these discussions because it leads to fewer painful conversations down the road.

HELPING YOUR ATTORNEY WITH AGREEMENT PREPARATION

Usually, your attorney won't be involved in business discussions and negotiations, but you may ask your attorney to draft or review final agreements. Be sure to give your attorney as much information as possible to aid in their document preparation or review. If possible, prepare term sheets or letters of understanding to cover the business terms.

A sample founders and formation documents term sheet is provided at the end of this chapter. By ensuring that you provide a summary of the key points of agreement, you will be sure that your attorney doesn't have to spend extra time trying to figure out the terms of the deal.

Keep Your Lawyer Informed

Sometimes, a client will decide to try to minimize legal bills by cutting off any communications with their attorney unless they have a specific legal need to discuss (and usually a rapid turnaround time). This means you go into radio silence until you plan to raise money from an angel investor, or you have a founder depart the company and need a separation agreement, or you get a cease-and-desist letter in the mail.

As with most of life, context is always helpful. So include your attorney on e-mails with updates on your product development, your sales wins, and your new hires. Your attorney should not bill you for the time he or she spends reading these updates (and if they do, find yourself another attorney). By providing this information, you may find that your attorney is better able to inform you of important information. When your attorney sees an e-mail about an important new hire, perhaps your attorney will shoot you a quick reminder to add the new hire to your directors' and officers' insurance, or if you e-mail about a new customer in China, your attorney may remind you of the tax or intellectual property implications.

Keeping your lawyer informed of nonlegal matters may add some billable time to your monthly legal bills, but it will also eliminate larger bills down the road. Just imagine if your attorney hadn't reminded you about paying local Chinese taxes and then you got slapped with a fine.

Board Meetings

Many of the most important discussions of your new business will be taking place during your regular board meetings. For some of these discussions, you'll need a seasoned legal advisor to walk you through key issues and help you avoid pitfalls. Therefore, you'll probably want to have your attorney present in person or via telephone.

What about the costs? Doesn't this really start to add up if my attorney is at a two hour board meeting every month? First, these costs are well spent in most cases. Having your attorney present will give your investors and other board members additional confidence in the organization (and hopefully you, too), but if costs still seem like a major problem, talk to your attorney. In some cases, you can frontload the meeting with discussions that would need counsel present and spend the later time focused on issues that are less likely to involve legal issues. In addition, some attorneys have a separate fee structure for meeting attendance that may help keep your bill lower.

Have Efficient Meetings

If you need to meet with your attorney or want to schedule a time to talk over a conference call, give your attorney a synopsis of the issue or question or send over an agenda before the meeting. Your attorney should know quite a bit, but if he doesn't know the answer, you've made your meeting an inefficient one. If you are planning to have a meeting to discuss a slew of hiring issues, your attorney may ask you whether he can ask his colleague who is an employment lawyer to join you for the first part of the meeting.

Only Make Your Attorney Ask You to Provide Them Something Once

A disorganized or forgetful client can make a relationship more costly than it needs to be. If your attorney asks you to send them the employment records of an employee to prepare a separation letter, send them the records. If you forget and they have to ask again, you've just paid twice for your forgetfulness.

Prevention Is the Best Cure

Most attorneys are happy to "turn off the meter" for certain types of meetings or discussions. Schedule a lunch every quarter. You may call these "relationship meetings" or a "business update meeting." In any case, go out to lunch with your attorney (and ask to have a nonbillable lunch, but you can take turns buying lunch). Spend the lunch updating your attorney on the status of the business and letting them know what could be coming down the pipe. These meetings are a great way to get advice on the key issues you may soon face. Going global? Find out how your attorney's other clients did it. Thinking about licensing technology from a local university? Find out what your attorney knows about the tech transfer office. Use these nonbillable lunches as a chance to pick your counsel's brain on trends and potential issues and to highlight challenges that will be coming up.

Some people view using their attorney like going to the doctors: "you won't see me at the doctors unless there is a good chance I could die or lose a limb because of the injury." Get regular checkups. These are an investment in your business and will prevent larger expenses down the road (like having to deal with a lawsuit, which is a surefire way to get stuck with a huge legal bill!).

Giving a Heads Up

Are you having second-round meetings with VCs and expect term sheets in the next month? Have you retained investment bankers and are planning an acquisition in the next few months? Are you looking into partnering with a European company to sell your products into the European Union? In each of these cases, you are likely to have a need for legal counsel for these transactions. Rather than letting your attorney hear about the deal when she receives the signed term sheet, provide a heads up so the attorney can consider how to staff the transaction and can perform certain preactivities when the deal approaches a certain level of certainty.

Dealing with the Last Minute

Your attorney never wants to tell you that they don't have time to address your problem or manage your deal. However, if you send them documents at noon and want them returned by five that same day, you could have problems getting your documents returned, or it could be that the partner will need to do this project himself rather than delegate to a more cost-effective associate.

Give your attorney ample time whenever possible and, when the project is going to have an extremely tight timeframe, give as much notice of the tight deadline as possible. This allows for proper staffing (so the highest billing attorney isn't stuck doing a project better suited for a junior associate).

Centralize Communications

In the beginning stages of your organization, you may find that each of the founders will be communicating with your attorney. As the organization grows, this approach probably will become much less efficient and effective. Be sure that you have a protocol in place for the use of legal services. You do not want the entire organization to have the ability to request work to be done by your attorney.

In many organizations, a key finance person will manage the relationship with your general counsel and a key engineering or research person will manage the relationship with your patent or intellectual property counsel. If anyone at your company needs to use legal services, you should have it go through this contact person, and you should put in place a backup in the event the primary point of contact in your organization is unavailable.

Paralegals and Junior Associates

Certain tasks are best performed in the hands of a paralegal or a junior associate. This includes managing typical stock option grants (unless you do these in house) and handling certain types of filings.

You should discuss work allocation up front with the lead attorney. Most firms will attempt to use the most cost-effective resources when they can, but if you have a rush project, those plans may go out the door and the most "available" person, rather than the most cost-effective person, will be doing the work. Plan ahead to avoid this type of scenario.

Don't Try to Do Legal Work Yourself

To save money, some clients will become amateur lawyers, using an old contract as a guide, failing to get a lawyer's eyes on a term sheet, or not informing their lawyer of a letter from a customer's counsel.

Trying to save money on legal services by doing them yourself nearly always costs you down the road. That "self-drafted" contract with a customer will make your potential investor's lawyer nervous and could sink the deal or make your lawyer spend hours redrafting the contract and trying to have the old contract voided. The improper separation letter could now raise age discrimination issues.

Legal fees are a cost of doing business, but you can find ways to minimize fees by developing an efficient relationship with your attorney. Ask your attorney to help prepare form contracts that only require minimal negotiations. Ask for a quick review of a contract to see whether any red flags are raised in the review. Call the attorney directly when an issue arises and ask whether they can inexpensively assist with the issue. Before you try to save money doing it yourself, attempt to use other methods to control the legal fees (there are several listed above). Don't be short-sighted in your approach; it will often cause well-intentioned companies more heartaches than they ever imagined.

SERIES A STICKER SHOCK

What could happen

You've hired a startup lawyer with a good reputation and a billing rate you can live with. Your attorney has given you an estimate on the amount of time it will take his firm to represent you in your Series A financing. Is there any reason why this could be higher?

Watch out for

You could be surprised. Each financing is different, but a good guess is that the costs for your attorney for a Series A financing will likely be between $15,000 and $40,000 in legal fees. This can be higher if the company has not kept good company records and you need assistance from your attorney to manage "company record cleanup." (So be sure to keep good records, even at the earliest days of your company.)

However, and here is the real shocker for some entrepreneurs, there will be more than just the bills of your attorney. It is typical that you will also pay the legal bill for your investors' attorney. That's right: you are paying to have two high-paid attorneys negotiate against one another. It may sound crazy, but this practice is now fairly standard. So in the event you have multiple investors, be sure that you are only responsible for paying for one investor counsel to prevent additional fees.

Be prepared to pay approximately $20,000 to $35,000 for each of the company and investor counsel for the financing. You should work with your attorneys to minimize the bill when possible, but know that venture capital financing will represent a large investment in legal fees, twice over.

Corporations as Clients

One interesting aspect of the modern law of corporations is the separate legal status of corporate entities. As entrepreneurs and businesspeople are well aware, a corporation is a business form that has a separate legal existence from its officers—a distinct corporate personhood—that allows it to own property and enter into contracts in its own name. This separate existence provides many of the benefits that enable corporations to operate so successfully in our modern economy, such as allowing corporations to continue in existence after a change of control and allowing shareholders' liability to be limited to their investments. See Chapter 7 for a discussion of different corporate entities and their business effects.

For entrepreneurs, an important consequence of the separate legal status of corporations is that attorneys retained to work for a company represent the company as their client. This means that, although you will routinely interact with the attorney in the course of your business, the company is the attorney's actual client. For the most part, the fact that the attorney technically represents the company leads to no significant problems. Both the attorney and you will be focused on making the company successful. Should a dispute ever arise between you and the company, however, the attorney will be required to work for the company's interest only. In such a case, you may need to retain separate counsel to make certain that you receive advice tailored to your personal situation.

One final note is in order if you are an entrepreneur in the earliest stages of building your business. When you hire an attorney to help you form your company, you will be the attorney's client until the company is formed, at which point the attorney will begin representing the company. This relationship will be clearly described in your engagement agreement, but be sure to speak with your attorney should you have any questions about the representation.

Confidentiality of Communications with Your Attorney

Attorneys are obligated to keep information about their work for you confidential, and it is worthwhile for you to have a basic understanding of the contours of this obligation.

Many people are familiar with the phrase "attorney-client privilege." Properly speaking, the attorney-client privilege is only one of three confidentiality-related doctrines that apply to attorney-client relations. The other two are the attorney's "professional duty of confidentiality" and the "work-product doctrine." The remainder of this section introduces each of these concepts. In reading the following materials, keep in mind that it is not necessary for you to understand these doctrines in detail. Rather, you should seek to have a general sense of how an attorney will treat the information gained during the course of working with you and know when to seek more information regarding the confidentiality of your communications.

Also, it is important to note that, although these doctrines are legally distinct, many lay persons and even some attorneys refer to the professional duty of confidentiality as attorney-client privilege. Thus, when you hear the phrase attorney-client privilege, remember that the speaker may be referring to an attorney's general obligation of confidentiality or to the narrower, strict legal definition of the attorney-client privilege.

Professional Duty of Confidentiality

Probably the most important of the confidentiality doctrines, at least for clients working with attorneys in a business setting, is the attorney's professional duty of confidentiality. The professional duty of confidentiality is a general duty of a lawyer to "not reveal information relating to the representation of a client." As the broad language of this command implies, this duty of confidentiality relates to any and all client information that the attorney might acquire. This means, for example, that an attorney must keep confidential any details of your business that you share, such as product designs, pricing strategies, or employee salary information. Additionally, the attorney's duty of confidentiality extends even after the attorney-client relationship ends.

There are several exceptions to this general duty. Attorneys may disclose client information when the client gives informed consent, when disclosure is implied for the attorney to be able to carry out the representation, and in certain other limited situations, such as when disclosure is necessary to prevent a crime or to comply with a court order. For example, your attorney does not violate the duty of confidentiality by disclosing information necessary to file a form that you requested to have filed. In this case, the disclosure is impliedly authorized so that the lawyer can accomplish the representation.

As a general rule, however, you may expect your attorney to keep the content of your relationship confidential. If you are in doubt about whether your attorney will have to reveal certain information to accomplish the goals of the representation, just ask. Attorneys understand that their ability to give their clients candid, valuable advice depends on their clients' ability to trust attorneys with confidential information, and attorneys take client trust very seriously. Your attorney will be happy to discuss with you the treatment of your sensitive information.

Attorney-Client Privilege

In contrast to the professional duty of confidentiality, which applies to attorney-client relations generally, the attorney-client privilege is a confidentiality doctrine that only applies in the limited circumstances of litigation. The privilege is, essentially, a rule of evidence. It is used when one party in litigation wants to discover information and the other party wants to keep the information secret. The party seeking to avoid disclosure will assert attorney-client privilege. Generally, a court will allow the party to keep the information secret if the party can show that the information sought was a communication, made between privileged persons, in confidence, for the purposes of obtaining legal assistance for the client.

It is worth making a few points about the application of this doctrine in the business setting. First, the attorney-client privilege only protects communications between a lawyer and client. As noted above, in corporate situations, the client is often the company and not the individual officers or employees of that company. This type of corporate representation can sometimes complicate application of the privilege, because the privilege technically belongs to the company and not the company's employees. Some employees will be covered by the employer-company's privilege, however, and historically, courts have used differing tests to discern which employees are covered by their company's privilege. When it addressed the question in 1981, the Supreme Court noted that, when employees communicated with corporate counsel, at the direction of the company, for the purpose of obtaining legal advice for the company and with the expectation of confidentiality, those employees' communications were covered by the corporation's attorney-client privilege. Although the court noted that it was only deciding the case before it (and not setting forth a general proposition), when employee communications follow these guidelines, it is reasonable to believe that they will be protected by the attorney-client privilege.

Second, the attorney-client privilege only protects communications made for the purposes of obtaining legal advice. This means that general business advice will not be protected by the privilege. Of course, the distinction between legal and business advice is an academic one. Every day, entrepreneurs face legal choices that have business consequences and business choices that have legal consequences. To increase the chance that a court will find communications with your corporate counsel within the attorney-client privilege, ask your attorney to respond to your legal questions separately, in either a separate section of a letter that contains both business and legal advice or a separate memorandum of law altogether.

As with the duty of confidentiality, the attorney-client privilege is also subject to some commonsense exceptions. For example, the attorney-client privilege does not apply to a communication when the client consults a lawyer with the purpose, later accomplished, of committing a crime or a fraud (or assisting a third party in doing so) or, regardless of purpose, actually uses the lawyer's advice to commit a crime or a fraud.

The Work-Product Doctrine

Like the attorney-client privilege, the work-product doctrine applies only in the context of litigation. Simply stated, this doctrine protects the work product of one side's lawyers against discovery by the other side. Work product is a legal term that refers to the materials prepared for a client by an attorney in anticipation of future litigation. The doctrine divides work product into "ordinary work product" and "opinion work product," although for the present purposes, it is not necessary for you to concern yourself with the distinction between these two categories.

A simple example will help you understand what the work-product doctrine protects. Imagine that you retain an attorney to defend you in an upcoming lawsuit and that the other party's lawyer requests your attorney's notes or other materials made in preparation for trial. The work-product doctrine provides your attorney with immunity from this discovery request, and your attorney need not provide these materials to the other side.

The basic thrust of the doctrine is to prevent the other side from using your attorney's work to gain advantage over you in trial or, in the words of Justice Jackson, concurring in the Supreme Court case that established the doctrine, to prevent the attorney from making her case "on wits borrowed from the adversary." Absent very unusual circumstances, attorneys must do their own work.

As with the other two confidentiality doctrines, there are exceptions and limitations to this general rule. For entrepreneurs, the crucial limitation here is that the work-product doctrine only applies to materials produced in anticipation of litigation. Little of the work corporate attorneys do for clients is done in anticipation of litigation, and so the work produced for you by your corporate attorneys will generally not be protected by this doctrine. (However, it will usually fall under the attorney's professional duty of confidentiality and may also be protected by attorney-client privilege.)

Specifics for Intellectual Property and Patent Counsel

Startup companies often need to retain separate counsel to assist with patenting and trademarks for their business. These legal specialties are important for any technology company and will oftentimes entail hiring separate counsel to assist with filings for any of you intellectual property assets.

The process to be a registered patent attorney requires that the attorney have a particular educational background and achieve a passing score on the U.S. Patent and Trademark Office registration exam. Candidates for the exam are required to have scientific and technical training, including a bachelor's degree in a field of natural science or technology, such as biology, chemistry, computer science, engineering, and physics, or technical training including certain engineering certifications, work experience, or sufficient relevant coursework. Patent attorneys may represent clients before the Patent Office, may prepare, file, and prosecute patent applications for their clients before the Patent Office, and may give patentability opinions.

When selecting patent counsel, you should attempt to find an attorney that has a scientific or technical background that is related to the field or fields of your products. In many cases, you can obtain referrals from other companies operating in your industry. A patent counsel will either use an independent patent search firm or may, in some cases, have resources at the firm to provide this service. An independent search firm may be preferable because the results of the search are often relied on by potential investors. Likewise, filing for trademark protections are not always an intuitive or easy process. Therefore, in many cases, hiring an attorney that specializes in trademark protections can assist with the process and help you to manage your ongoing obligations. Check with your corporate attorney for recommendations and cost estimates.

MEMORANDUM OF TERMS FOR FORMATION DOCUMENTS OF HIGH-TECH STARTUP INC.

Formation	Delaware Corporation 10,000,000 Shares Authorized, $0.0001 per share	
Founder Vesting	Vesting Time Period	Acceleration
Standard	• Time Period = **4 years** • No Cliff = Monthly over 4 year schedule	• Acceleration of Vesting on Change of Control ("CoC") = **50% of total** • Amount of Acceleration on termination after CoC = **100%** additional acceleration on unvested shares on Termination/Constructive Termination within 12 months of Change of Control
Customizations	Standard plus: • **Sally Founder**: 20% vested up front	Standard plus: • **Sally Founder**: 100% Acceleration on CoC
Other Founders	Standard	Standard
Employee Vesting	Vesting Time Period	Acceleration
Standard	Time Period = **4 year** Cliff = **25%** vests after 12 months; monthly over remaining 3 years	None
Capitalization	As set forth in capitalization table as provided	
Employer Identification Number (EIN)	To be obtained by the company	
Officers and Directors	Directors	Officers
	Sally Founder	Sally Founder – Chief Executive Officer, President, Secretary, Treasurer Mike Techie – Vice President of Engineering Jane Designer – Director of Product Development John C. Attorney – Assistant Secretary

Documents to be Drafted/Provided:	Corporate Governance 1. Bylaws 2. Organizational Board Consent 3. Form of Indemnification Agreement (for Directors and Officers) Capitalization Documents 1. Form of Founders Stock Purchase Agreement 2. 2009 Stock Option Plan 3. Form of Stock Option Notice of Grant Employment Documents 1. Form of Employee Offer Letter 2. Form of Invention Assignment Agreement 3. Form of Confidential Information Agreement 4. Form of Consultant Agreement Advisory Board Documents 1. Form of Advisory Board Offer Letter 2. Form of Advisory Board Agreement

MORE INFORMATION

To download a copy of the sample founders documents term sheet, visit the book's website at http://www.myhightechstartup.com.

6

Leaving Your Employer

We fail when we get distracted by tasks we don't have the guts to quit.

Seth Godin
Entrepreneur and Author

What to Watch For

Many established entrepreneurs can vividly describe that one moment when they'd finally decided to leave their employer. Some will reference this point as the first moment when they'd realized they had become an entrepreneur.

How do you decide when is the right moment to leave your job? There isn't one approach or strategy for leaving your job to launch your business full-time. So hopefully you've laid out all the options, discussed the business plans with your cofounders, your family, and your friends, and come to the conclusion that now represents that point in time. This point in time represents a closing of one door and the opening of an entirely different one.

However, before you send a "bridge-burning" e-mail to your bosses and coworkers, take a moment to understand how to carefully manage this transition. Nothing is worse for a new entrepreneur than to realize that your former employer has thrown up a number of roadblocks in your new business plans. Making your decision to leave your employer is an important one for high-tech entrepreneurs, but making a smart decision is even more important.

This chapter will discuss restrictive agreements, including noncompetition agreements, employee nonsolicitation agreements, and customer nonsolicitation agreements, an employee's duty of loyalty to his current employer, trade secrets law, and best practices when leaving your employer (particularly when competitive juices could start flowing!).

Thinking about It from "Their" Perspective

Leaving your employer to start a new business sometimes comes with its own share of hard feelings. The story below of Torrent Technologies and its founders is a useful example of what can happen to even the most well-intentioned startup founder.

A "TORRENT" DEPARTURE

Usually a startup welcomes a bit of free publicity, particularly when it comes from John Cook, the startup and venture capital reporter for the Seattle *Post-Intelligencer.* However, this wasn't the type of press Torrent Technologies founder Travis Pine had hoped for.

Pine and his cofounder Theresa Johnston started Torrent after each left the NFS. Torrent was featured in the *Post-Intelligencer* on this occasion because NFS, the former employer of over half of Torrent's employees as well as Torrent's largest competitor, filed suit against its former employees and their new company just two months after they'd incorporated. According to the *Post-Intelligencer,* NFS alleged that "the two former NFS vice presidents violated noncompete agreements, stole trade secrets, solicited NFS employees and interfered with customers" and the lawsuit described the conduct of Pine and Johnston as "willful, extreme and egregious."

Despite the mounting legal challenges, Torrent was still able to raise $3.3 million in its Series A round with the lawsuit outstanding, but not without its share of headaches in assuring potential investors that the lawsuit would not affect the company. Torrent finally settled the lawsuit in late 2006.

When Torrent opens its doors, the National Flood Service (NFS) had more to worry about than simply replacing departed employees. NFS lost two valued vice presidents and additional staff but gained a potentially significant competitor in Torrent. Torrent set up an office in the same city as NFS, and Johnstone and Pine wasted little time after leaving NFS to form the company and begin operations. It doesn't take a rocket scientist to understand why NFS became upset and ultimately sued Torrent and its founders.

All too often, employers depart their employer to set up a business in the same or in a derivative space as their former employer. It makes perfect sense: you know the market, the products, and the challenges, and you presumably have the special talents and skills that are useful in the industry.

However, from your employer's perspective, you've just become a competitor and hired away their talent. This can implicate certain legal risks, as well as create new headaches for a founder team. There has been an increase in lawsuits among competitors regarding employee recruiting and hiring, and former employees forming competing ventures. When a company has lost its employees to a competitor or a soon-to-be competitor, it may fear that the employees will use or disclose trade secrets or other confidential information in the course of working for the competitor. Even in the absence of specific proof of such wrongful activity, some companies may be inclined to file lawsuits for strategic purposes (i.e., to discourage the competitor from recruiting additional employees or to provide a disincentive for other employees to leave).

The purpose of this chapter is to provide you with legal information that you need to be aware of in the event that you leave your employer or if you begin recruiting talent from competitors. The first step in preventing conflict and the resulting legal issues is to put yourself in your employer's shoes. Strive to understand the implication of your departure, as well as the ramifications of moving in on some of their space.

Best Practices When Departing Your Employer (or When Hiring from a Competitor)

There are many legal issues that arise when departing your employer to start a competitive business or when recruiting candidates from a competitor (the analysis is very similar in each case). You and your former employer may have entered into agreements prohibiting solicitation of the employment services of your former coworkers. You may owe a legal duty of loyalty to your employer. You may even be restricted from disclosing certain forms of proprietary information or trade secrets belonging to your employer. All of these considerations may affect the timing of certain startup activities, as well as what sorts of legal problems a business may be facing.

CONSIDERATIONS FOR A "CLEAN" DEPARTURE: AVOIDING PROBLEMS BEFORE YOUR DEPARTURE

Before you depart your employer to start your new business, it is good practice to run through a series of questions and considerations.

Are you bound by an employee nonsolicitation agreement? Your new startup company should never initiate contact with your former coworkers about any job prospects if you or any other members of your founder team are under a nonsolicitation agreement. Disputes regarding employee nonsolicitation clauses often focus on who initiated contact and when. If the candidate initially contacted the company about employment prospects, then generally there will be no violation of the nonsolicitation clause. As a result, it is critical to maintain documentation establishing how the initial contact was made and when (e.g., an e-mail inquiry from the candidate, an Internet application, or written notes of a telephone inquiry).

Are you bound by a customer nonsolicitation agreement? Your new startup company should not initiate contact with the customers or clients of your former employer if you are subject to a nonsolicitation agreement for the customers of your previous employer. You may announce your new business and your participation, but be cautious about additional solicitation. If a customer does approach you (and not the other way), be certain to document the conversations and keep records as to the formation of the relationship.

Have you breached your duty of loyalty to your current employer before departing? For example, you should not solicit other employees to leave, undertake any work for the recruiting company, or inappropriately take or gain access to any information. If any such improper activity has occurred, you should promptly consult legal counsel to evaluate whether and how the situation may be corrected and whether it remains feasible to proceed with forming your business and hiring other employees.

Are there any restrictive agreements in your employment documentation and do you have copies in your files (including confidentiality agreements, noncompete agreements, nonsolicitation agreements, etc.)? Soon-to-be founders have a tendency to overlook or "forget about" restrictive agreements, so it is important to be particularly diligent in performing a search on any possible restrictions. Restrictive covenants are often found in confidentiality and invention assignment agreements, offer letters, employment agreements, and stock agreements.

Have you identified every contractual limitation you would need to observe? For example, if you are bound by a confidentiality agreement, how does the agreement define "confidential information"? What is the scope of any noncompetition agreement? What is the scope of any nonsolicitation clause?

Have you been careful to not reveal any trade secret information? You should identify, at a high level, the areas of trade secret information that may be considered sensitive and avoid tasks that would jeopardize the trade secret. You should also explore the time duration of the sensitivity and plan the timing of forming your business in conjunction with these durations. You should then evaluate whether any work duties you had contemplated undertaking in your new startup could create a risk of using or disclosing such sensitive information in the course of the new job.

After departing your employer to start a competing business, you should bring these issues to the attention of your cofounders immediately to discuss and establish any limitations on your activities to prevent potential issues. You and your team should also carefully document the agreed on limitations in a written memo to be placed in your files. One purpose of this memo is to document the company's good faith efforts to protect the competitor's rights.

Your company should consider providing a copy of the memo to your attorney and discussing these issues if you have any uncertainties. In this case, preventing lawsuits is critical: not all companies are able to sustain a substantial legal challenge like Torrent, so you should do everything to prevent those types of problems. Additionally, you should provide a copy of the memo to any other employees who need to know about these restrictions. In general, the company should be careful to brief any coworkers as necessary regarding the limitations. This will help to ensure that coworkers will honor these precautions and avoid (unwittingly or otherwise) putting the new employee in a high-risk situation.

More Information on Restrictive Agreements

The following sections contain additional information on restrictive agreements. It is helpful background information when examining restrictive agreements you have entered into with your former employer or you may find when hiring the employee of a competitor.

A variety of contractual agreements may restrict, or purport to restrict, an employee's ability to compete against a former employer. The most important types of these agreements are noncompetition agreements, employee nonsolicitation agreements, and customer nonsolicitation agreements.

Noncompetition Agreements

States outside California

A noncompetition agreement is an agreement between an employer and employee in which the employee agrees not to pursue a similar profession or trade in competition with the employer. Many states enforce agreements preventing employees from working for competitors, so long as the restriction is based on legitimate business need and is reasonable in scope, duration, and geographic limitation. When evaluating the enforceability of a noncompetition agreement outside of California, it is important to review legal authorities of the relevant state carefully.

New York

Although not as severe as in California, restrictive covenants in employment contracts are disfavored in New York. Restrictive covenants are recognized if carefully drafted. Contracts must be narrowly tailored, and restricting an employee for a nonunique skill will not be enforced. Restrictive covenants must also be limited in time and geographic area. Furthermore, noncompetition clauses in employment contracts are judged more strictly than noncompetition clauses in contracts for the sale of a business, in which the transfer of goodwill is involved. Restrictive covenants may be enforced if they involve trade secrets or a unique skill in the employment relationship or if the employer enabled the employee to obtain clients by using its proprietary information.

Washington

Within the state of Washington, noncompetition agreements are enforceable if reasonable. If entered into after employment, they will be enforced if they are independently supported by consideration or something of legal value.

California and Noncompetition Agreements

California law expressly voids contracts that restrain an individual from "engaging in a lawful profession, trade, or business" (California Business and Professions Code § 16600, often referred to as California 16600). The California Supreme Court has long recognized the strong public policy of employee mobility on which California 16600 is based.

This policy is so fundamental that an employer may be liable for the termination of an employee who refuses to sign an agreement that violates California 16600. Similarly, a company cannot interfere with a former employee's mere employment with a competitor, which does not implicate the employer's trade secrets. However, both the legislature and the courts have recognized exceptions to California 16600. Under sections 16601 and 16602, covenants not to compete are lawful when entered into as part of (1) the sale of the goodwill of a business, (2) the sale by a shareholder of all his shares in a corporation or a sale by the corporation of all or substantially all of its assets, together with the goodwill, or (3) the dissolution of a partnership, dissociation of a partner from a partnership, or sale or other disposition of a partner's interest in a partnership. These agreements must be limited to the geographic area where business was conducted, and they must otherwise be reasonable.

Nonsolicitation Agreements

Employee Nonsolicitation

These nonsolicitation agreements are aimed at preventing startup founders, such as yourself, from poaching your former colleagues. Yet, despite the strong public policy favoring employee mobility, most courts (including California courts) have upheld reasonable restrictions on a former employee's ability to solicit his former employer's employees or customers. Indeed, in *Loral Corp. v. Moyes*, the court enforced a provision in an employee's termination agreement that the employee would not "disrupt, damage, impair or interfere with his former employer by 'raiding' its work staff" for the one-year period following termination.

However, like noncompetition agreements, nonsolicitation agreements are construed narrowly by California courts. Thus, the *Loral* court found that antisolicitation clauses cannot prevent a former employee "from receiving and considering applications from" the employer's remaining employees or from entertaining inquiries from the employer's customers: "Equity will not enjoin a former employee from receiving and considering applications from employees of his former employer, even though the circumstances are such that he should be enjoined from soliciting their applications." Rather, antisolicitation clauses are construed narrowly to prohibit only active behavior on the part of the former employee.

Customer Nonsolicitation Agreements

Reasonable agreements not to solicit customers are often enforced by courts, but the use of such agreements has been narrowed by California precedents, including the decision in *Thompson v. Impaxx Inc.* In the *Thompson* case, an employee was terminated after he refused to sign an agreement not to solicit, for one year after leaving employment, any of the employers' customers with whom he had dealings as a result of his employment. The employee brought suit for wrongful termination, alleging that his termination for refusal to sign an unenforceable covenant not to compete was a wrongful termination in violation of public policy. The complaint also alleged that the identity of the employers' customers and potential customers did not constitute trade secrets. The trial court found no public policy violation, concluding that the employers were legitimately attempting to protect their proprietary interest in customer information. The Court of Appeals reversed the decision, holding that the determination of whether the customer information was a trade secret was a question of fact for the jury and that, in the absence of protectable trade secrets, the employee had a right under California 16600 to compete fairly with his former employers. In light of the *Thompson* case ruling, employers may prohibit solicitation of customers only when the identity of the customer is a trade secret.

However, if you have signed an agreement not to solicit your former employer's customers, you are allowed to announce your new affiliation and business. Thus, a departing employee may announce to her customers her new affiliation, regardless of whether the existence of those customers is her employer's trade secret. You may not, however, go further and solicit the customers, for example, by asserting the advantages to the customer of doing business with her new employer.

In New York, a nonsolicitation clause for customers of the former employer is unenforceable unless there is a showing that the particular employee's services are so extraordinary or unique that the employee would cause irreparable harm to the employer if he solicited the employer's customers. This standard is extremely difficult to prove and renders nonsolicitation covenants virtually purposeless.

More Information on the Duty of Loyalty

All employees have a duty to act loyally toward their current employer. This duty is a fundamental legal obligation that exists regardless of whether the employee has signed any form of agreement promising to act loyally.

Whether departing your employer or recruiting an employee of a competitor, you must be very careful not to interfere with the duty of loyalty owed to the existing employer. Your startup company could subject itself to legal liability by encouraging a recruiting candidate to violate the duty of loyalty.

HOW MIGHT YOU VIOLATE YOUR DUTY OF LOYALTY?

The most common mistakes that have been made that could infringe on the duty of loyalty include the following:

- Doing work for your new startup company while still employed by the competitor (also known as moonlighting).
- Soliciting other employees to leave your employer while still employed.
- Taking information or other property from your employer.
- Gaining access to information of your employer for improper reasons (such as customer addresses, pricing information, or research and development plans and goals).

More Information on Trade Secrets

Sometimes a company's most valuable asset can be its intangible, nonpatented trade secret information. The quintessential trade secret is the secret formula for Coca-Cola.

The formula has been kept secret for more than 100 years, and the company has profited enormously from the trade secret. No doubt, the company consciously chose not to patent the formula, because by doing so it would have had to publish the formula and allow it to be used by others after 17 years.

Likewise, some of your company's most valuable information may be trade secret information that your company has chosen not to patent. When an employee leaves one competitor to go to work for another, it is important that protective measures be taken to ensure that the employee does not intentionally or unintentionally use or disclose trade secret information of the other company.

USING WHAT YOU LEARNED FROM YOUR FORMER EMPLOYER

What could happen

While at your former employer, you learned a great deal of technical and market data that your former employer was not effectively leveraging. Can you use it in your new business venture?

Watch out for

Trade secrets learned on the job at your current employer that will be protected even after your departure, preventing you from using or disclosing them. Litigation over use of trade secrets can derail your new business. Courts have not been kind to former employees in these situations. If you were exposed to trade secrets at a former employer and left to work for another employer (including your own business), and your job will be substantially similar, you may increase your risk. Certain cases have found, under a doctrine known as "inevitable disclosure," that it is inevitable that you will use facts and information learned from your previous employer. A court could force an injunction to prevent you from starting your new business or working for your new employer for a number of months until the trade secrets at issue would be considered to have expired or no longer be useful.

Take the time to understand the information you will be using and its source. If you would not have known the information without your former employer and they took actions to protect that information, then tread lightly. Broad understandings of markets is typically of less concern than is specific customer lists or technical specifications. Don't derail your new business by acting careless with information from a former employer.

If there is information, technology, or market data that you want to leverage, you may consider negotiating a license from your former employer.

TIP: Don't risk an injunction by using trade secrets from a former employer.

What Is a Trade Secret?

Many states have enacted laws to protect such information as set forth in the Uniform Trade Secrets Act. Such trade secrets laws provide protection for such confidential and valuable information. Still, others have afforded protection to such information through case law.

Under the Uniform Trade Secrets Act as enacted in California, the term "trade secret" means "information, including a formula, pattern, compilation, program, device, method, technique, or process that: (1) derives independent economic value, actual or potential, from not being generally known to the public or to other persons who can obtain economic value from its disclosure or use; and (2) is the subject of efforts that are reasonable under the circumstances to maintain its secrecy."

WHAT IS MISAPPROPRIATION?

The Uniform Trade Secrets Act makes it unlawful to misappropriate trade secrets. Under the act, "misappropriation" is defined as follows:

1. acquisition of a trade secret of another by a person who knows or has reason to know that the trade secret was acquired by improper means; or
2. disclosure or use of a trade secret of another without express or implied consent by a person who:
   - A. used improper means to acquire knowledge of the trade secret; or
   - B. at the time of disclosure or use, knew or had reason to know that his or her knowledge of the trade secret was:
     - i. derived from or through a person who had utilized improper means to acquire it;
     - ii. acquired under circumstances giving rise to a duty to maintain its secrecy or limit its use; or
     - iii. derived from or through a person who owed a duty to the person seeking relief to maintain its secrecy or limit its use; or
   - C. before a material change of his or her position, knew or had reason to know that it was a trade secret and that knowledge of it had been acquired by accident or mistake.

The language in the box above highlights one of the key requirements of trade secret protection sometimes overlooked: it has to be a "secret" (or reasonably secretive) to qualify for protection. Many businesses lose this protection when they fail to take steps to ensure that their employees treat sensitive information as a secret. Again, to qualify as a trade secret, the information in question must not be "generally known to the public" and must be the "subject of efforts that are reasonable under the circumstances to maintain its secrecy."

How Could This Affect Me?

Companies take the protection of their trade secrets very seriously. Again, when it comes to trade secrets, be sure that you consider things from your former employer's perspective and are aware of the influence your competing startup could have. Look no farther than our friends at Torrent who were charged with stealing trade secrets to see what could happen to you.

PEPSI AND ITS TRADE SECRETS: AN EXAMPLE

What can happen when a company fears its employees may take their trade secrets to a new employer? Look no further than the story of the Pepsi employee and the Quaker Oats job.

Redmond had worked for PepsiCo for 10 years in various capacities, becoming general manager for noncarbonated drinks. Quaker Oats Co. offered, and Redmond accepted, the job of Vice President of Field Operations for Quaker's directly competing Gatorade product line. PepsiCo sued, claiming that Redmond was privy to trade secret information, such as strategic marketing, pricing and competition plans, as well as innovations in PepsiCo's selling and delivery systems. PepsiCo claimed that Redmond would inevitably use such information in his new job at Quaker.

The federal trial court issued a preliminary injunction prohibiting Redmond from assuming any duties with Quaker that related to beverage pricing, marketing, and distribution for six months. The federal court of appeals upheld this ruling and prevented Redmond from taking a position at Quaker. The court believed that Redmond possessed "extensive and intimate knowledge about [PepsiCo's] strategic goals for 1995 sports drinks and new age drinks" and reasoned that "unless Redmond possessed an uncanny ability to compartmentalize information, he [would necessarily be] making decisions about [Quaker's] Gatorade and Snapple by relying on his knowledge of [PepsiCo's] trade secrets."

Source: PepsiCo Inc. v. Redmond (7th Cir. 1995) 54 F.3d 1262.

Inevitable Disclosure Doctrine

The inevitable disclosure doctrine reflects the concern that, when an employee leaves a job to work for a competitor in a similar position, this employee, no matter how careful, vigilant, responsible, or trustworthy, will inevitably disclose the trade secrets of his previous employer, simply by performing the new job. In states that have adopted this rule, to protect the employer's trade secrets, a court may enjoin the employee from taking the new job, for some "reasonable" period of time, even if no actual disclosure or use of trade secret information has occurred.

Under this doctrine, entrepreneurs should be careful when timing the formation and operations of your startup competing business. When in doubt, check with your attorney before taking steps to start your new business.

A FEW PRACTICAL CONSIDERATIONS OF LEAVING YOUR EMPLOYERS

As you begin to consider leaving your employer, you should consider a few practical considerations:

- **Salary:** Although the concept of foregoing your salary may seem obvious, the reality is that many self-employed entrepreneurs fail to realize the fiscal impact and the likely time period it may be before they can begin drawing a salary. Be certain to plan appropriately. Use the personal financial summary found in Chapter 3 to determine the impact of leaving your employer.

- **Business equipment:** When you depart from your employer, remember that you will no longer have access to certain items of business equipment, such as photocopiers, printers, laptops, software tools, etc. In particular, many of the key tools necessary for operations such as software, laboratory equipment, or other technology will involve the incurrence of substantial expenses.
- **Insurance:** Insurance is one of the items that many entrepreneurs wrestle with after their departure from their employer. Obtaining insurance when you strike out on your own may be difficult. The first alternative to consider is obtaining coverage under a spouse, partner, or parent's health insurance policy. In the event you are unable to obtain coverage under another policy, one option is to obtain a high-deductible insurance plan (sometimes referred to as a catastrophic illness policy because insureds hope to use them only in the event of a catastrophic illness or injury). The premiums will be much lower when compared with low-deductible plans that would be more in line with the typical coverage an employer provides. However, it is important to keep in mind that these high-deductible plans do limit the benefits to major medical procedures. Some self-employed entrepreneurs will combine the high-deductible health plan with a health saving account (HSA). The benefit of the HSA is that it allows you to pay for medical expenses with pretax dollars. By combining the high-deductible plan with an HSA, you can pay for your routine medical expenses with money from the HSA (pretax dollars) and use the health plan for major medical procedures.

7

Forming the Business

Winning starts with beginning.

Robert H. Schuller

What to Watch For

Generally, your new venture will begin with a small, core group of individuals with an innovative idea or perhaps the Eureka! moment of a single entrepreneur. However, before you know it, your small venture could quickly grow in size with numerous new hires, rapid product development and exciting business opportunities, and new customers. In the world of startups, change happens in the blink of an eye, and being prepared for that change is just as important as the change itself. To be certain that the entity for your startup that you create will be the right one for both the small team of founders and the rapidly expanding organization you hope to become, it is important to consider how your entity will be structured from a legal perspective.

When should you form this new legal entity? How and under the laws of what state should it be organized? Once we are incorporated or formed, what are the next steps toward getting our business fully established to enter into contracts, hire employees, retain investors, issue stock, and do business?

Whether it is a corporation or an LLC, whether you form a C-corporation or an S-corporation for IRS purposes, whether you form the organization in Delaware, California, Washington, New York, Massachusetts, or another state, and the numerous other choices you have to make during this period, it is important to note that one size does not fit all. Structuring your business enterprise should be well researched and thought out. Remember that each decision should be made to fit your business and provide you the most optimal structure to succeed.

This chapter is set up to help answer a few of the questions and give you an overview into the formation process. If you have already hired an attorney or an accountant, you should discuss the formation with them to see what advice they may have for you and to see what type of support they may be able to offer in the formation of your company.

When to Legally Form Your Startup

DEL.ICIO.US, INC.: JUST A HOBBY?

Joshua Schachter started his collaborative bookmarking website del.icio.us in 2003 as a personal project simply to allow him (and a few friends) to tag and share webpages. Schachter had collected links to thousands of webpages he liked and didn't find anything that could help him organize and share them with others. So, he developed del.icio.us in his spare time and started tagging websites to share with others.

But Schachter didn't even consider himself to be an entrepreneur or a business owner. In fact, Schachter's full-time job was as a Morgan Stanley employee. However, the power of the Web made del.icio.us a huge hit, and by 2005, Schachter felt it was best to incorporate the company as del.icio.us, Inc. Shortly thereafter, Schachter sold the business for an unnamed amount. All this from a little hobby . . .

In her research, Professor Sue Birley of the Imperial College suggests that most new businesses follow a fairly standard path in their creation. According to Birley, the common path for a new business begins when the founder or founders do the following:

1. Decide to start a business
2. Quit their jobs
3. Incorporate the business

4. Establish a bank account
5. Lease or buy property and equipment
6. Receive their first order
7. Pay their first tax
8. Hire full-time employees

So, according to Professor Birley, once you quit your job, then is the time to incorporate your business, right? Well that might be the case, but it isn't always. Don't expect the path of your business to be the same path as another startup.

As has been the theme of this book, there isn't a one-size-fits-all approach to creating a high-tech business venture. Some high-tech companies will choose to incorporate immediately after an idea is drawn on the back of a napkin. Some companies are established while their founders continue in full-time jobs. Other companies operate fairly loosely, almost like a hobby for the founder, and don't get officially incorporated and "founded" for a number of years.

Why not just incorporate your business right away once you have your idea? The incorporation process itself is fairly simple: in fact, in many states, you can apply online for just a few dollars. The main reasons why you might consider waiting to form the business are as follows:

- **Still just an idea:** If you haven't decided whether or not this is just a hair-brained idea or if you can recruit an adequate team to execute on the idea, you may not be ready to incorporate or form the formal entity.
- **You are employed elsewhere:** To the extent forming a separate business could breach some of the obligations to your current employer or raise suspicions of your bosses or coworkers, you may decide to postpone incorporation.
- **You aren't sure the type of business you are forming:** Switching from one entity to another or one state of incorporation to another is not incredibly complicated, but it does involve some amount of work and possibly some costs to the extent you must involve your accountants and lawyers. If you are unsure about these choices, you may want to wait until you've settled on certain of these choices.

HOW TO KNOW WHEN IS THE RIGHT TIME TO INCORPORATE?

What are key considerations to help decide when you should incorporate your new business?

- Is your business currently operating? If your business is already operating and doing activities such as entering into contracts, providing services (even via a website), or hiring employees, you may want to incorporate or form an LLC sooner. If you do not operate under the protection of a corporation, limited partnership, or LLC, your personal assets may be exposed in the event of liability.

- Is your business not currently operating? If you are not operating the business (and are primarily in research, recruitment, and prestartup phase), there is little business risk in waiting to incorporate.
- Do you want to reserve a particular name for your entity? If there is a particular name you want for the company, you may consider incorporating to reserve that name. Otherwise, you may want to simply reserve the name without incorporating.
- Are you employed at another job? In particular, if your potential new business will have some overlap with your current employer, you may want to wait until after terminating your employment before incorporating or forming an LLC.
- Are you expecting tax losses in the near future? If you are planning to form a corporation rather than an LLC or another pass-through entity, you will treat taxable losses differently. Losses will not "pass-through" a corporation for tax purposes.
- Do you have a financing that will likely close in the near future? Problems can arise for certain new companies that want to issue stock to their founders at low prices if they wait until the company is nearing a financing. The problem arises because the company is selling its stock to investors at a much higher price than it hopes to issue stock to its founders.
- Are you applying for any grants, loans, or similar programs? Some grant and loan programs will require you to have established your entity to participate in the program application process.
- Are there cost implications? If you are not operating, when you form the company you'll likely have additional costs to pay your accountant for filing your annual taxes and to pay your attorney to assist with formation matters.
- Are there tax or fee implications? Some states have minimum tax filing amounts for a corporation even if you are not active.
- Are you currently paying self-employment tax? If you are currently operating as a sole proprietor or in a partnership, you will likely be subject to self-employment tax. This may mean you are currently paying more than 15% of your earnings for self-employment obligations to Medicare and Social Security. If you instead incorporated or formed an LLC, you would not pay the self-employment tax on any profit that remains in the corporation or LLC.
- Do you have any timing requirements? Depending on the time of year, you may find that you have to wait longer to receive confirmations of filings. This is particularly true at the beginning of a year and at the beginning or end of a month.

If you are certain you are going to start your business, you shouldn't wait too long before taking steps to incorporate your business as a corporation or to form an LLC. The costs to form the entity are fairly low, but it is always a good idea to protect your personal assets that may be exposed if you operate a business without proper protections of a legal entity.

CAN I WAIT TOO LONG TO INCORPORATE?

If you intend to issue founders stock at a nominal price (let's say $0.0001 per share), by waiting to issue the stock until you receive funding, you may create a "cheap stock" tax problem. Issuing stock at $0.0001 per share to founders a few days before you sell stock to outside investors at $1.00 per share raises some flags with the IRS. The IRS may consider this difference in stock price (between the price paid by the founder and the price paid by the investor) as income for the founder.

Forming the company may seem to be a formality, but this formality allows you to issue stock and to protect yourself and your cofounders from certain forms of liability, among other things.

Do These Choices Really Matter?

Perhaps you've heard the statistics that more than half of all U.S. publicly traded companies and more than 60% of the Fortune 500 are incorporated in Delaware. You might have also heard that many VCs prefer to fund Delaware corporations. Maybe someone once told you that "going public" is the easiest as a Delaware corporation, and, to heap on more evidence in favor of Delaware, you may have heard that Delaware has the most "business-friendly laws in the country."

So, because it seems like most everyone is a Delaware corporation and you would like to raise venture capital money and go public, then I guess all that evidence just means I should form a Delaware C-corporation too, right?

Not so fast . . . Before you rush to follow the pack, take a moment to understand why it matters:

1. Your choice affects your tax rates.
2. There are varied cost and fee structures in each state.
3. It may be costly to change entities at a later point.
4. Third parties may prefer or require your company to be organized a certain way.
5. The "pain in the butt" factor: how easy and simple is it?

Whether you decide to consult with your attorney and accountant to determine the best form for your company or whether you decide to forego the advice and go it alone, be sure you make a decision based on an understanding of how your choices affect the business and fit into your business strategy. Spending the time and money up front will help ensure that you make the right choices for your entity.

DOES IT REALLY MATTER HOW WE LEGALLY STRUCTURE OUR BUSINESS?

What could happen

We read somewhere that another successful entrepreneur said he was "double-taxed" on his C-corporation. He said that for his second company, he set up an LLC and then later converted the LLC to a C-corporation right before funding. Is this the right approach for us?

What to expect

Remember when considering how to structure your business, there is not one "right" approach. Every new business will have differing timetables, goals, and needs, so it is important to create a legal structure that accounts for this. Although this successful entrepreneur is correct that a C-corporation is subject to double taxation (whereby the corporation is taxed and the dividends distributed are also taxed), not every new venture would benefit from this approach.

Structuring your business is an important first consideration, so be sure to get advice customized to your personal and business plans. Although the upsides of using an LLC may avoid double taxation, conversions from an LLC to a C-corporation add additional legal and accounting expenses and may yield headaches for potential investors. Some new ventures begin as S-corporations (which primarily involve a C-corporation making a filing with the IRS, as well as other structural steps) and later change to C-corporations by revoking that filing. Other businesses may consider creating subsidiaries. Each option has its own set of benefits and drawbacks. Careful up-front planning on your business structure will aid in later efforts of fundraising, equity ownership, and costs for third-party service providers.

TIP: Proper planning for your business structure can save time, money, and headaches down the road.

"Rules of Thumb"

Although some startups will decide to create very complex structures with entities of various organizational forms located in various jurisdictions (such as setting up an entity in the Cayman Islands or splitting the company into a subsidiary to hold the intellectual property and license it to an "operating" subsidiary), you've decided to take a simpler approach. So, what next?

Here is a list of five basic concepts to consider. Once you've had a chance to review these general concepts and "rules of thumb," you can begin to get more insight into some of the implications of your choices.

Rule 1: If your business is a one-person company and you are making a profit that is in line with what your salary would be if you were not working for yourself, consider forming a single-member LLC. A single-member LLC is like being a sole proprietor for tax purposes (profits and losses will pass through the LLC directly to you), but it will give you protection from certain liabilities. An LLC has the simplest tax and accounting rules.

Rule 2: If your business is a one-person company and you are making big profits (in excess of a typical salary you'd earn working for someone else), consider forming an S-corporation. The benefit of forming an S-corporation in this scenario is that this structure will save you on self-employment taxes. You can form the S-corporation by incorporating as a standard C-corporation (or LLC in certain cases) and then making a filing with the IRS. The accounting and taxation are more complex in this setup.

Rule 3: If you are unsure whether you'll ever pursue venture capital financing and, if you were to pursue it, believe you'll wait a couple of years, consider forming either an S-corporation or an LLC. The most important and immediate benefit to incorporating or forming an LLC is liability protections. Both entities offer that protection and allow profits and losses to pass through to the individual for tax purposes.

Rule 4: If your business growth strategy involves raising money from investors in the upcoming 6–24 months, consider forming a C-corporation. An outside investor such as a VC will desire to purchase the stock of your company. The VC will also require that the profits and losses remain with the company and not pass through to the owners.

Rule 5: If you are planning to create a high growth company in need of substantial outside funding, consider incorporating in Delaware. A majority of venture-funded companies are incorporated in Delaware. There are a number of benefits, including greater comfort among investors, well-established corporate law, and fairly low incorporation fees. If you believe your company will become a venture-backed company, Delaware represents a logical choice.

Conversely, if the plans for your company are less clear and your growth horizon is longer term, plus you are planning to transact business in your home state, consider incorporating (or forming your LLC) in your home state.

Summary. These five broad concepts should offer you some rules of thumb for your choices. Obviously, your choice of entity is important, but don't be too concerned, as changing from one entity to another isn't uncommon. Ultimately, you should make your entity choice and jurisdiction based on research and critical thinking, but you should not let the decision allow you to lose sleep.

An "Inc.," "LLC," "S-Corporation," or Something Else? Which Entity Do I Choose?

By and large, most high-tech companies operate as standard C-corporations. The C-corporation is the entity venture capital investors are familiar with, and it offers the advantages of limited liability for its shareholders, centralized management, transferability of corporate interests, and perpetual existence, and it is not subject to the numerous eligibility restrictions applicable to S-corporations. So even if you choose a different entity for your company's early stages, there is a good chance you may be required to convert into the traditional C-corporation.

The type of business entity selected by a startup company will have an impact on such fundamental matters as the tax treatment of the business entity and its participants, form and methods of transfer of ownership interests in the entity, ease of obtaining financing for the enterprise, and liability for the obligations of the business. Although corporations are emphasized in this publication (and, generally in literature on startup ventures), businesses may also be structured as partnerships (general or limited), LLCs, sole proprietorships, or joint ventures, among other choices.

Current tax law creates substantial incentives for new businesses to structure themselves, if feasible, as partnerships, LLCs, or S-corporations, so as to permit shareholders to avoid double taxation of corporate earnings and to allow them to take a current deduction for losses incurred by the business. However, the C-corporation form of business is often favored. In addition to the reasons discussed previously, under the 1993 tax legislation, C-corporations have a tax advantage in that shareholders may qualify for a special tax benefit available on certain dispositions of "small business stock." (You can learn more about the tax treatment in Chapter 20.) In addition, many venture capital and other investment funds may not invest in partnerships or LLCs because of tax restrictions applicable to pension and profit-sharing trusts and other tax-exempt entities. As a result of these and other factors, most businesses historically have chosen the corporate form of business entity.

Comparison of Entity Types

Characteristics	Sole proprietorship	C-Corporation	S-Corporation	LLC
Formation	No state filing required	State filing required	State filing required	State filing required
Duration of existence	Dissolved if entity ceases doing business or on death of the sole proprietor	Perpetual	Perpetual	Dependent on the requirements imposed by the state of formation
Liability	Sole proprietor has unlimited liability	Shareholders are typically not responsible for the debts of the corporation	Shareholders are typically not personally liable for the debts of the corporation	Members are not typically liable for the debts of the LLC
Operational requirements	Relatively few legal requirements	Board of directors, annual meetings, and annual reporting required	Board of directors, annual meetings, and annual reporting required	Some formal requirements but less formal than corporations
Management	Sole proprietor has full control of management and operations	Managed by the directors, who are elected by the shareholders	Managed by the directors, who are elected by the shareholders	Members have an operating agreement that outlines management
Taxation	Not a taxable entity; sole proprietor pays all taxes	Taxed at the entity level; if dividends are distributed to shareholders, dividends are also taxed at the individual level	No tax at the entity level; income/loss is passed through to the shareholders	If properly structured, there is no tax at the entity level; income/loss is passed through to members
Pass through income/loss	Yes	No	Yes	Yes
Double taxation	No	Yes, if income is distributed to shareholders in the form of dividends	No	No
Cost of creation	None	State filing fee required	State filing fee required	State filing fee required
Raising capital	Often difficult unless individual contributes funds	Shares of stock are sold to raise capital	Shares of stock are sold to raise capital	Possible to sell interests, although subject to operating agreement restrictions
Transferability of interest	No	Shares of stock are easily transferred	Yes, but must observe IRS regulations on who can own stock	Possibly, depending on restrictions outlined in the operating agreement

A SEPARATE ENTITY FOR YOUR INTELLECTUAL PROPERTY

What could happen

You have developed some very interesting intellectual property, some of which doesn't relate to the business you are building. Someone has told you to set up a separate company for your intellectual property, which will "license" the intellectual property to your startup company. What will investors think of this?

What to expect

Many sophisticated entrepreneurs want to protect themselves from "giving away the farm" in their new business. Therefore, they consider setting up a separate holding company for their intellectual property, to keep the rights separate from the startup. The theory is that this will allow the entrepreneur to form multiple startups based on this intellectual property.

In practice, most investors are wary of such a structure. If an investor is pouring money into the business, they want to make sure that (1) the team is committed to the business and (2) the intellectual property is protected that will be used for the business. Unfortunately, setting up this separate entity for the intellectual property raises red flags for an investor.

Before you decide to structure your business with intellectual property located in a separate entity, be sure to speak with your attorney. If a future investor is likely to require that the structure be folded into the business, no need to waste valuable time, money, and energy to build this complex structure.

TIP: Sheltering intellectual property in a separate legal entity can raise red flags for potential investors.

Picking a "Pass-Through" Entity

LLC

Notwithstanding the historic preference for the corporate form, there has been considerable and growing interest of late in the LLC form of entity. LLCs are organizations that combine the business law advantage of limited liability for all of the owners of the business (such as shareholders of corporations and limited partners of limited partnerships) with the tax law advantages of a single level of tax on its income and pass-through of its losses to the owners of the business (such as partners of partnerships). The LLC form is now recognized in virtually every state, including California.

LLCs are treated as partnerships for tax purposes (unless the LLC makes an election to be treated as a corporation), as long as interests in the LLC are not publicly traded. An LLC normally will not be appropriate for an entity contemplating a public offering in the very near future or that otherwise is expected to have a large number of equity holders (although it usually is feasible to convert an LLC into corporate form later if a public offering is a long-term objective). In addition, LLCs will usually not be the entity form of choice for businesses that do not expect to derive substantial benefit from having a single level of taxation and pass-through of losses (such as a business that expects to realize most of its economic returns through a public offering or merger into a larger company rather than through substantial operating earnings or tax benefits).

However, the LLC form should be given serious consideration if the benefits of taxation as a partnership would be significant and the limitations noted above do not preclude using that form.

S-Corporation

In its simplest terms, the S-corporation is formed in the same way as all corporations (a filing in your state of choice), but, to become an S-corporation, you will make an additional filing with the IRS. It is just as simple to transition back to a C-corporation (except that you will be subject to certain filings for tax purposes).

If a corporation elects to be treated as an S-corporation under the Internal Revenue Code (I.R.C.), the corporation's shareholders retain the corporate form benefit of limited liability but are generally taxed in a manner similar to owners of pass-through entities like partnerships and LLCs. Consequently, profits earned by an S-corporation will generally be taxed only once for federal income tax purposes. Income, losses, and other tax attributes generally flow directly through to the shareholders (on a pro rata basis) without the "double taxation" of distributions to shareholders in C-corporations. The shareholders must include the profits of the S-corporation as income when earned, regardless of whether any amounts are distributed to shareholders. A distribution of earnings by an S-corporation to its shareholders is generally not taxed a second time. In contrast, a similar distribution by a C-corporation will be taxed twice: the corporation must pay federal corporate income tax on profits earned, and the shareholders must treat the distribution as a dividend subject to tax. Note, however, that one significant difference between the taxation of S-corporations on the one hand and pass-through entities like partnerships and LLCs on the other is that distribution of appreciated property by S-corporations results in the S-corporation recognizing taxable gain (which thus passes through to the shareholders), whereas such distributions are normally tax free in the case of partnerships or LLCs.

Shareholders generally elect S-corporation status when the corporation is profitable and distributes substantially all of its profits to the shareholders or when the corporation incurs losses and the shareholders are able to use the loss deductions in their personal income tax statements. However, there are substantial limitations on the eligibility of a corporation to make an S-corporation election and the allocation and deduction of S-corporation losses by its shareholders.

To qualify for S-corporation status, a corporation must be a U.S. corporation, it must have no more than 75 shareholders, all of whom are individuals (or certain qualifying trusts, estates, or tax-exempt organizations) and none of whom are nonresident aliens, it must have only one class of stock outstanding, and the corporation may not be an insurance company, a "section 936 corporation," or one of certain other types of ineligible corporations. Because of the significant consequences of the failure to meet these requirements, shareholders of an S-corporation usually enter into a written agreement to refrain from taking any action that would impair the election.

An eligible corporation may elect to be taxed as an S-corporation by filing a Form 2553 election with the IRS. To have a valid election, all shareholders must consent to the election by signing the shareholder consent statement that is part of Form 2553. The Form 2553 must be filed on or before the 15th day of the third month of the taxable year of the corporation for which S-corporation status is to be effective or at any time during the preceding taxable year. If a corporation does not meet all of the S-corporation requirements each day of the year, the election will not be effective until the following year. For purposes of the S-corporation election, the taxable year of a new corporation does not begin until the day the corporation (1) has shareholders, (2) acquires assets, or (3) begins doing business, whichever occurs first. Once the election is made, it continues until the corporation is no longer eligible for S-corporation status or until the election is revoked by shareholders holding a majority of the shares outstanding on the date of revocation.

Note that certain special taxes may be imposed at the corporate level on certain income of S-corporations that elect S status after having already operated as C-corporations.

There are also state tax rules to comply with. For example, California has an S-corporation provision in its Revenue and Taxation Code that allows a corporation and its shareholders to be taxed as an S-corporation for California state tax purposes but imposes a 1.5% tax at the corporate level on the corporation's net income. California has a separate election form, Form 3560, that also should be filed if California S-corporation status is desired (and that must be filed to elect California C-corporation status if the corporation elects federal S status but wants to be a California C-corporation).

The LLC versus the S-Corporation

If you've decided to take the plunge and operate your business as a pass-through entity, how do you choose between the LLC or the S-corporation? Many entities find that, if their business qualifies for the S-corporation election, it is usually the better choice in the long run. However, in the event that your company will be unable to comply with the IRS's S-corporation restrictions, you'll find that LLC represents the next best choice.

S-CORPPORATION AND THE LLC: THE DIFFERENCES THAT MATTER

What else differentiates these two different types of entities?

Downsides of an S-corporation:

- No more than 75 shareholders
- No foreign owners
- Only one class of stock
- Can easily fall out of compliance and lose tax benefits
- Less flexible structure

Downsides of an LLC:

- More complicated as its size increases
- More expensive state filing fees (typically)
- May create tax issues if acquired by a third party
- Unable to issue incentive stock options (although other mechanisms do exist)
- Unable to "write off" startup business losses
- Less "standard," which may increase compliance costs and attorney fees for preparation of operational documents
- Does not have stock, which is sometimes more difficult/confusing for investors or employees

Incorporating the Business

As you prepare to incorporate your business (or form an LLC, in which many of the steps below will also apply), there are a number of steps that you will take to legally form your entity. This section provides you with an overview to help offer some general insights into the process. Many startup attorneys will be able to assist their clients with this process:

1. Choose a state of incorporation
2. Select a corporate name
3. Prepare the appropriate state filing
4. Initiate formal corporate actions
5. Establish an employer identification number
6. Qualify to do business in another state
7. Other state and local filings

This section will go through each of these seven steps to provide background on some of these choices. When your corporation takes steps to legally form the business entity, you should be certain that you have completed the necessary actions required by each of these seven steps. Remember, as a new corporation, you will be subject to the laws of the federal, state, county, and local governments and agencies.

Choosing a State of Incorporation

In most cases for high-tech companies, the choice comes down to their home state or Delaware. As discussed previously, your choice affects your tax rates; each state has varied cost and fee structures; it may be costly to change entities at a later point; third parties (such as investors) may prefer or require your company to be organized a certain way; and you may find it simpler to operate in a certain state.

The principal advantage of incorporating in Delaware is Delaware's well-developed body of corporate law, which may favor management and the company's majority shareholders over minority shareholders and may provide significant flexibility in other areas of corporate governance and takeover defense. However, Delaware incorporation fees and franchise taxes can be significant, and expenses can be compounded by costs to qualify as a foreign corporation in your home state.

WHY DELAWARE?

What are some of the key reasons why so many corporations choose to incorporate in Delaware?

- Formation is fast and relatively easy
- Relatively low fees to incorporate
- Relatively low annual franchise taxes
- Separate court system for corporations that does not use juries
- Shares of a Delaware corporation are not subject to Delaware state personal income tax or inheritance taxes
- Delaware corporations that do not conduct business in Delaware are not required to pay Delaware state corporate income tax
- Only requires one officer (a single individual can hold all the offices of the corporation)
- Greater privacy for board members (Delaware corporations are not required to identify or give the addresses of individual board members)
- No need to be a state resident (you are required to have a registered agent in the state but not required to have any office or operations in the state)

A long-arm provision in your home state may also render the company unable to take advantage of key provisions of other states' corporations' laws. For California companies, Section 2115 of the California Corporations Code provides that certain provisions of the California Code will apply to any foreign corporation that has more than 50% of its stock held of record by California residents and more than 50% of its property, employees, and sales in California, unless the corporation has outstanding securities listed on the New York or American Stock Exchanges or has outstanding securities listed on the NASDAQ National Market.

In addition, a foreign corporation qualified to do business in its home state will likely be required to pay a corporate income tax on a portion of its income and certain of its affiliates' income in a manner similar to the taxes payable by corporations registered in the home state. Furthermore, the corporation can later decide to reincorporate outside of its home state if it decides to go public and wants to take advantage of Delaware's corporate law. However, reincorporation after the issuance of shares will be more burdensome than an initial decision to incorporate in your home state. The decision of where to incorporate is therefore not clear cut, and you should discuss any concerns you have with your advisors.

CASE STUDY

Comparing California and Delaware

Some of the most significant differences between California's and Delaware's corporate law (which are two of the more popular choices of entity location) are the following:

- **Cumulative voting:** Under California law, any shareholder may cumulate his or her votes for the election of directors at annual meetings, which may allow minority shareholders to achieve proportionate representation on the board of directors. Without cumulative voting, the holders of a majority of the voting shares can elect all of the directors. Cumulative voting is not permitted under Delaware law unless expressly authorized in the certificate of incorporation. The cumulative voting requirement applies to a Section 2115 Corporation (as discussed above), regardless of its state of incorporation.
- **Changing the number of directors:** California law permits only the shareholders of the company to change the number of directors, unless the shareholders approve a bylaw that permits the board of directors to fix its number within a variable range. In contrast, Delaware law permits the board of directors to change and fix the authorized number of directors by amendment of the bylaws, if the power to amend the bylaws is delegated to the board in the Certificate of Incorporation.
- **Staggered board of directors:** Delaware corporate law permits a staggered or classified board of directors in which the board is divided into one, two, or three classes. The California Code generally requires an annual election of directors unless the company is listed on the New York Stock Exchange, the American Stock Exchange, or the NASDAQ.

- **Promissory notes to purchase stock:** The California Code permits the use of promissory notes as payment for stock only if the stock is purchased by an employee or director under a stock purchase plan or agreement or if the stock is secured by collateral other than the shares purchased. Delaware law only prohibits the use of promissory notes to the extent of capital as defined in the Delaware General Corporation Law (which is also referred to as the Delaware Code), which requires that the purchaser pay for the capital value of the stock, typically a penny or fraction of a penny per share, with "cash, services rendered, personal property, real property, leases of real property, or a combination thereof."
- **Loans to officers:** California law prohibits loans to officers unless the loan is authorized by a majority of the outstanding voting stock of the corporation or the corporation has at least 100 shareholders and a bylaw has been approved by the holders of a majority of the outstanding voting stock that authorizes the directors alone to make such loans. The Delaware Code allows directors to authorize loans to officers.
- **Special meetings of the shareholders:** The California Code provides that special meetings can be called by the board, the chairman of the board, the president of the corporation, or the holders of at least 10% of the shares entitled to vote at the meeting. The Delaware Code limits the right to call a special meeting to the board of directors or "such person or persons as may be authorized by the certificate of incorporation or by the by-laws."
- **Shareholder voting on mergers:** With certain exceptions, the California Code requires that a merger or reorganization, certain sales of assets, and similar transactions be approved by a majority vote of each class of shares outstanding (such as the classes of common stock and preferred stock, voting separately). In contrast, the Delaware Code generally does not require class voting except in certain transactions involving an amendment to the certificate of incorporation that adversely affects a specific class of shares. Class voting applies to a Section 2115 Corporation, regardless of its state of incorporation.
- **Dividends:** The Delaware Code permits the payment of dividends out of paid-in and earned surplus or out of net profits for the current and preceding fiscal years. Under the California Code, any such distributions are limited to either retained earnings or an amount that would leave a corporation with tangible assets 1.25 times its tangible liabilities and current assets at least equal to its current liabilities. These dividend restrictions apply to a Section 2115 Corporation, regardless of its state of incorporation.

NOTE: Whereas the California Code refers to the "articles of incorporation" and "shareholders," and the Delaware Code refers to the "certificate of incorporation" and "stockholders." For convenience, we've used the California terms throughout this box.

Sample State Filing Fees and Expenses (as of February 2008)

	CA	CO	DE	FL	IL	MD	MA	NY	TX	WA
Incorporation (Corp.)										
Filing fee/state fee	$ 100	$ 125	$ 89	$ 70	$ 150	$ 120	$ 275	$ 125	$ 300	$ 175
Minimum franchise tax	800		35		75	300				59
Annual report filing fee	25	100	25	150	25		125	9		10
Total (Estimated)	$ 925	$ 225	$ 149	$ 220	$ 250	$ 420	$ 400	$ 134	$ 300	$ 244
Fees for expedited filings	*$350*	*$150*	*$119+*	*$70*	*$100*	*$50*	*4.5% fee*	*$25+*	*$25*	*$20*
Foreign Qualification (Corp.)										
Filing fee/state fee	$ 100	$ 125	$ 160	$ 88	$ 200	$ 100	$ 400	$ 225	$ 750	$ 175
Minimum franchise tax	800				75	300				59
Annual report filing fee	25	100			25		125	300		10
Total (Estimated)	$ 925	$ 225	$ 160	$ 88	$ 300	$ 400	$ 525	$ 525	$ 750	$ 244
Fees for expedited filings	*$350*	*$150*	*$119+*	*$70*	*$100*	*$50*	*4.5% fee*	*$25+*	*$25*	*$20*

	CA	CO	DE	FL	IL	MD	MA	NY	TX	WA
Formation (LLC)										
Filing fee/state fee	$ 70	$ 125	$ 120	$ 125	$ 500	$ 100	$ 500	$ 200	$ 300	$ 175
Minimum franchise tax	800		200		75					59
Annual report filing fee	20	100		139	25		500	325+		10
Total (Estimated)	$ 890	$ 225	$ 320	$ 264	$ 600	$ 100	$ 1,000	$ 575+	$ 300	$ 244
Fees for expedited filings	*$350*	*$150*	*$40+*	*n/a*	*$100*	*$50*	*4.5% fee*	*$25+*	*$25*	*$20*
Registration (LLC)										
Filing fee/state fee	$ 70	$ 125	$ 130	$ 125	$ 500	$ 100	$ 500	$ 250	$ 750	$ 175
Minimum franchise tax	800				75					59
Annual report filing fee	20	100			25		500	325+		10
Total (Estimated)	$ 890	$ 225	$ 130	$ 125	$ 600	$ 100	$ 1,000	$ 575+	$ 750	$ 244
Fees for expedited filings	*$350*	*$150*	*$40+*	*n/a*	*$100*	*$50*	*4.5% fee*	*$25+*	*$25*	*$20*

MORE INFORMATION

For a list of updated filing fees and additional information about various state filing requirements, visit the book's website at http://www.myhightechstartup.com.

Selecting a Corporate Name

Choosing your corporate name is important (although, as discussed below in the box, be sure to understand that a corporate name that is available doesn't mean that use of the name won't be restricted on other grounds such as trademark rules). You should carefully consider the name, whether a domain name is also available, and whether the name will violate any trademarks (you can do a check online through the U.S. Patent and Trademark Office website at http://uspto.gov). For helpful insights into name selection, check out "What to Consider When Selecting a Business Name" below.

You should also remember that there are no requirements that your corporate name must match the name you operate or do business under, or match a product name. Some startups, however, find that the expense of maintaining a different corporate name from their trade name adds unnecessary expense. Therefore, if possible, you may want to consider a corporate name that can be used for both corporate formalities and trade dress.

The general rule in nearly all states is that you may not form an entity or register as a foreign corporation with a corporate name that is already in use by a corporation in the state or one that closely resembles one already in use. You can contact the secretary of state's office in the state in question (many states now have the ability to check on corporate names online on their websites) or can hire a service to check the availability of a specific corporate name and to reserve such name, for a limited period of time, on a renewable basis.

The rules for naming conventions vary by state. Delaware generally requires the use of one of a selection of words (Corporation, Corp., Incorporated, Inc., etc.), although corporations with at least $10,000,000 in assets may, by filing a certificate with the secretary of state, avoid this requirement. California provides for more flexibility in the selection of a corporate name. In most cases, the name need not include words such as corporation or incorporated (or their abbreviations). The exception to this rule is when a natural person's individual name is used as the corporate name. In this situation, Corporation, Incorporated, Limited, or Company must appear in the name. California law prohibits the use of certain words in corporate names, such as bank, trust, or trustee, without the approval of the superintendent of banks.

WHEN IS YOUR CORPORATE NAME NOT REALLY YOUR CORPORATE NAME?

What could happen

You've picked out a name and found that it is available in your home state of Idaho. Once we incorporate under Idaho law, can we begin a national campaign under our new name?

Watch out for

It is important to note that the fact that a corporate name is available and that the secretary of state's office grants the right to use the name does not guarantee that the name actually can be used without restriction.

The fact that the secretary of state grants permission to use the name as the name of the new corporation does not in any way guarantee that the independent laws of unfair competition, fictitious business names, or federal and state trademarks, trade names, and service marks will also permit such name to be used. The permission granted by the secretary of state to use a particular name is fairly narrow in essence, a finding that the name is not confusingly similar to any other name on the consolidated list of corporate names maintained by the secretary of state. Another company may already have established rights to the name as a trademark, trade name, or service name.

The most fundamental right that a trademark, trade name, or service mark owner attains is the right to prevent others from trading on the owner's goodwill by confusing or deceiving third parties into purchasing a product or service through the use of a similar trademark or service mark. The basic question in assessing whether a trademark, trade name, or service mark has been infringed on is whether the use of a similar mark or name by another results in a likelihood of confusion among prospective purchasers of products or services. If the use of the similar name or mark by the new user is likely to confuse or deceive a substantial number of prospective purchasers into believing that products or services being purchased in connection with the similar name or mark come from the original trademark, trade name, or service mark owner or is somehow endorsed or sponsored by such original owner, then there is an infringement.

Where a corporation expects to try to establish a national reputation, a broad name search through trademark, trade name, and service mark registries may be desirable before significant investment is made in establishing good will in a name. If a particular Internet domain name is important to the corporation, the availability of that name can be confirmed through a search of Internet domain name registries.

TIP: An available name in your state of incorporation doesn't mean that your corporate name won't violate someone else's trademark.

Making the Appropriate State Filing

Once you've selected a state to incorporate under and found an available (and appropriate) name, the next step is making the appropriate filing with your state. Each state has unique rules for the filing and will reject your filing if you do not comply with the necessary provisions. Many states now allow you to file your incorporation documents online, further simplifying the process.

Your incorporation will oftentimes be handled by an individual known as your incorporator. The incorporator of the corporation has the power to do all things necessary for the initial organization of the corporation, including electing the corporation's directors if the initial directors are not named in the articles of incorporation. In many cases, the incorporator will also adopt the bylaws of the corporation (discussed below). Your attorney may be the initial incorporator to speed the process along until you are able to take certain corporate actions.

The filing of articles of incorporation or a certificate of incorporation, depending on the state, with the secretary of state establishes the existence of a corporation. In most cases, there are only a few items that must be included in your state filing to make the filing effective. Usually, you will be required to provide the following information: (1) the name of the corporation, (2) a brief description of the purpose of the corporation (which is generally a broad statement that the corporation may engage in any lawful activity for which corporations may be organized under the laws of the state), and (3) the authorized capital of the corporation.

Other than these standard items listed above, you will likely be required to provide other specific items depending on the state of choice. For example, Delaware requires the name of a registered agent for the corporation, the address of the corporation's registered office in Delaware, and name and address of the corporation's incorporator. The Delaware Division of Corporations maintains a list of individuals and entities qualified to act as registered agents. California requires that the name and address of the corporation's initial agent for service of process (the person to whom the state and the courts will direct papers and summonses) be set forth in the articles of incorporation.

Many new corporations will ask how many shares they should initially authorize. Although the corporation's initial authorized capital can be changed by amendment of these documents, it is prudent to authorize a number of shares of capital stock sufficient to cover not only the shares to be issued to the founders but also shares to be issued to prospective employees and investors.

You may also be required to include certain provisions relating to indemnification of directors, officers, employees, and other agents and limitation of the liability of directors with respect to certain matters. These provisions are generally of interest to all directors and potential directors of corporations with shareholders other than directors and may affect a corporation's ability to attract "outside" directors to its board.

Taking Formal Corporate Actions

Once you have arrived at this point in your formation, you will now have the shell of a corporation. At this point, you'll need to take certain steps to create a "functional" corporation. What do you need to do now?

Organizational Resolutions

First, you'll need to take your first set of corporate actions. This first set of organizational resolutions can be adopted at the first meeting of the board or through written consent of the board (a way to approve formal corporate actions without a meeting). These actions will handle many of the key formal matters of your new corporation.

WHAT TO INCLUDE IN YOUR ORGANIZATIONAL RESOLUTIONS

Generally, among other things, your organizational resolutions will take the following actions (many of which are formalities that are just important to get out of the way early on):

1. Call for the insertion into the minute book of the corporation's articles of incorporation/certificate of incorporation as filed with the secretary of state
2. Require the corporation to maintain a minute book including the articles and all amendments to the articles, the bylaws, and all amendments to the bylaws and the minutes of all board of director and shareholder meetings
3. Authorize the selection of a corporate seal
4. Authorize the election of the corporation's officers and specify which officers may sign on behalf of the corporation and/or obligate the corporation in matters relating to its business
5. Authorize the selection of a specimen stock certificate for the corporation's common stock
6. Authorize the issuance of stock to initial shareholders (if stock is to be issued)
7. Authorize the selection of the corporation's fiscal year
8. Designate the corporation's bank and accounting firm
9. Authorize the payment of the corporation's incorporation expenses
10. Authorize the corporation's officers to qualify the corporation to do business as a foreign corporation in states where such qualification is deemed necessary
11. Authorize the election of S-corporation status for the corporation, if desirable

Bylaws

Your bylaws typically set forth the rules and various general corporate procedures affecting the governance of the corporation. The bylaws are often adopted by your incorporator and ratified by the board.

Each state's corporate laws generally cover most of the topics normally addressed in the bylaws. Therefore, if the articles of incorporation set forth the authorized number of directors, there is no legal requirement that a corporation have bylaws. Bylaws are customary, nevertheless, even when the corporation does not desire to alter the rules established by applicable law.

If a corporation does choose to have bylaws, the bylaws must contain the number or range of number of directors, unless such a provision is contained in the articles of incorporation (although, generally, it is not advisable to put such a provision in the articles because they are more complicated to amend than the bylaws). The bylaws generally set forth, among other things, the powers and responsibilities of the corporation's directors and officers, the manner of calling meetings of shareholders and directors, the maintenance of corporate records, shareholder rights, voting and proxy solicitation procedures, regulation of transfer of shares, and indemnification of directors, officers, employees, and other corporate agents to the extent that indemnification is not covered in the articles. The bylaws may also contain any other provision not in conflict with law or the articles for the management of the business and for the conduct of the affairs of the corporation. However, all bylaw provisions must be consistent with the corporation's articles. If the two are in conflict, the articles' provisions will prevail.

Employer Identification Numbers

Any employer must have federal employer identification numbers to complete its federal and state tax returns. For your federal employee identification number number, it can be obtained by filing a Form SS-4 with the IRS. This can be filled out online in only a number of minutes.

Some states will also have similar filing requirements. In California, employers are required to fill out Form DE-1 with the California Employment Development Department and to register with the California Employment Development Department 15 days after becoming subject to the California Unemployment Insurance Code. A California employer is subject to this provision if it has had one or more employees and paid wages for employment in excess of $100 during any calendar quarter. In New York, employers must fill out Form NYS-100.

To find out some of the state filings that may be required, specifically whether your state has an employer identification number, visit http://business.gov. This site provides a service called "Permit Me" that provides links to appropriate state licensing authorities. Simply enter the state and the type of licensing, and it will provide you links to the appropriate state and local authorities.

Qualifying to Do Business in Another State (Other than the State of Incorporation)

If your corporation is considering transacting business in any other state, it should determine, before doing business, whether the company must be qualified or registered to do business in such other state. Most state laws provide that a "foreign corporation" (i.e., a corporation incorporated under the laws of another state) may not "do business" within the state unless it qualifies under appropriate statutory provisions. The scope and extent of the company's activities will govern whether qualifications will be necessary.

HOW DO I KNOW WHETHER I NEED TO REGISTER IN A STATE?

Typical activities that will require a corporation to qualify to do business in a state are as follows:

1. Transaction of a substantial amount of its ordinary business in the state
2. Maintaining an active office in the state
3. Manufacturing products in the state, but activities of substantially less magnitude may also require qualification

Although the penalties for failure to qualify vary from state to state, the following penalties may be applicable: (1) a denial of the right to enforce contracts in the state courts; (2) voidability by the other parties of all contracts entered into such state during the period when the company was required to qualify but did not; (3) monetary fines levied against the company; (4) monetary fines levied against agents or officers of the company; and (5) personal liability of the officers and/or agents of the company for the company's acts in the state.

Even if qualification is unnecessary, the company may be obligated to pay corporate income and other taxes (including sales and use taxes) as a consequence of operating in a state. For this purpose, "operating" in another state may include very limited and tenuous contacts; the states are becoming increasingly aggressive in treating foreign corporations as subject to their taxing jurisdiction based on virtually any activity within their borders. If the company employs persons located in other states, it may be subject to employer wage withholding requirements, worker's compensation requirements, and other regulatory requirements. Furthermore, if the corporation owns real or personal property in other states, it may be required to pay property taxes in such states.

Other State and Local Filings

Many trades, professions, businesses, and occupations are regulated by state law, which will often require that corporations meet various qualifications before granting certain certificates of registration or business licenses. Many cities also require that corporations doing business within the city limits obtain a local business license.

FINDING STATE AND LOCAL LICENSES

How can you find out about these various licenses?

Visit http://business.gov, which has a service called "Permit Me" that provides links to appropriate state licensing authorities. Simply enter the state and the type of licensing, and it will provide you links to the appropriate state and local authorities.

In addition, most states have a website dedicated to doing business in the state that contains this key information. A couple of examples include the following:

- Washington state has a Master Business License that is a single, simplified application for applying for many state licenses, registrations and permits, and some city licenses.
- California has the *California License Handbook*, published by the California Department of Economic and Business Development, which lists sources of California licensing requirements, the applicable regulatory agencies, and details the licensing process. The *California Permit Handbook* addresses permits that businesses may be required to obtain, such as certain environmental permits.

Formalities of the Corporation

Startups are oftentimes understaffed and have too little support staff for the founding team. This sometimes means that the founders struggle to follow proper corporate formalities. If you are concerned this may be you and your team, you should seriously consider hiring an attorney to assist you with following corporate formalities in the early stages of formation of your company.

The corporation is responsible for taking appropriate actions and following appropriate protocol for those actions. In particular, certain actions will require approval by your board of directors and others will require approval of your shareholders. Your company will also be required to maintain accurate records and follow proper rules and laws with respect to the rights of your shareholders.

Remember, the responsibilities of operating a new corporation are not intuitive and are not few. Therefore, be sure that you have a process in place to manage recordkeeping and ensure that your corporation complies with corporate formalities. There is no better way to torpedo a first financing, other acquisition, or sale event than by not being able to produce copies of records. Failing to keep records is a surefire way to prevent your company from growth (so for goodness sake, get a good assistant, controller, or other team member if you are worried you might not be keeping your records up to snuff!).

CORPORATE FORMALITIES FROM THE BEGINNING

What could happen

Because our company is in its early stages, is busy building the business, and has a board composed of just the founders, does it matter whether we keep corporate minutes, have formal board meetings, and do official option grants?

Watch out for

Absolutely. Companies that do not follow corporate formalities risk losing their corporate status and (as discussed below) could have a court rule to "pierce the corporate veil" and go after your personal assets.

In particular, you should be certain to follow corporate formalities with shareholder and director meetings, by signing appropriate documents as a corporation, with appropriate corporate record keeping, by appropriate state and federal filings such as corporate reports, franchise taxes, and federal and state corporate tax, maintaining proper bank accounts, and keeping proper financial statements and records.

In addition, if you are looking to raise funds, enter into certain partnership agreements, acquire an entity, merge, or be sold, you will risk sinking the deal if you cannot produce a clean set of your corporate records.

TIP: Follow proper corporate formalities or risk losing your corporate status.

Actions of the Board of Directors

Generally, the board of directors may only take action in meetings of the board at which a quorum is present (and which are either duly noticed or for which notice is waived) or by written consent without a meeting, as permitted by state law. You should always designate a corporate secretary to assist with recordkeeping and to assist with the publishing of minutes.

Meetings may be either regular meetings of the board or special meetings. Regular meetings are fixed by the board or the bylaws and require no notice. Special meetings, which require notice, may be called by the corporate secretary, chairman of the board, the president, any vice president, or any two directors, unless otherwise provided in the articles of incorporation or the bylaws of the company. Special meetings require four days notice by mail or 48 hours notice delivered personally by telephone or telegraph, and the articles of incorporation and bylaws may not dispense with such notice.

SAMPLE BOARD MINUTES

Go to Chapter 12 to find a sample of board minutes, or visit http://www.myhightechstartup.com to download a copy.

WHAT KIND OF THINGS DOES THE BOARD OF DIRECTORS NEED TO APPROVE?

Typical actions that may be necessary or desirable for the board of directors to approve include the following:

- Issuing securities and granting options, warrants, or other rights to purchase securities
- Adopting a stock option plan
- Amending the articles of incorporation
- Amending the bylaws
- Entering into major contracts, leases, or other obligations
- Declaring distributions, dividends, or stock splits
- Borrowing significant sums and the giving of security in connection with such borrowings
- Entering into employment agreements with key employees
- Electing officers of the company and setting or changing their compensation and terms of employment
- Adopting or amending employee benefit plans
- Forming subsidiaries
- Designating committees of the board and the powers of the committees
- Calling shareholders' meetings
- Buying or selling significant assets
- Electing directors to fill vacancies on the board
- Adopting company policies
- Merging or reorganizing the company
- Entering into a new line of business
- Changing the principal place of business
- Commencing or settling material litigation
- Taking any other actions material to the business

Neither notice nor a waiver of notice need specify the purpose of any meeting, and notice is not required for any director who signs a waiver of notice, a consent to holding the meeting, or an approval of the minutes of the meeting, or who attends the meeting without protesting the lack of notice either before the meeting or at its commencement. You should note that the foregoing requirements regarding notice and waivers are codified in many states including California, but they are not codified in Delaware. It is nevertheless customary and appropriate for Delaware corporations to impose such requirements in their bylaws.

PIERCING THE CORPORATE VEIL

Despite the general applicability of the rule that creditors of a corporation may not proceed against the assets of any shareholders of the corporation (the operational meaning of "limited liability"), specific circumstances may permit such creditors to "pierce the corporate veil" to satisfy corporate obligations by levying against assets of the shareholders. The cases generally have permitted the corporate veil to be pierced when fraud or similar malfeasance have occurred, and when it would be manifestly unfair to allow a shareholder to hide behind limited liability. Thus, it is important to note that the mere incorporation does not automatically prevent creditors of the corporation from reaching the assets of the corporation's shareholders.

The following are among the facts that courts have relied on in allowing the corporate veil to be pierced: (1) disregard of corporate formalities; (2) comingling of personal and corporate assets or diversion of corporate assets to personal use; (3) "holding out" to creditors by a shareholder that the shareholder is the obligor; (4) inadequate capitalization of the corporation; and (5) manipulation of corporate assets and liabilities by the shareholder.

To preserve limited liability for its shareholders, a corporation at a minimum should do the following to treat the corporation as a true separate entity:

- Obtain and record shareholder and board authorization for corporate actions
- Maintain complete and proper records for the corporation separate from the personal records of the corporation's owners
- Make it clear in all contracts with others that they are dealing with the corporation and not any particular individual, for example, by using the following signature block format on all contracts and agreements:

  [NAME OF CORPORATION]

  By:________________________________

  Title: ______________________________

- Conduct all transactions between the corporation and its shareholders, officers, and directors on an arms-length basis whenever possible. The board of directors should approve any such transaction by a vote of the disinterested directors (or, if all the directors are interested in the transaction, by a vote of the disinterested shareholders), after all the facts material to the transaction have been disclosed
- Start the business with a sufficient amount of equity in light of the future capital needs of the business

Actions of the Shareholders/Stockholders

Various actions of the corporation will require action by the shareholders, and these actions must be reflected in minutes of duly noticed meetings or by appropriate written consents. A corporation is typically required to hold annual meetings of shareholders, the principal purpose of which is to elect the members of the board of directors.

Unless prohibited by the articles of incorporation or state corporate law, any action that may be taken at a shareholders meeting may be taken by written consent if consents are signed by the holders of shares having at least the minimum number of votes that would be required to authorize the action at a duly held meeting at which all voting shares were present and voted. Many states, including both California and Delaware, have exceptions to this rule with respect to the election of directors. Generally, directors may not be elected by written consent except by unanimous written consent of all shares entitled to vote for election of directors or to fill vacancies on the board of directors. In addition, many states, including both California and Delaware, provide that notice must be given to those non-consenting shareholders entitled to vote of the taking of any corporate action approved by shareholders without a meeting by less than unanimous written consent.

WHAT KIND OF THINGS DO THE SHAREHOLDERS NEED TO APPROVE?

Typical actions that may be necessary or desirable for the shareholders to approve include the following:

- Merger or reorganization of the corporation
- Amendment of the articles of incorporation
- Amendment of the bylaws (other than an amendment setting the exact number of directors within the range established by the bylaws or articles of incorporation)
- Sale or transfer of all or substantially all of the corporation's assets
- Approval of contracts with interested directors
- Authorization of indemnity of a corporate agent for liability incurred when acting on behalf of the company
- Issuance of securities
- Adoption of stock option plans
- Winding up or dissolution of the corporation

Corporate Minute Book

The minute book of a company should contain its articles/certificate of incorporation, bylaws, and minutes or written consents covering all meetings and actions of the directors, committees of the board of directors, and shareholders. It is very important that the minute book be kept current and that it contain all necessary documentation. In financings, counsel for the investors or underwriters will often review the minute book carefully in connection with their due diligence investigations. Furthermore, up-to-date minute books will aid in establishing that corporate formalities were observed, which will be helpful in avoiding any shareholder liability problems.

Stock Ledger

The company must keep adequate records of stock issuances, showing the amount of stock issued, dates issued, and consideration received. A stock ledger can help the company organize this information. Usually, at least while the company is privately held, it is useful for the company to keep copies of all stock certificates issued.

Shareholder Inspection Rights

Many states, including California and Delaware, provide that the stock records of the corporation, the financial books, and the minutes of proceedings of the corporation's board of directors and shareholders must be open to inspection by any shareholder on written demand. However, the request must be for a purpose that is reasonably related to the shareholder's interest as a shareholder. These documents are not required to be kept at the corporation's principal office; often they are maintained by the corporation's legal counsel.

Additionally, in California, a corporation must keep a copy of its bylaws as most recently amended at its principal office. This copy should be open to shareholder inspection at all reasonable times.

Each director of the corporation has the right to inspect all books, records, documents, and physical properties of the corporation at any reasonable time. Any such inspection may be conducted in person by the director or by the director's agent or attorney. This right of inspection includes the right to copy and to make extracts of written materials.

WHAT TO CONSIDER WHEN SELECTING A BUSINESS NAME

Scott Trimble from Halfagain provides insights new businesses when considering their name. Halfagain LLC is a Portland, Oregon based creator of Search and Affiliate Marketing software.

Consider this: **The basic stuff.**

1. Be easy to pronounce and spell.
2. Make it memorable.
3. Don't pigeonhole yourself. (Being too specific in the naming of your company or product, e.g., Dave's 256k Flash Drives Inc. or Portland Flooring Inc., can hinder growth later.)
4. Go easy on the numbers.
5. Don't use names that could have a negative connotation in other languages. (Baka Software Inc. sounds OK in the US but won't fly in Japan.)
6. Stay away from negative connotations.
7. Make sure your name doesn't alienate any group (race, religion, etc.).
8. Search for existing trademarks on potential names.
9. Make sure the domain is either available or purchasable in the aftermarket. Use your favorite registrar OR use a bulk domain checker. (I've outlined one below.)

Consider this: **Domain availability.**

Domain availability is possibly the biggest hang-up to ever happen to naming. Sure, you can come up with great potential names, but can you come up with great potential domains that are available?

I won't spend much time on this because it's pretty simple. If you're creating a name for a product or business that will require a .com, be patient, keep trying and you'll start to get a feel for names that are more likely to be available than others. I've also listed some tools below that will help immensely with this.

Consider this: **Focused brainstorming.**

Every book out there prescribes brainstorming. However, instead of just sitting back and trying to come up with ANY words that describes your business, focus your brainstorming to answering a set of questions. Answer each by making as long of a list or words and phrases as you possibly can. Remember, the longer and more abstract your list, the better off you'll be. So go wild . . .

- What does your product do?
- What does your industry do, what's its purpose?
- What is your product's benefit to the consumer?
- What will happen for them?
- What will they get?
- What are the "ingredients" that go into your product or service?
- How are you different from the competition?
- What makes you unique?
- What's the lingo in your industry?
- What are the expressions that are unique to your offering and business? *(add your own as you see fit)*

Consider this: **Synonym search.**

It's pretty simple, really. Take each and every one of the words you brainstormed above and plug them into a thesaurus like Thesaurus.com. Run through each entry, keeping the words you like, trashing the ones you don't. Put these into a new list, paying attention to name possibilities.

Consider this: **Word combining + a cool name combining tool.**

After you've done some focused brainstorming and/or a synonym search, try word combining. Pop ALL of your words into a word combiner like My Tool (www.my-tool.com), tweak it's settings to reflect what you want it to show and combine. Depending on how many words you put into the system, you may get a massive list returned to you. To weed through them quickly, you can then hit the button at the bottom and check each domain for availability.

Consider this: **Name and word lists to get your juices flowing.**

Plenty of great product, company and website names have their roots in other, irrelevant names. Look up "list of ______" in Google and you'll get more than you can handle.

Geologic periods
Fruit or food names
Types of dinosaurs
Kinds of rocks
Latin or Greek roots
Place names
Historical figure names
Zoological names
Botanical names
Math or engineering terms
Astronomical terms
Animal, fish, or bug names

Think about this abstractly also. If your product is new and unique, what foods or plants have fresh connotations? Etc.

Consider this: **Punning and plays on words.**

I just tried a new beer recently specifically because of its name. It was called Tricerahops, a double IPA made by Ninkasi Brewery. Quite a beer, incidentally. But check out how you can create a name like that. Cruise your focused brainstorm and synonym lists for words that describe/define your product. In this beer example, we might find hops—one of the main ingredients in beer. Then, we can look through lists of animals, foods, places, etc. and see if we get any good combinations, where the words fit seamlessly. In this case, they chose a dinosaur name "Triceratops" and simply changed one letter. Here's an even easier way of doing it . . .

Consider this: **Groovy word tool.**

Use this tool More Words (www.wordlab.com) and search for any words that contain ____. You can search for anything—search for words that contain 'top', or words that have a double 'e'. Virtually any sound or letter combo you want to find in a word, this site will do it for you.

Consider this: **Meaningful or not?**

(Example: Dave's Rocket Repair Inc. has meaning, Simble Inc. does not)
Some say creating a name with built-in meaning is a must—new companies or products need to seem familiar and safe. Others say nonmeaningful names are the best—the name is completely yours, free of meaning (which you can then define), plus, newly coined word names connote innovation.

The jury, as they say, is out. Some things to keep in mind though:

Newly coined words CAN convey meaning. The most championed of these may be Acura, which was formed from the morpheme "Acu" and finishing with suffix "ra". Acu as a root connotes accuracy or precision, which fits nicely for a luxury car line.

The creator of the Acura name (Ira Bachrach of NameLabs) is purported to have a list of thousands of combinable morphemes. I, as of yet, have not found such a list. If you happen to run across one, I'd love to see it ☺.

Consider this: **A *truly* killer naming tool.**

Word Lab and specifically the Word Lab Tools pages.

This website I consider to be one of the single most powerful naming tools out there. With an absolutely massive list of company names, a morpheme name creator, name builder, and so on, this site is the juggernaut of idea generators. Every time I'm naming something new, I use this site.

Consider this: **Metaphorical naming [powerful stuff]**

I call it metaphorical or lateral naming, but no matter what you call it, it's a branch from the focused brainstorm and often, the coolest names come from this method. It'll take a more creative, abstract frame of mind, so whatever you need to do to break out of your linear comfort zone, do it.

So, after you've changed into your tie dye and stared at your Led Zeppelin poster for a while, grab your focused brainstorm. Here we're going to center on the question—"What does your product, business or industry DO". You're going to sequentially take each of the words and phrases you came up with, and come up with OTHER things in life that do these things too.

Let me repeat (or rewrite, as it were) that. You're going to take what your business does, and come up with other things in life that do the same thing. Make a list of everything you come up with. Here are some examples:

I have a software company and our newest product's function is to copy files (pretty high tech, I know). So I ask, "What else in life copies things?"
A copier – too logical.
A cell – might work, but a little "out there."
A mime – A HA!
Why not call the new software product – Mime.

Here's another:

My marketing company helps its clients' voices get heard above the competition.
So, what else gets voices heard or makes things louder?
A bullhorn.
A volume dial.
An Amplifier – A HA!
Why not call the company Amplify Interactive (happens to be a real company here in Portland). Volume Media wouldn't be bad either.

Consider this: **Misspellings.**

Misspellings of commonly used words can get you in familiarity's proverbial backdoor. Example – netflix.com. It's familiar, short and you instantly know what they do. Though, if looking for an available domain, you'll have to use some fancy combinations because common misspellings are already registered.

Consider this: **Industry lingo.**

Each industry has its lingo and you may have noticed that many taglines come from this, or more distinctly, those words and expressions that are used by your consumers.

For example, I've just developed the perfect fish hook. It never, and I mean NEVER lets a fish go. A common expression in fishing when you feel a fish take your bait is "Fish on". This great expression combined with something else, might make a nice tagline for my fail-safe hook. How about – **Fish on . . . never off**.

Consider this: **Ask your friends, but . . .**

Ask your friends' opinions, but take them with a grain of salt. First of all, your pool of test subjects is probably pretty small, leaving your results (ratio of yeas to nays) with little accuracy. Secondly, consider whether your friend is in your target market. If they're not, they may not 'get' a name which might be perfect for your market. Finally, people in general side with what's familiar. Finding your website, seeing an advertisement or having a friend suggest your product can have the unique ability of making your product's name sound good. The name or names you ask your friends to grade won't have the benefit of this.

Consider this: **How is the competition named? What are the trends?**

I've made the mistake (like an idiot, I might add) of not checking my competition first and creating a name, only to find out my name is JUST like a competitor's. Time wasted. Now, my general rule is to find out how my competitors are naming themselves and simply be different. Stepping out of the box is always a bit of a gamble, so make sure you're different in what will be seen as a positive way.

Consider this: **Name rhyming.**

Rhymed names are memorable and can work, as long as they're not too cute or overboard. Rhyme Zone (www.rhymezone.com) is fantastic for finding words that rhyme. More Words can also be good for this.

Consider this: **Web 2.0 name generators**

I'll be honest, they're generally crap. I've used this one – Web 2.0 Name Generator, but found that, for the most part, they return relatively useless gibberish. If you have a few extra minutes though, try popping some of your synonyms into the interface and seeing what it comes up with. At the very least it might give you some ideas and get your wheels turning.

Consider this: **Don't put TOO much stock in your name.**

They're certainly important, but naming can also be overemphasized. There are plenty of highly successful businesses and products out there with bad names. So, take your naming, like your friends' opinions, with a grain of salt. And, as with everything, the more you stress about obtaining perfection, the less likely you'll come up with that killer name that seamlessly fits your offering.

Source: Used with permission, Scott Trimble, Halfagain LLC.

8

Planning for Your Business

> Make sure you write a business plan because it will crystallize your thoughts to communicate your ideas with somebody else. Make sure that once you have written your business plan, you have somebody read and critique it and ask you questions. It doesn't have to be a cookie-cutter business plan with glossy pages and lots of information. Essentially it's a plan that says what the company is going to do, what problem it is going to solve, how big the market is, what the sources of revenue for the company are, what your exit strategy is for your investors, what amount of money is required, how you are going to market it, what kind of people you need, what the technology risks are, marketing risks, execution risks. Those are the fundamentals of what goes into a business plan, and many people have it in their heads but don't write it down.
>
> **Sabeer Bhatia**
> *Cofounder of Hotmail*

What to Watch For

At this stage, you most likely have an idea that has morphed into a business concept, one that you hope to make money at. The next step, as discussed by Sabeer Bhatia in the quote above, is to move this general concept into a usable action plan. Setting out the plan for the business points your business in the right direction and keeps you focused on the key business objectives.

Some entrepreneurs wonder whether it is worth the effort and time it will take to craft a business plan and instead opt to take a plan-as-you-go approach (perhaps only creating a business plan for fundraising purposes rather than actual business operations). Although this plan-as-you-go approach certainly does happen, researchers note that companies that have a formal business plan early in their lifecycle tend to produce better results.

According to Professor Scott Shane of Case Western Reserve University, companies that have a formal business plan in place are more likely to survive when compared with those that don't engage in formal planning. Again, no one approach is the right one, but most startups tend to engage in some form of business planning even if the planning never produces a formal business plan for the organization. Oftentimes, taking those informal plans and transforming them into a formal plan is less onerous than you might think.

The purpose of this chapter in the book is to help an entrepreneur develop a logically arranged, reasonably complete, and effective business plan that will provide financing sources the most useful and persuasive information about the proposed venture—information they need to make swift decisions. The goal of a business plan is to convince potential investors that, through a strong technical team, an experienced management team, and differentiation, the competitive advantages of the proposed technology or product or service will fill market voids and offer potential opportunities for revenue growth and investment return and maximization. Other than helping management drive business goals and decisions, the ultimate action objective of a business plan is to turn prospective investors into check-writing investors.

Business Planning

It can't be that difficult to write a business plan, can it? There are countless templates, preparation aids, and resources available to help an entrepreneur write a business plan. You just plug in some information about the business onto a template and you've got a business plan, right?

Business planning is much more than just getting words on paper. Planning for a startup company is a process that involves critical thinking, thorough research, and identifying the key opportunity for the business. It could involve highly charged discussions among the founders defining the short-term strategy and may involve several overhauls in the business focus. Those changes are part of the planning process. This planning stage should ultimately be a time in which the founders truly analyze the opportunity and the risks with the aim of being intimately aware of the requirements needed to build this successful business. This planning process and the challenge of a critical self-assessment is the biggest challenge in the preparation of any business plan.

HOW LONG WILL IT TAKE?

Experts suggest that the business planning process could take as few as 100 hours and up to 2,000 hours. The more common range given by experts is between 200 and 500 hours, oftentimes divided among a team of individuals building the plan. These hours can appear to be very deceptive. Much of the background and market research done in preparation of annual budgets or for other planning purposes can be used in the preparation of a business plan.

Surveys of entrepreneurs suggest that a business plan for a new business will take between 160 and 640 hours to research and prepare. For established businesses, the process will take less time to prepare, usually between 40 and 160 hours, depending on the amount of work that has been done previously.

There are primarily two types of business plans:

- **External plan:** Used as a tool to highlight the plans for the business to interested third parties, including potential investors.
- **Internal plan:** Used to set goals and objectives for the business and track progress. An internal plan is usually targeted at the board of directors, the management team, and others involved in the operations of the business.

For each of these business plans, the underlying information will generally be very similar. The difference is in the focus of each plan. An internal plan will be used as a tool for ongoing tracking of the business that is adjusted as changes arise in the business. This document is a roadmap for the business and should be used to guide the business as it matures. Conversely, an external plan is designed to be a solicitation tool. Here, the focus is on selling the business as a smart investment for any investor. The external plan will focus on the potential and promise of the company and lay out why external funds are crucial to the company's strategies.

WHAT ARE THREE MAIN TYPES BUSINESS PLANS?

A business plan is used for various purposes and will be prepared for various audiences. Listed below is a summary of three typical times a startup business will engage in business planning.

- **Business concept summary:** The founder's first attempts to lay out the reason he is creating the business. What is the market potential? What niche will our product fill? Who are the competitors? What is the "homerun" potential?
- **External business plan:** Depending on the needs of the third party, the company will prepare specific planning documents and resources to convince an outside investor to provide funds for the business. How does your investment fit our needs? What are the key metrics we judge our business by? Why is this approach the "right" move for the investor?

- **Internal (or strategic or operational) business plan:** The founders and management team will undertake a comprehensive evaluation of the business and prepare a detailed and thorough plan to be used to guide the business. What are key milestones? What are the hiring needs? What will the product development timetable be? What will the financial picture be today, this year, and three to five years from now?

In each case, the foundation of a thorough business planning process will be research, from both various reference sources available to the entrepreneur as well as anecdotal research available from key advisors and personal resources.

This chapter will focus on assisting an entrepreneur build an external plan used to solicit financing from third-party investors. However, many of the items discussed in this chapter can be used for the preparation of an internal plan. Additionally, energy spent developing an internal plan will oftentimes make the process of creating an investor-focused business plan easier and faster.

Where to Begin

Most entrepreneurs will initially start at the 30,000 foot level in their planning. This first stage is referred to as the business concept summary: a big picture analysis of the opportunity itself. (The business concept summary is discussed in greater detail in Chapter 2.) Usually, this phase will take all that is in your head (and the heads of your cofounders and collaborators), on the back of napkins, and in various e-mails, documents, and spreadsheets, and sprinkle in some initial research and data to try to identify whether an opportunity really does exist for a startup. This stage is the "could this work" stage, which is usually full of estimates, projections, anecdotal evidence, and experiences of the founder.

From this point, the founders will usually begin to expand into a "how could we make this work" stage and develop a more thorough understanding of the business (the basics that will ultimately become the external and internal business plans). This stage is most commonly understood to be the phase in which a basic outline for a business plan is prepared, including market research, product strategies, management team needs, funding requirements, competitive landscape reviews, and a roadmap to create a market leader.

Once a good foundation has been established for the business, the company will usually move into the "what will an investor need to know to fund this" stage. The founders will prepare an investor-focused business plan, designed to answer questions about the opportunity and identify why this is a smart investment.

TOO MUCH PLANNING

One common problem some first-time entrepreneurs face is focusing too much time and energy on creating the "perfect" business plan. Although the business plan is an important document, don't let it stop you from accomplishing other goals needed to build a business. Remember, your business plan is a work in progress and just one tool used to affect the path the organization takes. So don't let building a business plan stop you from building a business! Keep building the business; investors aren't investing in a business plan, they are investing in the business, the technology, and the people behind that plan.

Purposes of Business Planning

To be clear: the purpose of preparing a business plan is not simply to please a potential investor and get them to invest in the company. The purpose of preparing a business plan is to lay out the roadmap necessary to build the business. Ultimately, the goal is to prepare an entrepreneur to show the potential investor that the entrepreneur has thought about the problems, identified a set of solutions the business can provide, and identified a strategy to make money selling it. By showcasing this in-depth understanding of the business' opportunities and risks, the entrepreneur is aiming to convince a potential investor to invest in a team the investor can trust.

PLANNING TO SUCCEED

What could happen

Once you set up your business and received funding, your business plan has found its way into the scrap heap.

Watch out for

Successful businesses recognize the importance of strategic planning and use their initial business plan (with tweaks and modifications along the way) as a part of the regular operations. Many investors will ask for regular tracking of aspects set out in your business plan at board meetings or investor presentations. Strategic planning is an important part of a successful business strategy, so you should find ways to make goals tangible to all parts of your organization. Companies with a technical focus and founding team oftentimes struggle to maintain the focus and direction during their initial development stages.

TIP: Incorporate strategic planning into your business' fabric.

Again, this process of a self-assessment of the business is just as important as any single document and will result in a better product in the long run. The business planning process should prepare the founders to consider the critical challenges ahead and proactively consider the solutions. Who do we need on our team? What is the critical problem we can solve? How can we develop our product or service? Can we sell the product?

DOES PREPARATION OF A BUSINESS CONCEPT SUMMARY OR A BUSINESS PLAN VIOLATE THE OBLIGATIONS TO YOUR CURRENT EMPLOYER?

In most cases, no. Although it could seem underhanded to prepare a business plan for your startup while still employed, preparing your business plan or researching a startup business is not illegal or unethical while still employed. Here are some suggestions to eliminate potential problems when you depart:

- Avoid working on your business plan on company time
- Avoid recruiting coworkers to join your new company
- Try not to start a business that directly competes with your current employer before you depart
- Refrain from using company resources (computers, e-mail, copiers) in the preparation of your plan
- Continue to perform your job duties and provide the expected effort at your job

Building a Business Plan

Business planning is the process of developing a set of tools and resources for your business, tools that will change as the business grows and matures. Depending on the stage of your business, you'll need different tools and different levels of formality.

BUSINESS PLAN PREPARATION RESOURCES

Below are additional publications available online for preparing your business plan:

- *Outline for a Business Plan,* Ernst & Young LLP, 1997
- *Guide to Producing a Business Plan,* Ernst & Young LLP, 2001
- *Writing an Effective Business Plan,* Deloitte & Touche, Fourth Edition, 2003
- *Hurdle: The Book on Business Planning,* Tim Berry

Remember, you can literally find hundreds of resources available to assist entrepreneurs with business plan preparation, such as software, books, online publications, websites, consultants, and articles. Don't let information overload derail your business plan. Use applicable resources and information from personal contacts, but don't lose track of the purpose of the business plan: to assist you as an entrepreneur!

At the earliest stages of your business, you'll probably only have a few documents and spreadsheets laying out the business concept that you'll share among your cofounders and inner circle. However, as you begin to make strides toward establishing your business, you'll need a more detailed and thorough document. Eventually, these initial, internal plans will become tailored to attract money from investors, including angels, government grants, venture capitalists, and others.

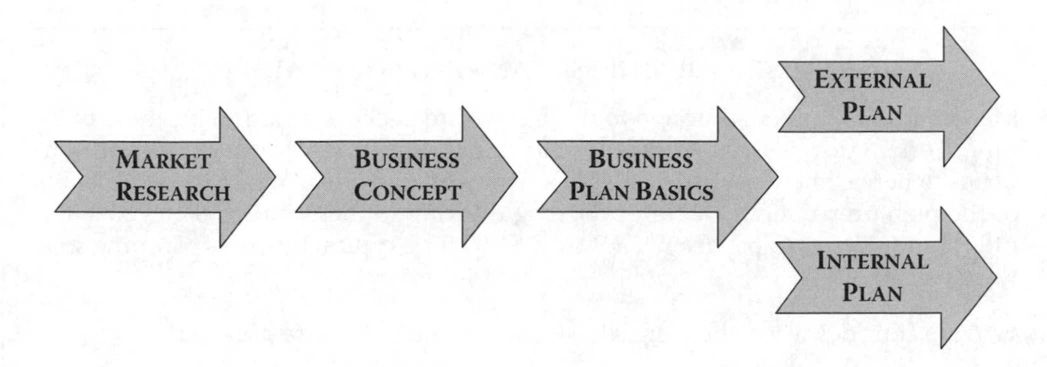

Building a thorough business plan will revolve around an in-depth understanding of the business, the industry, the competitors, and the market. As a result, research into each of these areas will form the foundation for the business plan itself. These research efforts will involve review and consolidation of various sources of data from personal business and industry knowledge to publicly available market research. Chapter 2 provides a number of key research resources to use the business planning process.

Our first step in the business planning process will be preparing a business concept summary. This step generally is done when the founder first begins to flush out his idea. Your business concept summary represents the earliest planning tools, including basic market analysis, high-level sales projections, and target customers. Generally, you'll find the business concept summary is very rough and is focused on very broad numbers, many of which are based on the founder's knowledge or commonly available research. From the business concept summary and the related tools, the basis for your budgeting process, your sales projections, your hiring goals, and much more will evolve. Hopefully, these tools will be used to prepare a full business plan for your company.

Once you have put together your business concept summary and the associated tools and documents, you'll then begin the more formal process of preparing the business plan basics. Preparing the business plan basics focuses on the following key areas: (1) product development, (2) marketing and sales, (3) personnel, and (4) financial projections.

Most high-tech startup companies will need to raise funds, and, to do so, management will need to prepare a business plan used to solicit financing from outside parties: an external business plan. These external business plans will generally be customized to the type of investment being sought. For instance, venture capitalists typically want to see documents look a certain way and may only ask to see an executive summary business plan in the beginning stages. Banks and grant programs may have different requirements that can be met through customizing your business plan.

Most businesses will put in place an internal business plan: an operational or strategic plan designed to set out a long-range strategy for the company. This internal business plan will represent a substantial document incorporating in-depth research into target customers or industries, laying out specific development timetables and objectives, identifying key inputs of additional team members as the business grows, and a providing deeper level of insight into the business. At the end of the process, you'll have a working document that the management team can use to guide the business and provide specific objectives for additional team members throughout the organization. The internal business plan is discussed in greater detail at the end of this chapter.

USING BUSINESS PLANNING SOFTWARE

Many entrepreneurs will decide to use a software package to aid in business planning. Certain software packages will simply provide business plan samples and templates, whereas more advanced systems will analyze your plan, assist with details of the plan preparation, and offer ongoing tracking to the business plan goals and milestones. Expect to pay from $39.99 to $299.99 for a typical business planning software package.

What to consider when choosing a business planning software package?

- Ease of installation and setup
- Price
- Time savings
- Simplicity
- Integration with other software tools
- Customer support
- Robust features and applications

Top-rated business planning software packages include the following:

- Business Plan Pro 11.0; Business Plan Pro Premier Edition 11.0; and Marketing Plan Pro 11.0 (by Palo Alto)
- Plan Write for Business; Plan Write Expert Edition; and Plan Write for Hi-Tech Marketing (by Business Resource Software)
- Fundable Plans (by Fundable Plans)
- BizPlan Builder 2008; and BizPlan Financials 10 (by Jian)
- PlanWrite; PlanWrite Expert Edition; and Plan Write Hi-Tech Marketing Planner (by PlanWare)
- Business 10.0 (by PlanMagic)

For a full review of more than 20 different business planning software titles, please see the table at the end of this chapter.

Finally, remember that business planning is ongoing. As you learn more information and continue to grow the business, your business plan will continue to evolve. Once you've put in place a plan, success will be measured by how you execute on that plan.

SAMPLE BUSINESS PLANS ONLINE

The following websites offer free sample business plans that you can download:

- Center for Business Planning (http://businessplans.org/businessplans.html)
- Bplans.com (http://www.bplans.com/sample_business_plans/index.cfm)
- Entrepreneur.com (http://entrepreneur.com/services/sampleplans/)

- About.com: Small Business Information (http://sbinformation.about.com/od/bizplansamples/Business_Plan_Samples.htm)

Most business planning software packages offer sample business plans with the purchase of the software.

Key Considerations for Your Business Plan

A well-prepared business plan tells the story of the company, both where it has been and where it will go. Although it may seem simple to a founder of the company, investors list a "failure to explain the opportunity" as one of the most common mistakes of a business plan. Therefore, the entrepreneur must focus her efforts on laying out a concise story for the readers.

KEYS FOR YOUR BUSINESS PLAN

The three most important considerations for a business plan are as follows:

- Be brief
- Be specific
- Be honest

When preparing a business plan for potential investors, remember who your audience is. That's right; it's the people you want to give you money to build your business. So focus on giving an investor reading your business plan the information he needs to make a decision. The business plan may be the first time a potential investor will hear about your business, which means it is important to provide him with a professional product.

ACCORDING TO VENTURE CAPITALISTS

What is the best way to make a good first impression with your business plan?

- The business opportunity is presented in a clear, exciting manner.
- The entrepreneur understands that projections are, at best, hopeful guesses and tries to base the projections on realistic assumptions.
- The entrepreneur makes as full disclosure as possible of the pitfalls of the business as well as its strengths.
- The plan is carefully proofread and edited until it does not contain any errors in grammar or math.
- The plan shows why the company and its products are different and significantly better than what is out there in the marketplace.
- The company has taken the time to study and understand its competitors and can address their strengths and possible weaknesses.

- The plan contains enough information to tell a complete story about the company but is presented in a concise, tight writing style.
- The plan does not make exaggerated claims about the product or the management.
- The entrepreneur knows the plan by heart before making a presentation to the potential investor.

Source: Profit Dynamics Inc.

ACCORDING TO VENTURE CAPITALISTS

What is the most common mistake entrepreneurs make when completing their company's business plan?

- Stating that the company had no competition or underestimating the strength of competitors — 32% of respondents
- Not clearly explaining the opportunity — 27%
- Disorganized, unfocused, or poor presentation — 12%
- Miscalculation of market share and market size — 9%
- Failing to describe a sustainable competitive advantage — 9%
- Failing to address the risks of a venture and failing to contain a contingency plans for coping with the risks — 9%

Source: Profit Dynamics Inc.

Building a Business Plan

You'll find experts who have a number of different structures for a business plan (you'll find four different outlines for business plans at the end of this chapter). In short, the business plan will lay out the product or service, the market and competitive landscape for the product or service, and the approach the company will take to tackle that market. All of this should fit together into a story of a founding team with a unique idea to tackle a problem. The key components of a business plan are as follows:

- Executive summary
- Product, service, or technology
- Market opportunities
- Competition
- Marketing and sales
- Financing and liquidity plan
- Management

To download samples of full business plans and executive summaries of business plans, visit the book's website at http://www.myhightechstartup.com.

Preparing a business plan for a high-tech company does involve certain differences from other industries. Many high-tech businesses will be unprofitable for a substantial period during product or technology development. As a result, the focus of the business plan will be on the sizeable opportunity for long-term profit growth on technology development. As a result, there is oftentimes less of an emphasis on short-term sales efforts and a greater emphasis on market opportunities in the long run. Additionally, expect to include a greater emphasis in the plan on topics such as development timetables and financing needs, protection of the technology, and retention of key research and development personnel.

WHAT ARE INVESTORS LOOKING FOR?

The following are key considerations for investors to consider when preparing a business plan for a high technology business:

- What level of financing is needed in the next 12 months?
- What level of financing does the company project to need over the next two to five years?
- Where is the company allocating money it raises and how does this contribute to the growth plans of the business?
- What is the expected return on investment (ROI) for the investor?
- What are the company's exit strategies (buyback, sale, or IPO) and how common is this strategy for companies in the market?
- What amount of ownership is the company willing to provide to an investor?
- Is the company willing to condition receipt of funds on achieving any milestones or targets?
- How with the investor be involved with the company (board of directors, management, advisory, etc.)?

Don't make a reader hunt for this information; be sure to include this in the executive summary that is readily available to a potential investor.

Executive Summary

The executive summary of the business plan is often called the most important section of the business plan. An investor that receives dozens of business plans in any given month may only skim the plan, focusing on the executive summary. For many investors, they will only give a few minutes to a business plan, and it is your job to hook them on the opportunity in those few minutes.

What are the primary objectives for your executive summary?

- Provide a clear statement of the business opportunity.
- Introduce the management team and highlight their combination of skills and experience.

- Detail financial projections based on clear assumptions.
- Identify the short- and long-range plans for the company.
- Explain the reason that an investor can achieve significant returns on an investment in the business.

The executive summary should be written to allow a reader to quickly understand the crux of the business plan without having to read the entire document. If a reader needs more information on a particular section or topic, this information can be found in the specific section of the plan.

GRABBING THEIR ATTENTION

According to Ernst & Young, some 85% of business plans are not seriously considered by investors. So what should you include in your executive summary that will capture the interest of a potential investor?

- Our product is targeting a sizeable market with unmet needs.
- Our business has an innovative and attractive approach for serving this market.
- We have already had some success with our approach.
- There are other companies similar to us that have had successful exit events.

A potential investor is looking for a business plan with tangible, specific, and clear evidence detailing each of these key statements. Build an executive summary that clearly answers these questions: "Can this business idea be a big success?" and "Why is this opportunity compelling?"

Usually, the executive summary will be the last section written and should be limited to one to two pages in length. Focus on what is most important, but be certain that the executive summary adequately explains the business and the opportunity in the marketplace. The primary objective of the executive summary is to offer sufficient information to the reader, who may only have time to read this section to make the initial decision to reject the plan. If the plan does not have a clear, concise, and compelling summary of the business concept, a plan will likely find its way into the reject pile.

GIVING THE VENTURE CAPITALIST THE INFORMATION NECESSARY TO MAKE A DECISION

According to Cindy Nemeth-Johannes, these are the things a venture capitalist wants to read in your executive summary:

- Here's our estimate of the market.
- The reason why you're going to be a market leader.

- How will you benefit your customers? How much will you benefit them? Why will they choose to do business with you?
- How are you going to prevent competition from going after your market and your customers?
- What does your management team look like? What have they already done? Do you have any advisors who will carry a lot of weight?
- How much money are you looking for, what are you going to do with it, and what is the current financial situation?
- What have you already accomplished? List up to eight things that you've done and don't restrict yourself to the product itself. Sure, they want to know that you can make the product work, but they also want to know what parts of your team are in place and whether you've come to agreements with any strategic partners.
- What are the milestones you intend to accomplish?
- When do you plan your exit strategy? What do you think the company will be worth then? Keep your numbers realistic or they'll figure that you're just a dreamer, not a doer.

Source: "How Do Venture Capitalists Evaluate Your Executive Summary?" (http://abcsmallbiz.com/bizbasics/gettingstarted/vencap_eval.html).

A well-written executive summary will include the following:

- An introductory paragraph to draw in the reader and identify the key opportunity the product or service will fill
- A company overview, including information on the organizational form, the date of formation, stage of funding/development, location, and milestones achieved to date
- A description of the core proprietary or licensed technological strengths
- A summary of short-, medium-, and long-term operational goals, implementation and growth strategies, and plans to attain and sustain market leadership
- A list of the factors necessary for the business to succeed
- Information on the market, the market size, and targeted or likely customers
- A clear description of the key benefits, features, and strengths of your product, service, or technology
- Profiles of key management and advisors
- Information on the financing needs of the business, including amounts raised thus far, future funds being raised, and the use for the funds
- A description of the most likely exit strategies for the company
- Financial highlights and projections
- Summary of the technology, team, and market opportunity

SAMPLE OF BUSINESS PLAN LANGUAGE

Executive Summary: Management Team

High-Tech Startup Inc. has a solid management team with years of e-commerce startup experience. Each of the founders has successfully founded and sold previous Internet startups. Each member of the founding team brings a wealth of experience to the company, including e-commerce, Internet strategy, application architecture, advertising models, content development and deployment, and business development.

Tina Talented, Cofounder/CEO

- Founded eSuccess Corporation
- Led product strategy, planning, and development of webSuccess, an Internet commerce product
- Sold eSuccess to MajorWeb Inc. (Nasdaq: MWI) in 2006
- Headed business development for MajorWeb's e-commerce efforts
- 16 years of relevant experience in operational and technical management in the communications arena

Ernie Engineered, Cofounder/Vice President of Engineering

- Founded vWorld Enterprises while a PhD student at State Tech University
- Helped vWorld attain market leadership position in tracking system for virtual reality applications
- Eight-year veteran of various technical product development

Seth Sellers, Cofounder/Vice President of Sales and Marketing

- Founded eMarket Inc., an online marketing software company
- Directed strategy and implementation of product marketing for eMark
- Sold award-winning OscarGrammy.com entertainment website to Newsite.com
- Developed comprehensive Internet strategy for World Telecom Inc.

Remember, the executive summary should be more than just a shortened version of the business plan. The executive summary must tell the story of the business and convince the reader (in perhaps as little as a few short minutes) that the opportunity does exist and the team is capable of capitalizing with this company.

SAMPLE OF BUSINESS PLAN LANGUAGE

Executive Summary: Critical Success Factors

The business plan that follows discusses the factors that make the company a well-positioned entrant in the competitive e-commerce arena:

- The technological breakthrough will overcome the inadequacies of alternative technical solutions

- Low-cost and high-performance proprietary technology delivering services adhering to/surpassing industry standards
- Technology is already operational (or in clinical trials)
- Management and advisory team with 60 years of combined management, consulting, research and development experiences and a proven track record for success
- Favorable regulatory environment
- A sizable local, national, and international market with tremendous potential for growth

Product, Service, or Technology

The aim of this section of the business plan is to focus on the technology, describing in detail the likely applications of the technology as well as the key opportunities for revenue and future growth. The focus should be on the benefits for the target markets, including why current products, services, or technologies are unable to provide comparable benefits to the end users.

What are the primary objectives for the product, service, or technology overview?

- Provide a clear description of the product, service, or technology
- Describe the current stage of product development
- Identify upcoming development milestones such as beta tests, clinical trials, or product releases
- Explain planned development for future products or services and identify the ongoing research and development efforts

One of the common problems faced by technology-focused entrepreneurs in preparation of a business plan is placing too much of an emphasis on the product or the technology in the business plan. This is common because many engineers or technology-focused entrepreneurs are most familiar with the technology and less familiar or comfortable with some business aspects of the plan. Remember that the purpose of this section is to explain the technology in a concise and clear manner but to focus on explaining the key aspects of the technology that will allow the company to differentiate it from outside competitors in the market.

A well-written product, service or technology overview will include the following:

- References to intellectual property found in your technology
- Illustrations, screen shots, or other images that can provide a visual representation of the technology or product
- A clear explanation of the features, contrasted with other competing technologies

Market Opportunities

The section of the business plan detailing the market opportunity should set forth the rationale in starting your business. This rationale will be laid out by highlighting the overall market opportunity, as well as the target markets for your technology. This section is crucial because it should offer a reader a view into the applicable market conditions and showcase why your business is well positioned to tackle the unmet market demands.

DEFINING YOUR MARKET

What could happen

You've been told that defining your market is a key component of the business plan. How do you define the market in a realistic manner?

Watch out for

Savvy investors don't want you to quantify the total available market. Focus on the target market for your product and provide a realistic estimate of that marketplace. Also provide the key assumptions used to reach your market estimate. Define the potential customers in the marketplace (be specific where appropriate) and identify the party within the organization that would make buying decisions. If you provide a five-year market-share forecast, identify what would be likely customers you would need to capture to reach that level. As one investor said, "I haven't seen a single business plan that was able to give me a realistic estimate of their target market." Use the help of potential investors to put together data for your market penetration and be willing to share your assumptions used.

TIP: Focus on a providing a target market and provide assumptions used to reach your market size.

What are the primary objectives for the market opportunities section?

- Describe the overall market conditions
  - Size, growth rates, trends, and key companies involved
  - Regulatory landscape for the market
  - Opportunities, unmet consumer needs, and weaknesses of existing products or technologies
- Identify the target markets for the technology
  - For each targeted end user, provide demographic information, trends in consumer activity, buying patterns, and price sensitivities
  - Identify the specific purchasing process for any competing products and for your potential technology

In this section, good research is not only important but crucial to identify the scope of the opportunity. Research can come from a variety of sources, including paid research services, government agencies, industry trade associations and publications, or information available over the Internet. Below are a few sources that provide research reports on specific companies and industries:

- **D&B's International Million Dollar Database** (http://www.dnbmdd.com) provides information on approximately 1,600,000 U.S. and Canadian leading public and private businesses. Fee for access to the database, but many libraries will have free access to the databases available for library patrons.
- **Hoovers** (http://hoovers.com) provides information on public and private companies from sources such as D&B and others. Subscriptions from $75 per month.

- **Hill Search** (http://www.hillsearch.org) provides detailed data on public and private companies in North America and access to current and archived articles from the national and regional newspapers, industry journals, trade magazines, and newswires. $59.95 per month; $650 annually.
- **Integra Information** (http://integrainfo.com) provides various private company research information. Costs range from $9.95 to $200 per report, with subscription fees available for access to multiple reports.

RESEARCHING INDUSTRY TRADE ASSOCIATIONS

Some freely available resources include the following:

- **Internet Public Library, Associations on the Net:** Offers a searchable database of various trade associations who maintain websites. Access at http://ipl.org/div/aon/
- **American Society of Association Executives:** Provides a searchable database known as Gateway to Associations for locating trade associations. The data may be filtered by state, city, country, association type and keyword. Access at http://asaecenter.org/Directories/AssociationSearch.cfm
- **Inc.com:** Provides a listing of industry trade associations broken down into various categories. Access at http://www/inc.com/articles/2001/02/22070.html
    - Consumer: advertising and marketing, consumer products, leisure, retail
    - Professional services: accounting, consulting, law
    - Technology: computer hardware, computer software, Internet and online, telecommunications
    - Health: biotech and pharmaceuticals
    - Finance: financial services, insurance, real estate, and construction
    - Industrial: industrial goods and services

Some fee-based resources include the following:

- **Directory of Associations:** Ability to download information to create mailing labels. Prices range from $165 to $695 for access. Access at http://www.marketingsource.com/associations/
- **Encyclopedia of Associations:** Leading resource with information on over 100,000 associations worldwide. Fee for use, but may be accessed at various public and university libraries. Access at http://www.gale.cengage.com

RESEARCHING INDUSTRY TRADE PUBLICATIONS

- **Yahoo! directories:** http://dir.yahoo.com/Business_and_Economy/Business_to_Business/News_and_Media/Magazines/Trade_Magazines
- **Wikipedia:** http://en.wikipedia.org/wiki/Category:Trade_magazines

USING U.S. GOVERNMENT INFORMATION

- **Bureau of Labor Statistics (http://www.bls.gov):** Identifies trends in hiring by occupation, demographic, geography and several other criteria. *Example*: If your product will be targeted at nurses, using the Bureau of Labor Statistics data, you would be able to show that there were 2,505,000 registered nurses in 2006, and the Bureau of Labor Statistics projects that these numbers will increase to 3,092,000 by 2016, an increase of 23.4%.
- **Census Bureau (http://www.census.gov/econ/www/index.html):** Provides data on the total number of establishments, employment, payrolls, and number of establishments by nine employment-size classes by detailed industry. *Example:* If you plan to market your product or service to physician's and dentist's offices and will market only to Chicago (your headquarters) in the first year, you can use this information to show that, in 2005, there were 6,843 physician's offices and 4,531 dentist's office in the Chicago metro area.

A well-written market opportunity section will include the following:

- Details on important trends in the industry that will affect your business
- Graphs, charts, and illustrations that visually explain the industry and its patterns
- Insights on purchaser demographics
- Reference to outside experts
- Information on each target market
    - Rationale for selecting a particular target market for initial efforts
    - Details on the size and demographic
    - Existing demand
    - Advantages and disadvantages your business has in the target market
    - Sales and distribution methods required to reach this market

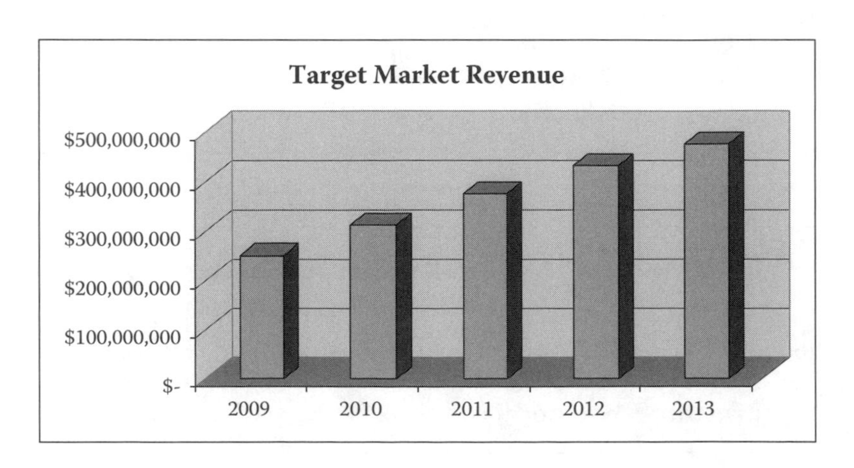

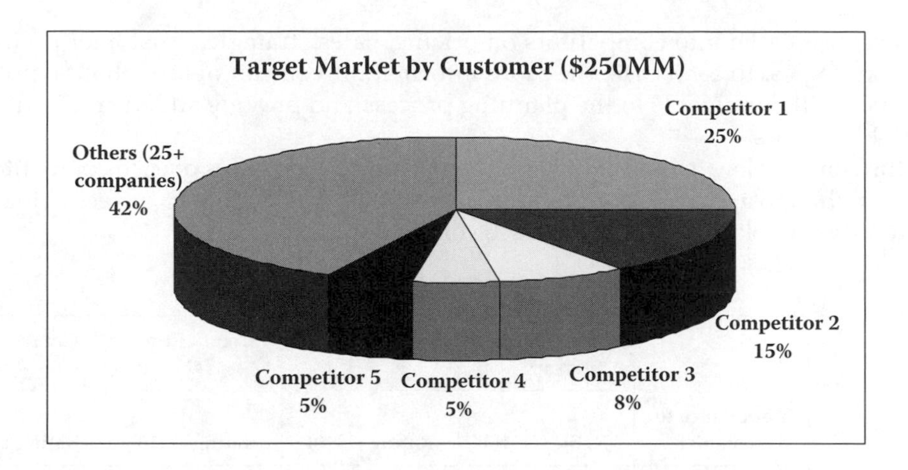

Competition

Failing to adequately identify and address the competitive landscape in the business plan is a key failing for rejected business plans. This section of the business plan should offer an overview of the landscape and explain how your technology will be cheaper, faster, more efficient, stronger, or simply superior to competitive products.

One of the key features in a well-prepared business plan is clearly differentiating your product or service from competitors. Identify factors such as price, payback period, quality, or features and compare your product across the competition. Be honest: if your graphs show that you blow away the competition on every category, a reader may doubt the company's true understanding of the marketplace.

Sometimes entrepreneurs will downplay the strength of a competitor because a technology without competition seems like it would be much more attractive. However, identifying competition and the risks associated with entering a market proves to an investor that the entrepreneur has critically examined the business landscape and is aware of key obstacles.

What are the primary objectives for the competition section?

- Offer insights into the competition
  - Direct competitors
  - Indirect competitors
  - Potential competitors
- Identify the barriers to entry for your technology that may protect you from competitors
  - Intellectual property
  - Partnerships
  - Regulatory obstacles

Much like the market opportunities section of the business plan, research data is a key to properly differentiating your product from the competition. Some of this research can be identified from sources used in the market research, such as competitor data from sources such as Hoovers, Hill Search, or Integra. Other information may require more direct research into competitors on pricing, sales strategies, customer policies, or market strategies. In each case, the more information you can obtain about a potential competitor will be valuable to the planning process and provide added credibility with potential investors.

Use the charts below to highlight key strengths and weaknesses of each competitor and to identify the most important criteria to customers for the product or technology, and then identify the relative strengths of each competitor.

		High-Tech Startup Inc.	**Competition #1**	**Competition #2**	**Competition #3**
	Importance to customer	**Rating (1–10)**	**Rating (1–10)**	**Rating (1–10)**	**Rating (1–10)**
Advertising					
Appearance					
Expertise					
Location					
Price					
Product					
Quality					
Reliability					
Reputation					
Sales strategies					
Selection					
Service					
Stability					

	High-Tech Startup Inc.		**Competition #1**		**Competition #2**		**Competition #3**	
	Strength	**Weakness**	**Strength**	**Weakness**	**Strength**	**Weakness**	**Strength**	**Weakness**
Advertising								
Appearance								
Expertise								
Location								
Price								
Product								
Quality								
Reliability								
Reputation								
Sales strategies								
Selection								
Service								
Stability								

MORE INFORMATION

To download copies of the competitive analysis worksheets, visit the book's website at http://www.myhightechstartup.com.

A well-written competition section will include the following:

- Market share data on each competitor, including graphical representations
- Comparisons of strengths and weaknesses of competitors
- Identification of current sales and marketing approach of each competitor
- Current and potential strategic partners in the target markets

Marketing and Sales

The first portion of the business plan emphasizes and focuses on the opportunity: how great the product is, how large the market is, and how your business will fill an unmet need of the competition. It is crucial to convince a potential investor that the opportunity really does exist. This next section focuses on how your business plans to go into the market, land customers, and make sales. A look at the video cassette wars between Betamax and VHS tells the story of how a product with a better marketing and sales strategy (VHS) beat out the product that many of the experts claimed was a superior technology (Betamax). (Some experts have claimed that the Blu-ray versus HD-DVD war for high-definition DVD players represented a similar battle, with the winner once again coming from a superior marketing strategy.)

The key components of any marketing and sales strategy (oftentimes called the "4 Ps of marketing") are as follows:

- Product
- Price
- Place (distribution)
- Promotions

Your business plan will need to address each of the 4 Ps and detail how your strategy will lead to sales of your technology.

What are the primary objectives for the marketing and sales section?

- Identify the sales process (how do customers purchase the product, by direct sales, retail, channel partners, etc.?)
- Detail the purposes of your marketing strategy
- Explain each of the 4 Ps and how the business will use each to generate revenue for the business

INTEGRATING THE 4 Ps INTO YOUR BUSINESS PLAN

Product: What are the key aspects in identifying the product or service to the end-user or consumer?

- Consumer benefits
- Competitive advantages
- Product positioning strategies

Price: What are the key aspects to the pricing strategies of your company?

- Price elasticity
- Comparison pricing
- Historical industry trends in pricing
- Your future pricing strategy
- Purposes of your pricing strategy: to gain market share, to establish a price-quality reputation, to increase cash flow, to gain major customers

Place (distribution): How will your product or service reach consumers?

- Industry sales channels
- Targeted sales channels
- Strategies for growth of distribution

Promotion: How will your product or service reach consumers?

- Identify target markets
- Offer marketing priorities
- Provide a marketing budget and short- and long-term marketing plan

A well-written marketing and sales section will include the following:

- A clear explanation of the customer sales process
- Specific targeted customers and a detailed approach to reach these customers
- Thorough explanation of the pricing strategies of competitors
- Explanation of the aims of your marketing strategies

Financing and Liquidity Plan

The financial section of your business plan represents one of the most important parts of the plan. However, many venture capitalists will discount this section of the plan because the projections are oftentimes based on a "best-case" scenario yielding the "hockey stick" growth plan. However, the underlying assumptions used to develop the projections in the business plan are highly valued by VCs that may use the assumptions to develop their own best- and worst-case models for the business.

These projections will drive much of the overall plan for the company. More than just impacting the plan document, this section of the report gives the entrepreneur and the startup company the opportunity to demonstrate to the investors that they understand the business, the product, the industry, the costs, and the market. The aim of this section is not to present a picture that you think an investor wants to see. Instead, it presents a detailed view that the founding team believes is attainable with the right mixture of hard work, foresight, and a bit of good fortune.

What are the primary objectives for the financing and liquidity plan section?

- Lay out the key financial projections
    - Projection for revenues
    - Projections of expenses by key category (research and development, sales, administrative)
    - Capital requirements
- Identify the exit strategy for the company

For entrepreneurs without a business background, it is a good idea to work with someone (accountants, advisors, retired executives, consultants, and lawyers) that has experience in building financial projections.

Your business plan should first provide information on the current status of the business. Identify the current ownership structure, the amount of funding obtained to date, the amount founders have contributed to the business, and the current development stage for the business.

Identify the following items in your projections:

- Short-term capital needs used to pay current operating expenses
- Projected costs for research and development
- Acquisition of equipment, buildings, and land
- Expected costs of various employees
- Timeline for go-to-market

SAMPLE OF BUSINESS PLAN LANGUAGE

Financing Section

The Company has received initial financing of $300,000 from the founders and private investors. The company expects to close the first round of financing in March of 2009 and to have raised $3.5 million. The company plans to begin a second round of financing in February of 2011, raising an additional $4.9 million.

The Company's technology is currently operational in a beta site for 150 customers. Expansion of the beta site is anticipated to begin in early 2009, with a broader and more intensive trial period for later in the year. Full-service commercial launch dates are being evaluated. The company expects the second financing round to fund the test phases and bring the company to the point of readiness for commercial service rollout.

SAMPLE OF BUSINESS PLAN LANGUAGE

Financing Section

We expect to raise $3 million in the current round of financing. These funds will be used primarily to fund product development and marketing of our product, eCom Success Software, and additional staffing in operations, testing, support, and administration. Beyond this, we will require additional capital to carry out our marketing, sales, and operations plans. The operating and financial forecasts attached to this document assume that we will close an additional $3 million round of financing by the end of calendar year 2009 and another $6 million (or more) in the following year. Market and competitive responses could accelerate our need for capital.

SAMPLE OF BUSINESS PLAN LANGUAGE

Exit Strategy Section

We expect to be able to achieve liquidity for our shareholders within 36–48 months, via an IPO, or a sale, or merger of the company. Although we are not actively seeking to enter into such a transaction, we are approached periodically by potential merger or acquisition candidates. Given the rapidly developing professional internet services industry and the proliferation of well-financed companies in this market space (e.g., Big Web Inc., ERP Software Corp., Softco Co., etc.), we think accelerating our success in the marketplace will make us a highly attractive target for acquisition by these firms who seek to add strategic components such as our e-commerce solutions to their product and service offerings.

SAMPLE OF BUSINESS PLAN LANGUAGE

Financial Assumptions

Revenue Assumptions

High-Tech Startup Inc. has segmented the anticipated market of e-commerce users into major e-tailers (defined as top 100 online retailers as reported by the Online E-tailing Review) and smaller e-tailers (all other e-commerce users). The relevant assumptions and estimates needed to estimate the revenue drivers are defined as follows *(truncated for example purposes only)*:

- **Number of major and smaller e-tailers:** The total number of e-commerce users adopting the HTWeb product and using it to create e-commerce transactions on a daily basis, segmented between major and smaller users. Based on information from the Online E-tailing Review, the top 100 e-tailers account for 62% of the transactions . . .

- **Average number of transactions per day per site:** The number of transactions per day per site, segmented between major and smaller users. The initial baseline estimate for the major sites (based on Online E-tailing's survey of 34 of the top 100 e-tailers) is 8,100 transactions . . .
- **Average dollar size per transaction:** The average dollar amount of each transaction, segmented between major and smaller users and segmented by industry categories is . . .

Expense Assumptions

The operating expenses are based on estimated quarterly salary amounts and headcount for the following functional areas:

- Executive
- Marketing
- Sales and Support
- Engineering
- Professional Services
- Administrative

Headcount increases represent the expected requirements to meet the anticipated growth of the company. Certain other cost areas are based on percentages of payroll costs (benefits, training, recruiting, etc.) or per-head costs (facilities and services, and equipment). All other costs are based on the estimated growth and needs of the company, such as nonsalary sales and marketing costs and other professional services.

SAMPLE OF BALANCE SHEET PROJECTIONS
High-Tech Startup Inc.

	Beginning Balance Sheet	End of Fiscal Year 1	End of Fiscal Year 2	End of Fiscal Year 3
ASSETS:				
Cash	$ 1,500,000	$ 1,601,727	$ 8,764,526	$ 26,911,533
Trade Accounts Receivable	$	$ 457,756	$ 1,949,045	$ 4,044,763
Other Current Assets	$	$ 25,000	$ 50,000	$ 75,000
Total Current Assets	$ 1,500,000	$ 2,084,483	$ 10,763,571	$ 31,031,296
Property, Plant, and Equipment	$ 2,000	$ 96,163	$ 284,813	598,263
Net Intangible Assets,	$ 15,000	$ 12,000	$ 9,000	$ 6,000
Net Total Assets	$ 1,517,000	$ 2,192,646	$ 11,057,383	$ 31,635,558
LIABILITIES AND SHAREHOLDERS' EQUITY:				
LIABILITIES:				
Accounts Payable	$	$ (165,896)	$ (319,623)	$ (563,647)
Accrued Expenses	$ (5,000)	$ (164,075)	$ (283,307)	$ (436,392)
Deferred Revenue	$	$ (70,000)	$ (175,000)	$ (245,000)
Total Current Liabilities	$ (5,000)	$ (399,971)	$ (777,930)	$ (1,245,039)
SHAREHOLDERS' EQUITY:				
Preferred Stock/Common Stock and Additional Bid-In Capital	$ (1,512,000)	$ (1,512,000)	$ (1,512,000)	$ (1,512,000)
Accumulated Deficit/(Earnings)	$	$ (280,675)	$ (8,767,453)	$ (28,878,519)
Total Shareholders' Equity	$ (1,512,000)	$ (1,792,675)	$ (10,279,453)	$ (30,390,519)
Total Liabilities and Equity	$ (1,517,000)	$ (2,192,646)	$ (11,057,383)	$ (31,635,558)

Balance Sheet Assumptions:

Accounts Receivable	Assume Days Sales Outstanding of 45 days
Property, Plants,	Assume an average useful life of four years and use the half year convention and Equipment
Intangible Assets	Estimated useful life of five years and assuming no capitalization of software development costs
Accounts Payable	Assume 45 days outstanding for accounts payable
Accrued Expenses	Assume salary and benefits paid in 24 periods, one period accrued and due at year end Additional accruals recorded in amounts of $75,000 in years 1 and 2 and $100,000 for the third year
Deferred Revenue	Deferred Revenue recognized over 12 month license contracts, assuming a half year convention
Equity	$1.5 million received today

SAMPLE INCOME STATEMENT
High-Tech Startup Inc.

	Fiscal Year 1	% of Total Revenue	Fiscal Year 2	% of Total Revenue	Fiscal Year 3	% of Total Revenue
Revenues:						
License Fees	$ 70,000	1.92%	$ 245,000	1.56%	$ 420,000	1.28%
Direct Customer Sales	$ 3,572,910	98.08%	$ 15,468,921	98.44%	$ 32,317,519	98.72%
Total Revenue	$ 3,642,910		$ 15,703,921		$ 32,737,519	
% Change	N/A		331.08%		108.47%	
Costs of Revenue:						
Cost of Goods Sold and Maintenance	$ 197,773	5.43%	$ 605,226	3.85%	$ 1,217,720	3.72%
Gross Profit	$ 3,445,137	94.57%	$ 15,098,695	96.15%	$ 31,519,798	96.28%
Operating Expenses						
Sales and Marketing	$ 1,548,004	42.49%	$ 4,089,019	26.04%	$ 7,553,735	23.07%
Research and Development	$ 709,107	19.47%	$ 1,021,141	6.50%	$ 1,730,471	5.29%
General and Administrative	$ 907,351	24.91%	$ 1,501,756	9.56%	$ 2,124,526	6.49%
Total Operating Expense	$ 3,164,462	86.87%	$ 6,611,916	42.10%	$ 11,408,732	34.85%
Income from Operations	$ 280,675	7.70%	$ 8,486,778	54.04%	$ 20,111,066	61.43%
Number of Employees	27		63		118	
Revenue per Employee	$ 134,923		$ 249,269		$ 277,437	

High-Tech Startup Inc.

January 1, 2009

CASH BURN BUDGET

	Opening Balance	Month 1	Month 2	Month 3	Month 4	Month 5	Month 6	Month 7	Month 8	Month 9	Month 10	Month 11	Month 12
OPENING CASH AMOUNT													

OUTGOING CASH													
Accounting and legal													
Car, delivery, and travel													
Insurance													
Interest													
Internet, Telephone													
Marketing, sales, advertising													
Outside services													
Payroll expenses (taxes, etc.)													
Purchases													
Rent													
Repairs and maintenance													
Supplies (office and operations)													
Taxes													
Utilities													
Wages													
Other expenses (specify)													
Other (specify)													
SUBTOTAL	$ -	$ -	$ -	$ -	$ -	$ -	$ -	$ -	$ -	$ -	$ -	$ -	$ -
Capital purchase (specify)													

Loan principal payment													
Other startup costs													
Owners' withdrawal													
Reserve and/or escrow													
TOTAL CASH PAID OUT	$ -	$ -	$ -	$ -	$ -	$ -	$ -	$ -	$ -	$ -	$ -	$ -	$ -
CASH RECEIPTS													
Cash sales													
Collections fm cash receipt accounts													
Loan/other cash inj.													
TOTAL CASH RECEIPTS	$ -	$ -	$ -	$ -	$ -	$ -	$ -	$ -	$ -	$ -	$ -	$ -	$ -
CASH AVAILABLE	$ -	$ -	$ -	$ -	$ -	$ -	$ -	$ -	$ -	$ -	$ -	$ -	$ -
CLOSING CASH AMOUNT	$ -	$ -	$ -	$ -	$ -	$ -	$ -	$ -	$ -	$ -	$ -	$ -	$ -

MORE INFORMATION

To download copies of the balance sheet projections, income statement or cash burn budget, visit the book's website at http://www.myhightechstartup.com.

Management

An investor is looking for a team that can implement the business plan and has a proven track record of success. The business plan should clearly identify and focus on successful business building and startup efforts. VCs know that historical evidence shows that a founder team that raised previous venture funding or executed an IPO is more likely to produce another successful result. Emphasize these successes and identify current recruiting efforts and the types of candidates being targeted.

What are the primary objectives for the management section?

- Profile key management team members and organizational responsibilities
- Highlight technical advisory board and strategic advisory board members
- Identify key support resources
  - Legal counsel
  - Patent/intellectual property counsel
  - Key consultants
  - Accountant
- List key management team members to be added and expected timetables

Venture capitalists also know that one of the most important factors in a successful venture-backed company is the scope of the networks of the management team and their advisors. The team itself will generally not be enough alone to obtain funding for the company. Showcasing a well-connected team does provide evidence that the business can use these networks to recruit additional talent for the organization, get connected to key business partners, and understand trends and challenges within the industry.

For many early-stage companies, your team will not be complete, and investors will usually not expect you to have a complete team in place when you are seeking funding. However, you should be sure to make it clear that you have identified the areas of need within the business for additional talent and have a plan in place to recruit individuals to fill these gaps.

A well-written marketing and sales section will include the following:

- Relevant experience of the management team
- Description of the management philosophy of the team
- Explanation of how current team members will complement one another
- Identifying key needs of the management team and timetable for hiring

SAMPLE OF BUSINESS PLAN LANGUAGE

Management Section

High-Tech Startup's management team combines exceptional engineering and product development experience with proven management and financial expertise. All members of the core team have been involved in startup ventures. The team is dynamic, cohesive, and well-balanced with strong technical, financial, and operations talents. The full team has been collaborating together for several months. Each member has broad capabilities and can successfully meet the diverse challenges of a technical startup.

Tina Talented, President and CEO

Tina has 16 years of experience in operational and technical management in the communications arena and is currently a consultant to venture-backed startup TechTop Inc., where she has participated in the company's development from concept through funding and initial product development. After founding eSuccess Inc. and spearheading its e-commerce product strategy, planning, and business development efforts, Tina sold the company to MajorWeb Inc. in 2006. Previously, Tina was the Division Manager and Executive Vice President of Product Development at a small privately funded startup in Singapore. In that position, Tina was responsible for the financial, operational, marketing, sales, and technical aspects of a design and development facility in Asia. Tina has also held various management and technical positions at HT&S Worldwide and Service Software. She holds undergraduate and advanced degrees from the College of Tech & Trade in Singapore and the University of Big State.

Appendix

Few investors want to see a book-sized business plan dropped on their desk or want to open a PDF file that is several hundred pages long. With that in mind, one area that entrepreneurs sometimes go overboard on is the appendix. Remember, the purpose of an appendix to support the information found within your business plan and reinforce what has been said throughout the plan. Don't just add additional information to add it; include materials, documents, spreadsheets, or biographies that add to the presentation and are important and compelling.

Items you may consider include the following:

- Financial projections and detailed financial statements
  - Income statement (profit and loss)
  - Cash flow/cash burn
  - Balance sheet
  - Financial ratios
- Resumes or expanded biographies
  - Key management team
  - Technical or scientific advisory board members

- Market research
- Marketing material
- References from customers or product users
- Description of patents, trademarks, copyrights, and licenses
- Detailed profiles of competitors
  - Company history
  - Detailed descriptions of products and services
  - Reviews of products or services
  - Pricing and sales strategies
- Media or press reporting

Other Details for a Business Plan

Here is a list of other details to consider for your business plan:

- On the cover page, include the following information:
  - Company name and logo
  - Contact information
  - Date of preparation
  - Confidentiality clause
- Triple check for any misspellings
- Limit the use of technical vocabulary or provide appropriate definitions
- Use graphs, pictures, images, and charts to make the business plan more appealing to the eye
- Provide detail on all the assumptions used in your business plan
- Use bold, italics, and underline to highlight text
- Extract key provisions and highlight in sidebars
- Be sure that the business plan is pleasing to the eye

"STRETCHING" THE TRUTH IN YOUR BUSINESS PLAN

What could happen

Your business plan initially contained projections and other information that was possible. However, you've now learned that these results cannot be obtained. Do you need to inform potential investors or is this just "puffery" and part of the sales process?

Watch out for

Making a statement that you know is untrue or a promise that you know you can't perform could be deemed to be fraud if an investor invested based on certain facts or information you know to be false. You have an obligation to correct inaccuracies if they arise. It is also good business practice to be truthful and forthcoming about changes in your assumptions or projections throughout the fundraising process.

For your business plan, you should be sure to include your key assumptions used for your projections to allow a reader to make her own judgment about the projections themselves. You should keep a list of parties who have received your business plan and update them with revised copies in the event things change in a material sense (certain intellectual property can no longer be protected, a founder leaves the team, or market conditions change dramatically). It is possible that a founder could be sued by the investors if fraud is discovered (while it somewhat unlikely to be the case, it is best to avoid the issue).

Many, if not all, new business ventures suffer a misstep in their first year, and investors know this. However, by setting impossible expectations, you may find that you wind up harming the relationship with these business partners. Remember, you are not only getting an investor for their money, but they likely represent a member of your board and oftentimes a source of funds for subsequent fundraising.

TIP: Update/revise your business plan in the event of changes in material information in your plan.

KEEPING YOUR BUSINESS PLAN CONFIDENTIAL AND PROPRIETARY

Because getting potential investors to sign an nondisclosure agreement may not always be an option, you should still take steps to provide some protection on your business plan itself. Use of the language "confidential and proprietary" may not provide the same level of protection as a nondisclosure agreement, but some courts have held that, in a scenario in which a person or persons were aware that they had been exposed to a trade secret, they would be barred from using it or disclosing it to others without permission.

Take steps when possible to protect your proprietary information and remember that, when you cannot get a nondisclosure agreement signed, to trust your gut: make sure to ask around before disclosing to a potential investor if you are wary of their reputation.

TIP: On the cover page of your business plan, state (in bold text) that the business plan is confidential and proprietary.

SAMPLE OF BUSINESS PLAN LANGUAGE

High-Tech Startup Business Plan

CONFIDENTIAL INFORMATION

This business plan is the property of High-Tech Startup Inc. and is strictly confidential. It contains information intended only for the person to whom it is transmitted. With receipt of this plan, recipient acknowledges and agrees that:

1. In the event recipient does not wish to pursue this matter, this document will be destroyed or returned to the address listed above as soon as possible;
2. The recipient will not copy, fax, reproduce, divulge, or distribute this confidential plan, in whole or in part, without the express written consent of SoftLink Inc.; and
3. All of the information therein will be treated as confidential material with no less care than that afforded to your own company confidential material.

This document does not constitute an offer to sell, or a solicitation of an offer to purchase.

Version: ______________

Copy number: ______________

Provided to: ______________

Signature: ______________

Company: ______________

Date: ______________

Principal Contact:

Name: ________________________

Title: ________________________

Address: ______________________

Phone, fax, e-mail: _____________

Business Plan Software Reviews

Software publisher	Website	Rating[1]	Relevant software titles	Price[2]	Notes
Palo Alto	http://paloalto.com	4.5	Business Plan Pro 11.0	$ 99.95	Provides largest suite of resources; best selling software.
			Business Plan Pro Premier Edition 11.0	$ 199.95	
			Marketing Plan Pro 11.0	$ 179.95	
Business Resource Software	http://brs-inc.com	4+	Plan Write for Business	$ 119.95	Offers unique business model evaluation function to identify the strengths and weaknesses of the plan; offers remote collaboration tools.
			Plan Write Expert Edition	$ 219.95	
			Plan Write for Hi-Tech Marketing	$ 299.95	
Fundable Plans	http://fundableplans.com	4	Fundable Plans	$ 39.95	Lots of links to appropriate industry-specific information
Jian	http://jian.com	4	BizPlan Builder 2008	$ 99.77	Contains numerous templates
			BizPlan Financials 10	$ 69.95	
PlanWare	http://planware.org	4	PlanWrite	$ 119.95	Best software to develop a business plan in an international format.
			PlanWrite Expert Edition	$ 219.95	
			Plan Write Hi-Tech Marketing Planner	$ 289.95	
PlanMagic	http://planmagic.com	4	Business 10.0	$ 99.95	Specific plans for hotels, bars, coffee bars, restaurants, resorts, construction, retail, bed and breakfasts
SmartOnline	http://smartonline.com	4	SmartOnline	$ 49.95/month	
Individual Software	http://individualsoftware.com	3.5+	Professional Business PlanMaker Professional 2008	$ 49.95	
KMT Software	http://send.onenetworkdirect.net	3.5+	OfficeReady Business Plans 2007	$ 79.95	
Nova Development	http://novadevelopment.com	3.5+	Business Plan Writer Deluxe 2006	$ 99.95	
Socrates	http://jdoqocy.com	3.5+	Winning Business Plans	$ 29.95	
Atlas Business Solutions Inc.	http://abs-usa.com	3.5	Ultimate Business Planner 4.0	$ 99.00	
Business-Plan-Success.Com	http://business-plan-success.com	3.5	Business Plan Success 5.0	$ 39.99	
My Business Kit	http://mybusinesskit.com	3.5	Complete Business Kit	$ 49.95	
NetEkspert	http://store.esellerate.net	3.5	iPlanner 2007	$ 39.95/year	Only does financials
VPS Pro	http://vpspro.com	3+	VPS Pro	$ 85.95/year	
Adarus	http://adarus.com	3	Adarus Business Plan	$ 55.95	
Village Software	http://villagesoft.com	3	Business Plan FastPlan	$ 99.95	

[1] Rating is out of 5, with 5 being highest and 1 being lowest. Ratings provided by Home Office Reports (http://homeofficereports.com/Business%20plan.htm).
[2] Prices as of January 2008.

INTERNAL BUSINESS PLANNING

Many early-stage companies will prepare an internal business plan (also referred to as the operational business plan or the strategic business plan).

The internal business plan is designed to serve as a detailed, strategic roadmap for the business. It is different from the business concept summary or a business plan prepared for third-party investors in that this document will represent a more substantial tool based on extensive research and a more detailed analysis of the company's plans and objectives. In most cases, an internal business plan will serve as the operational template for the organization to identify future hiring trends, sales targets, and key milestones.

Typical details included in the internal business plan:

- Capital purchases
- Headcount increases
- Department-level expenses
- Development schedules for new products and services
- Sales targets
- Customer and potential customer tracking
- Three- to five-year projections

An internal business plan should be structured to provide regular tracking against actual performance. Therefore, the accounting and financial systems should be set up to allow for performance measurement of actual data against the internal business plan.

One of the mistakes that many entrepreneurs make is that they focus solely on the preparation of their business plan for potential investors; the internal plan is usually more extensive, including specific headcount goals, specific customer targets, and detailed product development timelines. Many entrepreneurs find the process much more efficient to consider these distinct efforts. Although the internal business plan can serve as the foundation for information used in a third-party business plan, certain aspects of an internal business plan should not be shared with a third party.

Some startups will prepare an internal plan before seeking financing, whereas others will wait until after financing has been achieved to develop a plan with the aid of investors. The process may well take several months up to a full year.

Steps to Preparing an Internal Business Plan

To prepare the internal business plan, you'll need to the following:

- **Prestage:** When to begin? How long will it take? Who to include? What tools to use?
- **Stage 1:** Develop your steps to create an internal business plan

- **Stage 2:** Information gathering
  - Identify the current financial state of the business
  - Identify the financial and human capital needs for various business departments
  - Develop sales and marketing milestones
  - Create a list of potential customers
  - Collect competitor data
- **Stage 3:** Produce the internal business plan
- **Stage 4:** Initiate a critique and reworking of your business plan
- **Stage 5:** Management release of the business plan
- **Stage 6:** Ongoing tracking against the business plan

How long will it take?

The first rule of business planning is to give yourself more time that you think. Generally, the preparation will be done over a series of months; oftentimes, the planning is done while one or more of the founders continues to be employed elsewhere. As with the other business planning tools discussed previously, to prepare the documents themselves may only take a few days or weeks, but the research and discussions needed to gather and parse the underlying data driving the business plan will take much longer to collect and consolidate.

Much like the preparation of a business plan for potential investors, a typical range is between 200 and 500 hours for business planning. Depending on your experience and the complexity of your business, you may find yourself spending between 200 and 1,000 hours in the preparation of a full internal business plan.

MORE INFORMATION

To download copies or find links to any of the documents in this chapter, visit the book's website at http://www.myhightechstartup.com.

SAMPLE BUSINESS PLAN ORGANIZATION

Tim Berry of BPlans.com offers the following as a sample structure for a business plan.

1. **Executive Summary**
   - Objectives
   - Mission
   - Keys to success

2. **Company Summary**
    - Company ownership
    - Company history (for ongoing companies) or startup plan (for new companies)
    - Company locations and facilities
3. **Products and Services**
    - Product and service description
    - Competitive comparison
    - Sales literature
    - Sourcing and fulfillment
    - Technology
    - Future products and services
4. **Market Analysis Summary**
    - Market segmentation
    - Target market segment strategy
        - Market needs
        - Market trends
        - Market growth
    - Industry analysis
        - Industry participants
        - Distribution patterns
        - Competition and buying patterns
        - Main competitors
5. **Strategy and Implementation Summary**
    - Strategy pyramids
    - Value proposition
    - Competitive edge
    - Marketing strategy
        - Positioning statements
        - Pricing strategy
        - Promotion strategy
        - Distribution patterns
        - Marketing programs
    - Sales strategy
        - Sales forecast
        - Sales programs
    - Strategic alliances
    - Milestones
6. **Web Plan Summary** (if you are a Web-based business)
    - Website marketing strategy
    - Development requirements
7. **Management Summary**
    - Organizational structure
    - Management team
    - Management team gaps
    - Personnel plan

8. **Financial Plan**
   - Important assumptions
   - Key financial indicators
   - Break-even analysis
   - Projected profit and loss
   - Projected cash flow
   - Projected balance sheet
   - Business ratios
   - Long-term plan

Source: http://articles.bplans.com/index.php/business-articles/writing-a-business-plan/A-Standard-Business-Plan-Outline.

SAMPLE BUSINESS PLAN ORGANIZATION

Linda Pinson offers the following as a sample structure for a business plan.

1. **Cover Sheet**
2. **Executive Summary or Statement of Purpose**
3. **Table of Contents**
4. **Organizational Plan**
   - Summary description of the business
   - Products or services
   - Intellectual property
   - Location
   - Legal structure
   - Management
   - Personnel
   - Accounting and legal
   - Insurance
   - Security
5. **Marketing Plan**
   - Overview and goals of your marketing strategy
   - Market analysis (target market, competition, market trends, market research)
   - Marketing strategy
   - Customer service
   - Implementation of marketing strategy
   - Assessment of marketing effectiveness
6. **Financial Documents**
   - Summary of financial needs
   - Use of funds
   - Pro forma cash flow statement (budgeted)
   - Three-year income projection
   - Projected balance sheet

- Break-even analysis
- Profit and loss statement (income statement) (if applicable to existing business)
- Balance sheet (if applicable to existing business)
- Financial statement analysis (if applicable to existing business)
- Business financial history (if applicable to existing business)

7. Supporting Documents
- Personal resumes
- Owners' financial statements
- Credit reports
- Copies of leases, mortgages, purchase agreements, etc.
- Letters of reference
- Contracts
- Other legal documents
- Miscellaneous documents

Source: Anatomy of a Business Plan and *Automate Your Business Plan* by Linda Pinson.

SAMPLE BUSINESS PLAN ORGANIZATION

The small business resource, SCORE, "Counselors to America's Small Business," offers the following as a sample structure for a business plan.

1. Executive Summary

2. General Company Description
- Short description of the business
- Mission statement
- Company goals and objectives
- Business philosophy
- Target market
- Industry summary
- Core strengths and competencies
- Legal form of ownership

3. Products and Services
- Description of products and services
- Competitive advantages and disadvantages
- Pricing structures

4. Marketing Plan
- Economics
- Products
- Features and benefits
- Customers
- Competition

- Niche
- Marketing strategy
- Pricing
- Location
- Distribution channels
- Sales forecast

5. **Operational Plan**
   - Production
   - Location
   - Legal environment
   - Personnel
   - Inventory
   - Suppliers
   - Credit policies
   - Professional and advisory support (board of directors and management advisory board; attorney; accountant; insurance agent; banker; consultants; key advisors)
6. **Personal Financial Statement**
7. **Financial History and Analysis**
8. **Financial Plan**
   - 12-month profit and loss projection
   - Four-year profit projection
   - Projected cash flow
   - Projected balance sheet
   - Break-even analysis
9. **Appendices**
   - Brochures and advertising materials
   - Industry studies
   - Blueprints and plans
   - Maps and photos of location
   - Magazine or other articles
   - Detailed lists of equipment owned or to be purchased
   - Copies of leases and contracts
   - Letters of support from future customers
   - Any other materials needed to support the assumptions in this plan
   - Market research studies

Source: http://score.org/template_gallery.html.

SAMPLE BUSINESS PLAN ORGANIZATION

Tim Berry of BPlans.com offers the following as a sample structure for a business plan.

1. **Cover Page**
2. **Executive Summary**
3. **Operational Plan**
   - Company summary
   - Product or service, describing the details of the sale and customer benefits
   - Intellectual property
   - Legal structure
   - Management
   - Personnel
   - Accounting/legal
   - Insurance
   - Security
4. **Marketing Plan Analysis**
   - Overview
   - Market analysis
   - Marketing strategy
   - Customer services
   - Implementation of marketing strategy
   - Evaluation of marketing strategy
5. **Financial Analysis**
   - Summary of financial needs
   - Use of funds
   - Budgeted cash flow statement
   - Income forecast
   - Break-even analysis
   - Profit and loss statement
   - Balance sheet
   - Financial statement analysis
   - Business financial history
6. **Supporting Documents**
   - Management resumes
   - Credit reports
   - Copies of key agreements (leases, mortgages, contracts, etc.)
   - Legal documents
7. **Strategy and Implementation**
8. **Web Plan Summary**

Source: http://articles.bplans.com/index.php/business-articles/writing-a-business-plan/A-Standard-Business-Plan-Outline.

Part III

Building a Startup Success

The day you make it, call yourself a company and not a startup.

Anjali Gupta
Founder of Emphatech

9

Raising Money

When I finished school, I took my entire life savings—$5,000—and invested it in a business. I was young. I was inexperienced. But I was an entrepreneur, and I was proud. And in six weeks, I was broke.

Mark Warner
Cofounder of Capital Cellular Corporation,
Former Governor of Virginia

What to Watch For

Starting any new business will require money. According to research by Edward Roberts, approximately 74% of high-tech startups surveyed obtained their initial capital from the founders of the business. For high-technology businesses, there is a greater likelihood that the business will require money from a variety of sources as a result of increased capital needs. These sources may include initial monies invested by the founder or founders, money from family or friends, private investors (oftentimes called angel investors), bank loans, venture capital (VC) funds, government grants, bootstrapping (money earned from the business itself), or countless other sources.

The fact that successful companies raise money from so many different sources is a positive and a negative. The positive is that you can attempt to match funding sources to the company and the current needs of the business. The negative is that it may be difficult to sort through the numerous options to determine the right or best course of action for fundraising.

Fundraising serves as a key gating item for companies. Without necessary funding, a company may be unable to make strategic hires, conduct research and development initiatives, enter into license agreements, and exploit business opportunities. As a result, the fundraising process is integral to the success or failure of many businesses.

Raising money for your business is a process that may be new to an entrepreneur and may involve skills, knowledge, or experiences that a new business owner may not have. As a result, it is often a challenge.

WHY DO FUNDRAISING EFFORTS FAIL?

Perhaps the best way to approach the fundraising process is to consider what causes most fundraising efforts to be unsuccessful. The list below offers the top reasons why fundraising efforts for startup businesses most often fail:

- Little or no experience in fundraising
- Failure to find funding sources
- Lack of knowledge or understanding about a particular funding source
- Not matching your business to the most appropriate funding source
- Failure to focus on fundraising
- Focusing too much energy and effort on operating the business
- Starting fundraising too late or allowing too little time
- Lack of well-defined goals
- Failing to consult with outside assistance

What an entrepreneur should notice about the list above is that many of the items are within their control, from defining goals and consulting with experts, to understanding funding sources and focusing on the fundraising process. The key to a successful fundraising strategy is putting in place a set of fundraising goals and objectives and dedicating the time and energy to achieve these goals. With the right approach, reaching out toward the appropriate funding sources can provide necessary capital for the business to succeed.

Why Do You Need to Raise Funds for Your Business?

There may seem to be an obvious answer to this question, but many businesses fail to adequately understand the real needs of the business and the timetable required to raise the funds necessary for the operations of the business. Raising funds represents a marketing and sales effort for the business as a whole, and it represents an important effort for nearly every new business.

As discussed previously, Jessie Hagen of U.S. Bank found that, among small businesses that had gone out of business, 82% listed one of the causes as poor cash flow management skills or a poor understanding of cash flow, whereas 79% listed starting out with too little money. Many other researchers who have studied the success factors of high-technology businesses have found that an integrated business and fundraising strategy is crucial to the growth of a new business venture.

Determining what amounts or types of funding you need at various points is much more of an art than a science; many companies begin by casting a wide net with respect to various options and then begin to focus their efforts as they find which sources will best match the business.

What Are Investors Looking For?

Depending on the stage of your business, investors will be looking for different things. For instance, an investment by a family member into an early-stage business may be based largely on the reputation of the founder with little focus on the technology. A bank may look into credit histories of the founders and the cash flow of the business. An investment by an established venture fund may place a greater emphasis on your current market penetration and the experience of the management team in further exploiting the opportunities.

WHAT DO PROFESSIONAL INVESTORS LOOK FOR?

In general, the key aspects that most investors (from angels, to banks, to grant funds, to venture capitalists) evaluate in a potential investment are as follows:

- Team
- Technology
- Market potential

Investors will evaluate each of those three key areas to make a determination as to the likelihood that, given the right conditions, the business can be a significant success. More specifically, investors will look at the following:

- Experience of management team
- Relative skills of management team (including complementary nature of skills)
- Strategy of management team
- Stage of product development
- Protections of the technology
- Product market potential
- Current and likely customers
- Future financial needs of the business

The reality is that each investor will analyze the business and its future prospects through its own lens. What one investor sees as a futile idea without a strong management team, another investor may see as a surefire success. Therefore, businesses should cast a broad net to find investors that understand the technology, appreciate the relative experience of the management team, and believe in the market potential.

THE "REALITY CHECK"

One of the pitfalls for some technology entrepreneurs is falling in love with the eBay, Google, Yahoo!, etc. startup funding path. Let's just call it, the venture capital model. That is, take an interesting technology, get a couple angels to invest early, then find a venture capital firm to kick in several millions along the way until you can get acquired or go public. Although this model worked well for some notable successes (Apple, Google, and Microsoft), not every successful idea has grown from venture investments.

Nationwide, institutional venture capital firms typically make only between 2,500 and 3,500 investments annually (and only about one-third of those are initial investments in startups, whereas many of those are follow-on investments). With more than 1 million small businesses started each year, the chances of getting one of those coveted VC investments are extremely small.

Even still, technology entrepreneurs are oftentimes surprised to learn that their technology just isn't the type that traditionally gets funded by an institutional investor. Perhaps the company's market isn't large enough or the investment required is too great for VCs to be interested. Yet many technology entrepreneurs will spend 100% of their time looking to raise a Series A round from venture capital firms. For many companies, the venture capital model may be either unlikely or impossible, or perhaps the venture capital model won't work for the company until after several years of growth or development.

The reality is, many companies just don't fit the venture capital or institutional investor model. The venture capital model attempts to identify companies that will yield big returns in a short time period, usually in the form of revenues of at least $25 million in three years and $50 million in five years. Generally, the model involves a company with a technology familiar to the VC fund, a business that the VC can easily understand and explain to others, and has a revenue model that can be grown rapidly. VC firms usually want to see proprietary technology, substantial barriers to entry for potential competitors, and a management team with a proven track record.

As a company begins the process of considering the most likely funding alternatives, it is important to do a "reality check" to see whether the type of company being developed is a likely funding candidate for traditional institutional venture capital. If not, the company should identify alternative sources of funding for the company's growth plans.

> In the next chapter on venture capital, we've included a short questionnaire to help companies identify whether their idea and business concept is a likely VC funding candidate. Use it as a reality check to determine whether your company represents a viable candidate at your current stage and under your current model.

Determining What You Need

For most high-technology businesses, the company will usually require some amount of funding to move the business from the idea stage to a functioning business entity, albeit an entity that is still an early-stage venture. The early-stage funding is likely to come from personal savings, family and friends, a grant program, or perhaps angel funding. For most first time entrepreneurs, they will find out how quickly startup capital will evaporate and how fast expenses will be incurred.

In *Founders at Work* by Jessica Livingston, Arthur van Hoff, the cofounder of Marimba, which had grown into a 300 person software distribution company by the time of its IPO in 1999, highlighted the importance of living cheaply during the early days of a startup. Said van Hoff, "Initially we all put in a little bit of money, I think $25,000 each. If you don't take a salary, that can last you a long time. . . . We spent about $1,400 to furnish the entire office, including equipment like a fax machine and printer. We all used cell phones at first, and we had no Internet access for the first couple of weeks, just the whiteboard."

Once a business moves beyond the need for startup or seed fundraising into the need to secure more sizeable funding (typically from venture capital firms or other more significant investors), more questions will arise regarding the amount of funding to secure. There are different schools of thought regarding raising funds for a high-technology business. Some experienced entrepreneurs believe that the biggest risk to any company is undercapitalization and missing the narrow window of success a startup will have. Therefore, these entrepreneurs will encourage others to be certain to have adequate funds available for the business.

Other experienced entrepreneurs will urge caution in accepting funds from outside investors for fear of losing control of the business. For some founders that have taken significant payouts from investors, they have soon found that these funds were not without strings, strings that were much shorter than the founder initially realized. As such, these entrepreneurs will urge future entrepreneurs to be cautious with their fundraising and to explore nondilutive funding mechanisms such as bank loans and government grants and loans.

More considerations are discussed in the following chapter with regards to the process for determining the amount of funding to request from venture capital firms.

Matching Your Business with an Investor

Raising funds is a time-consuming process that will require you to "put many lines into the water to hook a few fish." Therefore, it is important for an entrepreneur to recognize the funding sources that are most available to the company depending on the growth stage of the company.

At different stages of a company's growth, certain funding sources will be more applicable than others. Early-stage companies most often rely on funds from personal savings, private investors, government grants, angels, and bootstrapping, whereas mid-stage and advanced-stage companies are more likely to pursue funds from venture capital sources or through investment bankers.

Business stage	Typical investors
Experienced founder and business plan	Founders; family and friends; government grants; angels
Experienced founder and prototype/beta product	Government grants; angels; early-stage VC
Experienced team and developed product and customers	Later-stage VC; strategic investor
Experienced team and sustained revenue growth and profitable in 12–24 months	IPO or strategic merger/acquisition

Startup Capital

Starting a business requires capital. The vast majority of new businesses require some initial investment from the founders to begin the business. Typically, this initial cash will be used to move the business from an idea or concept into a functioning business.

How Much Should a Founder Plan on for Startup Capital?

According to the 2004 financing report from the *Global Entrepreneurship Monitor,* the average amount needed to start a business was $53,673. For businesses that were necessity pushed (driven by a currently unmet need in the industry), the average startup only required $24,467. Depending on the industry and the specific growth plans, each business may require more or less to startup their business.

Where Do Founders Get This Startup Capital?

According to the *Global Entrepreneurship Monitor* report, more than 65% of the startup phase funding comes from personal savings, credit, and informal investors such as family and friends.

The perception may be that startups open their doors after receiving a big investment from an angel or a VC. However, this is far from the truth. In fact, it is very uncommon for a startup business to open its doors with venture capital funding in hand (fewer than 1 of 10,000 startups have VC money in the bank when they open for operations).

Seed and Angel Funding Rounds

Many emerging companies raise money in the first instance not from institutional investors but from "angels" (friends, family, and high-net-worth individuals) in a seed or angel round. In doing so, they need to decide whether to structure the seed round as a "bridge" (convertible debt) round or a "priced" (preferred or common stock) round.

Bridge Round versus Priced Round

If a company finds an early-stage investor who wants to invest capital into the business, how should the company decide what percentage of the company that investment is worth? It is a tough question to value a company without much to go on.

As a result of this uncertainty, a company will need to decide whether to issue convertible debt that will convert at a future point (usually when a larger financing occurs) or issue equity interest based on a current estimate of the value of the company. For a company, certain seed or angel funding is a first-step funding mechanism to be followed by a larger round that will involve a valuation of the company at that time, and a bridge round may be the right course of action. For a company that is uncertain whether or when it may raise a larger amount of money, potential investors may be wary to take convertible notes.

BENEFITS OF A "BRIDGE"

Some experts encourage early-stage startups to raise money involving seed financings with a convertible note with a discount that increases over time up to a certain cap. Others encourage the company to go with a priced round. So what are the pros and cons of doing a bridge round?

PROS

- **Doesn't require a valuation:** In early stage investments, the valuation is difficult to settle on or much lower than the company would like.
- **Preparing financing documents for a bridge financing are much simpler:** Simpler documents results in a quicker turnaround time (faster money in the door) and lower legal fees.

CONS

- **Misalignment of interests:** Strangely enough, a debt investor may actually want a *lower* premoney valuation for your eventual Series A round because it would *increase* their ownership percentage.
- **Unfavorable terms on the convertible notes:** Although the terms of the notes may not require a valuation, investors may insist on terms that are unfavorable to the company, such as founders' personal guarantees, heavy penalties in an event of default, grants of security interests in the company's assets, and others.

Discuss the pros and cons of the priced versus bridge round with your mentors, other entrepreneurs, and your attorney. There isn't a single "right" approach that will be the best approach for your business.

Bridge: Convertible Debt

At least for companies that anticipate doing a full-fledged financing in the near term (say, 6–12 months), a bridge round is considered to be more company favorable and is the easier, faster, and cheaper approach of the two. In this case, the investment is in the form of a promissory note that converts into equity on the terms set in a future "qualified financing" (in which the qualified financing typically is defined by having a minimum amount, say

$2 million, of total investment). The note will convert at a discount to the price per share set in the qualified financing (usually between 10 and 30%), will have warrant coverage (usually in the neighborhood of 20% of the dollar amount invested by each investor), or occasionally both. This discount and/or warrant coverage gives the angel investors some additional ownership in exchange for taking the early risk.

You can find a sample term sheet for a convertible note bridge financing at the end of this chapter.

Priced: Preferred or Common Stock

A priced seed or angel round is the more involved route, because it (1) requires that the company and the investors agree on a valuation during a time in the company's life cycle when pegging a valuation is inherently difficult and (2) involves drafting and negotiating a somewhat more comprehensive and complicated suite of investment documents. Occasionally, the investment is in the form of common stock, but more often it is known as a "light Series A": preferred stock that is similar to that a venture investor will get but usually with less complex, detailed, or complicated terms because of the relatively low valuation associated with it. If the company does not contemplate raising additional funds from venture or other institutional investors and will only be relying on additional small angel-type investments, the priced approach is usually more appropriate for the investors, because they'll more clearly be participating in the upside on terms that are agreed to early in the life of the company.

HOW TO DECIDE WHETHER TO RAISE A BRIDGE OR PRICED ROUND

Many early-stage investors lament the fact that valuations for an early-stage business are often unrealistic (or even downright "crazy" to some investors). Deciding to do a bridge round removes the requirement to set a valuation on an early-stage business. How can you determine if a bridge round is appropriate?

- Are you planning to raise additional funds from venture capital or other institutional investors?
- Are you planning to raise these funds in 6–18 months?
- Have you received positive responses or inquiries from potential institutional investors?

If you answered yes to each of these questions, a bridge round may be appropriate. Ultimately, a bridge round is an effective tool when it is a "bridge" to a later financing event. If you aren't planning on such an event or are unlikely to reach it, then consider a priced round.

Bridge Loan

The bridge loan can be used by angels, VCs as a short bridge until a pending VC round, or between VC financings to extend the financing until the next round is finalized. A bridge loan is an advance of funds by a proposed investor in a company, toward a planned future (usually between 3 and 12 months but sometimes as long as 18–24 months) equity closing, using an interest-bearing promissory note or similar device. In the event in which a bridge loan has been made and the equity financing never occurs, it is referred to as a "pier financing."

A bridge loan may be made in the case in which a company is actively pursuing venture capital funding or in the case in which indications are that a venture capital firm or firms are interested in funding a company but the process is going slower than expected or the venture capital firm and the company are unable to agree on the valuation and are looking to find another investor (a "validating investor"). In addition, a split Series A/B preferred round can serve the same function.

Typically, the transaction uses a straightforward promissory note, which may be secured or unsecured, has a due date 3–12 months later (or longer depending on the terms agreed on), and has a low interest rate. The promissory note provides that, if a company does an equity financing above a certain amount (say at least $2 million) before the due date of the note, the principal and interest under the note automatically convert at the equity closing into whatever is being issued (this is usually referred to as series next preferred or Series A preferred, if this represents a first round of funding), at a discount to the price per share as cash investors. If there are multiple lenders with notes outstanding, usually the notes will all rank equally as to payment and security priority.

Fundraising Process

Raising funds for a new business is usually not accomplished in a single event, nor do most businesses look to the same funding sources at the early stages as they would when they've reached a particular stage of maturity. For most entrepreneurs, you will spend personal savings or rely on friends and family to raise the startup capital. After this initial money, the business will likely need additional infusions and will need to incorporate fundraising plans into overall business planning. Therefore, entrepreneurs should look at fundraising as a process in which the business will acquire funds from a particular source or sources to allow the business to continue on its path of growth.

The following section provides an example of matching certain funding options with particular stages of the business. Most successful startups look to multiple sources to obtain funding. Different sources usually participate at different times in the startup process, and, to help understand the process of funding a new business, the following section follows the germination of an idea into a venture-backed enterprise. Not every company goes through each of these rounds, but it is a helpful exercise to consider. For example, if your company doesn't plan to pursue venture funding, replace that with another source in this exercise, perhaps an expanded angel financing strategy, Small Business Administration (SBA) loans and grants, or a slower growth plan. Ultimately, the aim of this exercise is meant to give you an example (albeit an example that is somewhat traditional for some in the technology space). More information on the specifics of certain types of fundraising methods can be found later in this chapter.

Founder Funding

Estimates show that nearly 90% of company founders will invest some amount of personal money into the business at the early stages of the business, with an average investment by each founder of $10,000. Startup capital tends to come from the initial founders. These investments can range from a few hundred dollars and the founder's time to several tens of thousands of dollars, with the average company requiring more than $50,000 in initial investments (oftentimes from founders or friends and family). Generally, this investment will be exchanged for initial ownership of the business.

SAMPLE FUNDRAISING PROCESS

High-Tech Startup Inc.: Founder Funding

The following boxes, each entitled SAMPLE FUNDRAISING PROCESS, represent a fictional example of the fundraising process for a fictional company: High-Tech Startup Inc. For the purpose of this exercise, assume that the founder funding is simply the initial investment by the founder or founders in deciding whether to pursue the business. These expenses may be incurred to determine whether it makes sense to start the business. The founder funding in this exercise could be used to purchase business planning software, to purchase research reports, to obtain licenses for development software or laboratory equipment, to attend conferences, or for business lunches with key contacts.

Seed Funding

Once the founder has decided to move forward with the business, the business may require some initial funding to move from an idea or concept to a business. This stage of fundraising requires the business to raise enough money to allow the entrepreneur and the initial founding team to commit the time and energy necessary to build a business. Seed money will generally involve raising additional funds from the founders or from outside parties, which include a variety of sources from family or friends to government grants or angel investors. (In this exercise, we've separated this "seed" round from an "angel" round, but these terms are interchangeable in practice and deal with similar issues.)

For some companies, raising seed funding may involve raising money from angel investors (discussed below in a separate section), but, for the purpose of this exercise, we'll assume that an effort to raise additional capital will follow. In this exercise, the business will just be raising a single seed investment necessary to move the business forward while the company is looking to raise a larger amount of money from angel or other seed investors.

After deciding to form a business and taking steps to begin the growth of the business, many entrepreneurs will need to understand how to migrate from the relative security of a job or schooling into the uncertain world of a new business venture, oftentimes with no assets apart from their ideas. Although some businesses are initially built by a team "on the side" or while continuing their current employment, most startups will eventually require more than a part-time effort to succeed in competitive markets. To most startups, the term "seed money" means an investment that will cover the first few months of startup costs and living expenses for the founders. Some businesses may only require a small seed investment from the founders, whereas others may require outside capital of several thousands of dollars to validate the business concept.

SAMPLE FUNDRAISING PROCESS

High-Tech Startup Inc.: Seed/Angel Funding

So, let's suppose that you and two cofounders start a business, High-Tech Startup Inc., and your initial injection of seed money comes from your wealthy brother Mike, who happens to be an accredited investor. Because Mike has done well for himself and wants his little brother or sister to have the same opportunities he did, Mike is willing to give you $20,000 for a 5% stake in common stock of the business. You and the other two founders will each receive 25% of the common stock of the company, and you agree to reserve 20% for an option pool to attract future employees.

Shareholder	Number of Shares	% Ownership
Founders	7,500,000	75%
Brother Mike	500,000	5%
Option pool	2,000,000	20%
Total	10,000,000	100%

The approach above is a priced round because you are essentially selling Mike 5% of the company in exchange for his $20,000. The sale of 5% at this price means the that overall value of the company after the investment would be $400,000 (to calculate this, simply divide the cash amount of $20,000 by the percentage purchased of 5%). This $400,000 represents the postmoney valuation, and the value less the $20,000 investment ($380,000) would be the premoney valuation.

Another alternative aside from issuing stock in the company in exchange for early-stage investments by friends or family is to issue promissory notes. These notes can be structured in such a way as to convert to equity of the company at a certain point or to be repaid at a particular date or time. The benefit is that a promissory note will provide the company with additional flexibility to pay off the promissory note or to convert the investment into stock of the company while ensuring that all the stock of the company remains in the hands of the founders.

In either case, the seed money from Mike pays for you to purchase some software and hardware that you'll likely need to build a prototype and test the concept and should give you some additional cash to help with the first six months of living expenses for you and your cofounders. For the early days, this will mean lots of Ramen noodles and living on the savings that each of you have without drawing much of a salary. If you are fortunate enough to have one or more of the founders continuing their employment, a spouse to provide support, or more extensive savings, you may be able to upgrade from Ramen noodles to a higher-end meal!

As a startup company at this stage, your business (and your life) will most likely be governed by three simple rules: live cheaply, work fast, and find more money. It may seem as though six months gives you plenty of time to get started on product development, but you'll need a constant stream of money for your business to grow, and it always takes longer than you think to find the right investors and close the deal. So you'll need to work hard both on the product you're developing and the financing necessary to make it a reality.

At this early stage, thriftiness is crucial. Remember, the more money you save, the more options you have. If you burn through money in the initial stages, you'll likely be forced into deals that you don't like later on. Working cheaply early in the process can actually preserve your equity stake, translating into huge rewards down the road if your company becomes a success. You may also discuss with Mike the opportunity for additional investments into the business as you meet certain milestones or continue to develop the business.

In this exercise, these first six months will need to be used to raise additional funds and to continue product development or undertake future research. To have the startup business primed to close the next round of funding (before the six months of seed funding disappears), fundraising for additional funding will need to begin as soon as possible. This means developing your business plan, researching potential funding sources, applying for loans or grants, and making contacts with angel and venture capital investors. The thought and care that goes into your business plan will partly determine your success in attracting investment in your startup, and an effective plan will guide management by focusing planning efforts and setting milestones. However, remember that a good business plan alone isn't enough; you will also need to network and build connections to get your business plan to the right investors and contacts.

Angel Funding

The purpose of the seed money provided by Mike has been to help move your business from a simple concept into a plan for a high-growth business with a plan for future product development. At this stage of the company's growth, you will need to begin looking for another outside infusion of cash to continue the growth of the business. After spending some time developing your product and preparing a business plan and presentation, you and your cofounders are running low on seed money, and your team has been out meeting with potential investors.

Some companies will look to grants or loans from government agencies for this additional infusion of cash. Other companies will look to more family and friends for additional investments or perhaps will discuss with Mike whether he is willing to make a larger investment into the business. Depending on the business concept, you may find that venture capital firms are hesitant to make an investment into a business at this early stage of growth; many firms are looking for more maturity in the companies in which they are willing to invest. Therefore, one of the key sources of capital for startup companies at a relatively early stage of maturity is angel funding.

At the point, you'll be looking for angel funding, you'll probably need more funding to hire some help, further your product development, and maybe even acquire some operating space. Angel funding represents a major source of capital for startup ventures. According to the Center for Venture Research at the University of New Hampshire and the MIT Entrepreneurship Center, more money is invested annually by angels ($23 billion) than venture capital firms ($21.9 billion).

CHECKING THE SERVICE PROVIDER BOX

What could happen

You want to get a meeting (or a follow-up meeting) with some top-tier venture capital investors or a local angel forum. Why would they be asking you to provide them the name of your bank, your accountant, your attorney, your intellectual property attorney, and your other outside service providers?

Watch out for

As your mother once told you, you are judged by the company you keep. Certain investors will take comfort in the service providers you have chosen to employ. They may be less comfortable that you have been properly accounting for revenues if you don't use an outside accountant. They may question your intellectual property protections if they aren't familiar with your intellectual property lawyers. They may wonder about your employment policies if you (as CEO/founder) run the human resources for your company.

If you haven't selected a bank, accountant, or human resources support provider yet, make sure you can name a list of parties that you are considering. Although it may simply be a perception issue, be prepared to show that you've considered each of these areas.

TIP: Be prepared to inform potential investors of possible third-party service providers for your business.

Angel investors are simply individuals who back emerging entrepreneurial companies. Generally, angels are willing to invest at an earlier stage of the business and will contribute money to help move the company to a stage at which it can attract venture capital investment or generate sustainable cash flows. Funding levels vary greatly but usually range from $50,000 to $2 million. Oftentimes, certain industries or regions will have groups of angels that meet together to listen to presentations of startup businesses to give angels opportunities to decide whether to invest in any of the businesses.

SAMPLE FUNDRAISING PROCESS

High-Tech Startup Inc.: Seed Angel Funding

You've been able to attend an angel investment forum and meet a group of angels interested in investing in your business. After a successful pitch, you find an angel willing to invest $250,000 at a premoney valuation of $1 million in exchange for shares of common stock.

So your company issues $250,000 in new shares of stock, which you give to the angel. If you had 10 million shares of common stock before the angel investment (with 500,000 to Mike, 2 million to future employees, and 7.5 million for the founders), this deal would generate 250,000 additional shares. Now the angel owns 20% of the shares of the company, and all of the previous shareholders' equity ownership is diluted by that 20%. Here is what your capitalization table would look like after the transaction:

Shareholder	Number of Shares	% Ownership
Founders	7,500,000	60%
Brother Mike	500,000	4%
Angel investor(s)	2,500,000	20%
Option pool	2,000,000	16%
Total	12,500,000	100%

As with Mike's deal before, this investment would be a priced round, in which the angel investor would be purchasing 20% of the company for $250,000 for a post-money valuation of $1.25 million.

A deal of this size will generally be more intricate than your brother Mike's investment. Although the angel might just pay for the stock in cash, the company or angel investor might instead choose to make the investment in the form of a promissory note that is convertible into stock of the company. This would provide the angel with more protection against equity dilution in later funding rounds. The angel might also demand preferred stock, which would give him certain additional rights over the owners of common stock, including vetoes over major decisions, the right to get the investment back in the absence of an exit strategy, and protection against equity dilution, and there's always a chance that the angel will want a seat on the board of directors. A deal of this size may require several weeks to close and could cost the company several thousands of dollars in legal fees (which is why it oftentimes makes sense to group the angel investors together into a single transaction to get all angels to invest on the same terms). Sometimes, an angel will agree to pay these fees for both sides if the startup is short on funds. If you find yourself in this situation, make sure that you are getting fair representation.

Once you have received an investment from an angel, you will be able to continue to grow the business and should be able to undertake key initiatives. With this $250,000 in the bank, you can afford to hire employees and purchase inputs that will bolster your product development process. Perhaps you can now hire a very talented engineer as your first employee. This employee may agree to a low salary plus 3% of the company in stock options or in restricted stock. Restricted stock is a lot like a stock option, only that instead of earning the right to buy the stock, you get the stock up front and earn the right not to relinquish it. Some startups have replaced stock options with offerings of restricted stock, yet the "option pool" would still be the source of these shares (so, after this deal, the option pool would have only 13% remaining to be issued and 3% outstanding).

Determining the proper amount of stock to provide to early employees is difficult for new entrepreneurs. Much more information is found in the later chapters on employees. In general, the stock option calculation is a function of the value of the employee and the stage at which they join the business. If you think someone is going to contribute as much to the success of your company as a founder (and you sign him or her really early), you might even give them an equivalent amount of stock. As your company ages, you should offer employees less and less stock.

Venture Capital Funding

Angel funding is oftentimes a precursor to venture funding. Angels may be able to assist you with introductions to venture capital firms and may provide added credibility with these firms. Unlike angel investors, venture capital firms tend to require companies to have greater maturity levels (some venture firms do invest in early-stage companies, but that is not the norm).

Venture capital has been an important source of funds for many high-technology companies, but not every startup will require venture capital funding. The decision to pursue venture funding should be based on a careful examination of the business and consultation with outside experts. In the event these initial discussions are positive and if your startup is undercapitalized and needs to invest in infrastructure, hire a sales or marketing staff, or put a product into mass production, you will probably need another big injection of capital that could come from the venture capital industry.

For a lucky few companies, venture capital firms might come to you, but most likely, you'll have to approach VCs through your contacts. You'll find more information on the process for finding and dealing with VCs in the following chapter.

SAMPLE FUNDRAISING PROCESS

High-Tech Startup Inc.: Venture Capital Funding

Let's suppose your company is one of the lucky few that has been able to find a VC firm or firms prepared to invest in your company. The VC will most likely offer you a term sheet summarizing what the deal terms will be. In some cases, VCs will want you to agree to avoid negotiations with other VCs for some period of time after you accept the term sheet. During this period, the VC will do its due diligence on your startup in an attempt to uncover any serious risks or barriers to success that might bubble up. If they don't find any fatal flaws, the VC will proceed with the deal.

Your VC has offered you a $2 million investment at a premoney valuation of $4 million dollars. Under this scenario, the VC will would get 6.25 million shares of stock of the company (generally, a VC investment would be for preferred stock, but that stock is convertible into common stock of the company). After this investment, everyone else's percentage of equity ownership would decrease by approximately one-third. Because the VC understands that you'll need to continue to hire new employees and will need additional stock options to attract quality candidates, the VC will permit you to expand the company's option pool by 1.25 million shares.

The postmoney valuation of the company after the transaction would be $6 million dollars ($4 million pre-money valuation plus $2 million of new money). The capitalization table and option pool would now look like this:

Shareholder	Number of Shares	% Ownership
Founders	7,500,000	37.50%
Brother Mike	500,000	2.50%
Angel investor(s)	2,500,000	12.50%
VC investor	6,250,000	31.25%
Total option pool	3,250,000	16.25%
Total	20,000,000	100.00%

Options	Number of Shares	% Ownership
Issued		
Engineer	375,000	1.88%
Other employees	625,000	3.12%
Available for issuance		
Original pool	1,000,000	5.00%
New pool	1,250,000	6.25%
Total	3,250,000	16.25%

Although this hypothetical deal helps give you an idea about the process of equity dilution, it is only simplified example. Companies may require several stages of investment from VCs to grow the business. More details about the specifics and process for VC investments follow this chapter.

At this stage, you will most likely do significant amounts of hiring while you further your product development. You also might prepare for more rounds of funding, hopefully at higher and higher valuations. If your company is extremely successful, you may even have the opportunity to go public through an IPO. Hopefully, your company will avoid a down round or a funding round at a valuation lower than the previous round. Common stock holders usually take a hit during down rounds, and many deals with VCs will include antidilution provisions in their deals to protect themselves.

The previous exercise is designed to provide you with an overview into the fundraising process and how each funding stage will affect your company and its allocations of ownership. Remember that timetables, trends, and funding courses will vary in the case of different industries, technologies, founder teams, and businesses. Many highly successful high-technology businesses will never go public and will focus their fundraising efforts on areas other than angels or VCs.

Funding Sources for Your Business

There are a number of sources from which a new startup can obtain the money that it will need to grow. There is not a one-size-fits-all strategy for funding a business; it often depends on numerous factors, many of which may be out of your control. Entrepreneurs can often self-finance the initial capital outlays for their business through personal savings, second mortgages, credit cards, and traditional bank loans. Some startup founders can also "bootstrap" their startup by using profits from early sales to grow the business. This approach works especially well in the service industry, in which startup expenses are sometimes low and the need for employees may initially be minimal. Most entrepreneurs, however, require at least some amount of additional money from alternative sources to fully capitalize their business, and this section explores some of the major sources of funding for new ventures:

- Friends and family
- Angel investors
- Government and public sector
- Seed funding firms
- Joint ventures and strategic alliances
- Venture capital firms

Each of these alternatives has its advantages and disadvantages, and they all require your company to form some sort of relationship with an investor or financier. While sorting through these potential sources for funding, you should consider the amount of control you want to retain, the amount of equity dilution you and your investors are willing to bear, and the rate of growth you want to achieve. You should also be mindful that the funding choices you make in one round can have downstream effects later. Without proper planning, due diligence, and careful negotiation of such partnerships, you may inadvertently miss opportunities or relinquish future rights to valuable assets in pursuit of immediate funds. All startup companies exploring these alternative forms of financing should obtain sound advice from an attorney, accountant, and financial advisor before entering into a definitive agreement.

Summary Details of Various Funding Sources

	Founder	Friends/family	Angel investors	Bootstrapping	Bank loans	Government grants/loans	Joint venture/ strategic	Venture capital
Typical maturity of companies	Early	Early	Early	Early/mid	Mid/mature	Early	Mid/mature	Mid/ mature
Time to obtain funds	Fast	Fast	Fast/medium	None	Medium/slow	Slow	Medium/slow	Medium
Amount of funds awarded	Small	Small	Small	N/A	Medium	Medium	Medium	Medium/ high
Likelihood of utilizing	High	Medium	Low/medium	Low	Low	Low/medium	Low	Low/ medium
Costs of funds	Low	Medium	Low	Low	Medium	Low	Medium	High
Comments	Typical for most startup companies	May be difficult to manage; may not provide strategic value	May require high degrees of effort to obtain	Unlikely in early stage companies	Unlikely in early stage companies; less favorable terms than equity	May be difficult and slow to obtain; not all companies will be eligible	May be difficult to obtain and be slow to close; can affect strategy choices	Typically requires a high growth company; may limit control of founders

Private Individual Investors

Private investors are a popular source of funding for new ventures in the early stages of their development process. Investments in this category range from small loans from friends and family to large injections of capital from angel investors. Stock offerings to private investors typically take the form of preferred stock, whereas founders traditionally hold shares of common stock.

CAN YOU HAVE TOO MANY INVESTORS?

What could happen

You've told a family friend about your business. This friend is a stockbroker and he's lined up 15 of his clients to each invest $5,000 into your venture.

Watch out for

Although it may seem counterintuitive to turn down any investors (or lenders if you arrange for small loans with a number of parties) into your business, remember that you add another relationship to manage with each new investor. Investor relations takes time and may cut into your ability to run your business. Be judicious about adding new partners to the venture, particularly if those investors are not savvy or you perceive may not be satisfied with infrequent communications. Consider the size of the investment relative to expectations this investment will create. If a new investor is not savvy about investing in startups and will create headaches with respect to getting out of the investment or getting information on the investment, you should consider whether it is the right source of investment dollars. In addition, always consider the securities law implications of adding nonaccredited investors.

TIP: Treat each investor or lender as a partner; each relationship will take ongoing time and energy to maintain and cultivate.

Founders

The majority of startup companies will require an initial investment from a founder or founders. This investment will oftentimes take the form of cash, purchase of company goods, services, or assets with personal funds or may represent a deferral of salary:

- For founders that own their own homes, some have taken out home equity loans to fund their startup business.
- Founders have also used personal credit cards to infuse initial capital into the business. There are some zero interest, deferred payment cards that can provide a founder with a mechanism to purchase goods or services for the business and defer payment (without interest) for 6–12 months. Use of personal credit cards should only be used as a short-term alternative and should be transitioned to a more traditional bank loan or other financing device as the business matures.

In some cases, these investments by the founders will be made in exchange for additional common stock of the company. In other cases, the founders may choose to issue promissory notes for these investment amounts to be converted into additional stock or returned on availability of capital to repay such amounts.

Friends and Family

Friends, family, neighbors, and colleagues might be some of the first sources you consider early in the funding process for loans and the purchase of stock. Garnering investments from the people close to you offers a few advantages: you won't need to expend valuable time and effort making connections and establishing trust, and friends and family are more likely to offer favorable rates on loans. These friends and family members also might be more forgiving if their investments fall through.

Yet there are several important considerations for your business to discuss before accepting investments from friends and family. First, unless these investors have more than $1 million in assets or an income of more than $200,000 a year ($300,000 if the person is married), they may not qualify as an "accredited investor." The SEC imposes lower regulatory burdens if a company's shareholders are accredited investors. Taking money from unaccredited investors can restrict your options and create major headaches down the road. If your startup goes public, the SEC will carefully study all previous issuances of stock and demand that your company take immediate action to cure any past violations of securities law arising out of investments from unaccredited investors. Although it may seem like a good idea to give stock to these unaccredited friends or family members initially, it may cause expensive and time-consuming problems later in your startup's life cycle. Even if you haven't violated any laws, SEC investigations can waste time and money. You should consult an attorney before accepting or soliciting money from unaccredited investors.

Second, in most circumstances, friends and family members will lack industry-specific connections and business sophistication, two prized qualities of investors. Every investor represents a potential wellspring of information that can help guide your company to success, and the knowledge and experience that your investors bring to the table can be very important assets. Other investors in future rounds of funding (such as angel investors and venture capital firms) will look closely on how you valued the equity given to your friends and family and may be turned off if they fear SEC complications or inexperienced decision-making.

INVOLVING FAMILY AND FRIENDS

What could happen

To get your business started, you are considering taking money from a collection of family and friends. You are also considering employing a family member in the business.

Watch out for

Family and friends can be a valuable source of initial startup capital or talent for your business, but remember that your family and friends may not understand the time horizons of a startup business or the applicable risk. Be careful about using money from family and friends unless you have a strategy for repayment and a time horizon that is realistic. Starting a business is stressful enough without adding interfamilial conflicts into the mix. In addition, be especially cautious with unaccredited investors (especially family and friends). Remember that unaccredited investors can cause future problems for compliance with securities law exemptions.

TIP: Be cautious about involving family and friends in your startup (especially in financing aspects).

Moreover, financial relationships with friends and family members carry extra emotional burdens. You may not want your loved ones' financial futures riding on the success of your startup or their constant input on how to run your business. To minimize any potential strain on these relationships, it is important to be as upfront as possible about your realistic expectations for growth and the inherent risks of investing. If you do want to accept money from friends and family, make sure to have an attorney draft a letter of agreement that outlines the funding terms and cancellation policies, if necessary.

Angel Investors

Angel investors (or simply angels) are affluent investors (typically accredited) who provide capital for startups early in the development process, usually in exchange for an equity stake. (The term came from individuals who would invest in Broadway theater productions as unnamed donors and became commonly referred to as "angels" for their roles in saving productions that had overrun their budgets.) Many successful startups in their infancy have been bankrolled by private sales of debt or equity securities to angel investors. Angels typically look for businesses that have solid management teams and strong growth potential in industries that they know well.

INVESTMENT BY ANGEL INVESTORS

According to the Center for Venture Research at the University of New Hampshire and the MIT Entrepreneurship Center, angels invest in nearly 50,000 ventures each year, representing annual investments of more than $23 billion.

Angels will normally invest larger sums of money than your friends and family, but these greater sums come with somewhat higher levels of expectations. Angels want to get their investments returned so will usually only consider funding companies that have valid exit strategies. Although it is unusual for a venture capital firm to invest amounts under $1 or $2 million, the majority of angel investments will fall under this level, providing a company with a source of funds to further develop the product and grow the company, without necessitating a $5 million outside investment.

An added benefit many companies look for in angel investors is relevant industry or investing experience. Angels are oftentimes wealthy from their own previous entrepreneurial or investment success, and the contacts and experience that they can bring to your startup can be as valuable as the capital. Taking money from an angel that has previously invested in successful startups within your industry is a particularly good idea. Besides being able to offer insights and advice, knowledgeable angels can make your company much more attractive to other private investors and VCs, as well as customers, suppliers, and employees. The ideal angel would have previous experience in marshalling young companies in your industry to success, would buy into your vision and long-term goals, and would bring both contacts and credibility to your business.

Some startups try to attract eminent angels who can provide guidance by letting them invest in their company when it has a low valuation. The lower the valuation for your company, the cheaper it is for an investor to acquire an equity share. Early in the funding process, it's hard to peg down an accurate valuation for a startup, because the number is just a byproduct of the respective investments of everyone involved. For example, if an angel pays $100,000 for a 10% stake of your startup, the startup has an implicit valuation, in theory, of $1,000,000. As a company gets more established, its valuation approaches its actual market value, and startups' valuations are expected to rise over time. It is impractical (and potentially illegal) to adjust your startup's valuation for the purposes of attracting individual investors, so if you want to entice eminent angels with cheap stock, do it early in the funding process when it's natural to have a low valuation. You and your angels will have to agree on a premoney valuation of your startup before they invest. The premoney valuation is as simple as it sounds: the agreed-on worth of your company before an investment. The lower the premoney valuation, the more equity an angel can acquire.

Finding the right angel or angels to invest in your startup can be a difficult, time-consuming process. Some angels organize themselves into groups to share research and pool capital, which makes them easier to find. The Angel Capital Association is a good source for finding a group near you. These groups can be found in most large cities, but the bulk of them reside in Silicon Valley, Seattle, Boston, Austin, Denver, New York City, and other similar cities or regions with a track record of developing new ventures. Bear in mind, however, that most angels (especially the prominent ones) don't belong to a group. Also note that you should carefully research any angel group that tries to charge you money just to pitch your idea. Many groups do not charge for presenting to their group, whereas some groups do have a fee. If the group has a fee, ask around to ensure they are reputable.

RESEARCHING ANGEL INVESTORS

- **Angel Capital Association (http://angelcapitalassociation.org):** North America's professional alliance of angel groups provides information on the more than 265 angel associations across the country.
- **vFinance Inc. (http://vfinance.com):** Offers paid searching tools to find angel investors based on net worth on the individual, industry that the angel will invest in, and location of the investor. Customized searches for $1.00 to $2.50 per contact.
- **Angelsoft (http://angelsoft.net):** Offers Web-based software tools for entrepreneurs to identify angels and angel groups and submit information to these groups online. More than 12,000 investors are standardized on this platform. Free trial of the tools; $250 for full access.

The best way to meet individual angels is through industry contacts. Although cold calling occasionally works, angels will pay more attention to investment opportunities recommended to them by someone they respect and trust. Your accountant, lawyer, or entrepreneurial friends might possess the contacts you need. Once you've established contact with an angel, you need to present your company in the best light possible.

Angels will look for a strong executive summary and management team, a "need-to-have" product or service, industry contacts to support your claims, and the aforementioned exit strategy. Angels will also be attracted to a legally sound business plan free of any significant downstream problems (such as intellectual property conflicts or securities issues).

There aren't any generally accepted standards for dealing with angels, so deal terms vary greatly. Some will require intricately structured arrangements rivaling those of venture capital firms, whereas others (especially those investing very early in the process) will be content with simpler agreements. Angels without significant investing experience may not even know exactly what terms they want. Although you might sometimes want to wait for the angel to draft the agreement on their terms, you can always have your lawyer draft a model agreement that meets your expectations. Many angels will appreciate this savings of time and expense.

Because angels are primarily concerned with getting a return on their investment, most will not demand an active role in the business. This means that, unlike most venture capital firms, angels will seldom insist on board representation or veto rights over employee decisions. Angels may only require the right to veto significant changes to the business plan, management salary levels, and the amount of equity available for employee incentive programs. Angels are also free from some of the restrictions that bind venture capital firms. For instance, angels will sometimes allow founders to cash out partially by selling some stock directly to investors during a funding round. Venture capital firms will rarely allow such a transaction because of concerns it will cause the founder to be less committed to the enterprise.

One of the risks of accepting investments from individual angels, rather than through an angel group or investment firm, is that you may have less knowledge about the potential investor. Angels may also be somewhat more likely to drag their feet before writing you the big check, insisting on lower investment amounts or a multistage investment. Sometimes, the only way to get one investor to commit is to have another investor lined up to invest to increase the sense of urgency.

Bootstrapping

Some startups will plan to avoid or postpone the need to find outside investors for as long as is possible. Such an approach allows the founders and other stockholders to avoid diluting their ownership interests and build the company with the profits of the company. Even startup companies that plan to take funds from outside investors will engage in some amount of bootstrapping to limit the amount of funds they need to take and limit dilution.

This approach may be difficult for some high-technology businesses that are capital intensive or require funding to get the product to market. However, companies may decide to bootstrap for a period of time to extend the burn rate of cash for the company.

How does a company bootstrap itself? Generally, this approach will be done by limiting expenses to maximize cash flow for the company. In the case of bootstrapping, cash really is king. A company may choose to offer a basic product or service for sale to customers while developing a richer product offering or providing consulting services while developing the ultimate product offering.

WHAT DOES "BOOTSTRAPPING" ENTAIL?

Here are a few examples of approaches to bootstrapping:

- Providing consulting services related to the product you are developing
- Licensing your technology for an alternative application
- Factoring your accounts receivable
- Leasing equipment or buildings
- Using trade credit (purchases made on net 30-, 60-, or 90-day terms, for example)
- Requiring upfront payment from customers
- Developing favorable relationships with key vendors
- Entering into a sale-leaseback arrangement in which a third party will purchase corporate assets such as computers, furniture, or other company equipment for cash and lease them back to you

Bank Loans

Although many startup companies seek equity financing instead of debt financing, entrepreneurs may still find it helpful to consider the various bank loans available.

Bank loans can be classified according to whether the loan is short term or long term and whether the loan is secured or unsecured. Short-term loans are generally used to finance the company's inventory needs, accounts payable, and general working capital. Interest rates are typically lower on short-term loans than long-term financing. Long-term loans typically require a larger amount of collateral to secure the financing far into the future. They are usually used to finance fixed assets, such as the company's property, plant, and equipment.

Loans may also be secured or unsecured. Unsecured loans are simply promises to pay a debt. If the borrower defaults on the loan, the lender's only recourse is to sue the borrower. In such a situation, the lender will not have any priority claim to a particular piece of the borrower's property. Consequently, businesses have a difficult time obtaining unsecured loans unless they have a strong credit history. Secured loans are also promises to pay a debt, but the promise is "secured" by the property of the debtor (called "collateral"). If the borrower defaults on the loan, the creditor can recover his losses by seizing the property that was collateral for the debt.

For companies that are unable to obtain banking loans from traditional large financial institutions, some startup companies will choose to work with community or local banks. These banks may offer a complete banking relationship and be willing to provide financial products with more varied minimum amounts, payment plans, or interest rates. Startups also may benefit from microloans, which have begun to be an important tool for newly established small businesses. These loans can range from several thousands of dollars to up to $35,000. Microloans are funded by the SBA through grants to nonprofit community lenders that oversee the lending process to business borrowers. The unique feature is that the lending and credit decision is made locally by the community lender. Each community lender will have individual credit and lending requirements, but the maximum term of these loans is six years. The microloans will require the borrower to provide a personal guarantee and some form of collateral. Additionally, the community lenders will require the borrower to complete a business planning and training program before issuance of the loan, but the tradeoff is that microloans are easier to obtain than a traditional bank loan.

LOAN PROGRAMS FOR STARTUPS

Some examples of the types of loans available to entrepreneurs and start-up businesses are as follows:

1. **Working capital lines of credit:** Under a line of credit, a party may borrow funds as needed, up to a specified maximum amount. The line of credit is used to fund the working capital and cash needs of the business. To secure the loan, the bank will sometimes use the company's accounts receivable or inventory as collateral. The term of the loan may vary and is often renewable. Borrowers will pay interest on the outstanding balance of the line of credit.
2. **Short-term commercial loans:** Short-term loans are usually given for a specific expenditure, such as a piece of equipment. Interest is paid on the lump sum of the loan and is often a fixed rate, so businesses usually do not face much risk of rising interest rates. Short-term loans may be as short as 90–120 days or may extend from one to three years. The loan is typically secured by collateral, such as accounts receivable, inventory, or a fixed asset of the business. Most loans to startup companies and new small businesses are short term, and the lending agency will review the company's cash flow and credit history before providing funds.
3. **Long-term commercial loans:** Long-term loans (those with terms longer than one to three years) are more difficult for new businesses to obtain because the risk that the new business will default increases with the length of the loan. The length of the loan usually ranges from five to seven years, although loans secured by real estate may extend much longer. Long-term loans are generally used for business expansions and to fund major plant and equipment purchases. Lenders usually require that the loan be secured by the asset being acquired. In addition, the lending agency will review the company's business plan and cash flow to determine whether the company will be able to repay the principal and interest over the term of the loan. Lenders may also require insurance to protect the collateral.
4. **Small business credit cards:** Small business credit cards offer an alternative to working capital lines of credit. They provide a quick source of limited funds when cash flow is tight. The interest rates are typically only slightly less than the interest rate on individual consumer credit cards and may not have very high spending limits. Small business owners who are considering this option should also be aware that they may have personal liability for the credit card, at least until the business has an established credit history of its own and the business owner can negotiate a new arrangement. If the credit history of either the business owner or the business itself is less than flawless, the credit card company may also require that the business deposit a specified amount of cash as collateral for securing a credit line.

5. **Letters of credit:** Businesses engaged in international trade frequently use letters of credit as a method for making payments. In these situations, the buyer and seller will arrange a contract for the sale and shipment of goods. The buyer will then deposit money (or take out a loan) at his local bank in the amount of the letter of credit. The buyer's bank will issue the letter of credit to the seller's bank in the foreign country. The letter of credit will specify certain documents that must be presented for buyer's bank to transfer the funds. Such documents frequently include a commercial invoice, the bill of lading for the shipment of the goods, and insurance documents. The seller's bank will then notify the seller that a letter of credit was opened in his favor and that the goods may be shipped. Once the goods have been sent, the seller will present the requisite documents to his local bank for approval. If the documents conform to the letter of credit, the issuing bank will transfer the money to seller's bank and the money will be deposited in the seller's account.

   Letters of credit are beneficial for both parties to the transaction for a couple of reasons. First, by requiring certain documents in the letter of credit, the buyer is offered some protection that the goods were sent on a particular date and that they were shipped in a particular condition. Second, the seller is offered protection that he will be paid, because the buyer's bank is required to honor the letter of credit on presentment of conforming documents. The flipside of these advantages, however, is that parties must be exceptionally precise when filling out the letter of credit and presenting the required documents. Even if the goods arrive on time and in the required condition, the issuing bank will not be obligated to make a payment if the documents do not conform to the letter of credit or if they are not received on time. Thus, the parties must pay very close attention to detail and must be as accurate as possible in their terminology.

Seed Funding Firms

Seed funding firms typically invest small amounts of money in startups, almost exclusively in the early stages of a company's development. Usually, the goal of seed funding is to cover the operating expenses of a startup until it reaches the point at which it produces something impressive enough to raise money on a larger scale. Seed funding firms come in many different forms. Some seed firms primarily provide funding and advice, whereas other firms will also provide operating space and even employee and legal help. This latter type of seed firm is sometimes referred to as an incubator, and, according to the National Association of Business Incubators, there are about 800 incubators operating in the United States.

Early-stage investments are particularly risky, and, in many instances, seed firms have very little information on which to base their funding decisions. Because many young startups radically change their business plan early on, seed firms care as much about the founders themselves as the strength of their ideas. Seed firms look for motivated, driven entrepreneurs that appear highly motivated. Occasionally, they will even invest money before a business plan has been developed or a management team has been put in place. Seed firms are conditioned to dealing with startups in their infancy, so they will be able to offer valuable advice about early-stage decision-making and later-stage financing. Given the risks involved, however, you shouldn't expect large amounts of funding from a seed firm.

One advantage of seed firms is that they are easier to find and reach than individual angels. Seed firms are established companies with websites and public contact information; reaching them is as simple as sending them an e-mail. Although a personal introduction to a seed firm through a trusted contact would certainly help your chances of getting your foot in the door, it may not be as important as it is with angel investors or venture capital firms. Arranging deals with seed firms might also be easier than with an angel, because many seed firms have standardized investment processes with set deal terms they use for every startup they fund. Standardized deal terms certainly expedite the process, but be aware they might not be so favorable to you, so consult with your advisors. If other successful startups have signed deals with a seed firm using the same terms offered to you and are satisfied with the results, it is a good sign that those terms are sufficient.

Government Funding and Public Sector Support

Federal and state governments offer grant, loan, and technical assistance programs designed to promote emerging businesses. Competition for these forms of public sector funding can be fierce, and the application process is oftentimes quite arduous. Additionally, the funds usually come with burdensome restrictions on what you can do with the money. Nevertheless, some startups effectively use government funds to advance their business development, so these funding sources shouldn't be dismissed without some research and review.

Many federal agencies participate in the government's SBIR and Small Business Technology Transfer (STTR) programs. The SBIR and STTR are funding programs designed to stimulate technological innovation and fulfill the research needs of the federal government. Businesses are required to meet several criteria to be eligible for grants under either program, including U.S. ownership, for-profit status, and restrictions on number of employees. The SBIR and STTR programs differ in two major ways. First, under the SBIR program, the principal investigator listed on the SBIR application must be employed by the small business at the time of the grant and for the duration of the project. Under the STTR program, there is no such employment requirement. Second, unlike the SBIR, the STTR program requires the small business to be engaged in a collaborative relationship with a nonprofit research institution located in the United States.

For SBA loan guarantee programs (the SBA will not actually give you the loan but will guarantee a loan made through a traditional commercial lender), there is not a maximum loan amount and companies are eligible for loan terms that can be up to 25 years. SBA loan guarantee applications tend to approve applicants in which the founders have good credit scores, a strong business plan detailing the use of the proceeds from the loan, and evidence that the founder or founders have made personal financial investments.

The SBA 504 loan program (sometimes called a "development loan") is designed to assist small businesses with financing of fixed assets, including the purchase of buildings, land, and certain types of equipment. Private institutions will provide the financing for the assets through certified development companies (CDCs). Loans made under the 504 program are generally made at a fixed-rate, long-term basis. The way the development loan works is that the private lender institution will lend the company 50% of the total project, a CDC will lend 40% (guaranteed by the SBA), and the company will be responsible for the remaining 10% of the total project cost. The uses are somewhat limited: inventory, debt service, short-lived equipment, and machinery aren't eligible. However, for various projects, including building, construction, and facility renovation or retrofitting, these programs will be an option.

Companies in the importation and exportation business may be eligible for import-export bank programs, supported by the SBA. Loans for working capital of up to $1.1 million ($1.25 million if combined with an international trade loan) can be guaranteed by the SBA. This program is typically only available for U.S. companies that have been in business for one full year, operate at a profit, and do not exclusively rely on the loan to support the business operations.

Some federal agencies also run venture funding groups. Rather than awarding grants, these groups make investments in emerging businesses. For example, In-Q-Tel is a venture fund run by the Central Intelligence Agency that invests in high-tech startups, typically that look to develop technology related to the intelligence community. These venture funds normally produce good returns, and approaching them may be easier than their private sector counterparts.

Many U.S. states also offer financial assistance and other support to new startups. These funding programs are usually run through specific agencies or departments based on the relevant subject matter. You should be cautious with state funding initiatives requiring your startup to locate to an area that lacks other entrepreneurial support organizations and companies. Although free money is enticing, it certainly shouldn't be considered free if it hamstrings your business by forcing you into an unsupportive location.

Joint Ventures and Strategic Alliances

Although less common than the other forms of funding mentioned previously, strategic alliances and joint ventures can be rewarding sources of financing for a startup. These sorts of arrangements typically occur when an established company and a startup share complementary needs and objectives.

Transactions like these can come in all sorts of shapes, sizes, and structures. For example, if a funding-hungry startup was developing a product that had a good chance of succeeding in an established company's business, the parties might be willing to share in the risks and rewards on an alliance in which both parties co-develop and co-sell the product, sharing in the proceeds equally. If a startup company has two good products, applications, or ideas, the startup may want to consider selling one idea or product to another party. After the sale, the startup company can use the proceeds to fund the second idea or application. Some larger companies have investment arms that provide direct funding to interesting companies or technologies. This allows the larger business to "outsource" research and development activities without weighing on profitability numbers for the entire corporation. As you can see, there are numerous ways for a startup to ally with an established company.

How do parties structure these relationships? One method is for the parties to come together and form a separate joint-venture entity, sharing resources, facilities, and information. This is a great way for a startup to get access to equipment, personnel, manufacturing capabilities, distribution channels, and sales forces that would be otherwise unavailable. Strategic alliances with a high-profile established business might also provide validation for the startup's vision and business acumen, facilitating future funding from other sources.

Established businesses look for joint venture partners that will provide significant business benefits. Typically, they will only be interested in your product if it can increase their sales 15% to 25% or if it provides them with a market advantage over their competitors. They will also prefer partners that have firmly established intellectual property rights. To find the right business partner, you must first develop your understanding of the market. Partnering with a larger entity may involve some creative thinking to ensure the parties' interests are aligned.

Where can you find potential corporate partners? Read the trade journals, talk to firms up and down the distribution chain, and find out which businesses in your industry are the early adopters of new advances. When you've found a potential suitor, make sure to push for a formal agreement as early as possible to establish your rights in the relationship. Above all, be reasonable in demands for profit sharing; asking for too much is an easy way to repel established businesses.

Finally, licensing arrangements may represent the simplest way for a startup to profitably ally with an established business. The parties might reach a licensing arrangement wherein the established company agrees to fund a startup's product development for the right to exploit that product for their own benefit. Alternatively, the parties might agree to cross-license each others' technologies. Startups can profit handsomely from these arrangements by collecting upfront or access payments, milestone payments and royalties, research funding, and loans and equity investments. It is crucial to consult with an attorney about your intellectual property rights and the optimal scope and span of your licenses before discussions with potential business partners.

Venture Capital Financing

There are a number of examples of highly successful companies that used venture financing: Amazon.com, America Online, Amgen, Apple Computer, Cisco Systems, Compaq, DEC, Federal Express, Genentech, Google, Intel, Lotus, Netscape, Oracle, Seagate, Sun Microsystems, 3Com, and Yahoo!. Because of the importance of venture capital as a financing source for many high-technology companies, Chapter 10 takes a closer and more thorough look into the process for obtaining venture financing, details on the venture capital financing transaction, and understanding a term sheet and other deal points in a venture capital deal. The section below provides an introduction and a broad overview into venture capital firms and financings.

Public perception is that VCs only fund high-tech companies. However, the truth is that VCs will fund a variety of companies that fit their investment profile and provide returns consistent with their internal metrics. Although firms may invest in industries outside of high-technology fields, it is still typical that a firm will focus its investments into certain fields, industries, or technologies. This focus is the result of the ability of the firm to understand the technology, market, and potential of any investment, as well as to allow the firm and its partners to offer its portfolio companies ongoing value as an outside advisor.

Likewise, the perception exists that venture capital firms only invest in mid-stage or advanced-stage companies. Although there are many firms that will focus their investments on mid-stage and advanced-stage companies, firms do choose to invest in early-stage companies. Matching the typical investment stage and technology or market focus of the venture capital firm is integral to obtaining funding from a venture firm.

HOW DOES THE VENTURE CAPITAL PROCESS WORK?

From a company's standpoint, here is how the whole venture financing transaction and relationship looks:

- The company starts up and needs money to grow. The company seeks venture capital firms to invest in the company.
- The founders of the company create a business plan that shows what they plan to do and what they think will happen to the company over time. The business plan should include how fast the company will grow, how much money it will make, who the key managerial leaders will be, and other relevant information.
- The VCs look at the plan, and, if they like what they see, they invest money in the company. Very often the most important aspect of the plan is a clear articulation of who is running the company. VCs deeply value leadership and management success when dealing with startup companies that all begin to look alike. Good management is a key differentiator.
- The first round of investment is typically called the Series A round, and the company will receive cash in exchange for equity ownership, which is usually given in the form of preferred stock of the company. Over time, a company will oftentimes receive three or four (more or less depending on the needs of the company) rounds of funding before going public, getting acquired, or going out of business.

In return for the money it receives, the company gives the VCs stock in the company as well as some control over the decisions the company makes. The company, for example, might give each VC firm a seat on its board of directors. The company might agree not to spend more than $X (say $250,000) without the VC's approval. The VCs might also need to approve certain people who are hired, loans that are made, and other key decisions.

In many cases, a VC firm offers more than just money. For example, it might have good contacts in the industry or it might have a lot of experience it can provide to the company. The value an experienced VC may add to a startup company often may transcend mere financing.

One big negotiating point that is discussed when a VC invests money in a company is, "how much stock should the VC firm get in return for the money it invests?" This question is answered by choosing a valuation for the company. The VC firm and the people in the company have to agree how much the company is worth. This is the premoney valuation of the company. The VC firm then invests the money in the company and creates a postmoney valuation. The percentage increase in the value determines how much stock the VC firm receives. A VC firm might typically receive anywhere from 10 to 50% of the company in return for its investment. More or less is possible, but this represents the typical range for a first-round investment. The original shareholders are diluted in the process. If the situation exists in which the shareholders own 100% of the company before the VC's investment, then following an investment by the VC firm in exchange for 50% of the company, the original shareholders' shares would now represent the remaining 50% ownership in the company.

After several rounds (each round involving a new series of stock) of financing, the company and the investors will usually be looking for liquidity for their investment. Private companies without a market to buy or sell their shares usually have few opportunities for an investor to "cash out" their stock. Therefore, most VC-backed companies will look to do some type of a liquidity event such as an IPO or an acquisition event. At or after either event, the VC firm and company will look to end or scale back the relationship. However, to be fair, reaching such a point often will involve three to seven years, multiple rounds of financings, and a substantial amount of time to find and finish such a transaction. To satisfy its investors, a VC will ultimately need to be able to extract its investment (plus a healthy return) to return the funds to its investors. For example, in many e-commerce companies or biotechnology, it isn't uncommon that a group of institutional investors have invested between $50 and $100 million before an IPO.

Most VCs will look at a highly successful return on their investment if they are able to return 10 times or more of their investment back to the fund. The odds of this happening are low, so having one homerun in the investment portfolio can pay off the numerous low-performing investments. VCs are still happy with returning two to three times the investment on a company. If a VC has 10 portfolio companies it has invested in with one big winner (10 times), one or two medium winners (two to three times), one or two break-evens, and the rest losers, the fund could wind up a success for the investors and the venture partners. Venture capital firms operate in a very risky game in which they hope to find one Google or Apple and avoid investing in too many Webvans (a famous Web-based grocery delivery company that once had about $800 million in venture capital but ended up with $830 million in losses and just $40 million on hand when it closed up shop) or Kozmo.com (a small-goods delivery service that raised more than $250 million only to be forced to liquidate in 2001).

To download a copy of the sample terms for a convertible note and warrant term sheet, visit the book's website at http://www.myhightechstartup.com.

SAMPLE TERMS FOR A TERM SHEET

Convertible Note and Warrant Financing

MEMORANDUM OF TERMS FOR THE PRIVATE PLACEMENT OF SECURITIES OF HIGH-TECH STARTUP INC.

This term sheet summarizes the principal terms of the proposed financing of High-Tech Startup Inc. (the "Company"). This term sheet is for discussion purposes only; there is no obligation on the part of any negotiating party until a definitive note and warrant purchase agreement is signed by all parties. This term sheet is subject to the satisfactory completion of due diligence. This term sheet does not constitute either an offer to sell or an offer to purchase securities.

Amount to be raised: $250,000

Type of security: Convertible promissory notes (the "Notes").

Warrants:

Warrants to purchase securities issued in the Company's next equity financing having an aggregate exercise price equal to 20% of the principal amount of the Notes.

- OR -

Warrants to purchase Common Stock having an aggregate exercise price equal to 20% of the principal amount of the Notes.

Interest rate:

Prime plus 2% per annum.

- OR -

6% per annum.

Maturity:

Principal and accrued interest shall be converted on or before December 31, 2009 into equity securities issued in the Company's next equity financing in an aggregate amount of at least $5,000,000 (including conversion of the Notes) (the "Next Equity Financing").

If the Next Equity Financing does not occur on or before December 31, 2009, principal and accrued interest shall be payable upon demand of the Holder.

- OR -

If the Next Equity Financing does not occur on or before December 31, 2009, principal and accrued interest shall be due and payable on such date.

- OR -

If the Next Equity Financing does not occur on or before December 31, 2009, principal and accrued interest shall be payable in four equal quarterly installments.

Conversion discount [only used if the Company will provide the Note Holder the ability to convert at a discount]:

Each Note will convert at a 10% discount to the price in the Next Equity Financing.

- OR -

Each Note will convert at a 10% discount (the "Conversion Discount") to the price per equity security paid by investors in the Next Equity Financing, as adjusted as follows: For each full month that the Notes are outstanding, the Conversion Discount shall be increased by 2.5% up to a maximum of 35%, as set forth in the table below.

Month 0 = 10.0%
Month 1 = 12.5%
Month 2 = 15.0%
Month 3 = 17.5%
Month 4 = 20.0%
Month 5 = 22.5%

Month 6 = 25.0%
Month 7 = 27.5%
Month 8 = 30.0%
Month 9 = 32.0%
Month 10 = 35.0%
Month 11+ = 35.0%

Subordination: The notes will be subordinate in right of payment to all current and future indebtedness to banks and other financial institutions.

- OR -

The notes will be subordinate in right of payment to certain indebtedness to banks and other financial institutions.

- OR -

None of the above (no subordination).

Security interest:

The Notes will be unsecured.

- OR -

The Notes will be secured by all of the Company's assets.

- OR -

The Notes will be secured by certain assets of the Company, including computer equipment, network equipment, and electronic storage equipment.

Investors:

	Principal Amount of Note
Jane Angelita	$100,000
John Familia	$75,000
Joe Entrepreneur	$25,000
ABC Ventures LP	$50,000
	$250,000

Closing Date: The closing of the sale of the Notes will occur on or before March 31, 2009.

10

Venture Capital

It's always a good time to be in venture capital.

Thomas J. Perkins
Cofounder of Kleiner Perkins Caufield & Byers

What to Watch For

Most of today's business people are familiar with the term venture capital, but the term itself has only been used publicly since the late 1930s. The first reported public use was by Jean Witter, the president-elect of the Investment Bankers Association of America, in his speech at the association's annual convention in 1939. The concept was formally adopted in 1946 but didn't become a major source of investment into entrepreneurial companies for some time, only reaching investment totals of less than $100 million annually by 1965.

From that point on, the venture capital community exploded, reaching highs of $48 billion invested in U.S. companies in 1999 and, according to some, fueling the Internet bubble of the late 1990s (investment totals fell after the highs of 1999 but have since rebounded up to $29.4 billion invested in 2007 according to the National Venture Capital Association). In 2007, venture capital firms invested more than $40 billion worldwide into startups, an increase over the $37.3 billion in 2006, according to market researcher Dow Jones VentureOne. Although these worldwide investments are down from the $56 billion invested in 2001, it still shows solid gains since 2003 and continued momentum going forward.

Numerous success stories from the venture investment community are referenced regularly. Just a few include Amazon.com, America Online, Amgen, Apple Computer, Cisco Systems, Compaq, DEC, Federal Express, Genentech, Google, Home Depot, Intel, Lotus, Netscape, Oracle, Outback Steakhouse, Seagate, Staples, Starbucks, Sun Microsystems, 3Com, Whole Foods, and Yahoo!. Some analysts note that, without the risk taken by the venture capital investing community for many of these success stories of recent business history, we might have seen a slower introduction and adoption of many technologies that today pervade our everyday living.

THE VENTURE INVESTMENT COMMUNITY

According to the National Venture Capital Association, U.S. companies that received venture capital between 1970 and 2005 accounted for approximately 10 million jobs and $2.1 trillion in revenue in 2005. Although venture capital represents a mere 0.2% of gross domestic product of the United States, venture-funded companies account for nearly 17% of gross domestic product.

As a result of the importance of venture investments to many of successful technology companies, many of today's new technology entrepreneurs also look to venture capital as a source of dollars to build their business, a source of prestige in the entrepreneurial community, and a source of insight from talented and well-connected venture fund partners. Although venture capital remains an important source of funds for technology startups, the odds of obtaining venture capital continue to be relatively small.

Many venture capital funds receive hundreds of business plans each month (for some funds, they may receive well over 1,000 in a single month), meet with only a handful of company founders in any given month, and perhaps choose to invest in a total of half a dozen companies each year. Although the odds of success may be low, a better understanding of the industry can help a company focus its energies and target the right firms. Finding ways for companies to increase the odds of obtaining venture financing is important in this game.

Is Venture Capital Right for Your Company?

Raising capital for a technology business represents one of the key milestones of the business. In particular, venture capital investments tend to be made into companies with revolutionary ideas filled with potential and high risk. For many businesses, this new source of funds allows the business to move toward a new level and begin to execute on the initiatives set forth in the business plan. Receiving funding serves as a confirmation of your business concept, the management team, and the potential product markets.

Venture capital is not a good fit for every company (in fact, probably not for most companies). Venture capital is expensive, has a fairly limited time horizon, and tends to operate in a selected group of industries, technologies, or fields. Before investing substantial time, resources, and energy into attracting venture capital, it is extremely important that an entrepreneur understand the market and the types of companies that make sense for venture capital. Countless excellent companies have been built without venture capital, because either other funding alternatives were available or the company was not a good fit for venture funding.

However, certain high-technology business ideas will require both substantial upfront time and money investments to perform research, development, and testing efforts. In many of these cases, venture capital offers the right mixture of capital and expertise to help an idea grow rapidly to a significant size. Moving a technology from a laboratory to a prototype may involve well over a year of time and millions of dollars. Finding a traditional bank loan program or living on savings alone is simply unrealistic for many startup companies. Therefore, obtaining venture capital financing is a necessity for certain businesses in the high-technology arena.

Benefits of Venture Capital

This fact is clear: venture capital is a key driver of innovation and entrepreneurship. Many of the most successful companies today simply wouldn't exist without this funding. Therefore, venture capital is instrumental for accelerating high-technology businesses, a model that other countries across the globe have begun to leverage to further technology development in their local economies. Below are a few of the key benefits that venture capital firms provide to their portfolio companies.

Money

Cash is king, and providing cash to startup businesses that, if successful, offer handsome rewards to the investors helps to align the incentives for the parties to these transactions. The reason for the crucial role of venture capital in high-technology businesses is that few other sources of capital have risk tolerances that allow investment into risky, early-stage companies. The venture model works because the high rewards realized from a relatively few investment "wins" offset the investments that are losers. High-technology businesses require cash to develop their technologies, scale their businesses, and reach new customer streams. Although there may be less expensive sources of financing, venture capital money remains an integral source of that cash necessary for many of today's technology startups.

Advisory

The partners of venture capital firms tend to be individuals that have a mix of entrepreneurial, business, and investing experiences. This skill set often uniquely positions a VC as an excellent resource to provide operational, strategic, or practical guidance for a new startup company. In addition, because a VC will often focus on an industry or technology, the VC may be uniquely situated to share knowledge, insight, and information across multiple portfolio companies facing shared challenges and obstacles.

Contact Networks

VCs provide a powerful resource in their personal and firm networks. From these contacts, VCs are able to tap into potential employees and management team members, find companies to be partners or customers, or can leverage market intelligence among their portfolio companies. Although a VC may be an early source of funds for the company, as the company looks to raise additional funding, many of the sources of fresh capital will result from introductions from the VC's network.

What Are My Chances of Getting Venture Money?

The easy answer is, it depends. It depends on the team you've assembled, the technology itself, the market for your technology, the right timing, and a little bit of luck. Although that may be the case, if you believe Sam Altman, the founder of Loopt, "If you've got a good idea, market, and team, raising money won't be your problem."

You can get a realistic sense of the odds of receiving a VC investment. Anecdotal evidence suggests that between 2% and 3% of businesses seeking venture capital financing will ever receive funding. As mentioned previously, the majority of these cases are because venture funding is not the right type of funding for the business. Sean Wise looked further and interviewed a number of VC to find out where their deal flow came from and the sources of deal flow that had the highest probability of closing. Although these are not scientific numbers, these figures should offer some insights into the way VCs see the sources of their deals.

What are the odds of getting a meeting based on:

- An unsolicited business plan submission? Approximately 1 in 100.
- A solicitation from an unknown agent? Approximately 1 in 50.
- Direct contact from venture fairs, financing forums, and other industry events? Approximately 1 in 15.
- Referral from professionals with fund relationships (accountants, lawyers, and market dealers)? Approximately 1 in 3.
- Referral from current and future investors? Approximately 1 in 2.
- Referral from executives of a portfolio company in the VC's fund or other stakeholder in the VC's funds? Nearly 1 in 1.

Even getting an initial face-to-face meeting with a VC only represents the first in a long series of events. An initial meeting will likely be followed by a second meeting, perhaps with other members of the venture firm. Due diligence, company visits, and meetings with other company management team members will be the next phase, with a formal presentation to the firm partners after what could be several months of discussions, meetings, and presentations. Unfortunately for the companies, at each of these points in the transaction, the venture capital firm may find a reason not to continue the investment process with the company. The information from Sean Wise also suggests that getting a meeting with the venture capital firm is only the first step toward closing the deal.

What are the odds of a deal getting done based on a solicitation from:

- An unsolicited business plan submission? Approximately 1 in 1000.
- A solicitation from an unknown agent? Approximately 1 in 100.

- Direct contact from venture fairs, financing forums, and other industry events? Approximately 1 in 20 (much higher if the company wins the competition).
- Referral from professionals with fund relationships (accountants, lawyers, and market dealers)? Approximately 1 in 5.
- Referral from current and future investors? Approximately 1 in 3.
- Referral from executive of a portfolio company in the VC's fund or other stakeholder in the VC's funds? Approximately 1 in 3.

LONG ODDS?

According to a study done by Profit Dynamics Inc. on more than 250 venture capital firms from 1998 through 2000, venture capital firms invest in less than 7 of 1,000 companies that contact them (0.7%).

Josh Lerner, a professor from the Harvard Business School, concurs with these findings, suggesting that historically less than 1% of business plans submitted to a venture capital firm will receive funding.

Because many business plans will be submitted to multiple venture capital firms, venture industry statistics suggest that somewhere between 2% and 3% of companies seeking VC funding will actually receive funding.

According to the MoneyTree Survey prepared by PricewaterhouseCoopers and National Venture Capital Association, each quarter, VCs initiate approximately 300 first-round financing deals for companies receiving venture capital, equating to approximately 1,200 "first financing sequence" deals annually. The average investment for each of these deals is nearly $5.5 million.

Investments by Financing Sequence (Amount in Millions)

	Quarter 3 of 2007		Quarter 2 of 2007		Quarter 3 of 2006	
	Amount	**Deals**	**Amount**	**Deals**	**Amount**	**Deals**
First	$1,729	273	$1,731	347	$1,605	309
Second and third	$2,060	285	$2,252	306	$2,434	303
Fourth, fifth, and sixth	$2,458	235	$2,359	245	$2,136	202
Seventh and beyond	$856	94	$865	102	$620	83

Source: PricewaterhouseCoopers/National Venture Capital Association MoneyTree Survey (http://www.pwcmoneytree.com).

WHO AM I COMPETING AGAINST FOR FUNDING?

Because few venture capital firms publish the number of business plans they receive annually and most companies seeking funding are likely to submit plans to multiple VCs, it is difficult to know the exact number of companies competing for first-sequence venture funding.

However, according to venture industry statistics used within the industry, only 2% to 3% of companies seeking venture capital are successful in receiving funding. What this tells us is that you can safely assume that your company may be competing with 40,000–60,000 other businesses also seeking these coveted 1,200–1,400 first financing deals.

Although these numbers may appear daunting, remember that, by focusing on a targeted approach that matches your business with VCs investing in your market, your region, and your business stage, you can increase your odds substantially.

Are You a Good Candidate for VC Funding?

Certain companies are considered to be "natural" fits for VCs; oftentimes, they involve a former CEO or technical lead from a successful startup company. These natural fit companies will often have a proven team that has been through the startup process and a high degree of familiarity in a given industry segment. To illustrate why these companies are such natural fits, consider the following example. If Marc Andreessen, the founder of Netscape (which sold for $4.2 billion) and Opsware (which sold for $1.6 billion), walked into a meeting with nearly any VC asking for money, the VC will probably pull out a blank check.

Why is that? Well, to put it simply, Andreessen has proven he is a good investment. Andreessen is a known quantity, having built successful companies that earned his investors boatloads of money. Of course, this would be a logical investment for any venture capital firm. Although Andreessen's next deal could tank, VCs can argue that they've found a guy to lead the company who helps to mitigate their risk. Andreessen has not only done it once, but twice. In venture finance, that's as close to a safe bet as they come.

Obviously, there are only so many guys like Andreessen. If you aren't the founder of a company that sold for a billion dollars, how can you tell whether you are a natural fit for venture funding? The same basic rules apply. VCs are looking to mitigate their risk. Venture capital firms are looking for a management team they can trust to execute on a business plan they believe can provide high returns. Does your business have a team that can execute on a business idea with the potential for a big reward? What does your team have that mitigates the risk for VCs?

The box below describes a few of the key factors that VCs point to as instrumental in their decision-making process. Management has been and continues to be a key factor for investors. Proven startup experience (and hopefully startup success) is valued by the venture community, but also note the weight the investors polled belowplaced on factors such as market size and growth potential.

WHAT DO VCs LOOK FOR IN A COMPANY?

Profit Dynamics Inc. surveyed venture capital firms and asked them to rank five important factors that influence their decision to invest, with 5 being the highest rank you could award a factor, 1 being the lowest. These factors were as follows:

- Quality of the management team
- Size of the company's market
- Proprietary, uniqueness, or brand strength of the company's product
- ROI
- Company's potential for growth

The percentage of respondents that awarded 5 to a factor were as follows:

- Management — 52%
- ROI — 42%
- Market size — 27%
- Growth — 25%
- Uniqueness — 19%

In overall average score, the factors were ranked as follows:

- Management — 4.1
- ROI — 3.5
- Market size — 3.3
- Uniqueness — 3.3
- Growth — 3.0

Source: Profit Dynamics Inc.

Although it may seem that all venture-backed companies have certain common traits, many companies who receive funding may seem atypical at first blush. Here are just a few myths:

- **All companies are in the "hot" industries such as software or biotechnology.** As recently as 2005, energy technology companies were considered by many in the industry to be a poor fit for venture capital, and only a select few VCs were making investments into the sector. However, by 2007, things had quickly changed and energy technology represented the fastest growing sector in venture capital investing (suddenly becoming a new "hot" industry).
- **All VC investments involve high-technology companies.** Not so. As an example, the House of Blues restaurant chain is a venture-backed company. It's the food, not the technology, that House of Blues is known for. Here are a few other famous venture-backed companies that you might never have thought of: FedEx, Starbucks, Home Depot, Staples, Outback Steakhouse, and Whole Foods, all terrific companies and all in industries that wouldn't be described as high tech.
- **Only companies in the "right" cities receive funding.** Numerous companies located outside of Silicon Valley or Boston (the top VC markets in the country) will receive funding. In fact, more than 50% of venture investments annually go to companies located outside of these two markets.
- **VCs will only fund companies with large intellectual property portfolios.** Barriers to entry come in many shapes and sizes (not always intellectual property portfolios). Perhaps your barrier is an exclusive contract, unique customer loyalty,

or a secured customer base; each represents a barrier to entry by companies following your lead.

- **VCs won't invest in companies attacking a market with a large, entrenched competition.** Not too long ago, a small company, at the time, called Google only decided to enter into the online search market dominated by Yahoo! and Microsoft after the founders were unable to license their technology. Seems like they did okay. Although a market without a dominant player may appear to be a more natural or easier market to enter, don't discount the power of a good technology and a strong management team to enter the fray.

The Venture Capital "Fit" Test

Many business plans received by venture capital firms will be immediately rejected because the company is not a good "fit" for that venture fund or for venture capital financing in general. This could mean the company is serving a market that is too small or is a niche market. It could be that the technology has few barriers to entry, or it could be that a company will have fairly low gross margins.

The above represent a few of the important criteria in the venture capital marketplace. Use the tool below to understand whether your company is ready to focus its fundraising efforts on venture capital. Take the test below to see whether your company would be a good fit for venture capital investment. Keep in mind that this test is simply to gauge a company's fit with a traditional VC model. Many companies may have a different model that fits the particular investment strategy or focus of certain venture firms. Therefore, this information should only be used as a guide to understand the initial perception you may receive from a traditional venture firm.

NOT A GOOD FIT FOR VENTURE CAPITAL

Many very successful businesses or businesses with the potential for success do not fit the venture capital funding model. For example, a business plan to provide technology consulting services may not be a typical business model funded by VCs. Here are some examples of markets, products, and technologies that tend not to be fits for venture capital funding:

- A market such as retail, mining, or banking may not be a market the venture firm will invest in or has contacts in and knowledge of.
- A product or service that is highly technical, extremely complicated, and difficult to explain to a lay person, or that the VC is unable to fully grasp may not be a technology that the VC is willing to try to convince her partners to invest in.
- A technology that will only serve a small niche and will involve an extensive process of education, selling, and distribution may not have the risk-reward profile for a venture firm.
- A business designed to sell a low-cost product at a low margin, taking market share from the industry leader may not be a model that a venture firm will find fits its strategy.

In each of the cases above, it is not that the business is not a good idea or will not be a successful business. Instead, it is simply a failure to match the business profile of the company with the investment profile of the venture firm. Perhaps some of these business concepts will eventually be funded by venture firms after they are established and better fit a firm investing in later-stage companies.

Companies may find that they are not a company that is a good fit for venture funding. In some cases, this will mean that the company's model and goals do not align with the venture capital funding model. In other cases, a company will recognize that they need to attempt to grow the business, further develop the product, or increase the experience level of the management team. In both cases, the company should focus their efforts on growing the business through other funding sources until the business more closely matches the business model funded by venture capital firms.

THE VC "FIT" TEST

Select 1 to 5 for each of the categories below. Total your score below.

1. Size of your market: ______________

1 Under $500 million
2 $500 million to $1 billion
3 $1–3 billion
4 $3–5 billion
5 Over $5 billion

2. Revenues in five years: ______________

1 Under $10 million
2 $10–20 million
3 $20–35 million
4 $35–50 million
5 Over $50 million

3. How much investment do you require now? ______________

1 Under $500,000 or over $20 million
2 $500,000 to $1 million; $10–20 million
3 $1–2 million; $7.5–10 million
4 $2–3 million; $5–7.5 million
5 $3–5 million

4. Product gross margins: ______________

1 Under 40%
2 40%–50%
3 50%–60%
4 60%–70%
5 Over 70%

5. Industry: ________________

1 Other
2 Consumer products and services; retailing/distribution; healthcare services
3 Information technology services; networking and equipment; computers and peripherals
4 Industrial/energy; telecommunications; semiconductors; media and entertainment
5 Software; biotechnology; medical devices and equipment

6. Location of company: ________________

1 Other
2 Within a two hour drive of any of below
3 Austin; Chicago; Denver; Philadelphia; San Diego; Seattle; Washington DC
4 Boston; New York City; Southern California
5 Silicon Valley

7. Management team business experience and success: ________________

1 Very low
2 Low
3 Medium
4 High
5 Very high

8. Costs to bring the product to market: ________________

(for example, sales, marketing, distribution, consumer education, etc.)
1 Very high
2 High
3 Medium
4 Low
5 Very low

9. How extensive are the barriers to entry you have? ________________

(for example, intellectual property, proprietary information, lead to the game)

1 Very low
2 Low
3 Medium
4 High
5 Very high

10. Current commitment to business: ______________

(for example, full-time founders, office space, development products, angel investors)

1 Very low
2 Low
3 Medium
4 High
5 Very high

Bonus points (1 point for each): ______________

- Former CEO
- Received VC funding at previous company
- 1 point for every five people you know personally at any venture capital firm

TOTAL: ______________

If you scored:

45 or higher	**Excellent fit.** Start sending out business plans.
40–44	**Good fit.** Try to expand your network to increase your fit.
35–39	**Okay fit.** Look to improve areas that are a 3 or below and expand size of network.
Under 35	**May not be a fit.** Determine whether the business is not currently at a stage ready for VC funding or whether the business model does not fit with the typical VC model.

In addition, some venture capital firms will not consider any company that does not score at least a 4 on each of the first four questions. For instance, your product may need to participate in a market at least $3 billion in size, a plan to reach sales of $50 million in five years, and a product with gross margins above 60%.

Venture Capital Industry

Venture Capital Funds

There are less than 800 venture capital firms in the United States. These firms create funds to serve as pooled investment vehicles (often in the form of a limited partnership). The partners (who are often referred to as venture capitalists, or VCs) are responsible for overseeing the funds in the firm and will make decisions to invest these funds by a vote of the general partners of the firm.

To begin financing, a venture capital firm will create a fund (for example, there were 55 new venture capital funds raised during 2007). The fund is a pool of money that the VC will invest in a variety of companies and businesses, and a firm can have a number of funds under management. The venture capital firm will obtain funds from wealthy individuals, companies, pension funds, and other sources to invest in this fund. Once the firm has raised funds to a particular level, the VC will begin to invest the funds. The fund managers, who are called "general partners," will receive approximately 2% of the fund each year as a management fee, in addition to approximately 15% to 25% of the long-term gains on the value of the fund.

The VC firm will then invest the fund in a number of companies, perhaps between 10 and 20, depending on the size of the fund. These investments generally will be made over a period of time rather than all at once, with follow-on investments in these companies as is necessary. Each individual VC firm and each individual fund managed by the firm will generally have a particular investment profile. For instance, the focus could be something as broadly defined as a technology fund, or the VC firm may manage a specific fund that will invest primarily in clean energy projects or biotechnology companies. Another fund could decide to invest in wireless companies looking for a second round of financing or could focus on all companies looking to enter IPOs within the coming six months. Many VC firms will specialize in investments in particular fields given the expertise within their firm and will use this specialized knowledge to assist the companies in which they invest. Each profile will be tailored to the investors and used as a selling point for investment into the funds themselves.

Broadly speaking, the venture capital firm invests the majority of the fund within a one to two year time horizon, anticipating liquidating the entire investment within a three to seven year time horizon, depending on the fund profile. As a result, the VC firm expects that the companies in which it has invested will engage in some type of transaction allowing for a liquidation of their investments, such as an IPO or another type of acquisition or refinancing. As a result, the cash generated by an IPO or refinancing will allow the firm to use the proceeds to return to their investors.

VC firms and their investors recognize that individual investments in these companies are likely to be a mixed bag. If a fund invests in 10 companies, the fund may see seven of those companies ultimately become unsuccessful. However, if the fund is able to realize a substantial success with one company and moderate successes with two more, the fund may have overall success. The ultimate success of any individual VC firm depends on its ability to select successful investments and develop the companies along a projected maturation timeline. The investors in venture capital funds are oftentimes looking for returns in the range of 20% per year on their investments into the fund. The traditional rule of thumb is venture firms only invest in firms that have the potential to provide a 10 times return on their investment, understanding that the odds of this occurring are quite low.

Investment Trends

Venture capital investment remains focused in a number of key industries, primarily technology and life sciences investments. Historically, life sciences (which are made up of biotechnology and medical devices and equipment investments) represent approximately 25% of the investment dollars made each quarter. Investments in software have historically been the next largest sector. However, investments are made in numerous fields from telecommunications, to financial services, to consumer products. For updated investment statistics and data, visit the book's website at http://www.myhightechstartup.com.

Investments by Industry (Amount in Millions)

	2007		2006	
	Amount	Deals	Amount	Deals
Software	$1,108	187	$1,521	253
Biotechnology	$1,091	99	$1,213	124
Industrial/energy	$ 921	83	$ 543	65
Medical devices and equipment	$ 825	76	$1,030	109
Telecommunications	$ 585	74	$ 515	84
Semiconductors	$ 513	55	$ 457	58
Media and entertainment	$ 508	96	$ 464	77
Information technology services	$ 354	51	$ 244	43
Networking and equipment	$ 289	29	$ 371	35
Financial services	$ 280	31	$ 54	16
Business products and services	$ 195	23	$ 200	36
Electronics/instrumentation	$ 153	21	$ 156	30
Healthcare services	$ 96	15	$ 27	9
Computers and peripherals	$ 88	13	$ 149	15
Consumer products and services	$ 57	20	$ 168	28
Retailing/distribution	$ 40	13	$ 93	17
Other	$	1	$	1

Investments by Industry (Amount in Millions)

	Quarter 2 of 2008		Quarter 1 of 2008		Quarter 4 of 2007		Quarter 3 of 2007	
	Amount	Deals	Amount	Deals	Amount	Deals	Amount	Deals
Software	$1,252	219	$1,355	245	$1,271	220	$1,108	187
Biotechnology	$1,079	111	$1,212	130	$1,286	129	$1,091	99
Industrial/energy	$1,152	89	$ 914	88	$ 648	76	$ 921	83
Medical devices and equipment	$ 833	98	$1,024	99	$ 901	95	$ 825	76
Telecommunications	$ 378	69	$ 378	57	$ 378	63	$ 585	74
Semiconductors	$ 328	39	$ 588	52	$ 441	46	$ 513	55
Media and entertainment	$ 586	110	$ 469	98	$ 407	83	$ 508	96
Information technology services	$ 553	74	$ 542	66	$ 364	59	$ 354	51
Networking and equipment	$ 142	22	$ 179	24	$ 222	29	$ 289	29
Financial services	$ 141	22	$ 166	19	$ 115	18	$ 280	31
Business products and services	$ 187	34	$ 158	31	$ 196	34	$ 195	23
Electronics/instrumentation	$ 132	23	$ 67	19	$ 150	17	$ 153	21
Healthcare services	$ 106	16	$ 44	9	$ 69	12	$ 96	15
Computers and peripherals	$ 147	18	$ 112	17	$ 212	26	$ 88	13
Consumer products and services	$ 130	30	$ 139	25	$ 156	35	$ 57	20
Retailing/distribution	$ 109	14	$ 73	14	$ 182	13	$ 40	13
Other	$ 6	2	$ 8	4	$ 12	8	$	1

Source: PricewaterhouseCoopers/National Venture Capital Association MoneyTree Survey (http://www.pwcmoneytree.com).

The largest single region for investments remains Silicon Valley, which consistently makes up 30% of the total U.S. venture capital investments each quarter. A large number of investors are located in Silicon Valley, which attracts new companies and entrepreneurs from other startup ventures to the area to try their hand. The entire state of California boasts large entrepreneurial centers of activity in other areas, including Los Angeles and Orange County, and San Diego. Outside of California, areas such as Boston, New York City, Washington DC, Austin, Seattle, Philadelphia, Chicago, and others have healthy numbers of investments made each quarter into technology-based local companies.

Even if you are not a company located in the United States, you could still potentially attract venture capital investment for U.S.-based venture funds. For instance, according to a MoneyTree Survey in 2007, United States-based VCs invested $1.1 billion in 91 deals in India and $1.4 billion in 133 deals in China. These totals were all-time highs for United States-based VCs investing in each country.

Investments by Region (Amount in Millions)

	2007		2006	
	Amount	**Deals**	**Amount**	**Deals**
Silicon Valley	$10,067	1,184	$9,528	1,169
New England	$3,695	478	$3,117	431
San Diego	$1,989	159	$1,306	131
Southeast	$1,856	243	$1,268	230
New York Metro	$1,715	254	$1,999	269
Los Angeles/Orange County	$1,648	205	$1,928	218
Northwest	$1,636	216	$1,194	174
Texas	$1,416	166	$1,450	186
Midwest	$1,298	236	$ 977	189
Washington DC/Metroplex	$1,242	207	$1,145	209
Philadelphia Metro	$ 861	125	$ 765	99
Colorado	$ 564	86	$ 661	100
Southwest	$ 541	90	$ 502	84
North Central	$ 526	80	$ 401	64
Upstate New York	$ 137	32	$ 156	37
Sacramento/Northern California	$ 100	16	$ 28	6
South Central	$ 92	27	$ 79	20
Alaska/Hawaii/Puerto Rico	$ 20	9	$ 47	14

Investments by Region (Amount in Millions)

	Quarter 2 of 2008		Quarter 1 of 2008		Quarter 4 of 2007		Quarter 3 of 2007	
	Amount	**Deals**	**Amount**	**Deals**	**Amount**	**Deals**	**Amount**	**Deals**
Silicon Valley	$2,956	306	$2,637	320	$2,399	301	$2,484	287
New England	$ 823	119	$ 780	113	$ 861	111	$ 998	119
Los Angeles/Orange County	$ 584	64	$ 532	61	$ 254	46	$ 425	56
New York Metro	$ 393	80	$ 569	69	$ 471	70	$ 385	57
Texas	$ 257	36	$ 397	39	$ 427	45	$ 386	39
San Diego	$ 366	38	$ 457	41	$ 424	32	$ 359	37
Southeast	$ 332	55	$ 406	52	$ 479	55	$ 353	52

(continued)

Investments by Region (Amount in Millions) *(continued)*

	Quarter 2 of 2008		Quarter 1 of 2008		Quarter 4 of 2007		Quarter 3 of 2007	
	Amount	Deals	Amount	Deals	Amount	Deals	Amount	Deals
Washington DC/Metroplex	$235	51	$257	47	$404	71	$331	34
Northwest	$333	58	$390	53	$438	54	$291	47
Midwest	$367	58	$203	60	$244	62	$343	45
Philadelphia Metro	$189	34	$162	30	$108	21	$238	26
Colorado	$184	26	$311	27	$102	20	$197	24
Southwest	$ 86	21	$184	24	$ 50	17	$134	23
North Central	$162	21	$119	19	$159	28	$111	17
Upstate NY	$ 21	8	$ 34	10	$ 75	12	$ 32	12
Sacramento/Northern California	$ 2	3	$ 8	5	$ 49	7	$ 15	2
South Central	$ 80	9	$ 52	6	$ 59	9	$ 19	7
Alaska/Hawaii/Puerto Rico	$ 18	3	$ 2	1	$ 9	2	$ 4	3

Source: PricewaterhouseCoopers/National Venture Capital Association MoneyTree Survey (http://www.pwcmoneytree.com).

The charts below offer information on the stages of companies receiving investments and the financing sequences.

Investments by Stage of Development (Amount in Millions)

	2007		2006	
	Amount	Deals	Amount	Deals
Early stage	$ 5,192	995	$ 4,102	923
Expansion	$10,845	1,235	$11,495	1,359
Later stage	$12,215	1,168	$ 9,797	1,006
Startup/seed	$ 1,153	415	$ 1,157	342

Investments by Stage of Development (Amount in Millions)

	Quarter 2 of 2008		Quarter 1 of 2008		Quarter 4 of 2007		Quarter 3 of 2007	
	Amount	Deals	Amount	Deals	Amount	Deals	Amount	Deals
Early stage	$1,294	257	$1,320	244	$1,363	251	$1,185	213
Expansion	$2,617	321	$3,064	331	$2,583	320	$2,718	294
Later stage	$3,123	318	$2,730	290	$2,637	281	$2,976	288
Startup/seed	$ 353	94	$ 386	112	$ 428	111	$ 224	92

Source: PricewaterhouseCoopers/National Venture Capital Association MoneyTree Survey (http://www.pwcmoneytree.com).

Investments by Financing Sequence (Amount in Millions)

	2007		2006	
	Amount	Deals	Amount	Deals
First	$7,230	1,267	$6,044	1,173
Second and third	$9,441	1,233	$9,311	1,232
Fourth, fifth, and sixth	$9,046	912	$8,694	869
Seventh and beyond	$3,689	401	$2,502	356

Investments by Financing Sequence (Amount in Millions)

	Quarter 2 of 2008		Quarter 1 of 2008		Quarter 4 of 2007		Quarter 3 of 2007	
	Amount	Deals	Amount	Deals	Amount	Deals	Amount	Deals
First	$1,583	301	$1,805	316	$1,917	343	$1,729	273
Second and third	$2,499	361	$2,726	347	$2,220	306	$2,060	285
Fourth, fifth, and sixth	$2,184	218	$2,042	217	$1,950	207	$2,458	235
Seventh and beyond	$1,122	110	$928	97	$924	107	$856	94

Source: PricewaterhouseCoopers/National Venture Capital Association MoneyTree Survey (http://www.pwcmoneytree.com).

Attracting Venture Capital

Once you've decided your company is a good fit for venture capital and the business is ready for a significant cash infusion, you and your business will begin an involved process of raising capital. This process of raising capital from institutional investors will likely prove to be time consuming, frustrating, intense, and, hopefully, successful. With nearly 800 venture capital firms, there are numerous opportunities to find funding. However, targeting the best choices for the business will provide the strongest opportunities.

ACCORDING TO VENTURE CAPITALISTS

What is the most common reason why you have declined to invest in a company?

- Lack of an experienced, complete management team — 40%
- Company does not fit VCs investment criteria, the industry(ies) on which the VC focuses, or the geographic area the VC invests within — 17%
- Size of the market the company was in, or the need in the market that the product serves, was too small — 13%
- Company has no competitive advantage or has a noncompelling technology — 13%
- Unclear business execution strategy — 10%
- Company is too small and will not grow large enough — 10%
- Company is too early stage — 8%
- Target industry has no barriers to entry for competitors — 6%
- Risk is too high relative to the projected return — 6%
- Company has low margins or the industry is facing margin pressure — 4%
- Entrepreneur put too high a valuation on his company's stock — 4%
- Other common reasons included the following:
    - Company is not profitable
    - Company faces huge, entrenched competitors
    - Entrepreneur was not referred to the VC firm

- Entrepreneur did a poor job presenting the company in a meeting
- Entrepreneur viewed the VC only as a source of money, not as a value added partner

Source: Profit Dynamics Inc.

The Fundraising Process

One of the most difficult decisions a business must make is deciding when to begin the fundraising process. Some may begin out of necessity (we've only got three months of cash on hand), others may begin when growth or progress has reached plateau and capital is needed for additional development (we've got to hire to grow), and still others may realize that their business has met the key criteria that investors look at (we've heard from our advisors that now is the time).

Once your business has decided to begin the fundraising process, what are the next steps to proceed? Here's the overview of the process going forward (the order may change depending on the business and the team):

1. Establish a fundraising team.
2. Evaluate the business as a candidate for VC funding.
3. Determine the funding needs of the business.
4. Analyze the marketplace for investor targets.
5. Develop a fundraising plan of action.
6. Create the necessary fundraising tools.
7. Solicit interest and meetings from targeted investors.
8. Investor presentations.
9. Receive feedback from investors.
10. Business due diligence.
11. Retool your "pitch."
12. And finally, the "Yes and here's a term sheet."

The process of readying the business to raise funds from venture capital firms (presolicitation) may take as little as a month or could take much longer without a heavy focus on the fundraising efforts. Ultimately, the effort put forth before sending out a single business plan is crucial to the success of the efforts.

ACCORDING TO VENTURE CAPITALISTS

How long will it typically take to close an initial investment in a company? (days measured from the receipt of the business plan to closing of the investment)

- Under 30 days — 1% of respondents
- 30–60 days — 18%

- 60–90 days — 45%
- 90–120 days — 26%
- More than 120 days — 10%

On average, according to VCs, the first-round venture financing transaction takes 80 days to complete from the receipt of the business plan to closing of the investment.

Source: Profit Dynamics Inc.

Once the business has prepared the fundraising tools and begun solicitation, the timeline itself will vary greatly depending on the level of interest generated by the potential investors. Below is a sample timetable of the key steps in the solicitation process, but the dates should be viewed as a "better" case scenario rather than the typical timetable. According to venture capital firms, once the venture firm receives a business plan that looks promising, the average time from the initial review of the business plan to the closing of the transaction is 80 days.

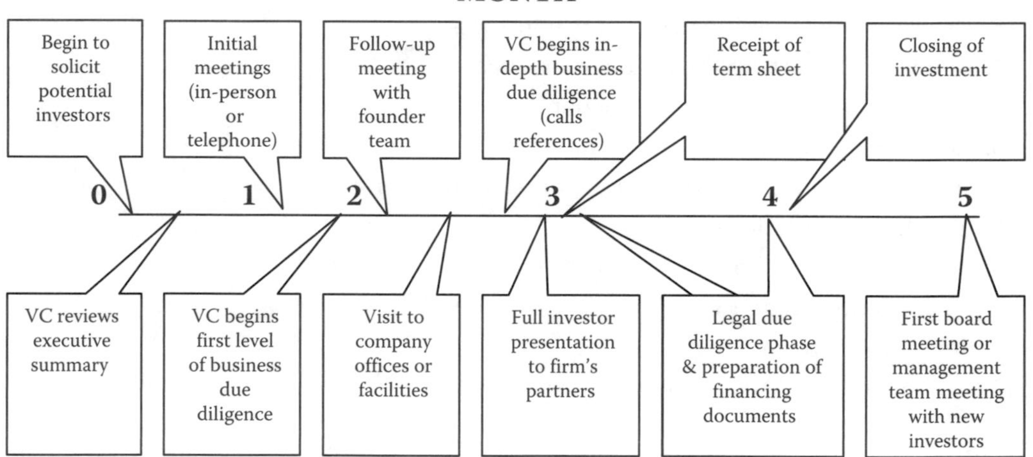

Fundraising Team

For many companies looking to begin fundraising, the CEO is the logical choice to lead the fundraising efforts. However, other companies may establish another individual within the company to lead the fundraising process such as the CFO or a founder of the company. In any case, the company will generally need to select one individual to be the lead for the process and set the tone and goals for the efforts. This lead individual will hold initial meetings, make personal contacts, and handle much of the "selling" process for the entity. Expect that, during the solicitation and negotiation phase of the process, the fundraising lead will devote a minimum of 20% of her time to these efforts and potentially much more if the process requires.

Generally, in establishing the fundraising team, the CEO will serve as the leader of the fundraising efforts and will be supported by others within the organization. The company will usually begin with the entire management team involved in the business evaluation as well as preparation of information and materials to provide to potential investors. However, as the process moves from preparation to solicitation, many companies will assign one individual to be responsible for managing the fundraising process (outside of the CEO) to coordinate contacts, timelines, provide materials, follow-up, and scheduling. As the process moves forward from the solicitation phase into evaluating term sheets and legal due diligence, companies may then decide to involve their outside counsel to a greater extent.

In later fundraising rounds for companies (after a seed/angel round or a Series A round), current venture capital investors of the company may play a role in the process, making additional introductions and solicitations. However, in most cases, the follow-on investment process will continue to be initiated and managed by the company itself.

Fundraising Plan of Action

One of the first steps in the fundraising process is to set a plan of action. Usually, this will involve setting a basic timeline, high-level discussions of the financial needs of the company, funding goals, and the acceptable level of dilution from a financing, and an early discussion of the networks of individuals familiar with the venture community to generate early contacts and leads in the process. The plan should involve a discussion of preparation of the key tools needed for solicitations.

The management of the company should also discuss the prospect that fundraising may not be successful. What amount of time should the company devote to fundraising before pursuing other alternatives or revising the business model?

Evaluation of the Status of the Business

This chapter provides a venture capital fit test. This should be a first step in the process of determining whether or not your business is a good candidate for funding. This evaluation should look into the management team, the product, or service you are offering, the potential in the marketplace, your business strategy, and the needs you will have going forward.

What will VCs look for? Ultimately, research points to two key factors that investors emphasize in their analyses: venture team quality (previous experience and success) and venture potential (favorable competitive conditions and market potential).

Research by Pankaj Patel and Rodney D'Souza of the University of Louisville found that each of these two criteria were key in the decision to accept or reject a company for funding by venture capital firms. According to Patel and D'Souza, team composition and team quality are major factors in explaining why a business plan gets rejected. However, the team is not significant in explaining why a business plan gets funded. (For instance, VCs will usually reject a business plan that has a weak management and so-so venture potential, but VCs won't fund a company with excellent management and so-so venture potential.)

However, this research did find that a VC may be willing to accept a marginal team if the venture potential is high. The logic could be that a VC believes they can leverage their networks and help fill a weak management team if the venture potential is high enough. (For instance, a VC may be willing to fund a company with a weak or so-so management but excellent venture potential. So a good idea with high potential is crucial.)

ACCORDING TO VENTURE CAPITALISTS

What are the essential characteristics of a quality management team?

1. Successful experience or proven track record (9.5 times*)
2. Integrity, honesty (3.0 times)
3. Dedication, commitment, passion, energy (3.0 times)
4. Vision and ability to articulate vision (2.3 times)
5. Knowledge, skill level, intelligence (2.0 times)
6. Leadership ability (1.0 times)
7. Ability to build a team (0.75 times)
8. Marketing focus (0.50 times)
9. Made investment in company (0.50 times)
10. Winning attitude (0.50 times)
11. Industry contacts (0.50 times)
12. Good references (0.25 times)

* Numbers in parentheses represent a scale of how frequently each characteristic was cited relative to each other.

83% of VCs surveyed included successful experience or proven track record as essential to a management team.

Source: Profit Dynamics Inc.

The main lesson here is that building a team with numerous smart, successful, skilled, and experienced people is very important to the fundraising process (it is the single largest factor cited as a reason a business plan is rejected), but a good team alone will usually not be enough. Potential investors want to see a business plan with a strategy to take advantage of favorable conditions in a large and growing marketplace. Building a strategy around both of these pillars is crucial; relying on a single pillar is a risk (although may be less risky if you can offer a compelling story on the venture potential and offer evidence that you can build out a team after receipt of funding).

How can you determine whether you have the appropriate mix of management team quality and venture potential? After a thorough self-assessment, you should begin to look for input from outside the business and the management team. Solicit feedback, insights, and thoughts from various internal and external parties. Many companies will solicit feedback from their board of directors and board of advisors, from their attorney, from experienced friends and family, or from informal connections they may have with current or former VCs.

ACCORDING TO VENTURE CAPITALISTS

What can entrepreneurs do if they don't have experience as a CEO or if their management team is not complete?

- The founder should step aside and bring in an experienced CEO — 40%

- The founder should ask the VC for management assistance or let the VC recruit a CEO — 35%
- Assemble a board of directors or an advisory board composed of CEO-type individuals — 15%
- Have such a winning business model or great idea that it outweighs any management team deficiencies — 7%
- Obtain CEO experience before seeking venture capital — 3%

Source: Profit Dynamics Inc.

Funding Needs

One of the most important questions you will hear from potential investors is, "How much money are you trying to raise and what will you use it for?" The fundraising process is predicated on the fact that your business needs capital to grow. Determining how much capital you will need, when you will need it, and where that capital will be spent needs to be addressed before soliciting investors. Generally, your business plan should set out the needs of the business for each year of a three- to five-year period, with an emphasis on the next 15 months of the business. In addition, the business plan should identify where the funds will be used, with a fair degree of specificity.

Venture capital firms do tend to have minimum investment levels to ensure the partners of the firm can service the company on the board, through strategic guidance, and through networking. As such, it is useful to see the average investment levels for venture capital deals. The average first-round investment by VCs was $6.33 million in the third quarter of 2007, $4.99 million in the second quarter of 2007, and $5.19 million in the third quarter of 2006 (an average of $5.48 million over those three quarters), according to the MoneyTree survey. The average early-stage company investment by VCs was $5.56 million in the third quarter of 2007, $5.04 million in the second quarter of 2007, and $4.69 million in the third quarter of 2006 (an average of $5.48 million over those three quarters), also according to the MoneyTree survey. The general rule of thumb is that venture firms prefer their investments to be within the range of $2 million to $6 million for an initial investment.

To better understand the issues, let's take an example at play. Assume that your business plan projections hold that the business will need $15 million over the next five years. In the first year, however, you are only projecting that the business will require $1.5 million of cash. Should you set out to raise $1.5 million, $15 million, or something in between?

There are numerous perspectives on the debate. The two extreme positions of this debate are as follows:

- Raise as much as you can (1) to avoid having to raise funds again when management least needs to be distracted from building the business and (2) allow the business to invest quickly and grow rapidly to tackle the small market windows.
- Only raise what you need because (1) few VCs are willing to go "all-in" in a startup still developing its product, (2) unused money sitting in a startup's bank makes VCs unhappy, (3) it increases dilution of the founders, and (4) raising the full complement of funds will set the company on a specific path and may not allow management to modify the business model from a "homerun" company to a "solid-double" company (to use baseball terminology).

The experts who say to raise as much as possible may encourage an entrepreneur to raise the funds needed for the first three years or perhaps even the first five years. Although this approach may increase the dilution you face (because you will be raising the funds at a lower valuation), these experts may argue that investing heavily in a solid business idea and team can create a much larger business than a more methodical approach (effectively increasing the total payout for all, even if the percentage decreases). If your product or service represents a major opportunity, it will require a rapid implementation and development strategy, this full go-to-market strategy may fit within your funding request.

Conversely, raising $10 million for a first-time entrepreneur or for a new idea represents a much more significant investment for a VC than $2 million. Most VCs will happily shell out the additional $8 million after year one is a success but may have doubts about handing over a blank check to most entrepreneurs. Therefore, some suggest that an entrepreneur pursue a more measured approach to fundraising, raising only what you will need for a 15 month aggressive development period. Although this type of an approach will require the company to raise additional rounds, it is generally easier to solicit larger investments (likely to be at higher prices per share) as the company continues to make positive progress. In the hypothetical above, raising only $1.5 million is likely to be lower than many VCs will prefer for an initial investment round, and therefore the requested amount may need to be higher to attract the interest of certain VCs.

The timetable to receive funding from a VC averages approximately three months but can be faster depending on the investment cycle of the venture firm. However, you should adequately prepare the business to have sufficient capital to continue to operate in the event the investment process takes a longer period to complete.

ACCORDING TO VENTURE CAPITALISTS

What is the fastest a company has ever received a first-round investment, from the time they received the business plan until the deal closed?

- Under 30 days — 41%
- 30–60 days — 39%
- Over 60 days — 20%

The average "fastest time period" for respondents was approximately 40 days.

Source: Profit Dynamics Inc.

Determining Investor Targets

Before beginning your fundraising process, it is important to identify likely investor targets. In most cases, through a simple review of the portfolio companies of any venture firm (oftentimes listed on the firm's website), you will notice an identifiable investment strategy. Focus your efforts on firms that have a strategy that most closely match your company. Look at comparable companies and find their investors to identify who you should also target. If you have contacts at the portfolio company, ask for a referral because those types of referrals are usually quite successful at generating meetings and interest. For a list of venture capital funds, review the tables at the end of this chapter or visit the book's website for updated information at http://www.myhightechstartup.com.

RESEARCHING VENTURE CAPITAL FIRMS AND ANGEL INVESTORS

- **National Venture Capital Association (http://www.nvca.org):** The trade association of the venture capital industry provides a list of the member-venture capital firms and links to their websites.
- **vFinance Inc. (http://www.vfinance.com):** Offers paid searching tools to find VCs based on the size of deal, industry that the VC will invest in, and location of investor. Customized searches for $3.00 per contact. Offers paid searching tools to find angel investors based on net worth on the individual, industry that the angel will invest in, and location of the investor. Customized searches for $1.00 to $2.50 per contact.
- **Capital Hunter (http://www.capitalhunter.com):** Provides a database of venture capital and active private equity investors and key executives, and downloads of funding event details. Subscription of $59.00 per quarter.
- **MoneyTree Report (http://www.pwcmoneytree.com):** Provides a database of venture capital transactions that can be customized by location, investment stage, industry, and keywords. Public access allows for searching the current quarter's results only. Free. (More information below)
- **Grow Think Research (http://www.growthinkresearch.com):** Provides database of venture capital activity detailing United States-based companies, funding transactions, executives, directors and advisors, and investors. Subscription of $499.00 per month or individual searches for $5.00 per record.
- **TheFunded Website (http://www.thefunded.com):** TheFunded.com is a community of founders and CEOs to research, rate, and review funding sources worldwide. Offers reviews, information, and actual term sheets from various funds. Free for entrepreneurs.

Other subscription-based services include Dow Jones VentureSource, Venture Xpert Web, and Capital IQ.

How should you customize your search? Look at the following characteristics of firms to develop a list of prospects:

- **Location:** Investors often prefer to invest in companies located in the investor's hometown. Otherwise, see whether the investor has invested in any other company's located in your town. If you are from out of town, most VCs will tell you to find a local VC and then approach them about the possibility of coinvesting.
- **Industry focus:** Identify which investors are investing in other companies in your industry or using similar technologies.
- **Investment stage:** Although a firm may invest in businesses at various investment stages, most firms have a preference for the stage of company they usually invest into. If you are an early-stage company, focus your efforts on companies that are investing in either seed- or Series A-stage businesses.

- **Size of investment:** If your business model requires a substantial investment or will need significant follow-on investments in a short period, look at firms that make investments that match those needs.
- **Activity level:** Only eight venture firms investing in early-stage companies did more than nine deals in 2006, according to Entrepreneur.com. No firms investing the late-stage companies did more than seven deals in 2006. See whether a potential investor is actively funding new companies or is only involved in follow-on investments.
- **Industry reputation:** Word of mouth is perhaps the best source of inside information about various firms, particularly from portfolio companies. If you do not have contacts in the field, many discussion boards or a website such as The Funded.com will offer some insights into the reputation of a firm or firms.

HOW TO FIND OUT WHO INVESTS IN YOUR TECHNOLOGY, INDUSTRY, OR AREA

Target your search for venture capital firms by finding out the investors of other startups in your industry or field of technology. A helpful, free tool is the MoneyTree "Custom Search" function. This search can be used to find out specific companies that received funding in the most recent quarter and their investors. Here is an example.

If you are a Texas-based software company and wanted to find out who had invested in a first round financing of other Texas-based software companies in the third quarter of 2007, simply input those parameters (Texas, First Round Financing, Software) and submit the report.

The report would tell you that CBG Holdings Inc. and Phurnace Software Inc. fit these search criteria.

If you then clicked on CBG, you'd find that the company is an early-stage software developer that "develops software for the banking industry," and CBG received a $10,250,000 first-round investment from Adams Street Partners LLC (formerly known as Brinson Private Equity). If you clicked on Phurnace, you'd find that the company is also an early-stage software developer that "operates as a Java EE application deployment company," and Phurnace received a $1,330,000 first-round investment from DFJ Mercury.

Using this information, you would know of two potential investors that invest in early-stage software companies located in Texas. Comb your personal network to see who you know at those investment firms or try to reach out to connections at these companies to find an introduction to these investors.

Visit the MoneyTree website (http://www.pwcmoneytree.com) from PricewaterhouseCoopers and the National Venture Capital Association for more information.

Fundraising Tools

There are a number of different tools that will be important for various stages of the fundraising process. Investors will prefer different sets of materials to review; some will only want to initially see an executive summary business plan over e-mail, whereas others prefer you to mail a full business plan. To be able to provide materials that an investor requests, you should prepare the tools that will be used throughout your fundraising process. These tools include the following:

- A prepared and practiced "elevator speech"
- A two- to three-page executive summary business plan
- A 10- to 15-page "mini" business plan
- A full investor-focused business plan
- A one-page letter, with a two-paragraph introduction into the business
- Management team biographies and resumes
- A list of references (each of whom have a full set of company materials)
- A list of customers (if you have any), likely customer targets, and testimonials of current customers when available
- A set of company financials (current and projections)
- A 15- to 20-slide investor presentation (in PowerPoint)
- A complete set of due diligence materials (company records, documents, and files)

Be certain to prepare materials that look professional, are accurate, and tell the story of your business. A company will only have one chance to make your initial introduction, and word-of-mouth will spread about companies that are unprepared for the process. Invest the time up front to truly prepare and vet materials before sending them to any potential investor targets.

ACCORDING TO VENTURE CAPITALISTS

What information would you like to receive with the initial contact?

- Two- to three-page executive summary — 56%
- 10- to 15-page "mini" business plan — 22%
- Full business plan — 14%
- Information by telephone only — 7%
- Cover letter describing the company — 1%

Source: Profit Dynamics Inc.

Your elevator speech should have the following characteristics:

- Should be less than 10 seconds
- Should focus on telling the listener why they want to know more

- In a few simple sentences:
  - describe how the business will position itself (and try to avoid clichés)
  - the problem you are addressing and opportunity you are tackling
  - the unique nature of your business in the industry
- Show (don't tell) why you are enthusiastic about your business

For you investor presentation, be sure to avoid any errors, typos, or mistakes. Practice the presentation multiple times and develop answers to the 20 most likely questions. The presentation should include the following:

- A memorable introduction and theme of the presentation
- Company background (name, location, structure, stages of growth, and names of presenters)
- Opportunity/problem addressed
- Marketplace
- Revenue sources
- Management team
- Go-to-market strategy
- Marketing
- Competition
- Competitive advantage
- Risks
- Financial summary
- Timeline
- Capital needs and funds utilization
- Reasons why the VC should invest

For your company financials, have your numbers checked by an accountant and by an individual familiar with making investor presentations. Follow these suggestions for your financials:

- Include bottom-up numbers with assumptions clearly spelled out
- Provide numbers that look five years out (and two years back, if applicable)
- Break down expenses by department
- Break down revenue by product category
- Avoid the common problem of underestimating expenses or overestimating revenue
- Be able to explain numbers; investors will "drill down"
- Avoid using the phrase, "This market is so large that we plan to take just 2% of this market to be a success." Investors have heard it before and want a company with specific targets in the market and an aim to be a market leader rather than taking a small niche of a larger market.

Soliciting Investors

E-mail makes it easy to mass-mail your business plan to every venture capital firm out there. Unfortunately, an unsolicited, untargeted e-mail blast is not necessarily the best approach to building a base of interested investors. The goal of these initial solicitations is to obtain a face-to-face meeting with a potential investor.

ACCORDING TO VENTURE CAPITALISTS

What's the most common way you have found the companies you have actually invested in?

- Referral by another VC — 34%
- Direct contact by the entrepreneur — 30%
- Referrals by intermediaries — 17%
- Referrals by accountants and attorneys — 13%
- Investor events (such as industry forums and venture capital conferences) — 7%

Source: Profit Dynamics Inc.

To understand how to best solicit venture capital firms, it is important to understand how the process of choosing to invest works. From the company's perspective, the key player in this investment drama is your champion. The champion is generally a general partner of the fund (or sometimes an associate, but nearly always someone with a certain level of clout within the fund). When a company gets funded, there is nearly always one firm partner who champions the investment and is willing to still his neck out to get the deal financed. Generally, in the investment process, the CEO of the company will first meet with a contact at the firm. The hope is that this meeting is with the person in the firm who is the "specialist" in your particular industry, has at least some availability of time to have another company under his wing, and has clout with the copartners of the firm. Ultimately, the hope is that this first meeting is with your potential champion or is a person who will introduce you to your champion.

GETTING A REFERRAL

Referrals are often a helpful way to get an initial meeting with a VC. How do you get a referral if you don't know which of your contacts knows anyone in the investment community? Try using social networking tools, such as LinkedIn (http://www.linkedin.com), to find out who in your network knows someone helpful to your fundraising process. Search your connections for contacts at the venture firm or firms you are considering soliciting. You may be surprised to find out which of your direct connections may have a useful contact. Then, call in a favor: get that referral.

The champion is the company's advocate and main point of contact. She looks at your business and your team and ensures that, with financing and a bit of input from the investor-director, your business could produce big returns. The champion will drive the internal process within the fund, setting up meetings, bringing in experts, finding additional investors to invest in the deal if necessary, and ultimately making the pitch to the other partners to make the investment.

ACCORDING TO VENTURE CAPITALISTS

How would you prefer the initial contact by the entrepreneur (or their representative) be made to you?

- E-mail — 42% of respondents
- Postal mail — 29%
- Telephone — 17%
- Fax — 8%
- Referral — 4%

Source: Profit Dynamics Inc.

As you can probably tell, the champion really is the lynchpin in this entire process, and that is why a targeted solicitation is so crucial. The hope is that you are not only finding the right set of venture capital firms to solicit, you're looking for the individual at the firm who will become your champion.

One certainty is that VCs will each have unique preferences, so the importance of matching their preference is important. Some prefer an introduction to come by referral. Others are comfortable with an unsolicited e-mail or telephone call. Most VCs appear to favor receiving an executive summary business plan of a couple pages that answers the key questions on the business. This gives them the opportunity to judge for themselves after a high-level review.

Use the resources available to you to identify these preferences and use your referral network if it is available. Not all VCs will be interested (that is a given), and many might not respond to your initial contact. Be diligent, follow up, and consider alternative approaches to build a connection to your points of contact.

DOES IT HELP TO HIRE AN AGENT?

What could happen

You are approached by an agent to assist with your efforts to get venture capital financing. Is there anything you should worry about?

Watch out for

Securities laws limit your ability to pay commissions (or similar types of compensation) for sales of your company's securities. Only a licensed broker-dealer may receive commissions under securities laws.

Referrals by agents can be successful (17% of VCs in the Profit Dynamics survey listed referrals by intermediaries as the most common source of deals). However, these referrals are only as good as the agent and their personal networks. Before hiring any agent, be certain that you obtain information on previous placements, including the names of the companies and the investment firms. You should request a list of references as well. Here are some questions you may consider asking of any potential agent:

- Are you a registered and licensed broker-dealer?
- Can you provide any references to venture capital firms who can offer information about your services and experience?
- Can you provide a list of companies and venture firms where you were successful in raising funds?
- Can you provide a list of companies where you were not successful in raising funds?

TIP: Do your due diligence on any agent hired to assist with finding venture capital investment for your company.

Investor Presentations

The first hurdle in the fundraising process is getting an initial meeting with a potential investor. Obviously, even getting a meeting is a challenge (perhaps a 1 of a 100 chance for an unsolicited submission), so one of the first keys to the process is ensuring that you take full advantage of the meeting. Landing a meeting is certainly a good step in the right direction, but initial meetings are just that, initial. The initial meeting represents an opportunity to sell the vision for the company to the investors and, hopefully, land a follow-up meeting.

For a firm that is interested in an investment, there will generally be a series of meetings designed to be a two-way communication between the venture capital firm and the startup company. Once you have been invited to make a presentation to the investor, you and your team will be asked to showcase the company and provide background on the company's vision, the management team, the broader market, and the details behind the approach you plan to use.

DO I NEED A VENTURE CAPITAL FIRM TO SIGN MY NONDISCLOSURE AGREEMENT?

In an ideal world, you would want everyone to sign a nondisclosure agreement if they will discuss the proprietary aspects of your technology.

However, common practice within the venture capital community is to refrain from signing nondisclosure agreements. The reason is that the VCs (and their attorneys) believe that there is some risk for VCs to sign a nondisclosure agreement when they see so many similar business plans on a daily basis, many of which may overlap in some capacity. Therefore, avoid the embarrassment of asking a VC to sign a nondisclosure agreement; industry practice is that VCs don't sign nondisclosure agreements.

What should you expect in your meeting?:

- You will meet with one or several individuals from the venture capital firm.
- A formal investment pitch will usually be one and a half to two hours in length (although an informal discussion before being asked to provide a formal pitch could be between a half hour and an hour or more, depending on the firm).
    - Prepare a presentation that will take 45 minutes to go through without interruption, but be aware that you'll likely be interrupted quite a bit.
    - Be sure to leave at least 20 minutes for questions at the end, even if it means going through some slides or information somewhat quickly at the end.
    - Immediately stop the presentation for a question.
- Prepare a PowerPoint slide presentation of no more than 15–20 slides (discussed above is a sample of slides you should consider including).
- Prepare answers to 15–20 of the most likely questions you expect to receive.
- Be prepared to discuss assumptions in creating your financial models and expect to receive questions about valuations even at an initial meeting.
- Be flexible. It is better to get off schedule if the investor is engaged and asking questions. The VC can review your prepared materials after you leave but can't engage you with questions then.
- Ask your contact what types of technology is available. Should you bring your laptop? A CD? Flash memory drive? Projector?
- Bring several hardcopy printouts of your materials to provide to your primary contact and any additional parties that may join the presentation.

What should you avoid in your meeting?:

- Don't have any misspellings or errors in your presentation materials.
- Don't look unprepared. Practice the presentation several times to varied audiences, particularly to individuals without an industry or technical background in your technology.
- Don't get stumped by a question. Prepare an answer to the 15–20 most-likely-to-be-asked questions.
- Don't "fudge" something you don't know. Offer to find out the answer and then quickly follow up once you've answered the question.
- Don't present proprietary information in the investment presentation. You may need to reveal such information at a later point, but early on you should avoid offering such information because VCs will not generally sign nondisclosure agreements.
- Don't ask the VC to sign a nondisclosure agreement.
- Don't bring team members to the meeting who will not be participating or presenting.
- Don't forget to ask questions about the VC as well. The VC wants to know you are as interested in them as they may be in you.
- Don't schedule investor presentations back-to-back. Take time between presentations to reflect on the previous presentation and modify items as appropriate.

- Don't change the presentation simply because one VC recommended that you add or change a particular portion. This process is highly subjective, so note trends of comments but don't overhaul the presentation based on the concerns of one person alone.

Receiving Feedback from Investors

Although a company soliciting investors will hear "No Thanks" or "Send us an update on your efforts" during the solicitation process, these negatives can be very helpful in eventually securing funding. One of the most important things you can receive from potential investors is feedback. Obviously, most companies that ultimately receive funding will hear "No" or "We're not sure" many more times than "Yes." Therefore, understanding why an investor just isn't sold is extremely helpful.

THINK VCs DON'T EVER MAKE MISTAKES?

The stories of successful investments into startups and big payouts to VCs are legendary, but remember that for every Sequoia returning $250 for every $1 invested in Google, there are many that didn't see that potential. Here are a couple examples:

- Geoffrey Y. Yang of Redpoint Ventures famously passed on Google (after successfully investing in Ask Jeeves, NetFlix, and Excite), and Redpoint wasn't alone because at least one other VC admitted to passing on Google (but did not want to give his firm's name).
- Stewart Alsop of New Enterprise Associates passed investing in X.com (after most of the company's employees quit). X.com was later acquired by PayPal, which was sold to eBay, netting a big return for the investors who did stay on with the founder of X.com.

Firms that have "missed" opportunities may have done so by virtue of different risk tolerances more than anything. Some of the larger venture capital funds can take more risks (this is why Kleiner Perkins Caufield & Byers and Sequoia Capital were more able to take a risk on a company such as Google: they each have a large enough fund to make more mistakes and still show strong returns).

First, expect to receive numerous refusals in the process; this generally is to be expected in the process (this means you are meeting with lots of potential investors, which is crucial). Second (and most importantly), use any outright "No" or even a lukewarm "Send us an update on your fundraising progress" to find out information to help you tailor your pitch and focus your business plan. Here are a few questions you may ask:

- What is the most/least interesting proposition we offer?
- Where do you see the needs to build out our team?
- What would you need to see in six months to invest?
- What should we reconsider in our business plan?
- Can you refer us to another firm that might be a better fit?

A "No," a "Yes," or a "Maybe"

Many entrepreneurs struggle to interpret the message from a VC they just met with. Oftentimes, the meeting goes well and the entrepreneur senses that the investor seemed engaged. Then things seem to unexpectedly cool down. Calls are less frequent. E-mails go unanswered. What gives?

If a venture capital firm partner is interested, the process will likely move fairly briskly. Generally, you can expect to hear things like the following:

- "I want to get you in front of the firm partners. Are you and your team available to present next week?"
- "What is your timing? I think we want to get involved."
- "What do you need from us to move forward?"
- "Let me have some of our people do some more analysis and diligence work."
- "Can we get a list of customers and references to interview?"
- "When can I come and meet the rest of your team?"

If the reception you receive is more lukewarm, you can expect that the VC is probably not entire convinced by some portion of your business plan.

When you've gotten in the door for a face-to-face meeting with a venture capital firm, you've obviously done something right. Generally, the company has at least "intrigued" someone at the firm if they've arranged for a meeting. By inviting you to a face-to-face meeting, you've passed the initial sniff test (an interesting business plan, a good management team, or a well-respected referral). Once you've passed that test, you are "fund-able."

HOW CAN I TELL WHETHER THEY ARE *REALLY* INTERESTED?

What are some of the signs that VCs give when they are interested in a potential investment?

- VC calls you (or quickly calls you back).
- VC sets up a follow-up meeting.
- VC wants to meet other members of the company and team.
- You receive an e-mail or document with a number of questions.
- VC provides a checklist of items they need to review.
- You are asked for a list of references or customers.
- VC asks you to speak with a technical specialist or expert.

Venture capital partners are busy individuals. Be persistent with efforts to reach out to your contact.

What does "fund-able" mean? A select few business plans are no-brainers for an investor (think Marc Andreessen's potential new company). The rest of the business plans a VC sees probably have some warts: perhaps an unseasoned management team, a large potential competitor, a complex technology, or a market that isn't a multibillion dollar space. In each case, there are some positives and negatives to weigh for any potential investor. Getting your foot in the door makes it likely that you are fundable, simply that you fall within a broad category of companies that the VC feels deserves a look.

However, a fundable company isn't a funded company, and that's the biggest difference. If you are still in the fundable category after the meeting, the VC may not be ready to commit but might not want to refuse the investment just yet. Perhaps you'll find another investor and will come back looking to fill out the round. Perhaps you'll have a large customer sign up in the next couple months. Perhaps you'll have a technological breakthrough. This means that you are likely in the realm of "Keep us updated."

WHY DON'T THEY JUST TELL US "NO" IF THEY AREN'T INTERESTED?

Once an entrepreneur has had an initial meeting with a potential investor, the entrepreneur may expect to hear the following:

"Yes, we're interested in investing in you. Let's move forward."
or
"No, but thanks for coming in. Good Luck."

Then, the entrepreneur can move forward or move on rather than be in limbo. Unfortunately, limbo may be more likely than not because an entrepreneur may not get an outright rejection and instead may hear something like the following:

"We're interested if you can find a lead investor."
or
"Let's see if you can build up some customers/hit some development milestones/make some significant sales and then come back to us."
or
"Like the idea—keep us posted on the fundraising process."
or
"Looks to be a bit too early-stage. Come back to us when you've got some additional traction."

Unfortunately, none of these responses are particularly helpful to the entrepreneur and the startup company. They aren't a "yes" or a "no." They are more like a "maybe."

So why don't they just tell you "no"?

- It is no fun to tell someone "no."
- What if they are wrong and you turn out to be the next Google, Apple, Genentech, or Cisco? How bad would they look as the VC firm who passed on that . . .
- Sometimes (but not very often) things do change and the business does become a much more attractive candidate.

How should you handle a lukewarm reception? First, don't wait on a firm to get back to you with a final answer. Be polite and try to gather as much feedback as you can to help in the process, but don't alienate a potential investor by pushing too hard for a yes or no response. In the meantime, continue to set up meetings, build relationships, and focus your message. The process of building the business shouldn't be put on hold. Momentum is a tough thing to regain once it is lost. Second, be persistent in your efforts to try to keep them updated. Let a potential investor know when you've reached certain goals you discussed in your presentation or when you've hired new talent into the organization. Don't just hope for a potential investor to change its mind at some point down the road; give the firm a reason to change it.

The biggest mistake an entrepreneur can make is wasting valuable time waiting on their top choice. Continue to build momentum and act on your business plan. Meet with many potential investors and keep communication lines open with all potential investors (even those that may have said no outright).

INSIGHTS FROM A VC ABOUT GETTING A SIMPLE "NO"

Guy Kawasaki is a managing director of Garage Technology Ventures, an early-stage venture capital firm. He offers some important insights about understanding why it may not be easy to get a simple no from a VC.

- "Venture capitalists are simple people: we've either decided to invest, and we are convincing ourselves that our gut is right (aka, "due diligence") or there's not a chance in hell. While we may be simple, we're not necessarily forthcoming, so if you think it's hard to get a 'yes' out of venture capitalist, you should try to get a conclusive 'no.'"
- For a VC, "there's no upside to communicating a negative decision. Entrepreneurs will simply hate us sooner—instead the game is to string along entrepreneurs in case something miraculous happens to make them look better. (An example of a miracle would be Boeing approving a $5 million purchase order.)"
- "Alas, entrepreneurs are also simple people: If they don't hear a conclusive 'no,' they assume the answer is 'yes.' This is an example of the kind of breakdown of communication between venture capitalists and entrepreneurs that causes much pain and frustration for entrepreneurs."

Source: http://blog.guykawasaki.com/2006/01/the_top_ten_lie.html.

Business Due Diligence

If the process continues to go well, the VC will begin its process of further validation of the business. The VC is looking to make a substantial investment in time and money into the business and therefore wants to be certain of the background of the business and the underlying assumptions on which the business is based.

Much like a job interview, expect that the VC will do its own homework on the company, speaking with your references, talking to your current or prospective customers, researching into the management, and consulting with contacts the VC may have within the industry. Ultimately, their goal is not to have anything surprise them. Therefore, talk with your references and customers before you give the list to the VC. Let them know what to expect and why they'll be receiving calls from potential investors. Provide your references with a copy of the business plan or certain materials to ensure they understand the business and can advocate on your behalf.

Be prepared to have your financial projections and current financial statements thoroughly reviewed by representatives of the potential investor. These requests for information may seem time consuming, but it is important to help minimize the time this research and review will take. In some cases, use your attorney, your accountant, or other contacts to ensure that the process goes smoothly.

Retooling Your Pitch

Once you've met with investors and heard "No" or "Keep us updated," you'll want to review your efforts and rework your pitch. Investors may have advice in all areas, such as target another market, hire an experienced CEO immediately, or get patents. Remember to use this information to help you, not to change the overall vision of the company founders.

USING FEEDBACK

Once an entrepreneur has received feedback from various meetings with potential investors, what should you do?

- Continue to focus on building your business (remember raising funds isn't an end in and of itself).
- Don't wait so long to fundraise that you have to shut down the business without financing from the first round of solicitations.
- Tweak your materials and presentation.
- Don't overhaul the overall vision or direction of the company.
- Get advice from many sources.
- Don't put too much stock in one person's advice (particularly if that person doesn't believe in the company enough to invest).
- Continue to look for talented individuals to join your team.
- Don't try to recruit an advisory board just to show that smart and successful people support you.
- Update firms that told you no or asked you to keep them updated.
- Don't burn bridges with a firm that wasn't interested.
- Look to new firms that you didn't initially consider.
- Don't scrap your fundraising strategy; rework it.

Once you've used the feedback, continue to reach out to new potential investors as well as keep those potential investors with whom you had previously met updated on your progress. Discuss new successes, including angel investments, technology enhancements, important hires, new customers, or changes in the marketing or sales approach. If you've proven that you are fundable, the VC may be looking for certain evidence to validate.

And Finally, The "Yes and Here's a Term Sheet"

The finish line is in sight: a term sheet. As the initial meetings and business diligence processes come to a conclusion, the final step in the solicitation process will culminate with a presentation to the firm partners (usually instigated by your champion). After this presentation, the partners will discuss the business and make a decision to invest.

Find out when the partners of the VC firm have their weekly meeting (traditionally, Silicon Valley firms have scheduled these meetings on Monday mornings). It may be a good idea to follow up with your champion the following day, after the fund has their all-partner meeting. Use this information to see where the partnership stands on your potential investment.

Then begins the process of turning that yes over the phone or in a meeting, into a financing contract and cash in the bank. The first step will be receipt of a term sheet from the investor, nearly always drafted by the lawyers for the venture capital firm.

Once you've received the term sheet, the work of the company is far from complete. The company must now drive the process to completion to secure the funding. One mistake many entrepreneurs make is failing to realize that, once a term sheet has been received, the company must manage the process to close the investment, including final negotiation of the term sheet and working with legal counsel to finalize the investment contracts. The company should work closely with the venture capital firm and partner, but don't assume that the VC will oversee the deal for the company. Traditionally, the company should use its advisors and counsel to set a timeline and manage completion of the transaction.

Understanding the VC Term Sheet

After a handshake and a few pats on the back, a partner at the venture capital firm will inform you to "Look for a term sheet from us very soon." This represents excellent news for the company and the fundraising team, but it doesn't mean the work of the company is done just yet. Now comes the final push to seal the deal and get the cash in the bank, which involves determining whether the investment meets the company's needs, doing your diligence on the investor, and getting a better understanding of the terms of the deal.

To see a sample Series A term sheet, go to the information at the end of this chapter or visit the book's website at http://www.myhightechstartup.com.

> **WHAT IS A TERM SHEET?**
>
> A "term sheet" or "memorandum of understanding" is basically an outline and a blueprint for the financing deal. It saves time and confusion later on by allowing the deal to be structured around the terms agreed on in the term sheet. It also serves as the easiest way for the principals to focus on and agree on the deal points.
>
> After the term sheet is negotiated, then the documents of the deal will be drafted to incorporate the points in the term sheet. These deal points may seem somewhat trivial (who cares about antidilution or liquidation preferences, or even knows what they mean for that matter?). The truth is, these seemingly trivial deal points influence the company over the long run and will need a thorough review before agreeing to the deal.

If you are fortunate enough, the company may even receive several term sheets over the coming days and weeks, and, in that case, you will need to evaluate the investor and the deals they are proposing.

The Process of Term Sheet Preparation

The term sheet will usually be drafted by the VC and primary negotiations will be between the company and the investors. However, the startup may be responsible for drafting the term sheet in seed round or Series A financing round if there is no "professional" (VC or corporate) or "lead" investor in the round. It is advisable to have your attorney review the term sheet before transmission to the investors to ensure that the terms are accurately reflective of the current marketplace.

WHY PREFERRED STOCK?

Venture financings typically involve the sale of convertible preferred stock. The special rights of the preferred stock support a two-tiered capital structure, with higher-priced preferred stock being sold to investors and lower-priced common stock being issued to employees. A typical price differential between preferred and common stock for an early stage company would be between approximately 2:1 and 10:1, respectively (this price or value differential is just a rule of thumb and is due to the addition rights provided to the preferred stock, including liquidation preferences, voting rights, participation rights, and veto powers). The convertibility of the preferred stock permits the investors to become common shareholders if it is in their economic interests to do so.

The chief disadvantage of issuing preferred stock is that, at least under California corporation law, preferred shareholders are entitled to a separate class vote on certain matters. For example, the preferred shareholders of a California corporation are entitled to veto a merger involving the company. Even if the company is incorporated in a state such as Delaware, where class voting does not apply to a merger, the investors will typically insist that a separate class vote on mergers and other matters be included in the financing documents.

The proposed transaction will generally begin with a term sheet drafted by one side (usually the investor). Once you receive the term sheet from the investor or his counsel, it becomes the company's turn to examine and revise the term sheet. This begins the process wherein the term sheet is revised and rerevised to reflect the ongoing negotiations of the parties. In other situations, the term sheet may reflect terms that have already been agreed on orally. Once a final term sheet has been formulated, the parties are ready to begin drafting documents that will reflect those terms, as well as the finer points to be worked out in the drafting stage of the transaction.

In negotiating the terms of an initial round of venture financing, the company and the investors will generally first try to agree on several key terms (which could be potential deal killers). These initial discussions will likely center on the premoney valuation of the company and the dollar value of the investment.

The premoney valuation of the company is usually based on the business plan, prospects, and any technology owned or developed by the company before the investment. (Valuation will be discussed in further detail below.) The size of the investment will then determine the percentage of the company being acquired by the investors. The per share price of the preferred stock issued in the financing is a function of the premoney valuation and the number of shares and options outstanding before the financing. Sophisticated investors will treat the shares reserved for future stock and option grants, if any, as being fully issued when calculating the per share price.

IS THE TERM SHEET BINDING?

For the most part, no. Usually a VC term sheet provides that it is nonbinding; that is, even after the terms have been agreed on for the term sheet, neither party is obliged to consummate the transaction until it has executed a purchase agreement.

However, as you can see below, this model term sheet does provide that a few items will be binding. These binding items are (1) the no-shop/confidentiality provisions that don't permit the company to shop the deal or communicate about the deal once the term sheet is signed, and (2) the legal fees of the investor will be paid by the company even if the deal doesn't happen.

"This Term Sheet summarizes the principal terms of the Series A Preferred Stock Financing of [__________], Inc., a [Delaware] corporation (the "Company"). In consideration of the time and expense devoted and to be devoted by the Investors with respect to this investment, the No Shop/Confidentiality and Counsel and Expenses provisions of this Term Sheet shall be binding obligations of the Company whether or not the financing is consummated. No other legally binding obligations will be created until definitive agreements are executed and delivered by all parties. This Term Sheet is not a commitment to invest, and is conditioned on the completion of due diligence, legal review and documentation that is satisfactory to the Investors. This Term Sheet shall be governed in all respects by the laws of the [State of Delaware]."

Source: Preamble to the "Model Term Sheet for Series A Financing," National Venture Capital Association.

Negotiating and revising a term sheet usually involves a number of tactical considerations by the company and its attorneys. One approach to the term sheet negotiation is to modify the term sheet to include terms that heavily favor the company. This may allow the company to "give" on certain issues down the road. Another approach is to offer revisions that would tend to be the "middle ground" to avoid lengthy back-and-forth negotiations. The strategy to use here will depend on the leverage of the company, the relationship of the parties, and the level of sophistication for each of the parties to the transaction. The company will also need to determine the level of detail it hopes to include in the term sheet. Attempting to include the bulk of the key details of the transaction in the term sheet will make the document drafting and closing process much simpler but could force discussions around issues that otherwise might not be discussed if simply included in the financing agreements.

TERM SHEET NEGOTIATION TIPS AND TECHNIQUES

Negotiating a term sheet with a venture capital firm may seem like a challenge, and there will be some challenging aspects to the negotiations. However, remember that the venture capital firm understands and expects some negotiating around the term sheet. Everyone hopes to get the deal done, so be pragmatic and consider the big picture when working to finalize the term sheet. Some key considerations of the negotiation process include the following:

- **Determine the lead investor and a single investor's counsel.** Negotiating with multiple investors and multiple lawyers can be time consuming and expensive. If you have multiple investors, ask to have a lead investor designated and shared investor's counsel.
- **Involve your attorney.** If you have hired an attorney that has done venture financing deals, your attorney likely knows what terms are reasonable and where you should focus your negotiations.
- **Model it out.** There are a lot of variables at play here and changes in some can drastically change the payouts down the road. Model out the investment over the five-year period in your projections and see the impact of changes in the valuations, liquidation preferences, etc. Look at hypothetical transactions down the road (What if we sold for $20 million in two years or $100 million in four years?) or see what occurs if the business goes belly up down the road. Find someone that is skilled in Excel and ask for their help to build the financing (and possible future financings) into a model.
- **Prioritize your negotiations.** The term sheet will have numerous terms that could be negotiated, but a heavily edited term sheet could kill the deal. Focus your comments on a few key items, such as company valuation, the size of the option pool used to attract future employees, the number of board members, compensation for the founders, vesting on founders stock, and liquidation preference.
- **VCs don't have "standard" deal terms.** A VC or attorney may tell you that certain deal points or provision are "standard" or "always the way we do it." However, that doesn't mean it isn't negotiable. Although it may be a preference or typical, each investment is different and few terms (if any) are never changed in the context of a deal.
- **Act professionally if you are evaluating multiple term sheets.** In the event you are considering multiple investment offers or are awaiting a term sheet from an investor who is likely to invest, be certain that you keep the potential investors informed of your timetable. Investors who feel like they are being played against one another will oftentimes balk at the deal and find another candidate for their investment.
- **Remember that you have to work with the VC after you close the financing.** Negotiating a term sheet (and closing the entire deal) can seem adversarial, but don't create a situation in which you can't work with the VC who will ultimately be serving on your board of directors. VCs expect to negotiate the deal, expect issues to arise in the process, but expect to get the deal done.

The Terms of a Term Sheet

To help you understand some of the provisions, a sample term sheet for a Series A preferred stock financing is included at the end of this chapter to help the entrepreneur see what to expect.

> **COMPARING YOUR TERM SHEET**
>
> Once you receive a term sheet, your attorney may be able to provide industry knowledge on the terms their clients have received in previous venture capital financings with a specific venture firm.
>
> In addition, the website http://www.thefunded.com provides members the ability to upload venture capital term sheets they have received previously (and either accepted or rejected) to allow other members to compare and contrast individual terms from each VC. The site is limited to members only, and applications for membership require a link to an online biography. The service is free for founders and entrepreneurs and allows users to review and rate funding sources.

In the context of a preferred stock financing, the term sheet is used to describe (1) the essential characteristics of the stock to be issued, (2) the capitalization of the company, (3) the main conditions to closing, and (4) registration rights, information rights, and any other rights and obligations that will survive the closing. Below is a list of some of the most typical terms in a financing term sheet:

- Short preamble or summary of the deal
- Type of stock being sold (usually preferred stock)
- Price per share and total funding
- Consideration (usually cash or converted bridge loans)
- Date of the deal's closing or the dates of multiple closings
- Names of the investors and the general breakdown in ownership
- Pre- and postfinancing valuation of the company
- Attributes of the stock in the deal, including (1) dividends, (2) liquidation preference, (3) redemption, (4) conversion, (5) antidilution, and (6) statutory voting rights
- Counsel fees
- Standstill and no-shop provisions

> **SUMMARY OF KEY TERM SHEET ISSUES**
>
> **The following represents a list of the key issues that are often negotiated and the various choices and alternatives found in a VC term sheet:**

Term	Choice or alternative
1. **Dividend preference**	a) Cumulative/noncumulative
	b) Rate
	c) When paid

2. **Liquidation preference**	a)	Amount of preference
	b)	Participating preferred
	c)	Caps
	d)	Treatment in mergers
3. **Conversion**	a)	Price
4. **Antidilution**	a)	Ratchet versus formula weighted average
	b)	Broad-based versus narrowly based formula
	c)	None; investor relies on rights of participation
5. **Voting**	a)	Class vote
	b)	Class versus series vote
	c)	Level of consent
	d)	Protective provisions/veto rights
6. **Redemption**	a)	Mandatory/optional/none
	b)	Initial date
	c)	Redemption period
7. **Registration rights**	a)	Threshold percentage of shareholdings to initiate a demand
	b)	Minimum number of shares to include
	c)	Maximum
	d)	S-3 expenses and threshold
	e)	Percentage in piggyback
	f)	Expenses
	g)	Transfer of rights
	h)	Expiration
	i)	Lockup
8. **Preemptive rights**	a)	Limit to holdings in company
	b)	Carve outs
9. **Pay to play**	a)	Necessary with full ratchet
10. **Cosale**	a)	Carve outs
11. **Board representation**	a)	Number
12. **Drag-along**	a)	Included
	b)	Approval percentage
13. **Stock option pool**	a)	Increase
	b)	Premoney or postmoney
14. **Founder/employee stock restrictions**	a)	Vesting
	b)	Time/cliff/acceleration
15. **Financing expenses**	a)	Amount of cap on fees

Investment Amount

Usually, the company won't be surprised at the investment amount when the term sheet arrives in their e-mail inbox. In most cases, the VC and the company will have discussed the investment amount and the investors are aware of the funds the company is seeking.

If the amount doesn't fit within the range of a typical investment for a fund, usually the venture partners will ask whether the entrepreneur is willing to increase or decrease the size of the round. If the entrepreneur isn't ready to change the amount of the financing, the VC may politely end the discussions or may see whether the fund can invite another investor to fund the entire round.

ACCORDING TO VENTURE CAPITALISTS

How much dilution should you expect?

A typical company will see its shares diluted by 40% to 50% after a Series A round. Therefore, expect the capitalization table to resemble something like the following after your Series A financing: 40% founders, 40% outside investors (VCs), and 20% option pool.

Average equity ownership by outside investors:

- After the Series A round — 42.9%
- After the Series B round — 58.4%
- After the Series C round — 64.8%

Median dilution after a Series A round:

- Investment of less than $5 million — 40%
- Investment of more than $5 million — 50%

Source: Private Company Financing Report, Cooley Godward Kronish LLP, September 2006 (based on information from 2004 to 2006).

Valuation

Perhaps the most fundamental issue in any financing is valuation. Valuation refers to the value of the company as established by agreement between the company and the investors. Investors typically refer to premoney valuation and postmoney valuation. Valuation is an art and the product of negotiation rather than a science.

Premoney valuation refers to the value of the company before the investment. It is calculated by multiplying (1) the price per share to be paid by the investors, and (2) the number of shares outstanding before the financing, calculated on a fully diluted basis so as to include the employee reserve pool. In the example given in the term sheet at the end of this chapter, the premoney valuation is $4 million ($1.00 per share × [3 million founders + 1 million employee reserve shares]).

Postmoney valuation refers to the value of the company immediately after the financing and is calculated by adding the amount of the new investment to the premoney valuation. The $2.0 million financing of "High-Tech Startup Inc." yields a post-money valuation of $6.0 million.

The "Valuation" box below discusses a number of tools that investors and entrepreneurs can use to identify the respective valuation for a company. In addition, as a general rule of thumb, many professional investors will automatically apply a substantial discount (say 30%–50%) to the company's valuation presented in the company's initial business plan. This rule of thumb tends to serve to discount an entrepreneur's oftentimes overly optimistic projections and should serve as a helpful guide to an entrepreneur as to what they should expect in early valuation negotiations.

ACCORDING TO VENTURE CAPITALISTS

What is the median company valuation for a financing?

Median premoney valuation (in millions):

	Series A	Series B
Quarter 1 of 2007	$9.0	$25.0
Quarter 2 of 2007	$7.2	$20.0

East Coast versus West Coast median premoney valuation 2006 to 2007 (in millions):

	Series A	Series B
East	$6.0	$10.0
West	$9.0	$28.0

Source: Private Company Financing Report, Cooley Godward Kronish LLP, 2006/2007.

VALUATION

Some industry experts will tell you that valuing an early-stage company is more a matter of intuition and experience than an actual empirical calculation (as discussed previously with the "30%–50% discount" rule of thumb). That said, there are a number of methods that an entrepreneur can look to for some guidance on the valuation figures to see whether the valuation is reasonable. Here are a few of the typical tools that VCs will use in their determination of a valuation:

- **Discounted cash flows:** Venture capital firms will often use a derivative of the discounted cash flows method to value the company. The approach estimates the earnings of the company at a point in the future where an exit is anticipated and then, using the price-to-earnings (P/E) ratio of comparable companies, determines the value of the company at that future point. Then, using a discount rate, the value of the company today is determined. Here are the steps to this valuation approach:

    1. Using the financial information provided to the potential investor, estimate the company's net income or EBITDA at the point in the future where the investor will exit the investment.
    2. Looking at comparable companies with similar economic characteristics, determine an appropriate P/E ratio.
    3. Multiply the P/E ratio with the net income or EBITDA value to determine the value of the company at the expected liquidity event.
    4. Discount this future value to the present, generally using a rate in the range from 30% to 40% (but perhaps as high as 80%).

    The obvious limitation of this method is the number of variables that can be changed slightly to change the entire calculation. Try various numbers in your calculations (including estimates on the high, middle, and lower end of your projections) to see the variations in valuation that will result.

- **Comparable valuations:** Comparison data about similarly situated companies can provide valuable information on appropriate valuations. Although the amount of funding for comparison companies is readily available on a variety of websites, including http://www.capitalhunter.com, http://www.growthinkresearch.com, http://www.localbusiness.com, and http://www.capitalgrowth.com, it is more difficult to obtain accurate valuation information. Your attorney may also have information on specific-industry deal points, including average valuation terms.

  Some companies are now providing more specific information such as pre-money valuations. The Private Equity Data Center (http://www.pedatacenter.com) offers valuation information on companies for $49.95 per company. VentureOne provides a "Comparable Valuations Report" analyzing comparable business fundraisings and valuations for $1,295.00.

- **Comparable exit events:** An additional approach that may provide another data point in determining the appropriateness of a valuation figure is looking at exit events of comparable companies. To find comparable companies that have recent merger and acquisition activity, you can look in industry trade press or review local news sources. The Private Equity Hub (http://www.pehub.com) provides news on exit events for venture-funded companies and has feature to search past news stories.

  Thomson, in its database Venture Economics (http://www.vx.thomsonib.com), offers a paid subscription service tracking venture-funded merger and acquisition activity.

A VALUATION CALCULATOR

You've heard that coming up with a company valuation is an art. So how then can you tell whether the valuation you receive is even reasonable? Cayenne Consulting provides a free valuation calculator tool (http://www.caycon.com/valuation.php) to assist entrepreneurs and investors with the difficult process of determining an approximate range. Based on your responses to 25 questions (each questions has four potential responses), the tool will provide you a valuation ranging from $0.5 million up to more than $40 million.

Dividend Preference

A typical term sheet will include a noncumulative dividend on the preferred stock, usually set at between 8% and 10% of the purchase price per annum. Although initially this may seem heavy-handed ("You are trying to tell me that they want dividends too?!?"), in practice, dividends are rarely a point of negotiation. The truth is that very few private companies pay (and few investors ever expect to receive) dividends.

ACCORDING TO VENTURE CAPITALISTS

Will the deal include cumulative dividends?

Percentage of financings that include cumulative dividends:

Q3'07	Q2'07	Q1'07	Q4'06	Q3'06	Q2'06	Q1'06	Q4'05
4%	6%	8%	4%	7%	8%	3%	4%

Source: Venture Capital Survey, Silicon Valley Third Quarter 2007, Fenwick & West LLP.

Sometimes, investors will seek cumulative dividends. Generally, these are not intended to be paid in cash because the startup company needs the cash. However, cumulative dividends may be paid in connection with a redemption or at the time of a merger, or, in the event of a public offering, would be added as a premium to the rate at which shares of preferred stock are converted into common stock (this rate is usually 1:1 at the time of the financing). Cumulative dividends can significantly increase the value of the investor's security, but they can also pose tax problems for the investor. As a practical matter, cumulative dividends aren't frequently requested. Dividends aren't a highly negotiated portion of the term sheet, although there may be some discussion regarding cumulative versus noncumulative dividends.

Liquidation Preference

The vast majority of term sheets will include a liquidation preference for the preferred stock, basically ensuring that the investors are paid out first if the company dissolves or is liquidated through a merger or acquisition. Nearly all venture financings have a liquidation preference, but there will often be negotiation around the details of how the preference works.

A liquidation preference comes into play in one of two situations: (1) the company fails and has to sell the remaining assets and liquidate the business, or (2) the company is acquired by or sold to another company (or also a merger transaction in which the current shareholders wind up no longer owning the company). In each of these cases, the shares will end up liquidated, and it will need to determine how to divide the proceeds from the transactions. The liquidation preference for the investors ensures that the investors are the shareholders that will be paid back first, allowing the investor to recoup their initial investment (and perhaps more). Then, after the preference is paid, any proceeds that remain will be allocated proportionally to the common stockholders.

As an example of why this provision matters, consider the following hypothetical. Company A raises $6 million on a $6 million premoney valuation for a postmoney valuation of $12 million. Now the founders of Company A own 50% of the company, and the investors own the other 50% in preferred stock. One year after the financing, an offer is made by a third party to buy the company for $10 million. In this simple example (and absent any liquidation preference), the founders would get $5 million of the proceeds (50% of $10 million) and the investors would get the other $5 million. Do you see the problem here? On this transaction, the investor just lost $1 million, whereas the founders just made $5 million. That's not a great business model, is it? So to prevent this from occurring, the investor will require a liquidation preference. This insures that the Series A investors get paid back first—their full $6 million investment—before anyone else gets paid.

This is the reason the liquidation preference is typically the most important economic right of the preferred stock (and the major reason why it is usually valued higher than the common stock, as discussed previously). Having this liquidation preference ensures that, even if things sour, the investors are going to be paid back first. Generally, the liquidation preference is written in such a way that preferred shareholders receive their purchase price before any payment to common shareholders in the event of a liquidation or dissolution of the company and, usually, also in the context of a merger or other acquisition of the company.

ACCORDING TO VENTURE CAPITALISTS

What should you expect for liquidation preferences?

Percentage of Series A deals that have liquidation preferences: 97%

Series A	Series B	Series C	Series D
97%	98%	100%	99%

Source: Private Company Financing Report, Cooley Godward Kronish LLP, August 2007.

However, there is still more to the concept of a liquidation preference. The rest of the discussion focuses on what happens when liquidation occurs, making the stockholders lots of money. How are the proceeds divided and who gets what first?

Preference Multiples

In some cases, the liquidation preference may be greater than just the return of the investor's initial investment, sometimes two or three times the investor's initial investment. This is referred to as a preference multiple. In this case, in the event of a sale or liquidation of the company, the investor will first receive two or three times its initial investment before the remaining funds are distributed.

ACCORDING TO VENTURE CAPITALISTS

What should you expect for liquidation preference multiples?

Percentage of Series A deals that have 1× liquidation preferences: 91%

	Series A	Series B	Series C	Series D
1×	91%	85%	87%	77%
1–2×	4%	10%	10%	17%
2–3×	2%	1%	2%	1%
Greater than 3×	0%	2%	1%	4%
None	3%	2%	0%	1%

Source: Private Company Financing Report, Cooley Godward Kronish LLP, August 2007.

Here is the example of preference multiples at play. This time, Company A again raises $6 million on a $6 million premoney valuation for a total valuation of $12 million post-money. The investors have a three times (3×) preference multiple, meaning that they would receive $18 million back in a liquidation or sale event, before any other shareholders are paid. One year later, the company receives an offer to buy the company for $20 million. The investors hold a three times preference on nonparticipating preferred stock (see below for the difference between nonparticipating and participating). Now, after the liquidation, the investors will receive their $18 million (the initial $6 million investment × 3), leaving $2 million to be allocated among the common stockholders.

The difference between using a one-time liquidation preference multiple and a three-times preference multiple equates to $8 million of additional proceeds as a result of the investors on this $20 million transaction ($18 million versus $10 million if the preferred converted into common and received 50% of the proceeds).

Participating Preferred Stock

In some transactions, after the preferred stock receives their initial liquidation preference (one, two, three times or more), then they also receive the right to receive a portion of the remaining proceeds. Enter the concept of participating preferred stock. In the event of liquidation such as a sale of the company, the liquidation preference ensures that the preferred stock is paid back first (or paid back two or three times). However (and here's the kicker), participating preferred stock are then entitled to share (participate) with the common stockholders in the remainder of the proceeds.

Here is the example of participating preferred at play. This time, Company A gets an outstanding offer to sell the company for $20 million just a year after receiving funding. The investors hold participating preferred stock. This means that the investors will first receive their $6 million back, leaving $14 million to be allocated. Now, the preferred will participate based on their ownership (here 50/50) in the remaining amount, with $7 million going to each. So the investors receive $13 million and the founders receive $7 million of the $20 million purchase price. If the investors had held nonparticipating preferred stock, then the investors would have a choice: (1) exercise the liquidation preference and receive only $6 million, or (2) convert the preferred stock into common stock and take their portion ($10 million).

The difference between using participating and nonparticipating preferred stock in the term sheet was $3 million for the investors ($13 million versus $10 million) and the reason that understanding a liquidation preference matters.

So hopefully you can now see why participation rights are a major negotiating point in venture financings. A nonparticipating preferred effectively requires that the preferred shareholders elect whether to retain their preferred stock and receive only their purchase price or convert their preferred stock to common stock to share in the proceeds remaining after payment is made to any nonconverted preferred stock. Participating preferred shareholders receive their original purchase price and then participate with the common shareholders in the distribution of the balance of the remaining proceeds.

Capped Participating Preferred

Based on the previous examples of preference multiples and participating preferred stock, it may seem that these terms are unfair. That's not exactly the case. Historically, most venture deals involved a nonparticipating preferred stock. In recent years, however, many investors have insisted on a participating preferred to avoid the situation wherein the company is acquired for approximately the postfinancing valuation (i.e., there is no appreciation in the value of the company) and (1) the preferred shareholders receive their invested amount but no return, and (2) the common shareholders receive a substantial return based on their lower cost basis. In negotiating this point, the company would argue that the investors should not expect a return if the company does not appreciate in value and that the investors should not be paid on the front end (the initial preference payment) and the back end (the distribution of the remaining proceeds) if the company is successful.

As a result, use of capped participating preferred has become much more common. With capped participating preferred, the preferred stock will be participating in mergers in which the return to the preferred shareholders would be less than a fixed multiple of the purchase price (typically between two and five times) on a straight pro rata sharing and nonparticipating in mergers resulting in proceeds above that price point.

Capped participating preferred is meant to reward everyone equally when the company has a very successful sale or merger event. At an acquisition above the "cap" price, investors would convert their preferred stock into common stock and share the distributed assets pro rata with all other holders. However, if the event is a medium success, the investors will receive a greater portion of the proceeds.

ACCORDING TO VENTURE CAPITALISTS

How likely is participating preferred and what are typical caps on participating preferred stock?

Only 50% of Series A financings have participating preferred. Of those 50%, half (25% of the total) have uncapped participation rights, whereas the other half (25% of the total) have caps at various levels.

	Series A	Series B
No participation beyond 1× liquidation preference	50%	41%
1–2×	7%	10%
2–3×	13%	13%
Greater than 3×	5%	4%
Uncapped (full participation)	25%	32%

Source: Private Company Financing Report, Cooley Godward Kronish LLP, November 2007 (based on financing transactions in the second quarter of 2007).

Here is the example of capped participating preferred at play. Again, Company A gets an outstanding offer after the first year to sell the company for $20 million. As before, the investors hold participating preferred stock with a one-time preference multiple, but this time there is a cap set at four times the original purchase price (often this also includes dividends, but we'll skip that here). So, if the management decided to take this deal, the investors would get $13 million ($6 million from the one-time liquidation preference and then 50% of the remaining $14 million) and the founders would get $7 million (their half of the $14 million after the initial $6 million to the investors). Although this would be a nice result, management believes there is more to the company and rejects the merger.

Fast forward to two years later. The company has taken off like a rocket and now receives an offer from a new suitor to purchase the company for $60 million. Because of the cap, the investors now have a choice. They can (1) take $24 million, which is four times their original purchase price and no more, or (2) convert their stock into common and receive $30 million, half of the total proceeds from the sale. Without the cap, the investors would receive $33 million ($6 million from the one-time liquidation preference and then 50% of whatever remains) and the founders $27 million (their half of the $54 remaining after the $6 million from the liquidation preference). Use of the capped participating preferred may provide for better alignment of priority among the investors and founders.

Special Shareholder Returns

In the alternative for companies that do not have a cap on participation, management often seeks a special return when its efforts have generated added value for investors (e.g., a twofold to fourfold return on the investment in which the liquidation event is a merger rather than a winding up of the business). This is oftentimes done to offset some of the preference rights of the investors in a highly successful event. Accordingly, some financings now include a liquidation payment to the common holders (e.g., $0.25 per share) before commencement of the "participating" preferred distribution.

Automatic Conversion

Preferred stock will generally have a voluntary conversion right, permitting any holder of preferred stock to convert their preferred shares into common shares at any time. However, the automatic conversion refers to certain situations in which the preferred stock will automatically convert. The events that will initiate the automatic conversion are usually (1) a vote by the majority or supermajority of all the preferred stockholders voting as a single class (sometimes each separate class will need to approve), or (2) a qualified IPO meeting certain agreed-on criteria.

The terms of the conversion can be quite confusing, but the general fact is that preferred stock will initially convert on a 1:1 basis into common stock. However, based on future financing events and other factors built into the conversion ratio, preferred stock may be converted at varying rates into common stock.

The primary reason that these conversion rights are built into the preferred stock is to force the preferred holders to convert their stock into common immediately before an IPO. As a result of the conversion, most of the negotiated rights and preferences of the preferred stock will disappear, creating a simple capitalization structure before the company goes public. Most underwriters will require that outstanding preferred stock convert to common stock in the event of an IPO. To avoid the need to negotiate this conversion at the time of an IPO, the company will negotiate a trigger price. Automatic conversion at three times the purchase price would be ideal for the company in a first-round financing, but it is likely that investors will expect an automatic conversion price of four to five times their purchase price.

Although the automatic conversion itself is rarely negotiated, the criteria for a "qualified" IPO may be negotiated as a part of the term sheet or during the document drafting process. These criteria require that the initial public offering be priced at a minimum per share price (generally set at two to five times the purchase price of the convertible preferred) and for a minimum aggregate amount of dollars raised by the company in the offering. The minimum aggregate offering amount is often set at $7.5 million (the minimum amount currently required for registration on Form S-1), but the investors may request a higher threshold ($10 million to $15 million). This requirement is intended to ensure that there will be adequate "float" (shares held by public investors) after the IPO.

Antidilution

Dilution prevention provisions are designed to protect the investor from dilution that may occur in subsequent financings in which stock is sold at a price lower than the investor originally paid. The antidilution formula provides for an increase in the conversion rate of the preferred stock (and therefore a lower effective per share price on a common stock equivalent basis) in the event of a subsequent financing at a price less than that paid for the preferred stock being adjusted. A future financing sold at a price lower than in the previous round is called a "down round" or a "washout financing" (see the textbox at the end of the chapter that discusses washout financings). Down rounds occur more frequently than you might think (approximately 20% of financings each year are down rounds). As a result, investors will generally insist on antidilution provisions to prevent dilution in the event of a future down round.

The question that entrepreneurs don't always ask is this: If the investors are protected from dilution in a down round, what happens to the rest of the company? Unfortunately, the common stockholders are the ones who will be affected the most. Common stockholders are actually diluted twice in a down round: first the normal dilution from raising additional funds and then additional dilution to cover the dilution that occurred to the investors as a result of the antidilution provisions.

Although nearly all venture financings will include an antidilution provision, the primary point of negotiation will focus on the formula to use. In most venture financings, a weighted-average adjustment is a more common type of antidilution protection over the full-ratchet adjustment, which tends to have a more severe effect on the common shareholders.

Weighted Average

Weighted average antidilution is not as severe as the full-ratchet antidilution provision discussed below. Instead of repricing the previous rounds at the same price of the down round, the weighted average dilution reduces the conversion price based on the investor's ratio of stock owned to the total amount of stock of the company. Using this provision will decrease or limit dilution in a down round for the investor but will not prevent it entirely.

There are two different types of weighted average calculations: broad based and narrow based, with broad based being the more frequently used calculation. The primary difference between these terms lies in the way "common stock outstanding" is defined.

Under the broad-based provision, both of the following are included: (1) the current number of common stock outstanding (which would include the number of shares of common stock that would be issued in a full conversion of all preferred stock), and (2) the number of shares of common stock that would be issued from the conversion of all options, stock rights, warrants, and other securities.

The narrow-based weighted average formula would not include the convertible securities in option 2, only using the current outstanding securities.

Full Ratchet

Full-ratchet dilution protection gives investors the ability to recalibrate a per share purchase price downward to the price at which shares are sold in the next round of financing. As an example, assume the first-round investor paid $1 per share to own 10% of the company. In a later down round priced at $0.50 per share, the investor's shares would be repriced to $0.50 per share. The investor wouldn't receive additional shares of preferred stock, but the conversion rate would be adjusted to double the number of shares of common stock that the investor would receive if the investor were to convert the shares. This would ultimately increase the dilution on the common stockholders. This provision then allows the investor to maintain the ownership percentage in the company.

If the company does provide a ratchet adjustment, the company should consider asking that the investors be required to participate in future dilutive financing or lose all antidilution protection: the so-called "pay-to-play" feature (discussed further below). The problem for the company with full-ratchet dilution protection without the pay-to-play feature is that the ratchet protection eliminates an incentive to participate in subsequent dilutive financings because an investor can receive the benefit of the adjustment without putting additional money at risk.

ACCORDING TO VENTURE CAPITALISTS

What type of antidilution formula should I expect?

	Series A	Series B	Series C	Series D
None	6%	5%	5%	5%
Broad-based weighted average	83%	85%	80%	77%
Narrow-based weighted average	3%	3%	6%	2%
Ratchet	8%	7%	9%	16%

Source: Private Company Financing Report, Cooley Godward Kronish LLP, April 2006 (based on financing transactions in 2004–2005).

Some investors will offer a derivative of the full ratchet called a "partial ratchet," including a "half ratchet" or "two-thirds ratchets." The concept is similar to a full ratchet but will have less of an impact on the common holders. However, these provisions are fairly uncommon.

One other wrinkle on antidilution clauses is the pay-to-play provisions, which require the investors to participate in the dilutive round to receive antidilution protection with respect to their higher-priced preferred stock.

ACCORDING TO VENTURE CAPITALISTS

What percentage of financings have "pay-to-play" provisions?

Series A	Series B	Series C	Series D
13%	15%	22%	27%

	% with Pay-to-play
Quarter 2 of 2007	14%
Quarter 1 of 2007	9%
Quarter 4 of 2006	11%
Quarter 3 of 2006	8%
Quarter 2 of 2006	11%
Quarter 1 of 2006	13%

Source: Private Company Financing Report, Cooley Godward Kronish LLP, November 2007 (by quarter), April 2006 (by series, transactions from 2004 to 2005).

Preemptive Rights

In addition to the provisions discussed previously to limit dilution, investors will often request preemptive rights or a right of first refusal as a tool to allow the investor to maintain its ownership percentage. Preemptive rights will generally involve a contractual right to purchase additional shares in the company's next offering to maintain the investor's percentage shareholdings, usually the pro rata share. However, in some cases, the provisions provide that the investor has the right to purchase any future shares the company will issue, even in excess of their pro rata amount.

Preemptive rights can cause some complications for a company raising additional funds, because the company must offer a portion of the additional investment to its current holders, which can raise certain hurdles in adding a strategic investor or a new investor. As a result, if the investor requires preemptive rights, it is usually in the interest of the company to try to limit this right to only the purchase of a portion of the shares offered in the next round sufficient to maintain an investor's prefinancing shareholdings on a fully diluted basis. Preemptive rights should include language terminating the right before an IPO to prevent any problems that would result in a general sale of securities.

One key point on the right of first offer is to preserve the company's flexibility to engage in transactions such as equipment lease lines with warrant coverage without having to get waivers from the investor group.

Board Representation

VCs will almost always have at least some representation on the company's board of directors: usually, the lead investor or investors of the round will receive a board seat as a result of their investment. This role within the company should be welcomed because a well-connected and experienced director from your venture fund can offer insight, industry knowledge, and key contacts. However, depending on the size of the investment and the number of investors, the investors may control a majority of the board after their investment. In the transaction, the parties will typically determine that the new series of preferred stock (or all of the preferred stock as a class) will have specific voting rights with respect to the election of directors that vary from the one-share-one-vote rule set forth in the certificate or articles of incorporation.

A key point of negotiation will center on issues on the overall board composition. For example, before a Series A investment, the company may have a three-director board that will be expanded to four or five directors (with one or two investor designees) after the closing of the transaction financing. In the event that management is concerned about a losing control over its board, the company may want to consider the size of the investment or insist on additional board members to be elected by the common stockholders.

LIMITATIONS OF BOARD REPRESENTATIVES

Expect the board representatives from the venture capital firm to be busy. According to the National Venture Capital Association, the average number of company boards that a VC sits on is four. Understand the limitations that a partner of a venture capital firm will have if he serves on multiple boards in addition to your company.

In the case in which an investor does not hold a majority of a series of preferred stock, the investor will often receive board observer rights to permit the investor to designate an individual to observe the board meetings but not hold a vote. Additionally, investors may also negotiate for certain informational rights to receive financial statements and other company documentation on a monthly or quarterly basis.

Voting Rights

Venture preferred stock typically has voting rights, computed based on the number of shares of common stock into which the preferred stock can be converted. In addition, the preferred shareholders will be entitled to a class vote as a matter of corporation law under certain circumstances, such as certain merger situations.

In addition to the rights above, the investors will typically negotiate for veto rights over certain actions of the corporation. These veto rights are also known as protective provisions, approval rights, or negative covenants. Usually, the investors will negotiate for the right to approve the following:

1. Any merger or acquisition of the company irrespective of whether state corporation law grants them such a right
2. The creation of any new securities senior to or on a parity with their preferred stock
3. Any amendment to the articles of incorporation that would affect the rights of the preferred stock
4. Changes to any of the rights and preferences agreed to in the preferred stock sale

Although the protective provisions mentioned above are usually included in the transaction agreements, both parties may disagree on the addition of other provisions such as those discussed below.

The investors will likely want a separate vote of the preferred stockholders on some matters in which such a vote is not required by law. In situations in which the investors are acquiring a minority position in the company, and as such cannot control the board of directors, the investors may also attempt to negotiate the right to approve the following:

- Any sale, merger, or acquisition of the company
- Purchasing stock of another business
- Certain other corporate transactions, such as the sale of a product line
- The sale or an exclusive license of significant technology
- Purchase of assets over a certain dollar threshold
- Incurring debt
- The issuance of any new securities senior to or equal to the preferred stock
- Redemption of stock
- Increases in management's cash compensation beyond historical practice
- Additional stock or option grants to management
- Other points negotiated by the parties

These items listed above are typically more negotiable than items 1–4 in the preceding paragraph. Therefore, these provisions tend to be more negotiated by the parties. Some of these items may tend to fall into the traditional decision-making realm of management and the board of directors. In the event these veto rights encroach too far into management's responsibilities, you should attempt to negotiate those provisions away.

These veto rights in the event of specified events or protective provisions in the form of special voting rights of preferred stock holders are often (but not always) included in the term sheet. From the company's perspective, it is preferable to have a class vote (e.g., Series A and B voting together) rather than a series vote.

Redemptive Rights

The term sheet may include a provision to require the company to redeem (buy back at the original purchase price plus an annual carrying cost) the preferred stock. Investors sometimes seek redemption rights as a way of obtaining a return on their investment if the company does not go public or get acquired. Redemption provisions are fairly common in transactions, although mandatory redemption provisions are somewhat rare. Generally, in transactions that include redemption provisions, the majority will have a redemption provision at the option of the investor. Typically, redemption occurs over three or four years beginning five years after the financing, although some investors insist on total redemption all at once.

A redemption feature allows the company to retire the preferred stock after a specified period of time (typically between five and seven years) by paying the purchase price and, possibly, a small premium (although anything more than 10% of the purchase price creates potential tax problems for the investors). A redemption call effectively requires the preferred shareholders to elect whether to receive their investment and little or no return, with a corresponding loss of the opportunity cost of having their money tied up for a number of years, or converting their shares into common stock, with a loss of their preferential rights. In the case in which the investors see little hope of a successful exit for the company after a fairly long investment period, the redemption feature may be an attractive alternative for a venture fund.

Mandatory redemption is more commonly requested by investors who are taking a minority position in the company. For these investors who may not have a strong position through the board or as a major stockholder, the notion is that the mandatory redemption feature will allow them to get their money out in the event the company becomes one of the "living dead" (i.e., no real prospects for going public or being acquired). Mandatory redemption is disadvantageous from the company's standpoint because of the potential for a significant outflow of cash. In this regard, if a mandatory redemption feature has to be included to get the deal done, a staged payout, perhaps spanning a three-year period, is often negotiated.

ACCORDING TO VENTURE CAPITALISTS

Will the deal include redemption provisions?

Percentage of financings that include mandatory redemption or redemption at the option of the VC:

Q3'07	Q2'07	Q1'07	Q4'06	Q3'06	Q2'06	Q1'06	Q4'05
26%	22%	26%	22%	29%	33%	27%	31%

Source: Venture Capital Survey, Silicon Valley Third Quarter 2007, Fenwick & West LLP.

With the exception of mandatory redemption clauses, the redemption feature, like the dividend provisions, tends not to be controversial. As a practical matter, use of the redemption provision is quite rare, particularly for companies within California, because (1) California corporate law prohibits payments to shareholders unless certain liquidity tests are met, and (2) a company that can meet the liquidity tests is probably strong enough from a financial standpoint to go public or be acquired at a favorable valuation.

Cosale Rights

To prevent a situation in which, after an investment, the founders jump ship and sell their shares to a third-party investor, most venture financings will include a cosale right. The cosale right gives the investor the right to sell its shares on the same terms as the investors, usually on a pro rata basis of its total ownership percentage in relation to the total sale.

More specifically, cosale rights, sometimes called tag-along rights, are the right to participate in any pre-IPO sale by the founders in proportion to the relative number of shares owned by the founders and the investors. In a company in which a founder controls a majority of the company, the investors will often request cosale rights. As a practical matter, the cosale right will limit the incentive and ability of the founders to dispose of a significant portion of their holdings in the company before the IPO. However, because the opportunities for separate liquidity for founders generally tend to be limited, cosale rights are not generally viewed to be significantly disadvantageous to founders in most situations and are less likely to be negotiated than other more meaningful terms, such as vesting on founders' shares.

Investors may want to participate in any pre-IPO stock sale by the founders. Cosale rights, if exercised, will result in less money to you on a private stock sale, and the notice and other requirements of a cosale right may impede a sale. Nevertheless, because the investors want to use equity to direct the company's management toward a long-term return rather than an early cash out, they are likely to insist on a cosale right.

Cosale rights are common in venture investments and typically will offer exceptions for certain types of transfers, such as for small amounts of stock, sales, or transfers to family members, or sales in the event of termination of employment or death of the founders. In addition, the cosale rights will often be accompanied by a right of first refusal for the investors. With a right of first refusal, founders will be restricted from selling their shares to any third party without first offering the shares to the company and/or the investors. The shares will be offered on the same terms as given to the proposed third-party purchaser.

Drag-Along Rights

The primary purpose of drag-along rights are to ensure that the investor's shares won't be held hostage in the event of a favorable acquisition or merger event. Drag-along rights permit the holder of the rights to force the other shareholders to sell their shares if there is a third-party offer for the purchase of the company that has been approved by a certain percentage of the shareholders, usually a majority or a supermajority such as 70 or 80%. The reason this right is important to investors is that many acquirers will not purchase a company unless they are able to acquire all outstanding stock of the company. If a few shareholders hold out approving the transaction, the drag-along rights will permit the investor to force a sale by the holdouts.

ACCORDING TO VENTURE CAPITALISTS

Will the deal include a drag-along provision?

Series A	Series B	Series C	Series D
48%	46%	42%	42%

Source: Private Company Financing Report, Cooley Godward Kronish LLP, April 2006 (based on financing transactions from 2004 and 2005).

Registration Rights

Registration rights are an important right to investors because they provide a contractual right for investors to demand that the company register their shares with the SEC. After a registration, the investor can sell their shares on the public market. The importance of this right stems from the U.S. securities laws that permit the company, and not the shareholder, to register the company securities. To ensure that liquidity will eventually become an option for the investor, registration rights are negotiated as a part of the purchase.

Investors are typically granted certain registration rights from their transaction, consisting of demand rights (the right to require the company to register their shares with the SEC), "piggyback" rights (the right to include their securities in the company's offerings registered with the SEC), and S-3 rights (the right to require the company to register their shares on the SEC's short-form registration statement; this applies only after the company has been public for one year and meets certain other tests).

Demand Registrations

Registration on Form S-1 is an expensive and time-consuming process. Therefore, the demand registration rights can be a very burdensome obligation for the company. Usually, the demand registration rights may not be exercised for three to seven years after the preferred stock purchase. If the company does make a registration on its own earlier than the three to five year minimum, there is generally a three to six month time period after the IPO until the investors can make their demand. From a company's perspective, the time limitations for the demand rights should be set at a reasonable time in the future to prevent the investors from requiring the company to go public before management and the company is convinced the company does represent an IPO candidate. In most cases, the shares that can be registered must (1) represent a specified percentage (perhaps as low as 20% and as high as two-thirds of the outstanding securities), or (2) their value must equal a minimum specified dollar amount (usually $5 million to $25 million). These limits are set to ensure that enough shares will be registered to make the effort and cost of the offering worthwhile. The company will usually limit the number of permitted demand registrations to a maximum of three.

S-3 Registrations

A registration on Form S-3 is generally much less expensive and time consuming than registration on Form S-1 (because this filing can incorporate information from previous filings with the SEC). As a result, the company will often grant unlimited S-3 registrations on request of holders of a minimum percentage of the shares that can be registered, oftentimes 20–30%. However, the S-3 registration right will usually be limited to one or two registrations in a single one-year period. As with demand registrations on Form S-1, some companies will negotiate to limit the total number of S-3 registrations.

Piggyback Rights

Piggyback rights permit the investor to participate in a public offering of securities by the company. These rights are not usually unlimited, providing the underwriters with the ability to limit the amount of shares that can be registered for sale if it will influence the public sale. These rights are generally fairly noncontroversial.

In addition to the registration and piggyback rights, the registration rights provisions oftentimes include information on responsibility of filing fees and transfer restrictions on registration rights. One key point on registration rights is to ensure that they terminate at some point after the company's IPO or when the investors can sell their shares under SEC Rule 144. Typical registration rights will provide the investor with two demand registrations that may be brought at any time commencing three to five years after the investment, along with piggyback registration rights for company offerings.

Financing Expenses

Typically with venture capital investments, the company will pay the legal expenses of the attorney for the investors; generally, the investors will select a single attorney to represent all the investors if multiple investors are part of the syndicate. The company's counsel nearly always prepares the financing documents. The advantage to the company of this arrangement is twofold. The financing can be closed more quickly because company counsel can deliver the initial document package shortly after agreement on a term sheet and as draftsman can better control the timing thereafter. As a result, the financing should be accomplished less expensively.

The company often pays the legal fees of both its own counsel and counsel, if any, for the investors (which typically is paid from the financing proceeds). A cap on legal fees for investors' counsel is often established and runs in the range of $15,000 to $35,000. Anywhere from $15,000 to $35,000 is typical, depending on the complexity of the transaction (for example, a Series A financing for a new startup will likely be more expensive than a Series E with a bridge loan and warrants for a long-established company) and which side will bear primary drafting responsibility. Some companies are able to negotiate that the payment of expenses is contingent on the consummation of the financing; however, this is not a typical practice.

Employee Stock Option Pool

Investors will usually insist on the company setting aside additional shares of stock or option grants for future hires. Moreover, this employee reserve will be assumed to be fully issued for purposes of setting the per share price for the venture investors. The size of the employee reserve will depend primarily on the perceived staffing needs of the company during the succeeding year or two. If the investors believe that a number of high-level employees must be hired, they may insist on a substantial reserve.

Employees in venture-backed companies who receive stock grants are almost invariably put on a vesting schedule. A typical vesting arrangement would often provide for a four-year schedule, with no shares vesting during an initial period of between six months and one year (a cliff). At the end of the probationary period, the employee would become vested in the same number of shares as if he had been on a linear schedule, with the remaining shares vesting in equal monthly installments.

Founders and Employee Stock Restrictions

Many sophisticated venture investors will ask that the shares held by founders and employees be subject to forfeiture on departing the company, with specified percentages becoming vested against forfeiture over time. There is no set custom on this; where you come out will depend in large part on your relative bargaining strength with the investor. Even in cases in which the founders have previously imposed some type of vesting schedule on their stock, the investors may require the schedule to be modified and acceleration provisions to be reset. The addition of these provisions will typically involve the founders entering into individual stock restriction agreements with the company as a condition to closing.

If the founders are subjected to a vesting schedule, there are a number of ways it could be negotiated to lessen its impact. For example, the vesting might cover only a portion of your shares, the vesting period might be relatively short (one or two years), you might receive credit for previous performance, and the repurchase right might only arise if you voluntarily terminate your employment or if your employment is terminated for cause. In the event the founders have previously subjected the founders stock to vesting terms, some investors will simply continue the vesting (or modify it in some capacity). However, generally an investor will want to ensure that the length of vesting is adequate to ensure that the founder is incentivized to continue with the business for a set period of time, typically between three and four years.

One important item that is oftentimes negotiated in the context of founder and employee vesting are acceleration provisions. Acceleration is primarily discussed in cases of a change of control, an event such as a merger or sale. In the event of a change of control event, usually two different approaches to acceleration are discussed:

- **Single trigger:** In the event of a qualified change of control (the trigger), the options held by the founder or employee will be accelerated. In some cases, all option will be accelerated or a portion may be accelerated.
- **Double trigger:** This acceleration is triggered when (1) there is a change of control event, and (2) the employment of the individual is terminated without cause.
- **Combination approach:** In this case, a portion of the options may be accelerated in the event of a change of control (say, 25% of the unvested option) and the remaining options will be accelerated in the event of a termination after a change of control (the double trigger).

In some cases, acceleration may also be given in the event of termination of the employee without cause (not in conjunction with a change of control event), to prevent a scenario in which members of the management team are terminated by the board of directors to bring in a more seasoned management team.

Other Possible Terms

No-shop provisions limit the ability of the company to solicit competing term sheets from other investors. A No-talk provision is a similar restriction but extends to negotiations with any rival investors. Usually these limitations will be restricted to a 30-day period, after which the restriction will no longer apply. Interestingly, unless there is a fee for terminating the deal (which is rare in VC financings), there is usually not a remedy for the breach of these provisions included in the term sheet.

Information rights or rights to financial information are fairly standard provisions. Generally, the only negotiation will center on the frequency or scope of the information, limiting the information rights to information that is usually prepared in the course of business (standard financial statements or other standard reports) and within the standard scope of materials reviewed by the board or other investors (not materials under the purview of management). In addition, information rights that provide the investor with unannounced access to the company can create an unmanageable relationship for the company.

Confidentiality agreements, such as a nondisclosure agreement, will generally not be signed by an investor during the financing process. However, investors may be willing to provide for confidentiality and nondisclosure at the time the term sheet is signed.

The following is a list of other terms sometimes incorporated into term sheets but which tend to be much more unusual than any of the prior terms discussed. For the most part, these terms are considered to be investor-favorable.

- Mandatory redemption (holders have the option to require the company to repurchase after a certain date, such as five years after the initial investment)
- Increased pro rata rights in follow-on rounds (one and a half to two times pro rata rights in any additional rounds to allow investors to increase ownership percentages)
- Most favored nation for future financing (guaranteeing terms for the initial investor that are equal to or better than any follow-on investors)
- Required sale provisions (company is required to liquidate if there is no initial public offering before a certain date, such as five years after the initial investment)
- Warrant coverage (usually associated with an investment round that has the same or similar price to a previous round)
- Increase in the size of the stock option pool on a premoney basis
- Payment of funds only on the achievement of certain company milestones
- Mandatory cumulative dividends
- Requiring the founders to individually make representations and warranties (often in the case of intellectual property matters)
- Investment takes the form of convertible debt rather than equity
- Preferred stock options for management (this is promanagement to better align investor and management incentives and is generally done when common shares become highly diluted or preferred have vastly superior rights)

In the case of the items above, many of these terms are more typical in a period of tighter venture money supply, in which the investor has greater leverage. Depending on market conditions, these terms may be more common as investors recognize an increased need to protect their initial investments.

VALUATION FIXATION

What could happen

You've narrowed the list of potential investors for your company to four firms. The range of premoney valuations is fairly large. Shouldn't you just select the firm with the largest premoney valuation (to reduce the dilution)? Doesn't a high premoney valuation mean that firm "gets" your business and understands the long-term potential?

Watch out for

A high valuation doesn't necessarily mean a good deal for your company. A company should consider the impact of additional terms, including the time period of redemption rights exercisable, the use of participating preferred with a cumulative dividend, as well as ratchet antidilution protection without a cap. You should discuss the entire term sheet with someone experienced in VC deals, such as other entrepreneurs, members of investor forums, or your attorney. Model out the impact of these additional terms on future events such as a sale, IPO, or other liquidation event, each at various proceeds levels for the company.

It is also important to consider the differences in the firms themselves when selecting an investor for your company. Look to other portfolio companies and the founders of those companies. Look at the industry expertise. Look at the location. Remember that, in addition to your money, the investors play an important role on your board and can be a source of important contacts for new hires, customers, and collaborators. A high valuation from a firm that is not a good fit won't be the best partnership.

TIP: Consider the entire term sheet and research the investor firm before selecting your investment partner.

Deal Documents in a VC Financing

Signing the term sheet is the first step in closing the venture capital financing. Expect several weeks (or more depending on the time it takes you to provide legal due diligence to investor's counsel and the time they need to review it) from the term sheet signing until the closing of the investment. Generally, after consultation with your attorney, you should set a targeted closing date for the investment to push the process forward.

Every year, at a minimum, thousands of venture financing deals of private companies occur. Countless hours of preparation, negotiation, and drafting of investment documentation occurs. Nevertheless, venture capital financing documents tend to consist of a fairly standard set of terms and documents given the recurring issues that arise. The most common venture capital financing documents include a term sheet, a certificate or articles of incorporation, a stock purchase agreement, an investors' rights agreement, a right of first refusal and cosale agreement, a voting agreement, and various ancillary documents. Transaction documents may vary depending on the venture capital firm, the stage of the company, and the location of the venture capital firm.

Familiarity with the documents that will be a part of the transaction will help the entrepreneur understand the steps and discussions taking place between your attorney and the investor's counsel.

Term Sheet

As discussed previously, a term sheet is the document outlining the material terms and conditions of a business agreement. After the term sheet has been executed, it can be used as a template to create the final set of venture financing agreements. The term sheet itself guides the agreement but is not necessarily binding. Many of the terms not present in the term sheet are standard industry conditions that present less significant negotiating obstacles.

SEALING THE DEAL (WHAT YOU'LL NEED TO PROVIDE IN DUE DILIGENCE)

What could happen

You've found a source of funds, which may be a traditional venture capital firm, a bank loan, or a corporate partnership of some form. They are prepared to invest and have told you "now it is time for the lawyers to become involved."

What to expect

Traditionally, you and your attorney can expect to receive a due diligence request list. This will include a list of items you'll be expected to produce. Some of the items may be held by your attorney or accountant, but others may be in your control. The following represents a high-level summary of information you would typically expect to be asked to provide:

1. **Business plan backup:** Details of financial projections; backup data and information for key assumptions; sources for any third-party data; and more detailed information related to the use of proceeds
2. **Employment information:** All employment paperwork for key employees; invention assignment agreements; noncompetition agreements; confidentiality agreements; severance provisions for management; references for the CEO, current board members, and key managers
3. **Corporate documentation:** All corporate formation documentation; state filings; copies of board and shareholder minutes or resolutions; evidence of proper corporate filings; share issuances; option plans, option grants and option exercises; all corporate security issuances and evidence of compliance with securities laws
4. **Intellectual property:** Patents; trademarks; copyrights; filings with state or federal agencies; communications with the patent office; license agreements; information disclosure statements; trade secrets; internal policies for company intellectual property; invention assignment agreements; open source licenses (if applicable); copies of all nondisclosure agreements
5. **Market information:** Customer records and agreements; letters of intents for potential customers; channel partners; testimonials; reports from third parties; research data and source data

6. **Research, development, and engineering:** Product information; specifications; product documentation; performance data; beta tests and results; test data; benchmarking; industry research

TIP: Instill a culture of recordkeeping and maintain proper corporate governance records to make the due diligence process run smoothly.

Certificate of Incorporation/Articles of Incorporation

The certificate/articles of incorporation is a legal document relating to the formation of a company or corporation. It contains basic provisions detailing the corporate name, registered address and agent, statement of purpose, and other relevant information about the business's identity. After a preferred financing, however, the certificate is amended to include the terms of the preferred stock, including the number of shares, dividend payments, liquidation preferences, redemption, conversion rights, and voting rights.

Stock Purchase Agreement

A stock purchase agreement sets forth the basic terms of the purchase and sale of stock to investors. The main terms of negotiation that are involved in this agreement are the price and number of shares being sold and the representations and warranties that the company and perhaps founders make to the investors. The stock purchase agreement does not set forth the characteristics of the stock being sold or the relationship among the parties after the closing.

Many venture capital industry players often speak about differences in the "East Coast" and "West Coast" approaches to venture financing found in the stock purchase agreement. Although some cultural differences between the more bank-driven East Coast culture and entrepreneur-driven West Coast culture exist, many of the fundamental considerations are the same. One of the main differences is that northeast investors tend to want personal liability for misrepresentations of founders. Nonnortheast entrepreneurs are generally less comfortable with personal liability in risky investments.

Investors' Rights Agreement

An investors' rights agreement can cover many different subjects. Frequently, it covers information rights (the right to receive or have access to financial or other information about the company and sometimes the right to attend meetings of the board of directors), registration rights, contractual preemptive rights (the right to purchase securities in subsequent equity financings conducted by the company), and various postclosing covenants.

Right of First Refusal and Cosale Agreement

A right of first refusal agreement requires a shareholder to grant the company a right to match any offers for their stock. This effectively preempts other buyers and may prevent loss of control of the company. The right of first refusal may also include the provision that the company must buy back the shares of a shareholder who dies or leaves the company.

A variation of the right of first refusal (and often a more appealing option from the company perspective) is the right of first offer agreement. The right of first offer agreement provides that a shareholder takes a proposed price and set of terms for shares to be sold to the company, but, if the company turns down the offer, the shareholder becomes free to sell the stock to a third party, if the price is the same or greater than offered to the company.

The purpose of the cosale agreement is to provide the investors with the right to participate in sales by a founder (or other major shareholder) of the company.

Voting Agreement

A voting agreement is an agreement among shareholders to vote their shares in a certain way. In the context of a financing, this usually means holders of preferred stock agree with the major holders of common stock to vote their shares in a specified manner for the election of directors. By entering a voting agreement, the shareholders can override the standard one-share-one-vote rule and maintain management control.

Legal Opinion

In the opinion letter, a law firm renders an opinion as to the qualification of the company to enter into a venture financing deal and the legal validity of the documents used in the process. These opinions generally cover whether the company is properly formed, whether the company can enter into the venture financing deal, and whether the other legal documents used in the venture financing deal are enforceable and duly executed. The legal opinion, however, does not act as a warranty that all assets and activities of the company are legal. For instance, the opinion will not assess whether the assets and intellectual property claimed by the company is all properly owned and noninfringing.

Indemnification Agreement

Indemnification is the act of being held not liable or being protected from costs of liability by shifting them to another party. An indemnification agreement, then, simply shifts the financial liability of certain acts onto another party. Indemnification agreements can cover both corporate officers and directors. Some investors ask to be covered by an indemnification agreement as well. Investors will only be covered by indemnification when they act as an agent of the corporation. If investors seek indemnification for actions other than those under the corporate veil, indemnification provisions would need to be specified in the stock purchase agreement.

Indemnification agreements have several important purposes. On the one hand, they may provide heightened protection above that provided in a certificate of incorporation or bylaw because they cannot be amended without the approval of the indemnitee. Additionally, they can make indemnification mandatory in which default laws may otherwise make it permissive.

Some companies provide mandatory indemnification for directors and discretionary indemnification for officers. There may be situations in which mandatory indemnification of officers would make a company look bad, such as in a discrimination case. Companies should therefore use care when deciding whether to indemnify their officers.

SAMPLE DOCUMENT

Series A Preferred Stock Term Sheet

MEMORANDUM OF TERMS FOR THE PRIVATE PLACEMENT OF SERIES A PREFERRED STOCK OF HIGH- TECH STARTUP INC. (the "Company"):

Amount to be raised: $2,000,000

Type of security: Series A Preferred Stock ("Series A Preferred")

Number of shares: 2,000,000 shares

Purchase price: $1.00 per share (the "Purchase Price")

Investors:

	Investment amount	Number of shares
TopTech Venture LP	$1,250,000	1,250,000
ABC Ventures LP	$750,000	750,000
	$2,000,000	2,000,000

Closing date: The closing of the sale of the Series A Preferred (the "Closing") will be on or before June 30, 2009.

Postfinancing capitalization:

Class	Number of shares	Percent
Common Stock	3,000,000	50.00%
Series A Preferred Stock	2,000,000	33.33%
Option pool:		
Outstanding	100,000	1.67%
Future grants	900,000	15.00%
Total	6,000,000	100%

Rights, preferences, and restrictions of preferred stock:

Dividends: The holders of Series A Preferred will be entitled to receive noncumulative dividends in preference to the holders of Common Stock at an annual rate of 8% of the Purchase Price per share from legally available funds and when, as and if declared by the Board of Directors.

Liquidation preference: In the event of any liquidation, dissolution or winding up of the Company, the holders of Series A Preferred Stock will be entitled to receive in preference to the holders of Common Stock, the amount of $__ per share plus declared and unpaid dividends, if any. Thereafter, the remaining assets of the Company will be distributed ratably to the holders of Common Stock. A Liquidation Transaction shall be treated as though it were a liquidation, for purposes of triggering an immediate obligation to pay an amount equal to the aggregate liquidation preference of the Preferred Stock. A "Liquidation Transaction" means an Acquisition of the Company, as defined immediately below, provided that if the holders of at least ____% of the Preferred Stock elect not to treat the transaction as a Liquidation Transaction, an Acquisition of the Company shall be deemed not to constitute a Liquidation Transaction.

An "Acquisition of the Company" means (1) a sale, conveyance or other disposition of all or substantially all of the property or business of the Company, (2) a merger or consolidation with or into any other entity, unless the stockholders of the Company immediately before the transaction own 50% or more of the voting stock of the acquiring or surviving corporation following the transaction (taking into account, in the numerator, only stock of the Company held by such stockholders before the transaction and stock issued in respect of such previously held Company stock), or (3) any other transaction which results in (assuming an immediate and maximum exercise/conversion of all derivative securities issued in the transaction) the holders of the Company's capital stock as of immediately before the transaction owning less than 50% of the voting power of the Company's capital stock as of immediately after the transaction, provided, however, that an equity financing transaction in which the Company is the surviving corporation and does not (directly or through a subsidiary) receive any assets other than cash and rights to receive cash shall be deemed not to constitute an Acquisition of the Company. A series of related transactions shall be deemed to constitute a single transaction, and where such transactions involve securities issuances, they shall be deemed "related" if under applicable securities laws they would be treated as integrated.

The liquidation preference of the Preferred Stock, if not specified in a merger agreement, will be paid 100% in the form of cash unless the Board of Directors elects to pay it in another form; if the Board does so elect, all stockholder of the same class must be treated equally.

Redemption: The Series A Preferred will not be redeemable, except to the extent the liquidation provisions specified above are deemed by any applicable law to constitute redemption.

Voluntary conversion: Each holder of Series A Preferred will have the right, at the option of the holder at any time, to convert shares of Series A Preferred into shares of Common Stock at an initial conversion ratio of one-to-one.

Automatic conversion: The Series A Preferred will be automatically converted into Common Stock, at the then applicable conversion rate, in the event of either (1) the consent of the holders of a majority of the then outstanding Preferred Stock, voting together as a class, or, if earlier, (2) immediately before the closing of an underwritten initial public offering of the Company's Common Stock pursuant to a Registration Statement under the Securities Act of 1933, as amended with aggregate proceeds of at least $__ million at a public offering price (adjusted for intervening common stock splits, reverse stock splits and stock dividends) of at least $_____ per share (a "Qualified IPO").

Antidilution provisions: The conversion price of the Series A Preferred will be subject to proportional adjustment for stock splits, stock dividends and the like, and to adjustments on a broad-based weighted average basis for issuances at a purchase price less than the then-effective conversion price, subject to the following carve outs:

(1) common stock issued pursuant to stock splits and common-stock-on-common-stock dividends,

(2) up to 1,000,000 shares of common stock issued or issuable to employees, officers, consultants or directors of the Company, or

other persons performing services for the Company, directly or pursuant to a stock option plan or restricted stock plan or agreement approved by the Board of Directors;

(3) capital stock, or options or warrants to purchase capital stock, issued to financial institutions with federal or state charters or to lessors in connection with commercial credit arrangements, equipment financings, commercial property lease transactions or similar transactions;

(4) capital stock issued or issuable to an entity as a component of any business relationship with such entity for the purpose of (a) joint venture, technology licensing or development activities, (b) distribution, supply or manufacture of the Company's products or services or (c) any other arrangements involving corporate partners that are primarily for purposes other than raising capital, the terms of which business relationship with such entity are approved by the Board of Directors;

(5) capital stock, or warrants or options to purchase capital stock, issued in connection with bona fide acquisitions, mergers or similar transactions, the terms of which are approved by the Board of Directors;

(6) common stock or other underlying security actually issued upon conversion, exchange or exercise of any derivative security;

(7) common stock issued or issuable in or under a Qualified IPO;

(8) common stock issued or issuable as a result of the antidilution provisions of any derivative securities; and

(9) common stock issued or issuable in or under a transaction for which the holders of at least a majority of the then outstanding shares of Preferred Stock, voting together as a class, adopt a resolution that expressly states that such Common Stock is not to be considered Additional Shares.

Voting rights: The holder of a share of Series A Preferred will be entitled to that number of votes on all matters presented to stockholders equal to the number of shares of Common Stock then issuable upon conversion of such share of Series A Preferred. Notwithstanding the foregoing, the Series A Preferred shall not be entitled to vote on any matter for which voting is expressly reserved, by law or the express provisions of the Company's Certificate of Incorporation, solely for a class or classes of stock other than the Preferred Stock, or for one or more series of Preferred Stock other than the Series A Preferred.

Protective provisions: So long as at least 1,000,000 shares of Preferred Stock remain outstanding, the Company will not (by amendment, merger, consolidation or otherwise, and either directly or indirectly by a subsidiary), without the approval of a majority of the Preferred Stock, voting together as a class.

(1) effect (a) a liquidation, dissolution or winding up of the Company (b) an Acquisition of the Company, or (c) any other merger or

consolidation of the Company, or a subsidiary of the Company, with or into any other entity;

(2) alter or change the rights, preferences or privileges of the Preferred Stock so as to materially and adversely affect such shares;

(3) increase or decrease the number of authorized shares of the Preferred Stock or the Series A Preferred;

(4) authorize the issuance of securities having a preference over or on a parity with the Series A Preferred as to dividends or liquidation;

(5) redeem, repurchase or otherwise acquire shares of Preferred Stock or Common Stock other than in accordance with **[the redemption provisions of the Preferred Stock or]** the repurchase of shares from employees, officers, directors, consultants or other persons providing services to the Company at no greater than cost pursuant to the original terms of such agreements, or such modified terms as have been agreed to by the Board of Directors **[including the Series A director representative then in office, if any].**

Registration rights:

Registrable Securities: All shares of Common Stock issuable upon conversion of the Preferred Stock shall be deemed "Registrable Securities."

Demand registration: Beginning December 31, 2012, or six months after a Qualified IPO, whichever is earlier, ______ demand registrations of at least $_________ each, upon initiation by holders of __% of the outstanding Registrable Securities.

Piggyback registration rights:

Unlimited piggyback registration rights, subject to pro rata cutback to a minimum of __% of the offering (complete cutbacks on the IPO) at the underwriter's discretion.

Registration on Form S-3: The holders of at least __% of the Registrable Securities will have the right to require the Company to register on Form S-3, if available for use by the Company, shares of Registrable Securities for an aggregate offering price of at least **[$500,000]**. The Company will not be obligated to effect more than two S-3 registration statement in any twelve month period.

Registration expenses: Registration expenses (exclusive of underwriting discounts and commissions, stock transfer taxes and fees of counsel to the selling stockholders) will be borne by the Company for all demand, piggyback and S-3 registrations. The Company will also pay the reasonable fees and expenses of one special counsel to the selling stockholders **[not to exceed $_______]**.

Assignment of registration rights: The registration rights may be transferred to a transferee who acquires at least _______ shares of the original purchaser's Registrable Securities (or all of such the transferring holder's shares, if less), provided that the Company is given prompt notice of the transfer and the transferee agrees to be bound by the terms and conditions of the Investors' Rights Agreement. Transfer of registration rights to a partner or affiliate of the transferee will be without restrictions as to minimum shareholdings.

Lockup agreement: In connection with a Qualified IPO, each holder of registration rights will be required not to sell or otherwise dispose of any securities of the Company (except for those securities being registered) for a period of 180 days following the effective date of the registration statement for such offering if so requested by the underwriters of such offering subject to such extensions of time as may be required by the underwriters under NASD Rule 2711 or a successor rule.

Termination of registration rights: The registration obligations of the Company will terminate on the earlier of (1) __ years after a Qualified IPO, (2) with respect to any holder of registration rights, at such time as all Registrable Securities of such holder may be sold within a three month period pursuant to Rule 144 or (3) upon an Acquisition of the Company.

Information rights:

So long as a holder of Series A Preferred continues to hold **[at least _____]** shares of Series A Preferred or Common Stock issued upon conversion of Series A Preferred, the Company will deliver to such holder annual, quarterly and monthly financial statements as well as an annual budget. The obligation of the Company to furnish such information will terminate at such time as the Company (1) consummates an IPO, (2) becomes subject to the reporting provisions of the Securities Exchange Act of 1934, as amended, or (3) upon an Acquisition of the Company.

Right of first offer:

Each holder of at least _____ shares of the Series A Preferred will have the right in the event the Company proposes to offer equity securities to any person other than:

(1) common stock issued pursuant to stock splits and common-stock-on-common-stock dividends,

(2) up to 1,000,000 shares of common stock issued or issuable to employees, officers, consultants or directors of the Company, or other persons performing services for the Company, directly or pursuant to a stock option plan or restricted stock plan or agreement approved by the Board of Directors;

(3) capital stock, or options or warrants to purchase capital stock, issued to financial institutions with federal or state charters or to lessors in connection with commercial credit arrangements, equipment financings, commercial property lease transactions or similar transactions;

(4) capital stock issued or issuable to an entity as a component of any business relationship with such entity for the purpose of (a) joint venture, technology licensing or development activities, (b) distribution, supply or manufacture of the Company's products or services or (c) any other arrangements involving corporate partners that are primarily for purposes other than raising capital, the terms of which business relationship with such entity are approved by the Board of Directors;

(5) capital stock, or warrants or options to purchase capital stock, issued in connection with bona fide acquisitions, mergers or similar transactions, the terms of which are approved by the Board of Directors;

	(6) common stock or other underlying security actually issued upon conversion, exchange or exercise of any derivative security;
	(7) common stock issued or issuable in or under a Qualified IPO;
	(8) common stock issued or issuable as a result of the antidilution provisions of any derivative securities; and
	(9) securities which, with the unanimous approval of the Board of Directors, are not offered to any existing shareholders, to purchase that portion of such equity securities equal to (a) the number of shares of Common Stock issued or issuable upon conversion of the Series A Preferred held by each holder of Series A Preferred divided by (b) all of the Company's Common Stock then outstanding or issuable upon exercise of options or warrants or conversion (on a fully-diluted basis) of Preferred Stock. Such equity securities shall be purchased within 15 days from notice by the Company and on the same terms as they are purchased by other third party purchasers of the equity securities. Such right of first offer will terminate upon an IPO or upon the acquisition of the Company. The right of first offer shall not be applicable with respect to any covered investor and any subsequent securities issuance, if (a) at the time of such subsequent securities issuance, such investor is not an accredited investor, and (b) such subsequent securities issuance is otherwise being offered only to accredited investors.
Cosale right and right of first refusal:	Holders of Preferred Stock shall have the right to participate on a pro rata basis in transfers of stock for value by _________ and ______ (the "Founders") (with customary exceptions for transfers in connection with estate planning and similar matters and an exception for up to 5% of each Founder's common stock). This right will terminate immediately prior to the IPO or an acquisition of the Company. The existing right of first refusal and vesting provisions with respect to the Founders' stock will remain in effect. The right of first refusal will be assigned to the holders of the Preferred Stock on a pro rata basis in the event it is not exercised by the Company.
Board of directors:	So long as ______ shares of Preferred Stock shall remain outstanding, the Preferred Stock, voting together as a class, shall be entitled to elect _____ member(s) of the Company's Board of Directors, and the remaining members will be elected by the Preferred Stock and Common Stock voting together as a class. Upon closing of the financing, the Board of Directors of the Company shall consist of _________.
Exclusivity:	In consideration of ___ investing substantial time, effort and expense in connection with the evaluation and execution of the transactions contemplated by this term sheet, the Company agrees that it will immediately cease any and all discussions or interactions with third parties concerning alternative financing transactions and, for a period of time expiring on the earlier of _______ or the date that ___________ shall have notified the Company in writing that it does not intend to proceed with the transactions contemplated hereby, (a) the Company and ___________ will work in good faith toward the

	consummation of the transactions contemplated by this term sheet and (b) the Company will not negotiate with, provide information to, or otherwise interact with any third party concerning any alternative transaction.
Purchase agreement:	The sale of the Series A Preferred will be made pursuant to a stock purchase agreement reasonably acceptable to the Company and the Investors, which agreement will contain, among other things, appropriate representations and warranties of the Company and the Investors, covenants of the Company reflecting the provisions set forth in this term sheet and appropriate conditions to closing which will include, among other things, qualification of the shares to be sold under applicable Blue Sky laws, and the filing of Amended and Restated Certificate of Incorporation.
Expenses:	If the Financing is consummated, the Company will pay the hourly fees and expenses of one special counsel to the Investors, not to exceed $[________].
Counsel to the company:	Cooley Godward Kronish LLP

BOARD OF DIRECTOR FIDUCIARY DUTIES IN A VENTURE FINANCING

Federal and state laws closely regulate the offer and sale of securities, which include stock, options, and warrants. On the corporate level, the board of directors must approve all offers and issuances of securities. When claims are brought for breach of fiduciary duties, however, VC directors are the most likely target. They are perceived to have the deepest pockets, both personally and as a result of supplemental directors' and officers' liability insurance maintained by their fund to cover claims at the portfolio company level.

In most jurisdictions, the directors of a company owe a fiduciary duty to both the company and its shareholders. This duty includes both a "duty of loyalty" and a "duty of care." Under the duty of loyalty, a director must act in a manner reasonably and honestly believed to be in the best interests of the company and its shareholders, seeking to avoid any conflict of interest situations. The director must place the interests of the company and its shareholders above the interests of the director or any entity that the director represents.

Directors are also required to exercise the amount of care that an ordinarily careful and prudent person would use in similar circumstances. A director's duty of care comprises care in both the decision-making process and overseeing the conduct of employees and advisors. To fulfill his duty of care in decision-making, a director must base his decisions on all material information reasonably available to him and he must critically assess that information. To fulfill his duty of care in oversight, a director must conduct adequate and necessary inquiries in which suspicions of misconduct should be, or are, aroused.

Adherence to one's fiduciary duties is critically important for a director because directors are personally liable for breaches of their duty of care. Some state laws, including Delaware, allow companies to limit or eliminate the personal financial liability of their directors for breaches of fiduciary duty. This director immunity does not apply in some circumstances, as follows: when the breach is of a duty of loyalty; when the director commits acts of omissions not in good faith or which involve intentional misconduct or a knowing violation of law; when directors make unlawful payments of dividends or unlawful stock purchases or redemptions; or when the director derives an improper personal benefit from transactions.

In addition to immunity from liability written into a corporate charter, the business judgment rule provides additional protection for directors. The rule creates a presumption that a director acted property in his/her decision-making process. The rule also places a difficult burden of proof to rebut this presumption on a plaintiff challenging it. For the business judgment rule to apply, generally the following criteria is needed: the decision was free from a conflict of interest; the directors must have acted or made a deliberate decision not to act; the directors must have considered all material information reasonably available to them; and the directors' decision must be based on a good faith belief that it is in the best interests of the corporation.

One particular aspect of a director's fiduciary duty with respect to venture financing is worth noting. When a company is insolvent or nearing insolvency, directors owe their primary duty to creditors rather than to stockholders. Directors are responsible for preserving asset value for eventual distribution to the creditors. A debtor becomes insolvent when its liabilities exceed its assets at a fair evaluation or it lacks sufficient property to pay its existing debts as they mature. After insolvency, directors' actions are judged by a trustee standard rather than by the business judgment rule. Under the trustee standard, directors are required to exercise such care and skill as a person of ordinary prudence would exercise in dealing with his own property. Directors may be subject to personal liability for harm to creditors resulting from mere negligence under the trustee standard.

WASHOUT FINANCINGS

Washout financing refers to financing at a late stage in venture financing when a once highly valued company is now valued at a much lower price (a "down round"). Venture capitalists are faced with the choice of either bailing on their investment or investing more money to save their struggling investment. Because the risk is high with washout financing, favorable terms are often extended in exchange for an equity stake.

According to the November 2007 Private Company Financing Report by Cooley Godward Kronish LLP, 78% of financings over the 2005–2007 period were "up rounds" (deals done at a higher price than the previous fundraising round), whereas 22% were considered to be "down rounds" (deals done at a lower price than the previous fundraising round).

To avoid accusations of breach of fiduciary duty, the VC firm (through its representative on the board of the company) will need to show that it has sought financing alternatives that were more favorable to the company and the nonparticipating existing investors. VC firms that have board representation in a portfolio company must remember that their representation on the board mandates owing a duty of fairness and good faith to all shareholders (including the minority shareholders). Often, controlling shareholders also have a fiduciary duty of disclosure to the minority. In the context of the financing, the VC investor is obliged to disclose all of the terms of the financing to the minority shareholders.

In lawsuits against VC firms and shareholders, courts tend to give deference to the decisions of controlling shareholders of the company. Courts believe that controlling shareholders can act in their own interest, even if at the expense of minority shareholders, as long as there is no fraud, bad faith, or malice. For Delaware companies, there is no statutory protection for minority shareholders. To protect against washout financing, a claim of breach of fiduciary duty would need to be brought. However, the business judgment rule provides directors with a presumption of good faith.

To avoid problems resulting from a washout financing, such as suits from disgruntled shareholders, investors must offer all shareholders the investment opportunity to purchase newly issued stock. Stockholders can purchase the percentage of stock issued that they already own in the company to preserve their ownership power. Of course, this offer may not mean much to the shareholders if they do not have the financial resources to buy the offered shares. Legally, however, it may prevent against breach of fiduciary duty claims.

When founders of a startup or other initial investors have negotiated antidilution clauses, washout financings are difficult. In most investments, however, the founders do not have antidilution protection. These provisions limit a company's ability to negotiate successive rounds of financing, and many VC firms do not permit such protection for founders. Also, during initial rounds of financing, investors and founders generally do not imagine scenarios of insolvency and are less likely to fight for antidilution provisions.

2007 Top Venture Capital Firms

Venture capital firm	Location	Number of deals in 2007
Draper Fisher Jurvetson	Menlo Park, CA	100
New Enterprise Associates	Baltimore, MD	87
Intel Capital	Santa Clara, CA	70
Polaris Venture Partners	Waltham, MA	61
Kleiner Perkins Caufield & Byers	Menlo Park, CA	56
Sequoia Capital	Menlo Park, CA	55
Menlo Ventures	Menlo Park, CA	54
Canaan Partners	Westport, CT	50
U.S. Venture Partners	Menlo Park, CA	49
Accel Partners	Palo Alto, CA	48

Foundation Capital	Menlo Park, CA	48
Highland Capital Partners LLC	Lexington, MA	47
InterWest Partners	Menlo Park, CA	47
Venrock Associates	New York, NY	47
Alta Partners	San Francisco, CA	44
Bessemer Venture Partners	Larchmont, NY	44
Domain Associates LLC	Princeton, NJ	43
Atlas Venture, Ltd.	Waltham, MA	42
Mohr Davidow Ventures	Menlo Park, CA	42
Austin Ventures, L.P.	Austin, TX	40
North Bridge Venture Partners	Waltham, MA	40
Sigma Partners	Menlo Park, CA	40
Morgenthaler Ventures	Menlo Park, CA	39
Versant Ventures	Menlo Park, CA	39
Benchmark Capital	Menlo Park, CA	38
Sevin Rosen Funds	Dallas, TX	38
Advantage Capital Partners	New Orleans, LA	37
Pequot Capital Management, Inc.	Westport, CT	37
Duff Ackerman & Goodrich LLC	San Francisco, CA	36
Frazier Healthcare and Technology Ventures	Seattle, WA	36
Lightspeed Venture Partners	Menlo Park, CA	36
VantagePoint Venture Partners	San Bruno, CA	36
Khosla Ventures	Menlo Park, CA	35
ARCH Venture Partners	Chicago, IL	34
Battery Ventures LP	Waltham, MA	34
First Round Capital	West Conshohocken, PA	34
Flagship Ventures	Cambridge, MA	32
Goldman, Sachs & Co.	New York, NY	32
Mayfield Fund	Menlo Park, CA	32
Sutter Hill Ventures	Palo Alto, CA	32
General Catalyst Partners	Cambridge, MA	31
Redpoint Ventures	Menlo Park, CA	31
SV Life Sciences Advisers	Boston, MA	31
Trident Capital	Palo Alto, CA	31
Oak Investment Partners	Westport, CT	30
Prism Venture Partners	Westwood, MA	30
Village Ventures	Williamstown, MA	30
Intersouth Partners	Durham, NC	29
Velocity Interactive Group	Palo Alto, CA	29
Oxford Bioscience Partners	Boston, MA	28
Sanderling Ventures	San Mateo, CA	28
Adams Street Partners LLC	Chicago, IL	27
DCM	Menlo Park, CA	27
Greylock Partners	Waltham, MA	27
Scale Venture Partners	Foster City, CA	27

Source: PricewaterhouseCoopers/National Venture Capital Association MoneyTree Survey (http://www.pwcmoneytree.com).

2007 Top 60 Venture Capital Firms for Early Stage Companies

Venture capital firm	Location	Number of deals in 2006
Maryland Technology Development Corporation	Columbia, MD	22
Draper Fisher Jurvetson	Menlo Park, CA	19
Tech Coast Angels	Laguna Hills, CA	17
New Enterprise Associates	Baltimore, MD	13
Khosla Ventures	Menlo Park, CA	12
Sequoia Capital	Menlo Park, CA	12
Village Ventures	Williamstown, MA	11
Austin Ventures	Austin, TX	10
Band of Angels	Menlo Park, CA	9
Omidyar Network	Redwood City, CA	9
Canaan Partners	Westport, CT	8
Foundation Capital	Menlo Park, CA	8
Illinois Ventures LLC	Chicago, IL	8
Intel Capital	Santa Clara, CA	8
JumpStart Inc.	Cleveland, OH	8
Mohr Davidow Ventures	Menlo Park, CA	8
Polaris Venture Partners LP	Waltham, MA	8
Atlas Venture	Waltham, MA	7
Ben Franklin Technology Partners Southeastern PA	Philadelphia, PA	7
Charles River Ventures	Waltham, MA	7
Draper Richards LP	San Francisco, CA	7
Hummer Winblad Venture Partners	San Francisco, CA	7
Venrock	New York City, NY	7
Advantage Capital Partners	New Orleans, LA	6
Bessemer Venture Partners	Larchmont, NY	6
Domain Associates LLC	Princeton, NJ	6
First Round Capital	West Conshohocken, PA	6
Kleiner Perkins Caufield & Byers	Menlo Park, CA	6
Lightspeed Venture Partners	Menlo Park, CA	6
Maryland DBED	Baltimore, MD	6
Morgenthaler Ventures	Menlo Park, CA	6
Prism Venture Partners	Westwood, MA	6
Spark Capital	Boston, MA	6
Sutter Hill Ventures	Palo Alto, CA	6
Trinity Ventures	Menlo Park, CA	6
U.S. Venture Partners	Menlo Park, CA	6
ARCH Venture Partners	Chicago, IL	5
Benchmark Capital	Menlo Park, CA	5
CMEA Ventures	San Francisco, CA	5
ComVentures	Palo Alto, CA	5
De Novo Ventures	Menlo Park, CA	5

DFJ Frontier	Santa Barbara, CA	5
Flagship Ventures	Cambridge, MA	5
General Catalyst Partners	Cambridge, MA	5
Greylock Partners	Waltham, MA	5
Highland Capital Partners LLC	Lexington, MA	5
Masthead Venture Partners	Cambridge, MA	5
Matrix Partners	Waltham, MA	5
New Jersey Technology Council	Mount Laurel, NJ	5
ONSET Ventures	Menlo Park, CA	5
Sevin Rosen Funds	Dallas, TX	5
Sigma Partners	Menlo Park, CA	5
Three Arch Partners	Portola Valley, CA	5
Versant Ventures	Menlo Park, CA	5
Accuitive Medical Ventures LLC	Duluth, GA	4
Alloy Ventures	Palo Alto, CA	4
ATA Ventures	Redwood City, CA	4
Avalon Ventures	La Jolla, CA	4
Battery Ventures LP	Waltham, MA	4
BlueRun Ventures	Menlo Park, CA	4

Source: http://entrepreneur.com.

2007 Top 30 Venture Capital Firms for Later-Stage Companies

Venture capital firm	Location	Number of deals in 2006
Tech Coast Angels	Laguna Hills, CA	7
Stonehenge Capital Company	Baton Rouge, LA	6
Edison Venture Fund	Lawrenceville, NJ	5
HIG Capital Management	Miami, FL	5
New Enterprise Associates	Baltimore, MD	5
Sequoia Capital	Menlo Park, CA	5
Benchmark Capital	Menlo Park, CA	4
Polaris Venture Partners	Waltham, MA	4
Redpoint Ventures	Menlo Park, CA	4
Safeguard Scientifics Inc.	Wayne, PA	4
Advantage Capital Partners	New Orleans, LA	3
Altos Ventures	Menlo Park, CA	3
Ascend Venture Group LLC	New York, NY	3
Capital Resource Partners	Boston, MA	3
Carlyle Group	Washington, DC	3
Core Capital Partners	Washington, DC	3
Draper Fisher Jurvetson	Menlo Park, CA	3
Expansion Capital Partners LLC	San Francisco, CA	3
Goldman Sachs & Co.	New York, NY	3
Intel Capital	Santa Clara, CA	3
Milestone Venture Partners	New York, NY	3

Oak Investment Partners	Westport, CT	3
Sierra Ventures	Menlo Park, CA	3
SV Life Sciences Advisers	Boston, MA	3
Ticonderoga Capital Inc.	Wellesley, MA	3
Trillium Group LLC	Pittsford, NY	3
Updata Partners	Reston, VA	3
Village Ventures	Williamstown, MA	3
Spring Capital	Salt Lake City, UT	3
Accel Partners	Palo Alto, CA	2

Source: http://entrepreneur.com.

11

Building Your Team

Frankly, I'm more of a researcher, teacher, motivator, and coach than I am an entrepreneur.

David Allen
Author of Getting Things Done

What to Watch For

For many high-technology startup companies, employees will represent one of the most crucial assets and one of the most substantial expenses.

The makeup of the team is one of the most important characteristics for potential investors considering an investment. Research has also found correlations between certain characteristics of a startup team and the likelihood of receiving venture capital financing and completing an IPO. These key team traits include previous startup experience (both successful and unsuccessful), years of management experience, numbers of patents filed, numbers of PhD degrees, previous team experience in the same or similar competitive marketplace, and countless other traits.

Building the startup team is a crucial piece of the success or failure of your high-technology startup venture. Hiring well is not a one-time event but represents an ongoing demand on the company to recruit, attract, hire, retain, and develop key talent for the organization.

Staffing a High-Technology Startup

Startup companies cannot afford to make too many hiring mistakes. Therefore, it is important to start with a top-tier management team and continue to expand the organization with high-level hires. This all sounds obvious, right?

However, high-level hires don't come along every day and very likely won't be looking for a new job at exactly the moment you come calling. Therefore, your company will need to balance the need to grow the organization with the limitations to attract and hire highest-quality employees. Rather, a high-tech startup should always be looking for good candidates, should hire candidates that fit the organization, and should be willing to fire employees that do not fit in with the company.

EMPLOYING SUCCESS (AND FIRING FAILURES)

What could happen

Hiring for startup success is crucial. In those first months, you may find yourself hiring quickly and you may find that some of those hires can't cut it in a startup.

Watch out for

You can't avoid mistakes in hiring. However, remember that, in a small and growing organization, the wrong personalities can quickly spoil the bunch. Successful employers will hire slowly but fire quickly.

TIP: For your startup, hire slowly but fire quickly.

As discussed previously in Chapter 4 on recruiting a founding team, the process to identify top talent is only the first step. After you've identified potential team members, you've got to convince them to join your team. To repeat the words of investor and author Guy Kawasaki included in Chapter 4 on recruiting founders, "The art of recruiting is the purest form of evangelism because you're not simply asking people to try your product, buy your product, or partner with you. Instead, you are asking them to bet their lives on your organization." In particular, the people you want to join your organization will have other opportunities. Remember, when someone says, "I'm not interested" followed by a litany of questions about the business, they may well be interested but not yet ready to commit. In those cases, be ready to evangelize.

You should overlay a hiring plan with your business plan to match your needs as you grow with the recruitment of talent, but remember that good candidates may not show up in line with your hiring plan. Be flexible and adapt as talent becomes available.

Finally, for most startups, salaries represent a major source of expense for the company. New hires are expensive and will cost you more than a salary and stock options. As such, startup companies should consider the use of independent contractors, temporary workers, and third-party service providers to balance the business needs.

A Startup Persona

Not everyone is startup material. Employees that succeed at a startup company are not necessarily the same types of people that succeed in a large, well-established company. As such, you will be looking to identify key talent to involve in your venture that brings the right mixture of experience and skills, as well as the personality to succeed at a startup company.

Researchers Edward Roberts and Alan Fusfeld studied the key traits found in individuals undertaking high-technology projects. They found that a successful high-technology project required a team of people in which the mixture of talent includes individuals with at least above-average skills in (1) idea generation, (2) entrepreneurship or championing, (3) project leadership, (4) gatekeeping, and (5) sponsorship or coaching.

This research emphasizes the importance of developing a team that not only includes technical skills but a mixture of interpersonal skills needed to accomplish tasks and meet goals. You should be aware of these interpersonal skills and remember when staffing project teams to find ways to incorporate a mixture of people that share these skills.

Technical Talent

By their very nature, high-tech companies require technical talent. Hiring technical talent is not as easy as it may seem; simply having the right credentials, certifications, and references doesn't make your potential hire a good programmer, scientist, or engineer. Identifying technical talent most likely requires a broad level of experience in the particular field to know when a potential hire has the right skills or simply has lots of degrees and certifications. To be blunt, good programmers, engineers, or scientists know what other programmers, engineers, or scientists need to know to be great. Use the resources within your company, including board members, advisory board members, current employees, and third-party information to made determinations on the technical talent of potential hires.

Finding this talent requires consistency (you should always be looking), an eye and experience for the right hires (credentials and experience don't always mean the right fit), and a bit of luck (because you can never get a full sense of a potential hire).

MISSING TECHNICAL TALENT

What could happen

You have a great idea, tremendous market potential, and some key business and entrepreneurial talent, but you are missing key technical hires at this stage of your business. How can you fix that?

What to expect

If you aren't a technical whiz in a particular area (perhaps you only got a B minus in quantum nanophysics or barely passed "science-for-dummies"). Don't fret. Oftentimes, a nontechnical founder or a founder without experience in one technical aspect will rely on the support of an advisory board member or a key technical founder. The phrase "it takes one to know one" is very true in identifying true technical talent. Be sure to rely on more than degrees and certification when making technical hires.

In the late 1990s, some have argued that many of the failures from the dotcom bust resulted from a dearth of talented programmers or other technical individuals. Business-savvy individuals came up with exciting and innovative ideas and brilliant marketing strategies but missed the boat when it came time to implement them. In a high-tech business, technical talent is at a premium, so surround yourself with individuals who can help you select the right candidates for your business.

TIP: Identifying true technical talent often requires a technical background and understanding.

The Hiring Process

The business planning process will oftentimes involve determining an employee headcount for the upcoming period. The hiring process will follow certain generalities, but some features of the process will depend on the timing for the hire and the seniority of the potential position. In any case, you should be aware of common problems that can occur in the hiring process. Some examples follow.

Legal Considerations

Federal and state laws provide limits to an employer's right to hire in any way it chooses. This means that you should be aware of these limitations and be certain that all other individuals who will be a part of the hiring process are aware of the limits. In particular, there are limitations set on the types of questions asked of a candidate. Other considerations include certain restrictions that may be imposed by former employers of your employees. These nonsolicitation restrictions may limit your ability to actively recruit candidates from this company.

Additionally, you should ask to review the current employment agreements of any potential hire to be certain that they are not subject to a noncompetition agreement barring them from working for your company. Although some states may not enforce noncompetition agreements or the agreements may have limited enforceability in others, it is important to be made aware of these issues before you hire an employee to have these provisions reviewed.

Forgetting Formalities

Although you may know members of your management team well, many problems can be prevented by formalizing key employment and formation terms with your new hires. Deal up front with issues such as salary, severance, stock or option issuance, vesting terms, equity transfer restrictions, bonuses, invention assignment, confidentiality, nondisclosure, and noncompetition. Prepare offer letters and related agreements before a new team member begins. This will help to avoid problems down the road and help you focus on the more important issue: running your new business.

Proper Candidate Screening

Screening applicants is oftentimes one of the most important phases of hiring. Either over the phone or in person, screening questions must avoid all information regarding an applicant's race, religion, gender, national origin, age, or marital status. An applicant may be asked whether they are over the age of 18 or whether they have any criminal conviction related to the job. Any questions asked of a reference must also avoid questions prohibited to ask the applicant directly. An exception to asking for information related to an applicant is when a protected characteristic is a bona fide occupational qualification.

Overselling a New Venture

With any brand new business, optimism will likely abound. However, when recruiting new talent to join the company (particularly in the earliest stages), avoid focusing too heavily on the positives of the technology and failing to provide a balanced view of the risks. Although potential hires with previous startup experience may not need these reminders, individuals that have only held positions with well-established companies may not have a full grasp of the potential downsides. If a new hire is looking for structure and guidance, your new company may not be the best place. Some new high-tech companies have found that bringing in new members of the management team on a "test" basis for three to six months is an effective way to prevent serious mistakes.

Failing to Reach Out to New Talent

It is human nature to be most comfortable working with individuals we know and trust. However, this mentality may not be the ideal approach for a startup. Investors may be wary of a team that is composed primarily or entirely of individuals with a similar working background. Look to establish a team with unique talents to tackle the business issues you will face. This diversity of talent may lead the group to new solutions.

Consider a Potential Investor's Perspective

When you begin looking for funds from potential investors, you are being judged on the management team for the company. Analyze your team as a potential investor would. Who will provide financial, sales, or marketing expertise? In some cases, it may make sense to bring these hires on earlier to cement the team and provide a more complete vision of your management before reaching out to investors. In other cases, be prepared to discuss likely or potential hires. If a hire has stated she is willing to start after funding the venture, you may be able to highlight the background of this individual without revealing her name.

NONCOMPETES FOR NEW HIRES

What could happen

You've just interviewed a potential employee who works at a company operating in a similar space. What should you do to prevent any issues from arising?

Watch out for

If you are serious about this hire, you should check the potential hire's paperwork with their existing or former employer. Specifically, many companies include non-compete provisions in employment agreements. Although not all states will enforce such provisions, you should check state employment rules in the new hire's state of employment. Most often, there are some limitations on noncompete provisions that may delay, rather than prevent, an employee from joining your company. You may also encourage the employee to discuss waiving these restrictions directly with their former employer.

When looking to hire a new employee, you should ask to review the employment and personnel records of the new hire. Noncompetition, nonsolicitation (limitations by former employees to recruit from their former employer), and assignments of inventions may raise certain restrictions with respect to your new hire. Also, review all nondisclosure agreements your new hire signed with her former employer where these provisions sometimes may be found.

TIP: Ask to review all agreements of a new or potential hire with her former employer.

Recruiting Top Talent

Depending on the position that needs to be filled and the stage in the cycle of your startup company, you will use a variety of methods to recruit new employees. For many positions, you may find that your personal network will be able to provide you a wealth of referrals to consider for positions. The downside may be that this referral network will consist of a number of people already in positions and convincing them to leave the position might be a challenge. For other positions, you'll likely find that job postings made on websites are a good source of applicants for a position. And, for certain key positions, you may find that retaining a search firm is the best approach for attracting the right talent.

RECRUITING TECHNIQUES

Most successful organizations will use a variety of theses methods depending on the time in the life of the organization, the job title, the timetable for hiring, and other factors.

Referral Search

- **How it works:** By leveraging your network, you may be able to find direct referrals for talented new hires. Use your current employees, your accountant, your attorney, and other service providers. Reach out at industry trade associations and other entrepreneur support organizations.
- **Costs:** None, other than your time.
- **Risks:** Depending on the size and scope of your network, you may be casting your net too narrowly. Be careful with particularly important positions that you do not miss opportunities to reach beyond your network to bring in new talent.

Retained Search

- **How it works:** Typically the use of a third-party retained search is used for high-level employees such as the CEO, president, CFO, or any of your vice presidents. The search firm will likely manage the majority of the process from finding candidates, to initial screenings, to setting up interviews with the company. This is a full-service process, and the firm will likely coordinate areas such as travel, reference checking, and evaluation processes. The process will likely take between 60 and 90 days.
- **Costs:** The placement fee will be negotiated but expect between 25% and 50% of the budgeted amount of the employee's first-year salary and bonus (including stock compensation). Typically, you will pay a portion up front and will be responsible for the remainder after final placement. You will also be responsible for other standard expenses such as travel, copies, etc.
- **Risks:** Some search firms do not work with startup or emerging companies. Be certain that any organization you work with has experience with emerging companies and in your industry. Your search firm should create a job description and a list of companies that are most likely to have the applicable talent. Be certain to review the description and the list of companies to prevent the search firm from reaching out to companies you may want them to avoid (such as a potential competitor, a business partner, or a customer).

Contingency Search

- **How it works:** For contingency searches, you will only pay the search firm after successful placement of a candidate (unlike the retained search process). The agency will be less likely to prepare job descriptions, to determine target companies, or to provide reference checking. In addition, they oftentimes do not provide a first screening and may provide you with a much higher volume of candidate resumes.
- **Costs:** The placement fee will be negotiated but expect between 10% and 25% of the first-year salary and bonus of the employee. However, you will not pay until the search is successful.

- **Risks:** Contingency searches may not provide you with the highest caliber of potential hires because a greater number of unemployed individuals or others that are looking to leave their employer will likely come through this process. If you would like the search firm to contact potential hires at specific companies or in specific jobs via phone or e-mail, this may be an additional charge.

Contract Search Staff

- **How it works:** At the early stages of your organization, you may find it is necessary to hire an individual or team to provide contract search staffing. This may involve employing a search coordinator on a limited, part-time basis to handle your internal search efforts.
- **Costs:** Contract employees to provide search servicing may likely cost between $25 and $100 per hour depending on the experience level and support required.

Traditional Advertising

- **How it works:** Using traditional advertising in local newspapers, placement websites (including places such as http://www.craigslist.org), and industry periodicals.
- **Costs:** Individual advertising rates will apply. These can run from the hundreds of dollars for an individual ad to thousands for a multiple ad listing.

Management Team Hiring

Early on, one of the key considerations a new company has is the timetable for filling its management team. Typically, a company will be formed by a single founder or a small group of founders, each filling a number of key roles within the company. As the venture continues to grow, the business will need to add key members to the management team to oversee areas of finance, marketing, sales, business development, manufacturing, and, in some cases, an experienced CEO or chief operating officer.

Although your business may require a unique timetable for executive hiring, John Nesheim in his book *High Tech Start Up* noted a general pattern in the process by which founders added to the executive team. According to Nesheim, a company would typically be founded by the inventor and a core group but would begin by adding an experienced engineer, preferably someone with previous experience as a vice president level of a startup venture. Next, founding teams would add an experienced CEO, particularly near to the time of venture fundraising. After fundraising, companies often first hired a vice president of marketing to help with creation of the first products of the company, followed by a vice president of business development who may oversee sales or marketing early in the life of the company. Tech startups would usually add a vice president of sales approximately two months before product launch date and a vice president of manufacturing or operations near the time of product launch. Nesheim found that companies were less consistent in their hiring of a CFO or vice president of finance and administration. In some cases, this position would be filled just after financing, whereas other times investors would encourage the company to wait to, according to Nesheim, give the investors greater input and oversight of the company's financial policies.

HIRING A HEADHUNTER

What could happen

You've just received funding, and it seems like headhunters are coming out of the woodwork offering you their services to assist with hiring your new employees. Should you consider using a headhunter?

Watch out for

Much of your initial successes (or failures) will depend on the employees brought into your organization. Finding the most effective and efficient ways to recruit new talent into your organization is critical.

Many new business ventures will use some type of headhunter or search firm to assist in the initial hiring efforts. Depending on the type of hire you are looking to make, you may find that using one or more types of services offered by search firms is appropriate for your organization. Be aware that, for their fees, search firms oftentimes charge between 15% and 50% of the salary and bonus for the employee's first year of service. Also, be cautious of hiring multiple search firms, as they may refer the same individuals to you, and each may demand payment for their fees for the same candidate.

As with any third-party service provider, be certain to get referrals, interview the search firm, and develop a recruitment strategy that fits within your budget. Oftentimes, search firms will specialize in a particular industry or market, so be sure to determine what types of hiring the search firm is best suited for.

TIP: Search firms can be a valuable third-party service provider, but be aware of the costs of using them.

Practical Considerations of Hiring an Employee

Hiring employees implicates a number of practical considerations from payroll and taxes to confidentiality and invention assignment. Therefore, startup companies should be aware of the necessary restrictions and requirements that are undertaken with a new hire.

Employees or Independent Contractors

For early-stage companies, hiring full-time employees may represent a substantial undertaking from a financial and operational perspective. An independent contractor is responsible for payment of the taxes that an employee would typically pay: payroll, Federal Insurance Contributions Act (FICA), Social Security, and unemployment taxes. Therefore, many companies will initially decide to use their team of founders and retain independent contractors on a project-by-project basis, for specific tasks, or on a less than full-time basis.

If you decide to use independent contractors, be aware that simply calling someone an independent contractor may not be enough in the eyes of the IRS to avoid paying certain taxes. The IRS recognizes that companies that rely on independent contractors have lower effective tax rates, and therefore the IRS carefully scrutinizes these relationships to determine whether the relationship is, in fact, an employment relationship and subject to taxes. There are multiple factors the IRS examines in these scenarios, but factors include the level of control the company has over the contractor's work, the number of other clients the independent contractor has, and the employment-like activities of the contractor (e.g., Does he have a desk? Does he have an identical name badge? Does he have an e-mail address?). If you are uncertain about these issues, you should discuss them with an attorney familiar with employment matters.

Independent contractors may be important resources for your company. Be aware that independent contractors that are engaged with your company should sign agreements governing nondisclosure of confidential information and assignment of inventions.

In the event you do hire any employees, the company is responsible for payment of taxes and proper employee withholdings. Be aware that many companies get themselves into trouble as a result of improper management of employment taxes and withholdings. You should carefully structure your accounting system to track payroll, taxes due, and withholdings. Many companies will outsource their payroll and employee taxes to a third-party provider.

RECORDKEEPING FROM THE BEGINNING

What could happen

Your company was in its early stages and didn't ask or require employees of the company to maintain records, files, and backup of key information. What happens if a key employee developing your technology departs or you have a dispute that arises with a competitor regarding development of your technology?

Watch out for

Although it seems like a low priority to enforce diligent recordkeeping, to ensure redundancy of technical knowledge, and to track activities of your development teams, forgetting these activities can lead to catastrophic problems down the road. Insist on proper recordkeeping from the beginning rather than attempt to instill discipline down the road.

You will be responsible for maintaining proper records for the following: accounting and bookkeeping records; bank records; contracts; corporate records; all corporate correspondence; employee records; business forms and form agreements; debt documentation; intellectual property records; marketing and advertising records; permits and licenses; stock, option, and other equity records; and tax records. By keeping these records from the beginning and maintaining proper procedures for ongoing recordkeeping compliance, you will ensure that a future transaction will not be derailed as a result of a lack of proper records.

TIP: Requiring proper documentation and recordkeeping may prevent a departing employee from sinking your ship.

Offer Letters and Employment Agreements

Offer letters and employment agreements represent a written understanding of the key terms of the employee-employer relationship. These agreements represent an important tool for clarifying the understanding of the parties and ensuring that the parties accept any restrictions or responsibilities.

Today, the trend is for companies to use offer letters, although some companies and some employees will insist on an employment agreement. The difference is a technical one: an offer letter is an offer for an employment contract that is accepted upon the employee beginning work, whereas an employment contract is accepted upon signing of the parties. Generally, an employment agreement is perceived to be a bit more formal, but both agreements will offer sufficient evidence of the employment relationship.

WHAT TO INCLUDE IN AN OFFER LETTER

The offer letter will set out the key provisions governing the employment relationship. In many cases, a startup company should consider creating a form offer letter that can be customized for each employee and having that form agreement reviewed by an attorney. Key provisions of an offer letter will include the following:

- **Duties:** An offer letter will typically state the job title of the position of the new hire and a summary of the key duties of the position in terms that allow the position to be modified if necessary (oftentimes a simple one- to two-line statement). Additionally, the company may include the title of the position that will supervise the employee.
- **Compensation:** An offer letter will lay out the base salary for the employee and a general description of how the salary will be paid (i.e., monthly, biweekly, or quarterly). The company should also include information on any potential signing bonuses, year-end bonuses, commissions, or other incentive-based awards, with specific details on the terms of the bonuses and commissions. In the event the bonus or commission is based on an incentive plan or based on specific terms located elsewhere, the offer letter should make reference to those documents.
- **Benefits:** An offer letter will set out a list of eligible benefits for the employee. As with the bonus or commission plans, you may reference the benefit documentation in the letter for specific details and terms. In addition, the offer letter will usually include language stating that the company may alter the benefit plans at its sole discretion, to prevent misunderstandings and potential ongoing obligations.
- **Severance:** In the event that the company wants to provide certain benefits after termination, such as payment of certain insurance premiums, payment of salary, acceleration of stock options, or other severance, these items should be set out in the offer letter.
- **Equity awards:** In the event that an employee will receive equity awards such as stock options or restricted stock, the company should include information on the awards. Because grants of stock options typically require the approval of the board of directors of the company, the language should be clear that the company will recommend the board to issue such a grant, but it is based on the final decision of the board. In addition, the company will want to include information about the specifics of the award, such as the number of

options or amount of restricted stock, the vesting terms, any acceleration of vesting, and the terms of repurchase after the termination of the employment relationship.

- **Duration of employment:** For most employees, the company will want to enter into an "at-will" employment relationship that allows the company to terminate the employee for any reason (as long as the reason does not result in a violation of certain employment laws and rules). The company should be careful about using time periods or implying an ongoing relationship. In the event the company wants to provide for a longer term of employment (perhaps in the case of your chief executive or other management team member), you should consult with an experienced employment attorney to ensure that the proper language is included for protection of the company.
- **Right to work:** An offer letter will include language providing that the employee must comply with certain rules and restrictions under federal immigration laws. The letter typically requires that the employee provide verification of citizenship, residency, or rights under a work visa.
- **Requirements to enter certain other agreements:** In the event the company has other agreements, such as confidentiality or nondisclosure agreements or invention assignment agreements, the offer letter should include language that the employee is required to execute these agreements before beginning employment.
- **Arbitration:** Companies will often include language in offer letters requiring disputes to be settled by mandatory arbitration to limit costly and time-consuming litigation. Although these provisions have been successful in some situations, courts have rejected the enforceability of the clauses in a number of cases.

Employee Handbooks

Although employers are not legally required to have an employee handbook, many misunderstandings about the employer-employee relationship may be avoided through use of an employee handbook. An employee handbook provides a clear statement of an employer's policies and procedures. Minimally, an employee handbook might include a statement describing "at-will" employment, an equal opportunity statement, a sexual or more general harassment policy, policies regarding Internet, e-mail, and voice mail access, and the Family Medical Leave Act.

- In California, courts have adopted the handbook exception in which provisions of an employee handbook or personnel provisions can form contractual rights between an employer and its employee.
- In New York, a very narrow handbook exception exists. To show an employment contract existed because of an employee handbook, an employee must demonstrate that the employer issued the handbook to the employee and that the handbook expressly limited the employer's right to terminate its employees at will.
- In Washington, assurances of job security can alter the at-will employment relationship and form an implied employment contract.

Restrictive Agreements

Companies will often include certain restrictions on their employees in their ability to join competing ventures, to solicit your current employees after they've departed, or to solicit your company's customers after they depart. The enforceability of these provisions depends on the state where the company is located or where the employee resides or does business, and you should only include the provisions if you have researched their enforceability in the state. More information on these restrictive agreements can be found in Chapter 6.

Noncompetition

A noncompetition agreement is an agreement between an employer and employee in which the employee agrees not to pursue a similar profession or trade in competition with the employer. Generally, these provisions will have a limited time period after departing the company. Many states enforce agreements preventing employees from working for competitors, so long as the restriction is based on legitimate business need and is reasonable in scope, duration, and geographic limitation. When evaluating the enforceability of a noncompetition agreement outside of California, it is important to review legal authorities of the relevant state carefully.

Employee Nonsolicitation

These nonsolicitation clauses are aimed to prevent a mass exodus from one company by limiting the ability of a former employee to recruit former colleagues. Yet, despite the strong public policy favoring employee mobility, most courts (including California courts) have upheld reasonable restrictions on a former employee's ability to solicit his former employer's employees or customers. Many companies will include nonsolicitation language in employee agreements to restrict the ability of a former employee from solicitation efforts.

Customer Nonsolicitation

Reasonable agreements not to solicit customers are often enforced by courts, but the use of such agreements has been narrowed by many states. These restrictions should be limited to a particular time period, and the company should recognize that these provisions will only forbid solicitation by the former employee but will not prevent customers from working with the departed employee.

Protecting Intellectual Assets

For most high-technology companies, disclosure of confidential or proprietary information, inventions produced by employees, and employee use of trade secrets represent potential risks to the company. Therefore, it is important that the company take actions at the time of hiring, throughout the employee's tenure, and after termination to protect the company.

Many technology companies will develop an agreement that includes matters of confidential information, invention assignments, works made for hire, and may include references to noncompetition or nonsolicitation restrictions (which are discussed in the previous section). In such an agreement, an employee will often agree, among other things, to the following:

- The employee will not use any confidential information of the company (including inventions, discoveries, concepts, and ideas that are useful or related to the business of the company and that are conceived by the employee during the period of his or her employment).

- The employee will disclose promptly to the company any inventions he may make, develop, or conceive during the period of his employment, with an agreement that all such inventions shall be and remain the property of the company.
- All records and other materials pertaining to confidential information, and all other records or materials developed by the employee during the course of employment, will be and remain the property of the company and, after termination of his employment, will be returned to the company.
- The employee's employment with the company does not and will not breach any agreement or duty that the employee has with anyone else, nor will the employee disclose to the company or use in its behalf any confidential information belonging to others.

These terms and restrictions can be included in a single agreement (sometimes called a confidential information and invention assignment agreement) or may consist of a number of separate agreements. In these cases, an employer may use confidentiality or nondisclosure agreements to protect confidential information and an intention assignment agreement to assign rights to employee-developed inventions.

Confidentiality and Nondisclosure Agreements

Your employees have a general duty to the company to keep information of the company confidential. However, you should still insist on a separate agreement written broadly to cover third-party disclosures made to your company, which may not necessarily be covered under this common law duty. In addition, for consultants or independent contractors who do not have this duty to the company, it is even more crucial to insist that they enter into confidentiality or nondisclosure agreements.

Invention Assignment Agreements

Most high-tech startup companies will have products and services involving numerous intellectual property, proprietary, and confidential information implications. Intellectual property assignment agreements are important tools to help sort out ownership rights of intellectual property in employment and consulting contexts. Absent these agreements, intellectual property law rests on a series of default ownership rules that often leave ownership claims unclear. Invention assignment agreements and other intellectual property assignment agreements clarify who owns what. These agreements must conform to standard principles of contract, intellectual property, and employment law. Such agreements are common in technical companies, academic institutions, and government, with the majority requiring employees and consultants to preassign their intellectual property creations to their employers.

Trade Secrets

Employees play a crucial role in limiting the disclosure of company trade secrets. Reported court decisions suggest numerous ways a company can act to ensure that its sensitive information qualifies for trade secret protection.

EMPLOYEES AND PROTECTION OF TRADE SECRETS

One of the most common sources of disclosure of trade secrets is by employees. Taking practical steps to protect the company from these disclosures can save time, money, and headaches down the road.

- **Employment agreements:** Requiring all employees (when hired) to sign an agreement in which they acknowledge that they will receive confidential information and in which they agree to maintain the secrecy of such information. Whenever an employee is promoted to a new position in which he may acquire access to new or different confidential information, the employer should check to make sure that the employee has signed such an agreement. Companies should also include similar language in independent contractor agreements.
- **Policies for protection of sensitive information:** Requiring employees to secure access to sensitive information. Sensitive hardcopy documents should be kept in locked storage space. Sensitive information contained on computer files should be accessible only to those provided with a password. Access to all confidential information should be limited to those who legitimately "need to know."
- **Properly mark documents and files:** Asking all employees to mark documents or files containing confidential information as "confidential and proprietary information."
- **Restrict access:** Restricting access to company premises to employees and contractors. Providing for additional restrictions to certain areas of the company's properties in which particularly sensitive information can be found.
- **Document retention policies:** Implementing a document retention policy that involves shredding paper documents containing sensitive information (i.e., instead of simply disposing of such papers with other trash) and consider periodic reviews of paper waste to determine whether employees are properly shredding documents containing sensitive information.
- **Nondisclosure agreements:** Putting procedures in place to discourage employees from providing sensitive information to outsiders unless the third party has a legitimate need to know and has executed nondisclosure agreements.

Additional information about a trade secret protection program can be found in Chapter 13.

Employee Costs

Hiring an employee involves a substantial investment, not only in a salary for the employee but costs ranging from bonuses and commissions for salespeople to various benefit plans, employment taxes, and costs to provide equipment and technology to the employee.

Employee Salaries, Bonus Programs, and Commissions

Salaries and bonus programs for startup companies vary based on the region, the industry, and the position. Startup salaries tend to be somewhat lower than established companies, but the salary is balanced by equity grants and the opportunity for the company to produce a sizeable appreciation in stock value.

FINDING MORE INFORMATION

Need to find out the going rate for a chief scientific officer for an early-stage (one to five employees) life sciences company located on the West Coast?

Visit http://www.compstudy.com to find out detailed information used to set salaries, bonuses, equity compensation, and more for specific positions. CompStudy offers guidance based on the size of your organization, the fundraising status, and the title or position of the employee.

Industry trade associations and local entrepreneurial support organizations may provide anecdotal evidence of salary ranges. You may be able to research comparable salaries on various job posting websites such as http://www.monster.com or http://www.careerbuilder.com. In addition, http://www.payscale.com and http://www.salary.com provide detailed salary and benefit reports you can purchase based on the position, the region, the size of the organization, and various other criteria.

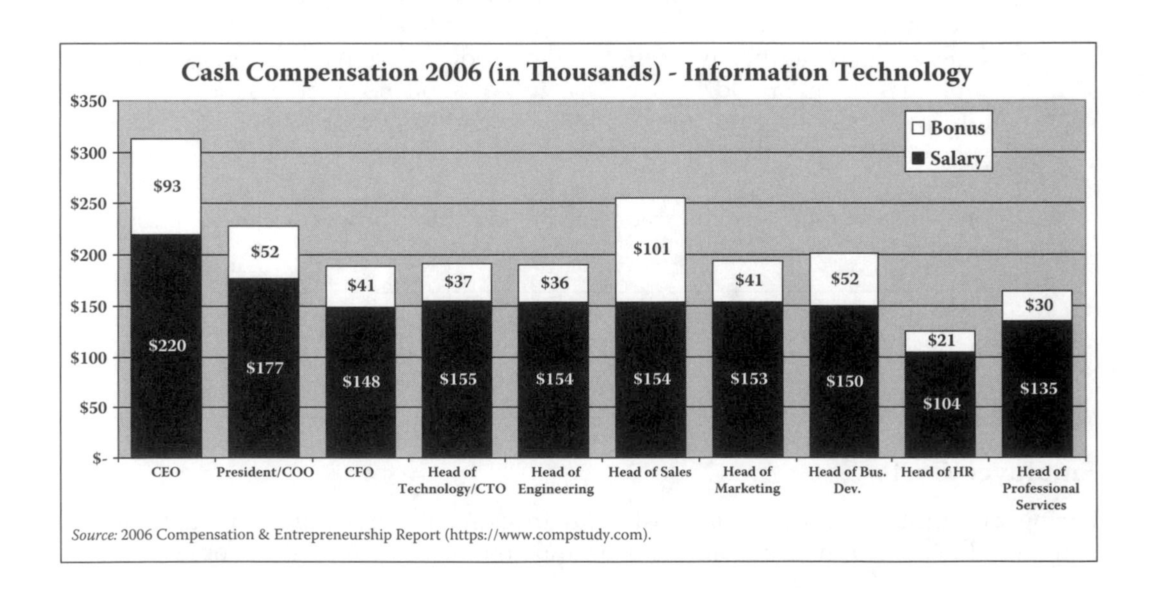

Source: 2006 Compensation & Entrepreneurship Report (https://www.compstudy.com).

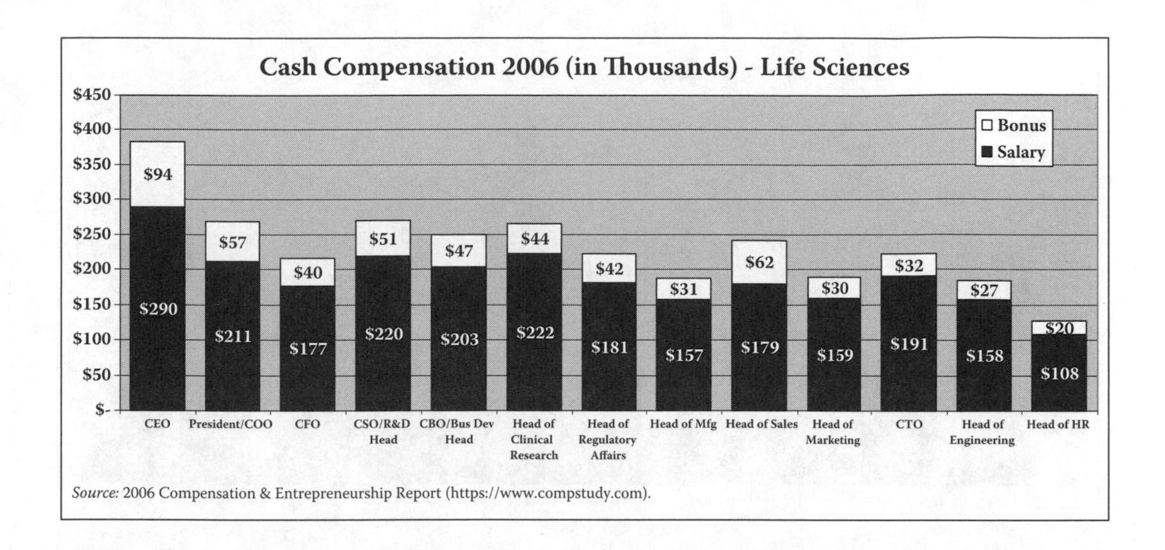

Source: 2006 Compensation & Entrepreneurship Report (https://www.compstudy.com).

One of the benefits of a cash bonus program (as opposed to an equity bonus program) is that it does not dilute the ownership of the company and may provide greater flexibility. Bonus programs can be based on sales, profits, or, in some cases, only payable in the event of an acquisition, IPO or other similar milestones. The company should be aware of accounting and tax implications of certain types of bonus programs to ensure that the company does not need to accrue for these expenses in the current period.

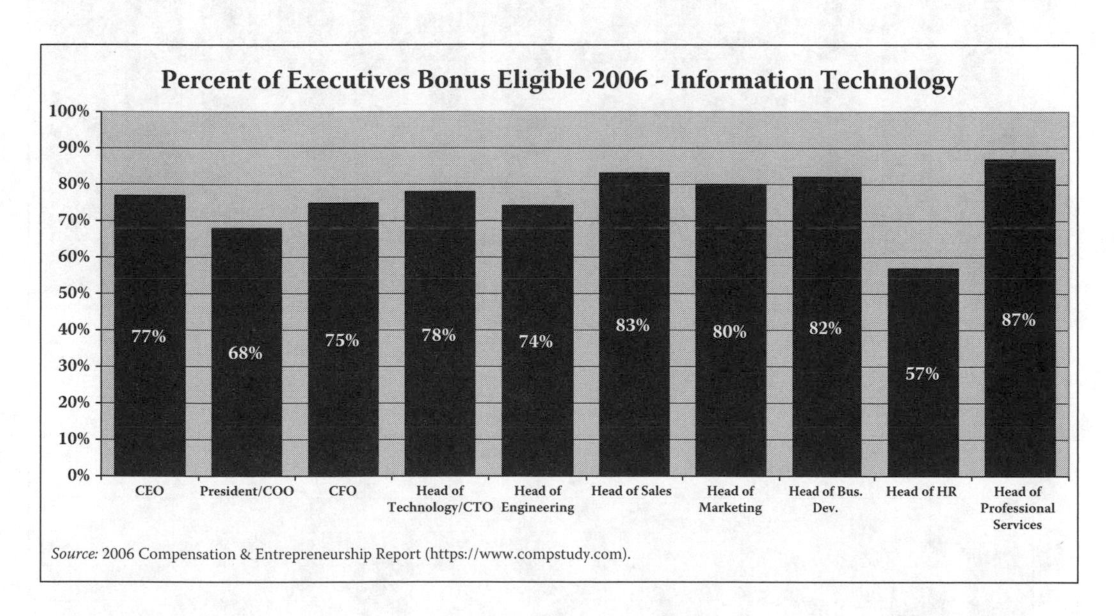

Source: 2006 Compensation & Entrepreneurship Report (https://www.compstudy.com).

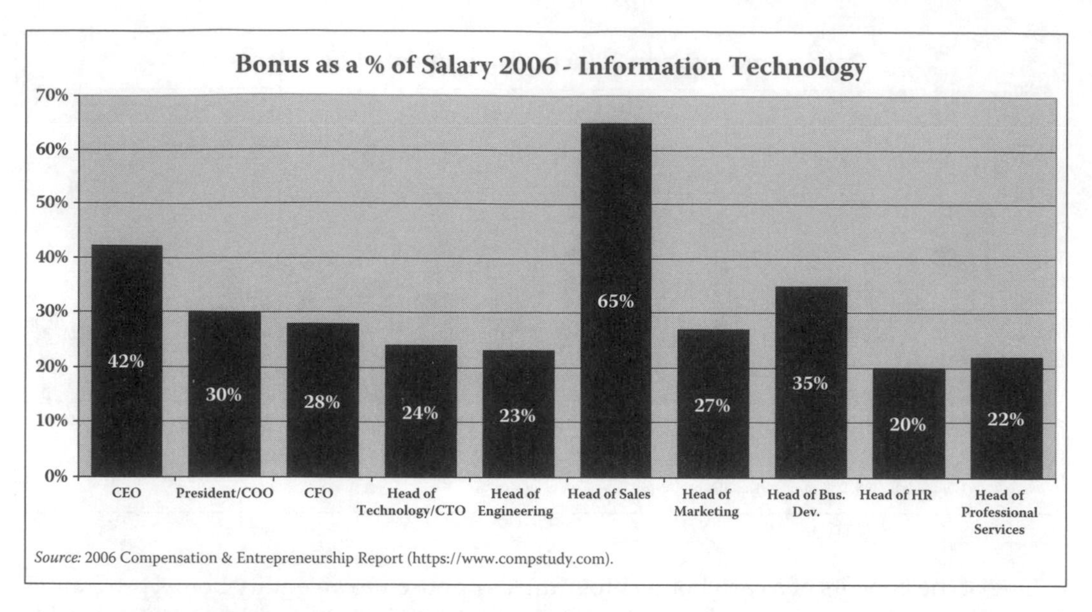

Source: 2006 Compensation & Entrepreneurship Report (https://www.compstudy.com).

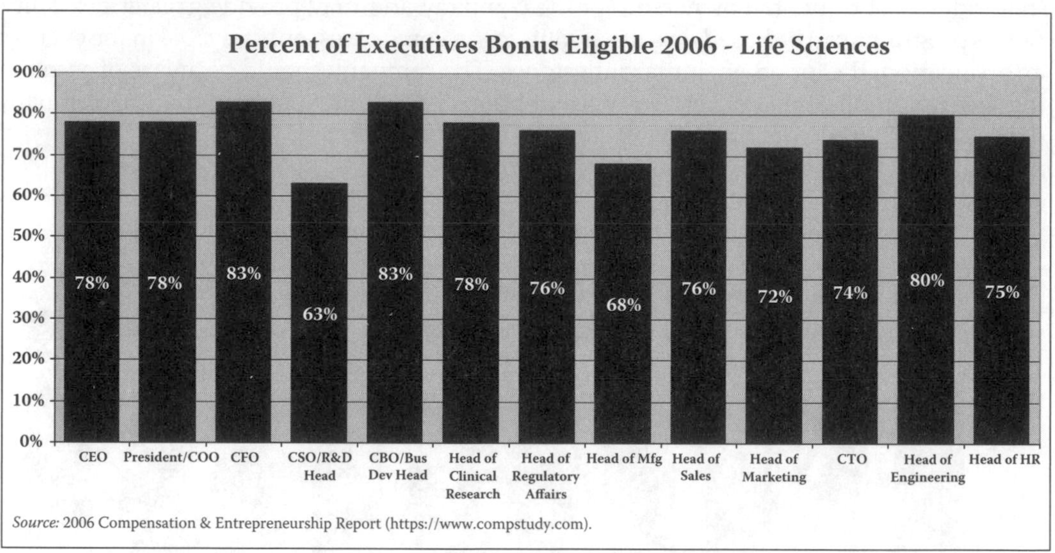

Source: 2006 Compensation & Entrepreneurship Report (https://www.compstudy.com).

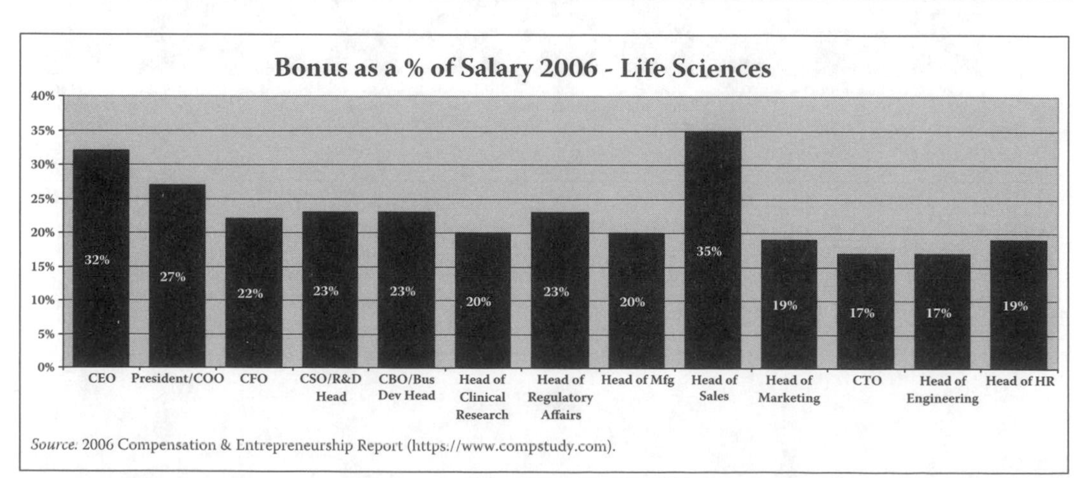

Source: 2006 Compensation & Entrepreneurship Report (https://www.compstudy.com).

Commission programs are typical for sales and business development employees. Program details vary according to the industry and the base compensation amounts. Amounts, schedules, and timing of the payments should be provided within the program. Companies should be aware that commissions are key drivers for many top sales performers and should be adjusted accordingly to stay competitive within the industry and prevent defections from occurring.

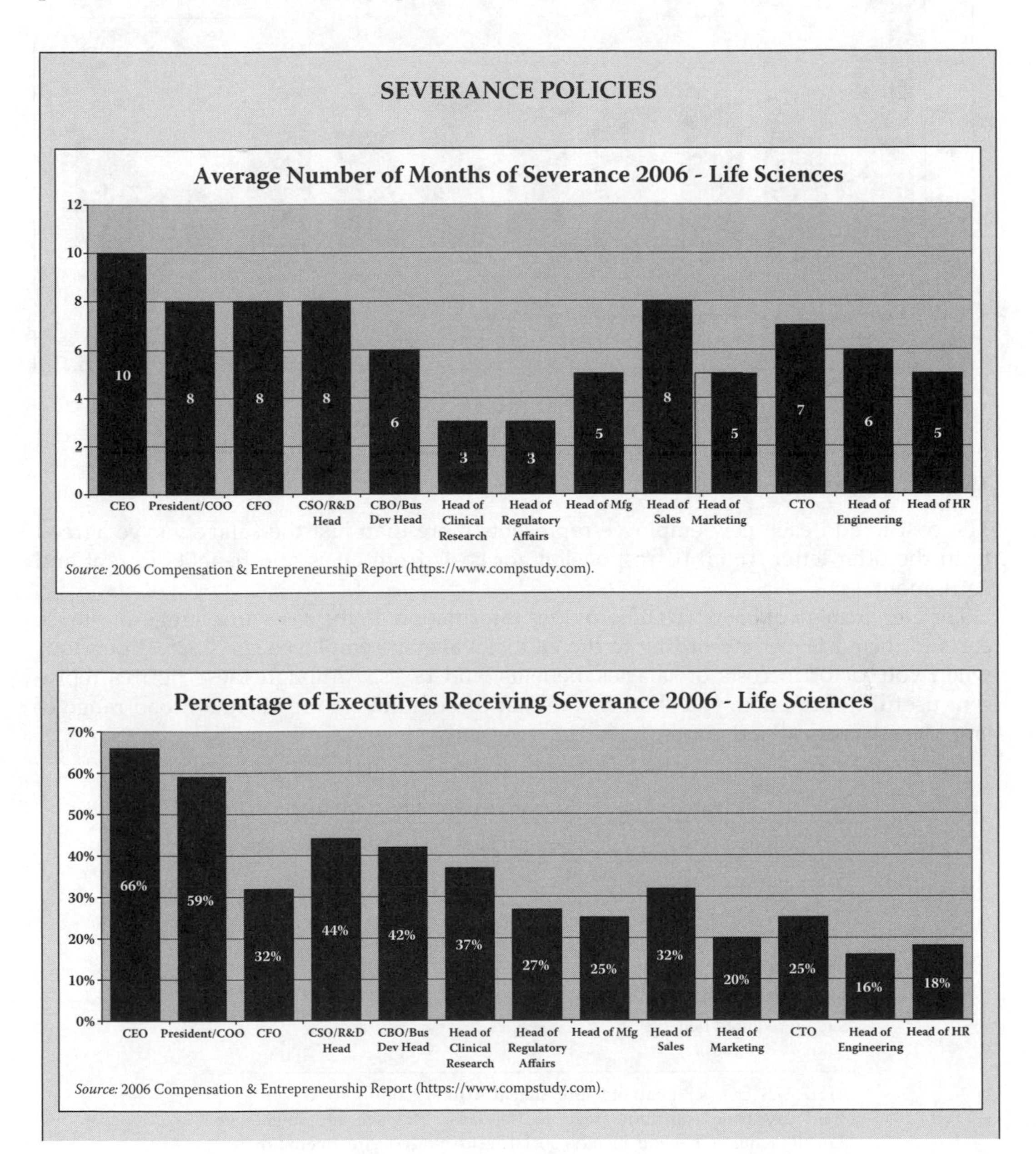

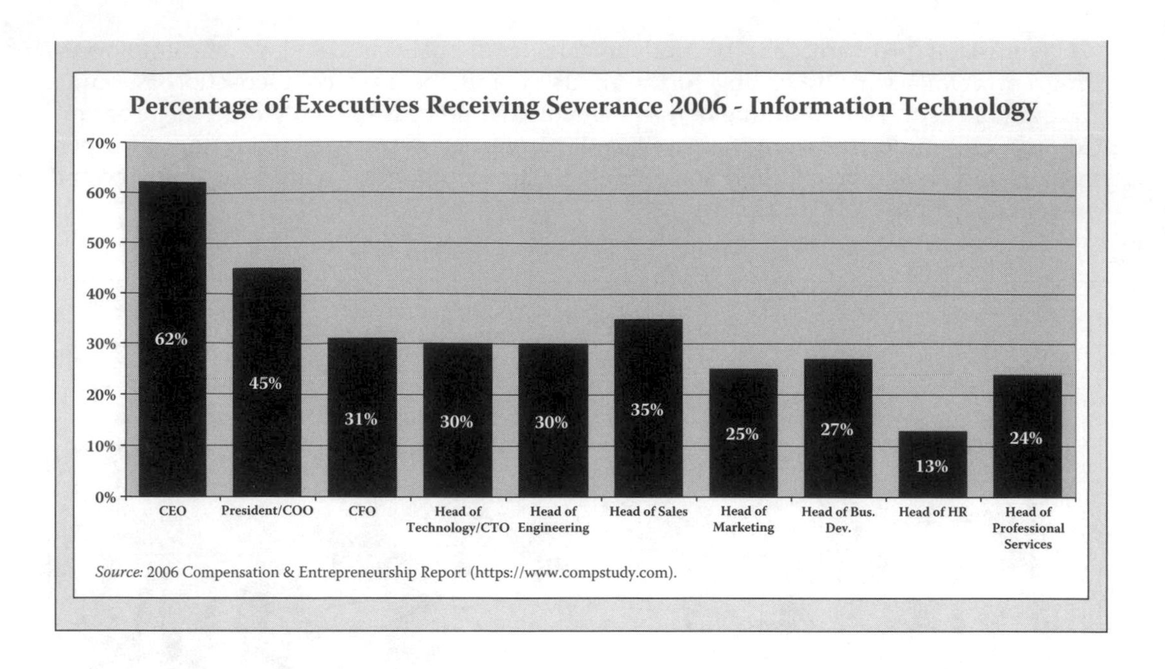

Source: 2006 Compensation & Entrepreneurship Report (https://www.compstudy.com).

The Costs of an Employee

The cost to add each new employee represents more than just the salary you've agreed to in the offer letter. You'll be responsible for costs from taxes and benefits to rent and equipment.

The Department of Labor (DOL) provides information on the costs an average employee costs to the employer. According to the DOL, an average employee costs $25.93 per hour when you factor in costs of salaries, benefits, and taxes. Although these figures represent useful information, you should note that these numbers represent a broad range of employees across all industries in the U.S. economy.

Private Industry Employer Compensation Costs

Employer cost	Cost per hour	Total costs
Wages and salaries	$18.32	70%
Paid leave benefits[1]	1.77	7%
Supplemental pay	0.78	3%
Insurance benefits	1.97	8%
Retirement and savings	0.88	3%
Legally required benefits[2]	2.21	9%
Total	$25.93	100%

Source: U.S. DOL's Bureau of Labor Statistics (June 2007).

[1] Paid leave benefits include vacations, holidays, sick leave, and other leave.

[2] Legally required benefits include Social Security, Medicare, unemployment insurance, and workers' compensation.

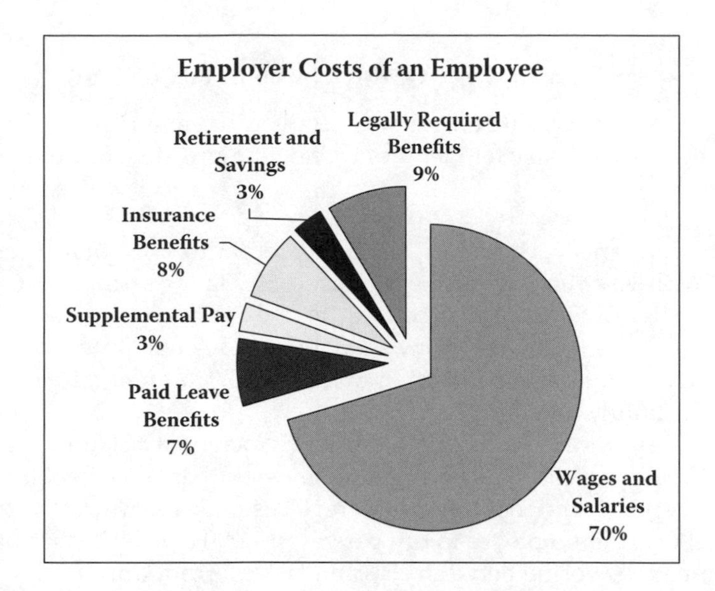

Although the DOL information represents aggregate data, the information can be quite helpful to gauge what it will really cost to hire another employee. If you'd like to estimate the real cost of adding a new employee, you may consider multiplying the employee's base salary by a multiplier that would reflect their salary, benefits, rent, equipment, training, and other general expenses associated with another team member.

For instance, looking at the chart below, you could see that, if you planned to add two programmers at salaries of $50,000 each and one manager at a salary of $100,000 for the following year, you could be looking at an increase of expenses of up to $540,000. When making your budget to add headcount, it is important to include costs associated with the employee in addition to their salaries. This is just a helpful big-picture tool and each company will likely need to adjust its calculations to fit its own operations, but it is helpful in gaining a quick sense of the true costs associated with increasing headcount.

Employee cost	Multiplier	Salary of $50,000	Salary of $100,000
Salary	1.0	$50,000	$100,000
Benefits	0.2–0.4	$10,000–20,000	$20,000–40,000
Rent, equipment, training, etc.	0.5–1.3	$25,000–65,000	$50,000–130,000
Total	1.7–2.7	$85,000–135,000	$170,000–270,000

Benefits

The type of benefits provided to your employees and the costs to your business will depend on specific choices the business makes. Many companies will provide medical insurance, dental insurance, and life insurance. Other companies will offer long-term disability insurance, tuition reimbursement, retirement programs, dependent care assistance, home-buying assistance, banking programs, and many others.

DETAILS ON COMMON BENEFIT PLANS

In general, providing certain benefits will cost your company an additional 20% to 40% of the employee's salary to offer these benefits. Here are a few of the key benefit plans offered:

- **Health insurance premiums:** In 2006, premiums for small group plans (plans with less than 50 people) averaged $3,732 for a single coverage and $9,768 for family coverage, according to the research of America's Health Insurance Plans. Although many small businesses will only cover a portion of the costs, these could still be in excess of $3,000 for single coverage and $8,000 for family coverage.
- **Dental insurance:** According to a 1999 Society for Human Resources Management survey, 83% of small businesses that offer health insurance also will offer dental insurance. Coverage costs will vary greatly, but expect the total cost per employee to run between $1,000 and $1,500 annually, with the company covering between 25% and 50% of premiums.
- **Vision insurance:** According to the same Society for Human Resource Management study, 58% of small businesses offering health insurance also offer vision insurance. The costs for vision insurance on the employer are generally very low, typically ranging from $50 to $100 per employee.
- **Life insurance and long-term disability insurance:** Expect costs for providing life insurance to cost between $150 and $250 annually per employee and long-term disability insurance to cost between $200 and $350 per employee each year.
- **Retirement plans:** A 401(k) plan will generally be funded by employee withholdings, although the company will bear the administrative costs of the program.

Additional information on many of the insurance programs can be found in Chapter 17.

Employment Tax

The specific amounts for certain required employment taxes for Social Security, Medicare, unemployment insurance, and workers' compensation, described as "legally required benefits" above, can be found on various government websites. FICA consists of both a Social Security (retirement) payroll tax and a Medicare (hospital insurance) tax. Individuals who are self-employed have their own tax that is similar to FICA referred to as the Self-Employment Contributions Act. The Federal Unemployment Tax (FUTA) finances the administrative costs of unemployment insurance. Finally, each state will have an individual state unemployment tax.

Under FICA rules for 2008, the company is responsible for payment of 6.2% of gross compensation up to a limit of $102,000 of compensation for each employee for the Social Security portion of FICA. The cap of $102,000 for 2008 increases annually based on average national wages (this cap is known as the Social Security Wage Base). In addition, the company will also be responsible for payment of 1.45% of each employee's wages for the Medicare portion of FICA, without a limit. Furthermore, the company will withhold the employee's portions of each of these taxes under FICA: 6.2% for Social Security taxes and 1.45% for Medicare taxes.

Under FUTA rules for 2008, the company is responsible for a 6.2% tax on the first $7,000 of gross compensation for each employee.

State unemployment taxes vary depending on the state and the category of the employee:

- In California, rates for 2007 range from 1.5% to 6.2% of the first $7,000 of gross compensation.
- In New York, rates for 2007 range from 1.1% to 9.5% of the first $8,500 of gross compensation (normal rates with the subsidiary tax and Reemployment Service Fund for 2007 factored in).
- In Texas, rates for 2007 range from 0.29% to 7.70% of the first $9,000 of gross compensation.
- In Massachusetts, rates for 2007 range from 1.12% to 10.96% of the first $14,000 of gross compensation.
- In Washington, rates for 2007 range from 0% to 5.4% of the first $31,400 of gross compensation.

FINDING INFORMATION ON EMPLOYMENT TAXES AND RATES

You can find updated information and rates for the following:

- **FICA:** http://www.ssa.gov
- **Medicare:** http://www.ssa.gov
- **FUTA:** http://www.workforcesecurity.doleta.gov/unemploy
- **Individual state unemployment taxes:** http://www.toolkit.com/small_business_guide/ (summary of individual state rates)

Rent, Equipment, and Training

In addition to costs for benefits, taxes, and wages, your company will also undertake additional costs for each employee related to office space, technology and equipment, training, travel, and administrative costs. Be prepared to factor in these costs into annual budgeting.

Equity Compensation

For many entrepreneurs and early employees of startup companies, one of the main reasons they join the venture is the lure of receiving stock or options to purchase stock that may one day skyrocket in value. Stock grants and stock options represent important tools to attract and motivate talented employees. Managing equity compensation is a complex issue to juggle for most entrepreneurs. The most common forms of equity incentives for the employees of young, growing companies include the following:

- Stock option plans
- Stock grants
- Stock purchase plans

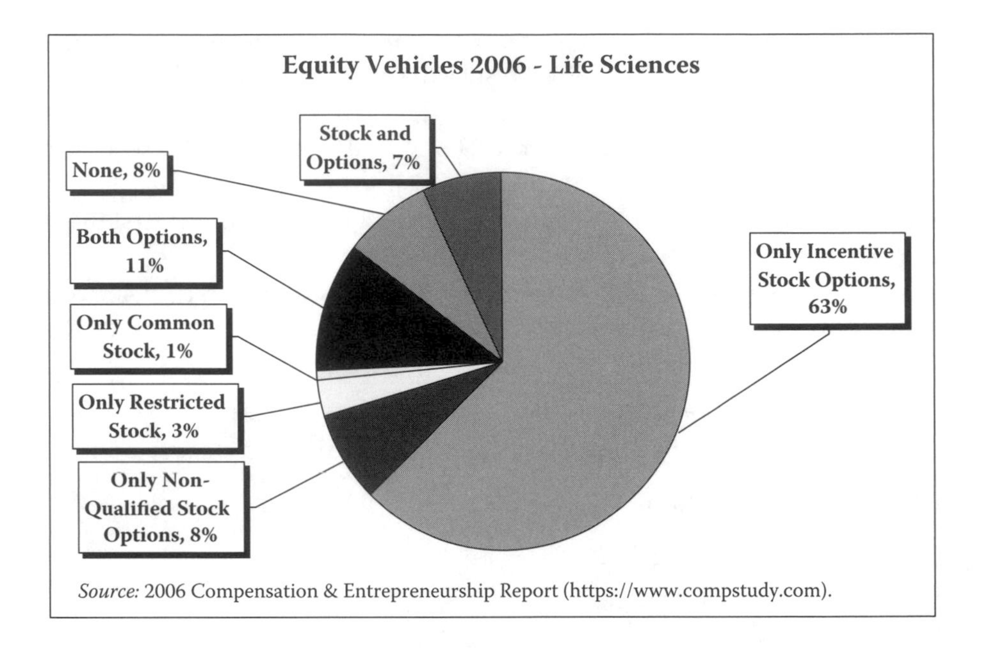

Source: 2006 Compensation & Entrepreneurship Report (https://www.compstudy.com).

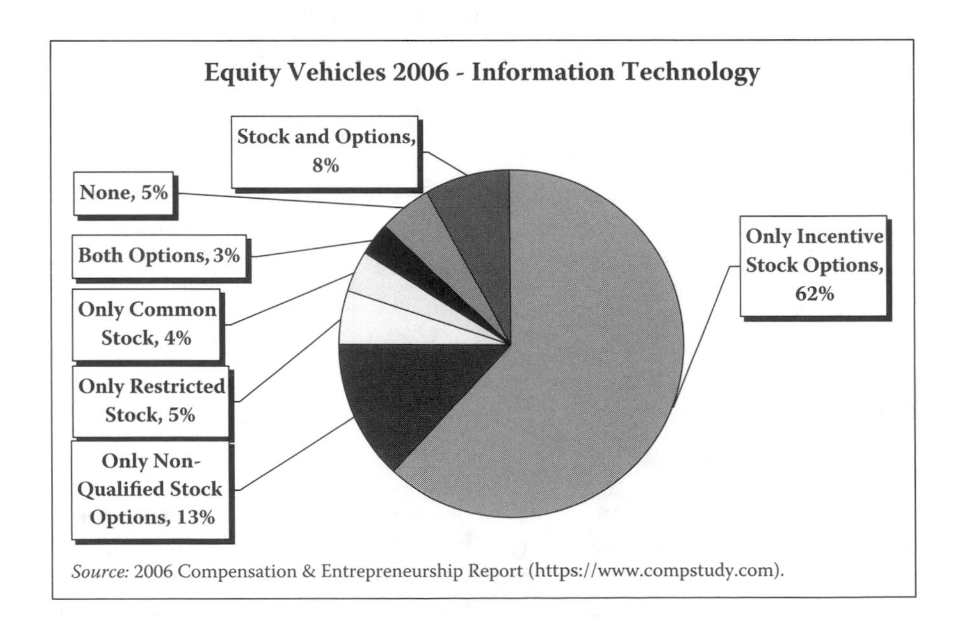

Source: 2006 Compensation & Entrepreneurship Report (https://www.compstudy.com).

Many companies will use each of these programs to provide incentives to their employees but may rely more heavily on one type of plan over the others at various points in the company's growth.

What Is the Difference between Stock and Stock Options?

Stock grants are the simpler of the two to explain. If you receive a grant of stock, you receive a portion of ownership (shares) of the company. At the time you receive the stock grant, you will actually own the stock, have certain voting rights in the company, and may (unless restrictions are applied) be able to transfer your stock to others.

If you receive stock options, you are not receiving stock. Instead, stock options are contracts that allow you to buy a certain number of shares of the stock of the company at a certain price within a set period of time. The price that you can buy the stock at is known the exercise price or the strike price. To purchase the shares of stock set forth on your stock option, you will pay the company the exercise or strike price in exchange for the shares. This exchange is referred to as an exercise of your option. You may choose to exercise all or a portion of your option.

HOW DOES A STOCK OPTION WORK?

High-Tech Startup Inc. grants one of its employees an option to purchase 100 shares of stock of the company. High-Tech Startup believes its stock is currently worth $1.00 per share. Therefore, High-Tech Startup sets the exercise price or strike price at $1.00 per share. This means that the employee may give the company $100.00 and will receive 100 shares, but High-Tech Startup doesn't ask for the money now. The stock option can be exercised for 10 years. Therefore, the employee doesn't have to exercise until he has the cash to exercise (at $1.00 per share) or believes the value of the stock exceeds $1.00 per share.

For some employees, this is the best of both worlds: the employee has the right to buy the stock for $1.00 but is not obligated to buy the stock. The employee keeps his $100.00 but knows that he can purchase the stock at any time until the option expires. High-Tech Startup is able to grant an option to the employee, which may motivate the employee, but has not had to issue actual stock to the employee.

Fast forward to several years later... High-Tech Startup has gone gangbusters, and now High-Tech Startup has had experts value its stock at $10.00 per share. The employee is holding an option to purchase the stock at $1.00 per share. When the option is exercised, the employee will pay $100.00 but will receive stock that is valued at $1,000, a gain of $900.00.

Instead, if the High-Tech Startup stock had decreased to $0.10 per share, the employee can just hold the option (and not exercise the option) until it expires and hope that its value increases. The employee keeps the $100. If the employee had purchased the stock for $1.00 per share, it would now only be worth $0.10 or $10, a loss of $90. This is why options continue to be attractive to startups and their employees.

Why Would You Grant Stock or Options to Employees?

The grant of stock or options may serve multiple purposes. Generally, options and stock are designed to align the interests of the employee with the company by providing a direct reward to the employee in the event the company is successful and the stock price appreciates. As a result of this potential, options may serve as a tool to balance a lower salary for a new employee. Ownership of stock or options to purchase stock of the company can be used to offset a lower salary than what could be earned working for a larger, more established company.

HOW COULD OPTIONS COMPENSATE FOR A LOWER SALARY?

The basic calculations

If an employee is offered a salary that is 25% lower than a comparable position at an established company, you may be able to convince the employee that this salary differential is offset by a stock or option grant. If you were to provide the employee 2% of the company with that grant, this could equate to between 1% and 0.5% of the company in three to five years after expected dilution from outside investments. In the event that the company is able to raise the value of the company in that five-year period to an amount that exceeds the decreased salary over this period, the employee would wind up ahead.

Applying real numbers

You offer Jane Programmer a $75,000 salary, which is less than the $100,000 she could earn at a large software company. You also offer Jane Programmer stock options equal to 2.0% of the company (which you tell her you expect to be diluted by 50% to 70% in the next three to five years). If the company becomes a $20 million company (0.6% of $20 million = $120,000), then she's made up the difference in the salaries. If the company becomes a $100 million company (0.6% of $100 million = $600,000), she's made up that difference five times over.

Remember that this calculation is not quite this simple because there are risk premiums and taxes involved as well, but it can be a helpful tool to explain the mixture of options and salary offered to a potential employee.

Under these plans, employees are given the opportunity to acquire stock of the company on favorable terms; in return, the employees are generally required to remain employed by the company for a certain period of time before obtaining unrestricted ownership of the stock or the right to exercise the options. There are several major types of stock plans, and the structure of individual plans may vary widely even within a single type of plan. For this reason, it may be beneficial for the company to use a number of different stock plans, tailoring each to specific circumstances and specific employees. Because each type of plan has strengths and weaknesses, a company should carefully examine its incentive goals and the characteristics of its employees before adopting any plan.

It is important to realize that issuances of securities to employees, like securities issuances to any other person, are subject to federal and state securities regulations. Consequently, neither an issuer nor an employee may sell unregistered securities without either registering the security or qualifying under an exemption.

Stock Option Plans

Stock option plans provide an important mechanism for high-technology companies. The purpose of granting options to acquire stock in the company at a designated price is to attract, retain, and motivate employees by providing for or increasing the proprietary interest of such employees in the company. Option plans also include the ability of the employee to receive compensation with favorable tax treatment, to give the employees compensation without a cash outlay by the company (but often still receiving a tax deduction), and to provide the company with a compensation plan that will generally provide favorable accounting treatment to the issuing company.

MANAGING A STOCK OPTION PLAN

After your company sets up an employee stock option plan and creates the appropriate documents, you should determine who will oversee and manage the employee stock option program. For many early-stage companies, your outside legal counsel can provide a system to properly process stock option grants and exercises. You should consider using your outside legal counsel to assist you with management of the stock option plan until you are able to have experienced human resources and finance departments.

TIP: Have experienced human resources and finance personnel before managing your stock option plan "in house."

An employee will generally receive one of two types of stock options: incentive stock options (ISOs) and nonstatutory stock options (NSOs), which are discussed in more detail below. In particular, ISOs require a stock option plan to be in place and that the company follows certain steps to properly issue its options. Remember that improper option issuances may lead to unintended tax liabilities for both the company and the employee. To properly enact and maintain your stock option plan, the company should follow certain rules to properly grant stock options to its employees.

RULES FOR A PROPER ISO STOCK OPTION PLAN

- The stock option plan must be in writing.
- The stock option plan must be approved by the shareholders of the company within 12 months of the plan's adoption by the board of directors (the plan may also be approved up to 12 months before adoption by the board).
- Options must be granted within 10 years of the formal approval of the option plan.
- Options must expire less than 10 years from issuance (or five years from issuance for any holders of more than 10% of the company's stock).
- Options must be granted only to employees of the company (not to directors or consultants).
- Options must be exercised within 90 days of termination of employee status or one year after the death or disability of the employee.
- The value of the stock to vest in any one year under the option (based on the value at the grant date) shall not exceed $100,000.
- Options may not be transferable except in the event of death by will or laws of distribution of assets.

The company should have a stock option plan in place that meets the criteria above before issuing any options to employees.

ISOs

ISOs are options that satisfy the requirements of Section 422 of the I.R.C. of 1986 (the "Code"). The Code provides special tax treatment in connection with the exercise of the option and the disposition of shares subject to the option. Key requirements of Section 422 are that ISOs be granted only to employees of the company or its parent or subsidiary and that the exercise price of an ISO be no less than the fair market value of the stock (as determined in good faith by the company's board of directors) on the date of grant.

There are no federal income tax consequences on grant of an ISO. After exercise, the employee incurs no tax liability unless he or she is subject to the alternative minimum tax (AMT) under Section 55 of the Code. After sale of the shares (assuming that the sale does not occur within one year after the date of exercise nor within two years after the date of grant), any gain is taxed to the employee as long-term capital gain. If the shares are disposed of within one year after the date of exercise or within two years from the date of grant, the employee will recognize ordinary income to the extent of the lesser of the excess, if any, of (1) the fair market value of the shares on the date of exercise or (2) the sales proceeds, over the exercise price. "Disposition" includes not only a sale but also gifts and certain other transfers.

The company is entitled to a federal income tax deduction only to the extent that the employee recognizes ordinary income on the disposition of the shares acquired pursuant to the ISO. No deduction for the company arises from any AMT liability incurred by the employee. Therefore, in situations in which the option qualifies as an ISO at all times, the company has no federal tax consequences with respect to that option.

NSOs

NSOs are stock options that do not satisfy the requirements of Section 422 of the Code and are not eligible for special tax treatment. NSOs are often issued to nonemployees such as consultants, who are not eligible to receive ISOs or participate in statutory employee stock purchase plans (ESPPs), and to key employees or directors to whom the company wants to grant options containing terms not permitted by Sections 422 and 423 of the Code.

Assuming that the NSO does not have a "readily ascertainable value" at the time of grant (and virtually no NSOs do), there are no tax consequences for the optionee at the time of grant. After exercise of the NSO, the employee will recognize ordinary income in the amount of the excess, if any, of the fair market value of the shares at the time of exercise over the exercise price. This ordinary income will be subject to withholding by the company if the optionee is an employee, from the current earnings paid to the employee, by an out-of-pocket direct payment to the company, or through other means the company may choose to allow. After sale of the shares, the employee will recognize capital gain or loss in an amount equal to the difference between the sale price and the fair market value of the shares on the date of exercise. If the shares have been held for more than one year before the sale, the gain or loss will be treated as long-term capital gain or loss to the optionee.

The company is entitled to a federal income tax deduction to the extent of the ordinary income recognized by the optionee after exercise of the NSO.

NSOs and ISOs: Tax Impacts

	Employer		Employee	
	ISO	**NSO**	**ISO**	**NSO**
Option is granted	No tax deduction	No tax deduction (unless option has a readily ascertainable market value)	No tax on grant	No tax on grant (unless option has a readily ascertainable market value)
Option is exercised	No tax deduction	Tax deduction on spread between exercise price and market value at time of exercise	No tax on exercise other than AMT, which may apply	Tax on spread between exercise price and market value at time of exercise
Sale of stock from option exercise	Tax deduction for a disqualifying disposition. No tax deduction for a qualifying sale.	No tax deduction	Tax spread from exercise price and sale price, which is taxed as ordinary income or capital gain/loss depending on holding period	Tax spread from exercise price and sale price as capital gain/loss (long or short term)

Stock Grants

More startup technology companies have begun issuing restricted stock to employees than ever before. What differs from options is that the employee holds the restricted stock outright but is restricted in his or her ability to transfer the stock and the company often holds a right to repurchase the stock in the event the employee departs before an agreed-on time period or fails to meet certain milestones. Some companies have found such grants to be easier to administer than a stock option plan.

An employee restricted stock purchase agreement establishes the terms and conditions under which an employee can purchase and hold stock in the company. In contrast to stock purchase agreements used for the purchase of stock by outside investors or venture capitalists, employee restricted stock purchase agreements generally provide the company with an option, which lapses over time pursuant to a "vesting schedule," to reacquire the employee's stock (generally at the price paid by the employee for the stock) if his employment with the company terminates before a specified period of service has elapsed. A typical vesting scenario is a 48-month vesting schedule, with $\frac{1}{48}$ of the shares vesting at the end of each month of continued employment, but the length and terms of the vesting schedule may be the subject of substantial negotiation between the company's founders and investors. In addition, a "blackout" period or "cliff" period on vesting is often imposed such that no shares will vest during the first 6 or 12 months of employment, at which time $\frac{6}{48}$ or $\frac{12}{48}$ of the shares will vest at once. This is sometimes called "cliff vesting." The purchase price for shares purchased pursuant to an employee restricted stock purchase agreement is typically payable in cash or by a full recourse promissory note in favor of the company.

WHAT THE HECK IS AN 83(b) ELECTION?

It is a tax-saving mechanism. Basically, an 83(b) election is made to try to reduce taxes when you sell stock down the road. If you didn't file the 83(b), that gain would be taxed as income rather than capital gains. As an example,

- Assume you are in the 33% tax bracket and assume the long-term capital gains rate is 15%.
- Let's say your stock appreciates by $100,000 when you eventually sell it (from zero to $100,000).
- Assuming you did file an 83(b) election, you would owe $15,000 in taxes (taxed at long-term capital gains rates).
- If you did not file and 83(b) election, you would owe $33,000 in taxes (taxed as income).

That's why filing can matter: it would be additional taxes of $18,000. Some additional details . . . At the time of purchase of the stock, your tax treatment depends on whether you timely file an election under Section 83(b) of the I.R.C. If you make the Section 83(b) election, any appreciation in the value of the shares after the time of receipt and before vesting will not be subject to taxation until the shares are sold. In the absence of a Section 83(b) election, the employee must include as compensation income in his or her income for each year in which any shares vest, the excess of the fair market value of the vested shares on the vesting date over the amount paid for such shares. In addition, the employee (and the company) may be required to satisfy certain withholding tax obligations at the time of vesting, which may require an immediate cash outlay by the employee.

This obligation can be potentially disastrous to an employee in a situation in which the shares have increased in value, and the employee is unable to sell any of the shares to offset the tax obligations. The Section 83(b) election must be made no later than 30 days after the date of transfer of the shares to the employee.

After sale of the shares, the employee will recognize capital gain or loss in an amount equal to the difference between the sale price of the shares and the fair market value of the shares on the date of purchase (or, if no Section 83(b) election is made, the vesting dates). If the shares have been held for more than one year before the date of sale, the gain or loss will be treated as long-term capital gain or loss to the employee.

The company is entitled to a federal income tax deduction equal to the amount of ordinary compensation income recognized by the employee.

Stock Purchase Plans

An ESPP provides another mechanism for employees to purchase stock in the company. These plans are mainly used by public companies, although some private companies do use these programs. Generally, an ESPP would be structured as a tax-qualified Section 423 plan that provides for participation by nearly all full-time employees with at least two years service.

Like a stock option plan, an ESPP must be approved by the shareholders and the board of directors. The plan itself provides employees with the opportunity to purchase company stock at a discount from the current market price of the stock, as much as 15% below the value of the stock at the time the price is set. Generally, the company would automatically deduct a portion of the employee's salary through payroll deductions. At the end of a period, known as an "offering period," the money that was set aside will be used to purchase the stock at a discount. After the purchase, the employee would own the stock and be able to sell it at that time.

Amount of Stock or Options to Grant

Most companies will set aside an amount of stock or options to grant to employees. Managing that stock or option pool is a difficult job for startup companies. Providing sufficient numbers of options or stock to attract and retain top employees while attempting to stretch the pool over a number of years is a difficult job.

The number of options or amount of stock themselves are less important than the percentage of ownership to which these options or stock are equal.

PROMISING A PERCENTAGE

What could happen

You've agreed to give a new hire "1% of the stock of the company." Could this cause a problem?

Watch out for

You were most likely implying that the employee would receive 1% of the currently issued and outstanding stock, or perhaps 1% of the fully diluted stock of the company (which would include authorized options). However, if you promise a percentage of the company, you may obligate yourself to provide the employee additional stock if the company issues more stock down the road. As more stock is issued, the percentages are diluted.

You should make grants of stock or options that are for a particular number of shares of stock, not solely for a percentage of the stock. In the event you want to be clear that the stock granted currently equals the a certain percentage, use the phrase "which represents __% of the current issued and outstanding stock."

TIP: Be careful of promising a "percentage" of the company to avoid ongoing grants of stock or option to maintain that ownership.

Companies generally offer initial stock or option grants at the time of hire, annual option or stock grants, or a combination of both these methodologies. More established companies are likely to give reoccurring grants, whereas startup companies tend to offer initial grants and may choose to approve subsequent grants to employees in the event of a change in position, the attainment of a milestone, or to better distribute options. Additionally, companies will tend to give reoccurring grants to more senior positions.

Option Pool Size

After an investment from a venture capitalist, the size of the option pool will typically be set at between 10 and 18% of the fully diluted stock ownership. These options are usually intended to last for two to three years, depending on the expected growth of the company. This means that the company must carefully plan to allocate the option pool for new hires and for additional grants for continuing employees.

As the company continues to grow, the board and the stockholders may need to replenish the option pool to continue to attract additional talent.

Grants

Many companies will attempt to structure a general rule to allocate initial hire stock grants that depend on the seniority of the individual. Companies will identify the general tiers of the organization, which would range from the CEO and the senior vice presidents through the entry-level employees. In general, the number of options granted to a new hire will decrease with the level of seniority; usually an employee will receive between one-half to one-quarter of the options granted to their supervisor. An example of a sample option distribution plan is below.

Tier/position	Option/share grant	Percentage[1]
CEO	600,000	3.000%
Chief operating officer/CFO/ senior vice president	200,000	1.000%
Vice president	100,000	0.500%
Director	50,000	0.250%
Manager	25,000	0.125%
Staff	10,000	0.050%
Entry level	5,000	0.025%

[1] Assumes 20 million shares of stock issued and outstanding (including option pool).

The following charts represent helpful guides to help determine appropriate initial hire grant levels for employees based on their relative position and the approximate ongoing grants to be made for positions.

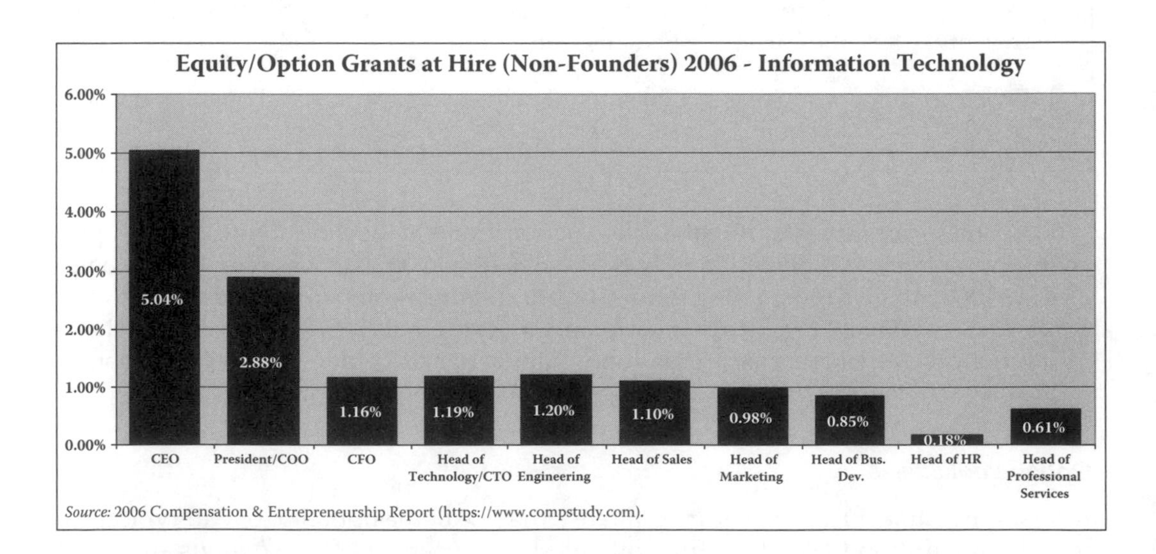

Source: 2006 Compensation & Entrepreneurship Report (https://www.compstudy.com).

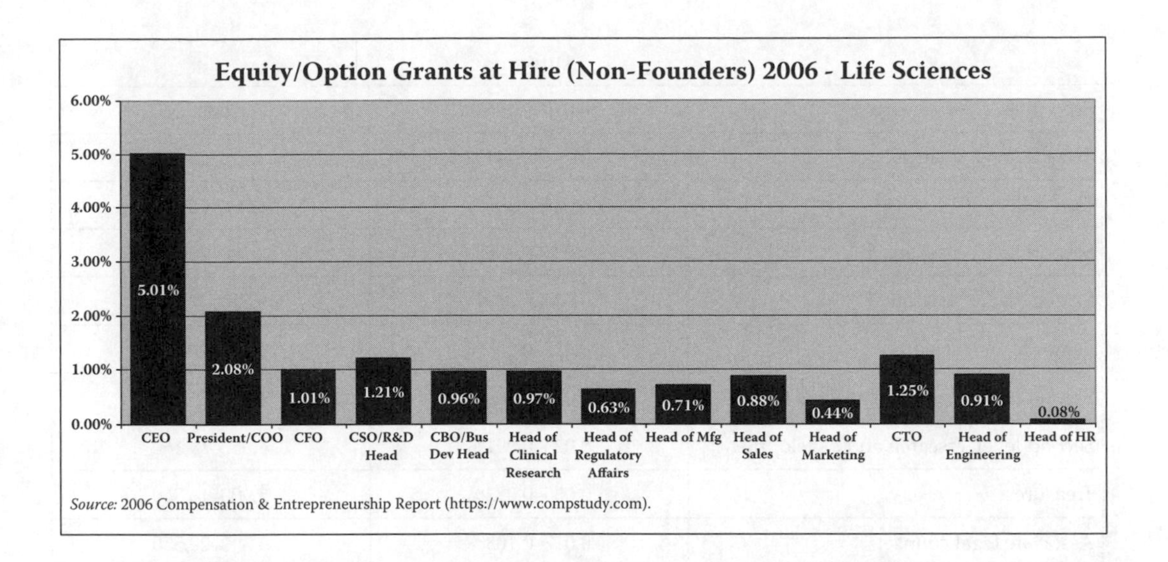

Source: 2006 Compensation & Entrepreneurship Report (https://www.compstudy.com).

Stock Option Hire Grants in the High-Technology Industry

Level	Hire grants as a percentage of shares outstanding	Options based on 20 million shares outstanding
	Officers	
CEO	0.781–2.513%	156,270–502,650
Chief operating officer	0.322–0.926%	64,340–185,190
CFO	0.204–0.616%	40,830–123,160
Chief technology officer	0.110–0.495%	21,950–99,050
Senior executive vice president	0.205–0.751%	41,070–150,280
Senior sales	0.121–0.572%	24,280–114,450
Senior marketing	0.138–0.482%	27,530–96,470
Senior operations	0.125–0.485%	24,910–96,940
Senior professional services	0.091–0.386%	18,160–77,180
Senior research and development	0.121–0.463%	24,190–92,680
Senior engineering	0.224–0.615%	44,800–123,090
SBU executive	0.100–0.287%	20,030–57,360
Legal counsel	0.064–0.236%	12,830–47,190
Senior administration	0.071–0.137%	14,200–27,330
Senior human resources	0.048–0.183%	9,670–36,580
Senior information systems	0.047–0.106%	9,330–21,230
Senior manufacturing	0.032–0.102%	6,410–20,390

Directors		
Second level, engineering	0.040–0.126%	7,910–25,160
Second level, financial	0.045–0.141%	8,970–28,250
Controller	0.044–0.142%	8,790–28,470
Second level, marketing	0.039–0.105%	7,700–21,000
Senior quality assurance	0.034–0.089%	6,810–17,820
Senior technical staff	0.025–0.105%	5,030–20,910
Second level, professional services	0.023–0.106%	4,660–21,140
Second level, research and development	0.023–0.091%	4,650–18,160
Treasurer	0.030–0.082%	5,900–16,360
Associate legal counsel	0.017–0.103%	3,390–20,550
Second level, information systems	0.015–0.065%	2,910–12,940
Facilities/real estate	0.016–0.045%	3,140–8,900
Managers		
Third level, engineering	0.017–0.071%	3,430–14,100
Third level, marketing	0.018–0.064%	3,650–12,870
Human resources, compensation/benefits	0.012–0.034%	2,320–6,740
Accounting manager, entry	0.006–0.031%	1,200–6,110
Lead – tech		
Test engineer	0.020–0.070%	3,930–14,070
Manufacturing engineer	0.017–0.061%	3,370–12,120
Technical support	0.020–0.060%	3,960–11,900
Sales reps		
Senior	0.009–0.034%	1,750–6,810
Entry	0.003–0.011%	580–2,100
Other exempt		
Exempt technical (senior)	0.012–0.045%	2,310–8,977
Exempt technical (intermediate)	0.006–0.025%	1,170–5,020
Exempt technical (entry)	0.003–0.011%	630–2,210
Exempt nontechnical (senior)	0.004–0.016%	830–3,100
Exempt nontechnical (intermediate)	0.003–0.010%	527–2,070
Exempt nontechnical (entry)	0.002–0.005%	350–1,060

Source: http://www.salary.com, based on data compiled from published surveys as of January 2000.

Ownership Levels at a Liquidity Event in the High-Tech Industry

Level	Ownership levels as a percentage of shares outstanding	Ownership based on 20 million shares outstanding
President and CEO	2.40–4.60%	480,000–920,000
Vice presidents (all)	0.25–1.95%	50,000–390,000
Vice president, CFO	0.80–1.60%	160,000–320,000
Vice president, engineering	1.15–1.95%	230,000–390,000
Vice president, operations	0.70–1.50%	140,000–300,000
Vice president, human resources	0.25–1.05%	50,000–210,000
Vice president, general counsel	0.50–1.30%	100,000–260,000
Vice president, marketing	0.40–1.20%	80,000–240,000
Vice president, sales	0.80–1.60%	160,000–320,000
Directors	0.10–1.00%	20,000–200,000
Managers	0.03–0.50%	6,000–100,000
Senior contributors	0.03–0.20%	6,000–40,000
Intermediate contributors	0.02–0.07%	4,000–14,000
Entry contributors	0.01–0.05%	2,000–10,000
Nonexempt	0.01–0.03%	2,000–6,000

Source: http://www.salary.com, based on data compiled from published surveys and analysis of actual practices from S-1 filings as of January 2000. Information excludes founder's holdings.

MORE INFORMATION

To find updated information on equity compensation figures, visit the book's website at http://www.myhightechstartup.com.

Vesting and Acceleration of Employee Stock and Options

For stock grants, restricted stock grants, and stock options, the company may include vesting terms on the equity. Under the terms of the agreements, the company may grant stock or options that are subject to repurchase by the company. As time passes or events occur, the repurchase rights of the company may be released for all or a portion of the equity. The benefit of this structure is that the company will be able to make a smaller, up-front grant of equity and the amount of equity held free and clear by the employee will increase over time or as milestones are met.

COMMON APPROACHES TO VESTING

- **Time-based (straight line):** Stock or options are released from vesting in equal amounts each month over a particular time period (say monthly, quarterly, or annually over a number of years, usually between two and five years).
- **Time-based (cliff):** No vesting for a particular time period (for instance, the first six months or first year or until financing occurs). Then, once that initial period is completed, the company will then release a certain portion from vesting. Once the cliff period has passed, the rest will typically vest on a straight-line basis afterward (monthly, quarterly, or annually).
- **Milestone-based:** Stock or options will be released after the achievement of particular milestones rather than based on time periods.
- **A combination-based approach:** Stock or options will be released on a combination of milestone-based and time-based vesting.

In addition, companies may also decide to add acceleration provisions onto the vesting. In these cases, the vesting will automatically accelerate on the occurrence of a certain event or events. The company may provide that, if the company is acquired, then the unvested stock or options will automatically accelerate. This is oftentimes referred to as a "single-trigger acceleration" provision. In other cases, the company may provide that, if the company is acquired and the employee is terminated within 12 months of the acquisition, then the unvested stock or options will accelerate. This is oftentimes referred to as a double-trigger acceleration provision.

Sometimes, the company will accelerate 100% of the unvested stock or options, whereas other times the company will accelerate a particular portion (say 25% of the unvested stock/option or 25% of the total amount of stock/option), and still other times, a company will only give acceleration to employees that have a certain tenure with the company or a certain level (vice president and above) with the company.

EXAMPLES OF VESTING IN PRACTICE

- **Time-based (straight line):** An employee has straight monthly vesting over a four year period. After being employed for 12 months, the employee leaves the company. At this point, the departed employee would only own one-quarter of the original stock issued or options granted to her. In the case of options, the employee would be able to exercise the options to purchase stock; the stock grant would be owned outright. The rest of the unvested options or shares (three-fourths of the original amount issued) would then be repurchased by the company at the par value for the stock grant.
- **Time-based (cliff):** An employee's stock or options are vested over four years. The vesting will be a one year (12 months) cliff, followed by straight-line monthly vesting over the remaining three years. If the employee departs after six months, she will have no shares or options vested and the company will repurchase the entire amount of original shares issued or cancel the option grant.

However, if the employee departs after 18 months, then one-quarter of the stock or options will have vested after the cliff and six more months of vesting would have occurred. So the founder would have 37.5% of her stock or options vested at that time.

- **Milestone-based:** A certain portion of unvested stock or options will be released from vesting when (1) the company receives at least $1 million in funding (to incentivize fundraising efforts), (2) the company reaches $250,000 in annual revenue (to incentivize sales), and (3) when the company releases its second-generation product (to incentivize product development).
- **A combination-based approach:** As an example, half of the stock or options will vest monthly over a four-year period and the other half will vest based on achievement of certain milestones.

Labor and Employment Laws

Although the entrepreneur may be focused on financing and product innovation, he should also be focused on whether his organization is complying with relevant labor and employment laws. Employment litigation is on the rise, and the entrepreneur does not need to deal with charges of harassment or discrimination amid trying to create a successful business venture.

Labor and employment law is both federal and state in nature, often with federal laws providing a minimum layer of protection for employees and state law adding further to those rights. Labor and employment laws in California are particularly protective of employee's rights. The following are some of the major areas for concern for employers generally; many of the issues involve specific state laws, and this provides information for several states.

Employment Status

An employee is generally retained on either a contractual or at-will basis. Employment contracts usually specify the terms of employment, including duration, salary, bonuses, benefits, and incentives. Small businesses rarely have employment contracts with their support staff, although employment contracts with licensed professionals are not uncommon. Absent a contract between an employer and its employee, most employees are hired on an at-will basis. Employment at will allows either the employer or its employee on notice to the other party to terminate employment for any reason.

- In California, both employment contracts and employment at-will conditions must adhere to California public policy and implied practices of good faith and fair dealing.
- In New York, the at-will doctrine is firmly established, with one exception. An employer may be bound to obligations or promises contained in an employee handbook or personnel manual.
- In Washington, courts have recognized public policy as regulating both the employment contract and at-will relationships.

Minimum Wage and Overtime

Startup companies are required to comply with federal and state minimum wage and overtime requirements for their employees. The Fair Labor Standards Act does exempt small businesses with annual sales less than $500,000 from paying the federal minimum wage to employees. However, your business must still comply with state minimum wage requirements.

In addition, be aware of the impact of minimum wage and overtime requirements on salaried employees. Employers have been required to compensate salaried employees when it was found that the hours worked by the salaried employee reduced their hourly compensation below the required minimum wage or when overtime pay was obligated to be paid. You should be certain that you are aware of the hours employees are working to avoid this issue from arising (which oftentimes is raised when an employee departs and decides to sue the company for back pay and overtime).

Discrimination

Contrary to popular belief, not all forms of discrimination are prohibited. Rather, discrimination is prohibited if it falls within a protected category defined by the law. These protected categories have developed over time in response to society's realization that certain categories of individuals have been mistreated for unfair reasons and require protection under the law. Under federal law, these protected categories include age, race, creed, color, national origin, gender, religion, military status, and disability.

An exception to discrimination law arises when a job requires a bona fide occupational qualification (BFOQ). A BFOQ might be discriminating against nonfemales in the casting of a role for a female character in a movie. The BFOQ exception is an affirmative defense and the burden is on the employer to prove its validity. Because of the potential for misuse, BFOQ exceptions are narrowly interpreted.

STATE LAWS: DISCRIMINATION

- **California:** California law, like federal law, prohibits discrimination against employees or job applicants based on race, gender, age, national origin, religion, and disability. California law goes further than federal law in several respects, however. Marital status is also a prohibited ground for discrimination. Courts have construed the California Labor Code as prohibiting discrimination based on sexual orientation, and certain local governments have enacted antigay bias laws. Plaintiffs may also sue for discrimination as a violation of public policy without first pursuing administrative remedies under the California Fair Employment and Housing Act. In addition, if an employee can prove unlawful discrimination, he may seek reinstatement or economic damages, as well as potentially significant damages for emotional distress caused by the discrimination and punitive damages. Such remedies are not currently available for all forms of discrimination under federal law. California's family-leave law requires employers to grant up to four months leave of absence to employees who need to care for a spouse, child, or elderly relative, or who want time off in connection with the birth or adoption of a child. The law is not clearly drafted, however, and there are many questions about its scope that remain unanswered. The law does clearly require that employers make available to employees on their return the same position or a position comparable with the one they held before taking family leave.

- **New York:** New York law contains the same protections as federal law and adds marital status, conviction of a crime, and genetic predisposition or carrier status to its protected categories. Certain local New York laws include sexual orientation. New York City prohibits employers, including private ones, from discriminating based on sexual orientation. Sexual orientation is defined to include heterosexuality, bisexuality, homosexuality, and transsexuality.

  New York Human Rights Law prohibits discrimination on the basis of a person's pregnancy status or birth control methods. Unlike in federal law, in New York, pregnancy is considered a disability, and the pregnant person can collect disability pay under the short-term disability provisions contained in the Worker's Compensation Law.
- **Washington**: Washington's antidiscrimination law, the Law Against Discrimination, prohibits discrimination on the basis of race, creed, color, national origin, ancestry, sex, sexual orientation, marital status, age, disability, use of a trained guide dog or service dog by a disabled person, and results of HIV or hepatitis C tests. The Law Against Discrimination, together with the blind and disabled person's law, make discrimination because of a person's disability unlawful.

  Washington's civil service law prohibits employment discrimination and mandates affirmative action in state government, as does the public education gender equality law, with respect to higher educational institution employment. Affirmative action is also required by the state patrol affirmative action law, a sexual harassment executive order, and an affirmative action executive order.

  Washington law also provides protection for employees who need time off or extended leave to care for themselves or family members. The family leave law provides leave for employees who must care for a newborn, an adopted or foster child, or an ill relative, as well as employees with their own medical problems. Employees are allowed to use sick leave to care for a child, whereas the parental leave discrimination law protects employees who need leave for the birth or adoption of a child.

Wrongful Discharge

When an employer terminates the employment of an employee for reasons prohibited by the law, the employer may be held liable for wrongful discharge. Generally, an employer may not terminate an employee based on any of the following: (1) a breach of an implied contract made by the employer, (2) the employee's refusal to perform illegal acts or for exercising a legal right, (3) an employee's whistle-blowing activities, or (4) the employer's violation of public policy. Wrongful discharge claims can entitle an employee to significant damage awards. An employer can prevent many wrongful discharge suits through thorough documentation of all personnel and termination-related decisions.

STATE LAWS: WRONGFUL DISCHARGE

- **California:** California law permits an employee in many circumstances to sue for wrongful discharge even if he does not have an express contract with the employer requiring cause for discharge. The employee can sue for breach of implied contract if he can establish factors such as longevity of employment, promotions, pay raises, and commendations, and employer promulgated personnel policies that promote job security. Employers may try to forestall such claims by having employees sign express agreements that they may be terminated at will or without cause. These statements may also appear in employment applications and personnel manuals or handbooks.

  However, several California appellate courts have held that, although such writings may be considered evidence of the employee's at-will status, unless they are part of a complete employment agreement, they may not be determinative of the issue. In addition, if the employee can show that he was discharged in retaliation for exercising a legal right, for refusing to break the law or for reporting or complaining about employer misconduct, or in violation of a statutory protection (like the laws against discrimination), the employee may sue for termination in violation of a public policy.

  Wrongful discharge claims, if not anticipated and handled properly, can lead to substantial jury verdicts and litigation expenses. Employers who are used to exercising virtual absolute discretion in firing employees can be in for a rude awakening when it comes to dealing with California wrongful discharge law. As a result, prudent employers will take precautions, including following progressive disciplinary programs, to help avoid such liability.
- **New York:** In New York, attempts have failed for the passage of a wrongful discharge law. However, other proposals continue to be announced. Typically, the proposed bills contain the following provisions:
  - Employers must give and employees are entitled to 15 days' notice of discharge.
  - Employers must give and employees are entitled to a statement of reasons for discharge.
  - An employee complaining of unjust dismissal can request arbitration; the arbitrator can order reinstatement and lost earnings.
  - A prevailing employee can be awarded expenses accruing from outside assistance, but the employee can also be required to pay costs if the claim is frivolous, vexatious, or made for the purposes of harassment.
- **Washington:** In Washington, wrongful discharge is regulated by case law, statute, and regulations promulgated by the Washington Department of Labor and Industries. Washington courts have adopted a public policy exception to the at-will doctrine, in which a termination of an employee in violation of a constitutional provision, statute, regulatory provision, or previous judicial decision may be actionable.

Workers' Compensation

Workers' compensation is a type of insurance that provides medical care and compensation for employees who are injured in the course of employment. Workers' compensation schemes differ between jurisdictions, varying from weekly payments in place of wages, compensation for economic loss, reimbursement or payment of medical and like expenses, general damages for pain and suffering, and benefits payable to the dependents of workers killed during employment. Benefits are administered on the state level.

STATE LAWS: WORKERS' COMPENSATION

- **California:** Every employer in California must provide workers' compensation insurance for employees who sustain injuries arising from employment. Compensable injuries are not only physical but also mental or emotional when they cause disability. Generally speaking, the workers' compensation insurance carrier will handle an employer's worker compensation claims, although certain claims relating to workers' compensation, such as liability for "serious and willful" misconduct causing injury, are not usually covered by insurance.

  In addition, Section 123a of the California Labor Code, which prohibits discrimination or retaliation against employees who file worker's compensation claims or participate in the Workers' Compensation System, has been construed very broadly to protect injured employees.
- **New York:** Employers in New York must pay medical costs of compensable injuries that occur in the course of employment. The list of compensable injuries is constantly expanding and includes disability benefits. Under New York's Disability Benefits Law, a nonoccupational disability qualifies even if the employee is not totally disabled, as long as he cannot perform regular job duties or other duties proffered by the employer for which the employee is qualified.

  To entice businesses to relocate to New York, New York law limits third-party suits against employers in workers' compensation cases. A third party may not receive a contribution from an employer for an injury caused to an employee unless the employee suffers grave injury.
- **Washington:** Washington workers' compensation is different from most other states. The Washington Act establishes two funds, the accident and medical aid funds, which are wholly administered by the state. Employers who are not self-insured pay into the state fund based on the degree of hazard involved in their occupation.

Immigration Laws

All companies must obtain an Immigration and Naturalization Services Form I-9 from each new hire. In the event your company is planning to involve foreign nationals as key employees in the early stages, you will need to be acutely aware of the limitations of immigration laws. Be aware that complying with immigration laws is often a difficult process when hiring a foreign national for your startup company. The entire process takes between three and five years or longer depending on the potential employee's qualifications and the backlog of work at U.S. labor, consular, and immigration offices.

For companies that would like to hire foreign nationals, there is basically a three-step process to obtain employment-based lawful permanent residence status. The steps are as follows:

1. The labor certification process conducted by DOL
2. The submission of the I-140 Immigrant Petition with U.S. Citizenship and Immigration Services
3. Finally, the employee's application for permanent residency, filed either as an adjustment of status application with the U.S. Citizenship and Immigration Services or through the appropriate U.S. consulate abroad

The process will be lengthy and require work with an immigration attorney. In the event you plan to involve hiring foreign nationals in your hiring plans, you should consult with an immigration attorney in advance of your hiring efforts that may be able to provide guidance to help the process move more quickly.

Workplace Safety

The federal Occupational Safety and Health Act governs workplace safety and deals with both traditional and current issues.

California recently adopted a comprehensive workplace safety law that requires virtually every employer in the state to adopt and communicate detailed employee injury and illness prevention plans. In addition, California law imposes criminal liability on corporate managers who knowingly expose employees to workplace hazards.

Government Contracting

Many startup companies will apply for government grants or may decide to apply for a government contract, which oftentimes have favorable rules designed to give preferences to small businesses. Although these programs may be very beneficial to a startup company, you should also be aware that these grants may also bring additional procedures for your company to undertake. Many of the rules at issue are typically not applicable for small businesses but may be required to comply with the rules of the government contract.

Be aware of both direct government contracts with your business and when your business will be a subcontractor. There are oftentimes very particular reporting requirements that are part of the contract.

Departing Employees

There are numerous legal issues that may arise when an employee departs for a competitor. The most important task an employer can do is done before an employee leaves, which is checking to make sure that every employee has an offer letter, invention assignment, nondisclosure agreement, and other agreements in place with the company. All employees should have agreements in place to maintain confidential information of the employer after the employee leaves the company, to fully assign inventions, and, in certain circumstances, to prohibit joining a competitive business, soliciting employees, or soliciting customers. The duration of this obligation is dependent on the nature of the company's business and should be considered when drafting the agreements.

PRACTICAL CONSIDERATIONS FOR A DEPARTING EMPLOYEE

After an employee leaves, the following steps should be taken by a business to help minimize legal problems it might face:

- **Consider whether it should waive any notice of the employee's resignation.** Once the employee gives notice, the employer may want to ask the employee to leave employment immediately. An employee with divided loyalties can be a risky proposition for the employer.
- **Conduct a thorough exit interview of the employee.** During the exit interview, the employer should identify the new employer, determine the employee's new job description, and understand the reasons the employee is leaving to join a competitor. The employer should remind the employee of any confidentiality and nonsolicitation obligations. The employer should also identify and list any confidential information in the employee's possession. Once the exit interview is complete, the employer should document the contents of the exit interview in a memorandum to the employee's file. Often, the employee's suspicious behavior during the exit interview, or lack thereof, will reveal how concerned the employer should be regarding the employee's departure.
- **Ask the employee to sign an exit memorandum or letter.** The purpose of such a memorandum or letter is that the employee will acknowledge his or her access to trade secrets and confidential business information of the employer and will reaffirm her confidentiality obligations to the employer. The exit memorandum or letter should also indicate and remind the employee that the obligations survive termination of the employment. If the employee refuses to sign the exit memorandum or letter, the employer should send it to the employee at the employee's home address.
- **Take physical action to minimize the risks that trade secrets will be misappropriated.** The employer should check the inventory accessible to the employee and, specifically, the employee's workspace. The employer should consider whether the employee should clean out his workspace with a human resources representative present. The employer should also lock out the employee from the network as soon as practical. After the employee's departure, the employer should maintain the employee's computer, including electronic document and e-mail accounts, for 30–60 days. This preservation will ensure that, if issues arise after the employee's departure, records are maintained that can be used to assess whether the employee has engaged in any wrongful conduct. Employees are often caught stealing employer's intellectual property after computer forensic experts analyze the former employee's computer or e-mail accounts.

- **Consider sending a letter to the new employer.** The letter should warn the new employer that the employee had access to or possesses knowledge of the company's trade secrets. The letter should also disclose the employee's confidentiality and nonsolicitation agreements with the employer. The letter should remind the new employer of its obligation to avoid placing the employee in a situation in which misappropriation is likely. Finally, the letter should seek assurances from the new employer that the company's trade secrets will not be misappropriated.
- **Monitor the activities of the employee throughout the period of transition while the employee leaves the company for a new employer.** If there are any suspicious activities, for example, potential nonsolicitation violations, these activities should be documented. If any of these circumstances surrounding the employee's departure appears suspicious, it is imperative that the employer raises the issues with its lawyer. A small amount of counseling at the beginning of this type of situation can often avoid irreversible errors in the process that result in the destruction of critical evidence and increase the risks that a company's trade secrets will walk out the door.

Trade Secrets

Employers are generally very concerned about the leaking of trade secrets when an employee departs the company. There are numerous ways a company can act to ensure that its sensitive information qualifies for trade secret protection when an employee is planning to depart.

PRACTICAL CONSIDERATIONS FOR PROTECTING TRADE SECRETS

What steps should you take to protect your trade secrets when an employee departs?

- **Exit interviews:** Conducting exit interviews with departing employees to remind them of their obligations to refrain from using or disclosing the company's confidential and proprietary information (preferably in the form of a written acknowledgment signed by the employee). The company should ask the departing employee (1) to identify any documents, computer disks, or other materials he possesses that contain confidential or proprietary information of the company, (2) to commit in writing as to when he will return the material to the company, and (3) to confirm what the departing employee will be doing at his or her new place of employment (i.e., to get a sense of whether the employee may be in a position to misappropriate the company's trade secrets).
- **Consult your attorney:** Working with legal counsel in cases in which a departing employer accepts employment with a new employer and it appears there may be a risk that the employee will misappropriate the company's

trade secrets in connection with his or her new employment. Legal counsel may determine it is appropriate (1) to send a letter to the employee's new employer to warn the new employer of the existence of trade secrets within the employee's knowledge, (2) to remind the new employer of its obligation not to make use of the company's trade secrets, and (3) to ask for assurances that the company's trade secrets will not be misappropriated. if the company does not receive a satisfactory response, it may need to initiate a lawsuit to protect its rights.

12

Your Boards and Board Members

All start up companies have one thing in common—they all get in trouble. It's how they and you on the board handle their trouble that separates the winners from the losers.

Thomas J. Perkins
Cofounder of Kleiner Perkins Caufield & Byers

What to Watch For

For many entrepreneurs, thinking about your board of directors may seem to be low on your list of priorities. Some entrepreneurs will decide to have a single-member board of directors, whereas others will have a board made up of just the management team and perhaps a few confidants.

Yet, according to investment banker Bill Vogelgesang, failing to focus some energy on developing a strong board represents a significant missed opportunity for a startup company. Said Vogelgesang, a managing director and principal in the Cleveland office of Brown, Gibbons, Lang & Co., an investment-banking firm, in an interview with *Inc.* magazine, "What entrepreneurs often don't realize is that a strong board can help a young or growing company build credibility in the outside world. A good board—especially one with heavy involvement from other CEOs and decision makers—reflects well on a chief executive because it shows that he or she can take criticism and doesn't just want to impose one business vision on the company."

Recognizing the importance of these outside advisors, many startup companies choose to have a board of directors, as well as an advisory board or a scientific advisory board. These boards and the board members can represent important sources of strategic thinking, networking, and experience for a young company.

The key feature distinguishing your board of directors and advisory board is the legal role the board of directors plays in the corporation. Although each of your directors and advisors may provide varying levels of strategic insight, industry contacts, and sage advice, the role defined for your board by state law and in your organizational documents will tend to become more important as the company matures.

Roles of the Board of Directors

A board of directors is responsible for overseeing the management of the company. The directors represent the shareholders to ensure that the company acts in the best interest of its shareholders. More than just providing oversight, a board provides assistance for a new company and may offer their experience to assist new entrepreneurs in their business development.

The board of directors serves to provide strategic thinking based on an "outsider's" perspective. It is important to consider each appointment to your board as an important resource and asset for the company. Board members may be able to provide you with key contacts and access to individuals to assist in the development of your business. For high-technology companies in particular, research has shown that your board plays an important role in successful innovation and commercialization efforts.

For a newly formed or emerging company, the board of directors can also act to assure key constituents of the company: investors, employees, and customers. Some VCs and other sophisticated investors will consider the directors as part of the knowledge base of a company and may evaluate a company based on the composition of the board of directors. In addition, some investors feel that a board that has at least some directors that have no direct relationship with the company will be better able to protect shareholders' interests with constructive, unbiased guidance.

The role of the corporate board of directors has changed in the past century. Corporations had been understood to be primarily governed by the will of shareholders, and the board of directors was the mechanism that acted out that will. By the early twentieth century, that perception had changed and the role of the board began to increase as corporations provided greater direct decision-making authority to the board elected by the shareholders.

This role has continued to evolve, and, today, more than ever before, the board of directors serves a critical role in corporate governance and decision-making.

Boards of directors have seen the importance of their position within corporations increase dramatically, particularly since the inception of the Sarbanes-Oxley Act and its creation of a growing emphasis on effective corporate governance. Although this growing scrutiny and regulation is overtly affecting public companies, many private companies are placing important oversight roles in the hands of their boards at earlier stages than before.

Although the legal role of the board to provide oversight and make certain decisions is a key part of the board's duties, a board of directors plays multiple roles for a startup company. Understanding these different roles represents an important first step in working effectively with a board and its directors.

For companies that are organized as an LLC, oftentimes the LLC will define a board of managers in their operating agreement. Generally, much of the information about a corporation's board of directors is applicable for an LLC's board of managers.

VARYING ROLES OF THE BOARD OF DIRECTORS

There are five key roles a board of directors play within the company, and particularly play with respect to a high technology startup company. Those roles are as follows:

- Legal role
- Representative role
- Strategic role
- Advisory role
- Service role

The legal role played by the board of directors is determined by federal and state laws (primarily state laws governing the state of incorporation, but other laws may also be implicated), rules and procedures set forth in your corporate articles or certificate of incorporation and corporate bylaws, and additional rules and statutes from the SEC, Financial Industry Regulatory Authority, or other entities that regulate certain stock exchanges.

The representative role relates to the relationship between the corporate shareholders and the board of directors. As discussed previously, the board of directors historically had been elected by the shareholders as representatives and acted out the will of the shareholders. Although that relationship does remain in some senses, particularly in that the shareholders continue to elect the board, the power has shifted in many cases to the hands of the directors. Even so, it is still important to recognize that the board does represent the interests of the shareholders and is responsible for providing oversight in the interest of all shareholders of the corporation. This relationship can be at issue in cases in which a shareholder (oftentimes a holder of a sizeable stake of stock) believes the board is not acting in the best interests of the shareholders, but courts have tended to give substantial deference to the board.

The strategic role primarily stems from the role the board plays in corporate decision-making. Because of the fact that a board will be responsible for approving decisions from potential acquisitions to option grants to new hires, the board will hold an important role with respect to certain strategic decisions. The challenge that some corporations face is the balance between the board and management in these strategic decisions. However, there has been research that suggests there is a positive relationship between a board's involvement in corporate strategy and the performance of the company.

The advisory role for a board of directors is most often seen in the earliest stages of corporate development. Early investors in the company, oftentimes who represent venture capital funds or other serial startup investors, will provide capital for the corporation but may also view their role to guide and assist the company in its earliest stages. The experience many of these directors will have of working with numerous companies in a similar market, at a similar development stage, or facing common market conditions gives the board the ability to provide advice and insights to the company.

The service role of the board and its members come as a result of the fact that many small and early-stage companies will not pay their board members until the company matures or will only pay a nominal fee for board service. Compensation will usually be in the form of stock options, reimbursement for travel expenses, or stipend for meeting attendance. At least in some senses, the board and its members will be providing service to the company and its contributions will be aimed at adding long-run value to the entity.

Forming the Board of Directors

Understanding the structure and makeup of the boards of startup and emerging companies has been the subject of a great deal of research since the Sarbanes-Oxley Act went into effect, because it has been suggested that corporate governance before going public represents an important insight into the governance practices to expect after the company has completed its initial public offering. The National Association of Corporate Directors (NACD) has conducted research on more than 200 companies with $50 million to $200 million of revenues, and TopFive Data Services researched 177 high-tech companies with revenues under $250 million. The results of these data reveal some important insights into the size, structure, and compensation practices of small and emerging companies.

How Big Should the Board Be?

For public companies, the National Association of Corporate Directors recommends a board of no less than nine and no more than 15 members. For smaller private companies, board sizes are typically between five and seven members with one to two inside directors and the rest outside directors. For a new startup company, the minimum number of board members is typically set by the state laws, but oftentimes can be just one individual. In the event that the company is an early-stage company and has not received any investments from outsiders, many companies choose to operate with a single director for ease of operations. For companies that are founded by a group of individuals, oftentimes the founders will decide that each of them will begin as directors with the understanding that, as the company matures, some of the initial founders will most likely be removed from the board.

SIZE OF THE BOARD

According to research by Noam Wasserman, the size of the board increased or decreased based on certain events.

- As the number of rounds of financing increase, a startup company has a *larger board of directors.*
- As the dollars raised in the most recent financing round increases, a startup company has a *larger board of directors.*
- As the number of employees increases, a startup company has a *larger board of directors.*

What Is the Makeup of the Board?

The most effective board of directors will grow and change as your business develops and matures. A good director during the initial startup of an organization may not be the most effective director when the business is contemplating an IPO or making future merger or acquisition decisions. Therefore, it is important to continue to evaluate your board and determine the best directors as the company develops.

Many early-stage companies will have less than five directors and, as the company matures, will add additional seats as the company takes on additional investors and continues to grow. In many cases, it may be required to add seats on its board for its large investors.

DIFFERENT TYPES OF DIRECTORS

Your board will likely be made up of some composition of the following types of directors:

- **Inside directors:** These are individuals that are most likely part of the day-to-day operations of the company or were at one point involved in the operations. This would include having the CEO or a founder serving as a director.
- **Outside directors:** The directors are individuals that do not have any role with the organization other than as director and do not serve as an officer of the organization. These individuals may be current or former executives of industry companies, professionals from accountancy, finance, academia, or legal careers, or may have experience as a former entrepreneur of another successful company. These individuals are valued for their contacts and outsider perspectives.

- **Investor directors:** As part of their role as a source of financial capital for your business, investors may serve as directors on your board. They too can provide similar qualities as the other non-investor outside directors through contacts and strategic advising. However, it is also important to note that these investor directors are also responsible for monitoring their investment and may not have their interest completely aligned with the company.
- **Board observers:** Because of constraints of size or maintaining a balanced board, you may want to select individuals to serve as board observers. These individuals will not vote in matters but will be able to participate in discussions and provide guidance. These individuals can serve a valuable role and help the functioning of the board.

Although it remains important to have a board that provides key functions for your organization, it is also important to consider the size of your board in light of your needs for efficiency during the early stages of your organization. Selecting a board that is not only the right composition but also the right size is important.

The makeup of the board is particularly important for a company looking to raise money. The caliber of the board members reflects on the caliber of the company, so be cautious of placing individuals on your board that do not portray the image of the company that you want to display. Generally, a board that is made up of 60% to 80% of outsiders is preferred to prevent a scenario in which the board is dominated by insiders and unable to allow the board to have a fresh perspective and provide an observers view.

Do You Need to Appoint a "Chairman" or "Lead Director"?

Ultimately, appointing a lead director or a presiding director is designed to provide greater structure to the board and its operations. Additionally, selection of the lead director is recognized as a best practice in corporate governance circles.

For many early-stage companies, the direction of the board and the leadership of the meeting will be handled by the CEO. As a result, a separate lead director may not be necessary. Research shows that only between 10% and 30% of companies with revenues between $50 million and $200 million have a lead or presiding director.

Should the Board Have Committees?

The decision whether to employ the use of board committees oftentimes depends on the size and involvement of the board members. Many-early stage companies with five or fewer directors will not necessarily find the need to divide certain responsibilities of the board. However, as the company continues to mature and add new investors, it may become necessary to create board committees to oversee audit, compensation, and executive transitions.

The use of committees ultimately will allow directors to focus their backgrounds and expertise into key areas of need for the company and assist the board to be more efficient and functional. The decision to add board committees may be required after investors add board seats. The decisions of the committees must still be brought before the entire board and voted on.

Although a board may not initially determine the need for committees, the board should begin to undertake certain activities to help ease the transition into a more formal board operating structure as the company matures. The company and the board should identify individuals of the board to serve in the lead capacity or role over the key areas of audit, nominations, and compensation. This will help make the transition to formal committees much simpler and more effective. In addition, the board should meet at least biannually in executive session without inside directors present. The board and the company should select one outside director to organize, plan, and moderate these sessions. This activity will assist with determinations that require independence and put in place a practice that will become important going forward.

BOARD OF DIRECTORS COMMITTEES

The following are committees that are most often used by companies or required by investors:

- **Audit committee:** The major U.S. stock markets require all listed companies to have an audit committee. The audit committee is primarily responsible for oversight (in cooperation with independent auditors) of the accounting financial controls of a company, the financial reporting process, and the internal audit function.
- **Nominating/governance committee:** The nominating or governance committee is responsible for the recruitment of potential candidates to fill initial board positions and vacancies. As a company grows and matures, it is important to continue to fit directors into the underlying needs of the company itself. It is important to engage in these discussions for the long-term health of the company.
- **Compensation committee:** This committee will serve to advise the board on matters related to compensation and is responsible for establishing an overall compensation policy for board members and key executives.

Even for smaller technology companies, the number of board committees has been increasing since the Sarbanes-Oxley Act was enacted. In 2003, the National Association of Corporate Directors found that the smaller technology companies in its survey had an average of 2.7 board committees, up from 2.1 in 2001. Although it has been quite common for smaller companies to have an audit and compensation committee, more companies are adding a nominating or governance committee.

Recruiting Directors

The decisions to select the directors of a new entity are some of the most important for long-term successes. Recruiting dedicated, intelligence, creative, and connected members of your board of directors will assist the company during its initial growth and as the company undergoes future corporate transitions including future financings, partnerships, mergers, and possibly an IPO.

Once you've identified a top-tier candidate, you may first be told no (which is often the response from the candidates you want the most). Find opportunities to get a potential board member to interact with your CEO or other board members. Building personal connections may be one of the best ways to convince a prospect to join your company over a period of time.

Who Should You Recruit to Serve on Your Board?

Ultimately, you should recruit people for your board that will be able to provide assistance to the company, from guidance attributable to industry or previous company experience to important contacts in the marketplace. You should look to create a team that will work well together and advance the aims of the company.

WHAT TO LOOK FOR IN POTENTIAL DIRECTORS

Identifying board members is a difficult process for many startup companies. So what are important things to consider when you are looking at a potential director?

- Relevant industry experience
- Ability to provide relationships and "open doors"
- Background in entrepreneurial activities, marketing, sales, or strategy
- Interpersonal and communications skills
- Ethical reputation
- Ability to work in a team
- Good relationship with the CEO or founders
- No conflicts with your business
- Adequate time to devote to your business
- Reasons for joining the board

Directors can be recruited from various backgrounds, including management executives, industry experts, professionals, academics, or retirees. Ultimately, your board should be selected on each individual's ability to work with a diversified group of people toward a defined objective, their willingness to commit to a company's goals and plans, and a history of good judgment.

Where to Find Potential Directors

Recruiting directors can be a tedious process but use references, contacts, and even search firms in some cases to find the right people. Allow plenty of time to find the right board members and don't hold up your business plan because you may not have the right team in place right away. Look for references from your current board members, attorney, accountant, or other professional services providers, from other entrepreneurs, from local professional associations or groups, from directors and executives of successful companies in your area, and from former colleagues or executives from previous employment situations. Finding the right team is important and deserves a substantial commitment of time and energy from the new business.

Selecting directors that can open doors for the company is an important characteristic. Research has shown that the number of contacts your business has with VCs is correlated to your ability to get funded (proving that one aspect of getting financing is simply a numbers game: finding as many potential investors that you can get access to). Look for potential investors that can offer you access to a personal network for business opportunities, sales leads, potential employees, interesting investors, and other board members. Particularly for early-stage companies, you will have the luxury to wait longer to find high-impact board candidates, so set the bar high for your ideal candidates and work to fill the seats with top-tier candidates.

UNIQUE PLACES TO FIND POTENTIAL BOARD MEMBERS

- **Academia:** Recruit former professors that may have expertise in a key aspect of your business.
- **Retired executives:** Look to executives, former partners of accounting or law firms, or other successful individuals who have or will soon retire.
- **Former employers:** Recruit executives from former employers who will know you personally and may offer an outsider's perspective.
- **Board search firms:** Use a search firm to fill important board seats.
- **Board members of other companies:** An individual who currently serves on a board may be more willing to serve on another board.
- **Industry and trade associations:** Find out leaders and members of key trade associations and use those connections as a way to gain access to potential board members.

Compensation of the Board

There are various incentive compensation alternatives for directors, including a particular emphasis for many companies on equity compensation plans and cash compensation arrangements tied to the employer corporation's business performance that have similar qualities to equity compensation plans. Companies will also use compensation packages that include cash retainers, meeting fees, and travel reimbursements.

	Restricted stock	Option	Phantom stock	SARs	Cash bonus
Payment form	Stock	Stock	Usually cash	Usually cash	Cash
Economic benefit to director	Value of equity	Stock value increase	Value of equity or other performance measure	Increase in value of equity	Set dollar amount if goals met
Legal rights of director	Equity ownership	Equity ownership	Contract right	Contract right	Contract right
Vesting	Usually	Usually	Usually	Usually	None
Tax treatment of director	Income on receipt based on then value, if election made	Income generally on exercise unless ISO (in which case may be AMT)	Income when phantom stock cashed out	Income when exercised	Income when bonus paid
Tax treatment of company	Deduction equal to director's income	Same, except no deduction if ISO and holding periods met	Deduction equal to director's income	Deduction equal to director's income	Deduction equal to director's income
Accounting treatment	Compensation expense over vesting period	No compensation expense if exercise price not less than stock's FMV at grant	Compensation expense as value increases	Same as phantom stock	Compensation expense when paid

Equity Compensation Arrangements

Restricted Stock

Restricted stock plans provide for the grant or sale to directors of actual equity of the company. Directors may receive the stock subject to restrictions that require them to serve as a director for the company for a specified number of years to vest in the stock. If the director terminates her role on the board before the service period is completed, the unvested portion of the stock typically must be sold back to the company at the purchase price originally paid by the director, if any. The director is generally treated as the legal and beneficial owner of the shares for all periods unless the shares are returned to the company by reason of failure to satisfy the vesting requirement or, for certain purposes, unless the director fails to make a Section 83(b) election (see below). Restricted stock is typically sold to a director at a purchase price equal to the fair market value of the stock at the time of grant, although in certain circumstances the company may choose to issue the stock at a discount to its fair market value (usually a discount of not more than 15% of the stock's fair market value).

Directors are taxed on restricted stock at ordinary income rates based on the difference between the value of the stock (disregarding the restrictions) and the amount paid for it, if any. Provided that the director makes a Section 83(b) election, the amount of income recognized by the director is determined at the time of receipt based on the stock's value at that time. If that election is not made, then the director is taxed on the value of the stock as it vests over the service period, based on the excess of the stock's value on each vesting date over the purchase price. The company receives deductions at the time and in the amounts of the income recognized by the director and is required to withhold income taxes due from the director with respect to such income. The company generally recognizes compensation expense for financial accounting purposes equal to the excess of the value of the stock at the time of grant over the purchase price, if any, whose expense is recorded over the vesting period of the stock.

Stock Options

Stock options are a common form of equity-based compensation: nearly 80% of companies surveyed by NCAB in 2003 offered an initial option grant and ongoing annual grants to directors. A stock option gives the director the right to purchase stock of the company or its parent corporation for a prescribed price (generally the fair market value of the stock on the date the option is granted). The director therefore receives the benefit after exercise of the option of the increase in the value of the stock above the exercise price. Depending on how the stock option plan is structured, non-employees such as directors may or may not be eligible to participate. You should check on the specifics of your plan before issuing options, particularly if you intend to issue ISOs.

Stock options generally take one of two forms: (1) ISOs, which may provide certain income tax advantages to directors but must meet certain requirements to do so, and (2) nonqualified options, which do not provide such director income tax advantages. Like most of the other forms of compensation discussed in this memorandum, stock options typically are granted to directors subject to vesting requirements that prohibit exercise of the unvested portion of the option before completion of specified employment or service requirements (or may permit immediate exercise but with the stock subject to a repurchase right on the company's part that lapses over the vesting period in a manner similar to restricted stock).

Directors are not taxed on receipt of a nonqualified option but recognize ordinary compensation income at the time of exercise of the nonqualified option based on the excess of the value of the stock at that time over the exercise price, with the company receiving a deduction in the same amount (assuming no repurchase right applies to the stock at the time of exercise or, if such a right does apply, a Section 83(b) election is made). Directors recognize no regular taxable income at the time of grant or exercise of an incentive stock option (although they may incur AMT in connection with the exercise of such an option) but instead generally are taxed at the time the stock is sold at long-term capital gain rates, provided certain holding periods are satisfied. The company receives no deduction with respect to incentive stock options for which such holding periods are met. The company recognizes no compensation expense for financial accounting purposes at any time under current rules as long as the exercise price is not less than the value of the stock on the date the option is granted.

Equity-Based Cash Compensation Alternatives

Phantom Stock

Phantom stock is a form of incentive compensation that gives the director a contractual right to receive amounts linked to the value of the company's equity (or some other measure of the value of its business or a portion thereof) but does not involve the actual issuance of stock or options to the director. In a typical phantom stock plan, participants are granted phantom stock rights or units that give them solely a right to receive cash compensation on a specified future date or event (generally a number of years in the future) when the units are redeemed by the company. The amount of the cash received on redemption may be determined by reference to the trading price of the company's stock (in the event the stock is traded over an exchange), a formula that reflects the performance of the company's overall business, a formula tied to the value of a particular business unit, or other similar measures. Phantom stock units are usually subject to vesting requirements that provide for forfeiture of nonvested units after termination of the director before completion of specified service periods. Some phantom stock plans (particularly if they are tied to the value of the company's stock) provide for payment of "dividend equivalents" to reflect any dividends paid on the company's actual stock during the period the phantom stock units are outstanding.

Phantom stock plans are most often used when the company prefers not to issue actual stock to directors but wants to incentivize and compensate directors in a manner similar to issuance of restricted stock to them. The term "phantom stock" is used because the units are intended to simulate the economic returns the director would have received had he owned actual stock of the company but without in fact giving the directors the legal rights associated with equity ownership. Phantom stock also is commonly used when the company would be willing to issue actual stock to directors but wants to tie the compensation directly to the performance of a particular business unit rather than to the performance of the company's overall business. Although phantom stock normally is settled in cash, some phantom stock plans provide for directors to receive the value of their units in the form of either cash or stock. Some phantom stock plans permit participants to defer receipt of cash settlements for a period of years or until retirement.

In general, plan participants will not recognize taxable income unless and until they receive actual payments under the plan (whether in the form of cash or stock). At that time, the directors will recognize ordinary compensation income in the amount of the cash (or value of the stock) received, whose income is subject to applicable income tax withholding. The company will receive a deduction equal to the amount of income recognized by the director. The principal disadvantage of phantom stock plans from the company's perspective is that the company must record compensation expense for financial accounting purposes equal to the excess of the value of the outstanding phantom units as of the end of each accounting period over the value at the start of the period.

Stock Appreciation Rights

Stock appreciation rights (SARs) are similar to phantom stock in that an SAR gives the director a contractual right to receive compensation based on the value of the company's stock or some other performance measure but does not involve the issuance of any actual equity or options of the company. An SAR typically entitles the director to receive, after exercise of the SAR, the difference between the value of the company's equity (or other performance measure) at the time the SAR is exercised over the value at the time the SAR is first granted. SARs also are typically subject to vesting requirements under which non-vested SARs are forfeited after termination of employment before completion of specified service periods. Like phantom stock units, SARs typically are settled in cash (although plans sometimes provide that they may be settled in cash or stock) and may provide for deferral of payout for a period of years or until retirement.

The key difference between SARs and phantom stock is that the former entitles the holder to the appreciation in the value of the company's equity or other performance measure after the date the SAR is granted, whereas the latter entitles the holder to the entire value of the equity or other performance measure on which it is based. For example, if the company's equity had a value of $10 on Date 1, at which time a phantom stock unit and an SAR were issued, and each was cashed out on Date 2 when the value of that equity had increased to $20, the holder of the SAR would receive $10 and the holder of the phantom stock unit would receive $20. SARs, therefore, are generally analogous to stock options, whereas phantom stock units are analogous to restricted stock. In each case, they tend to be used for the same reasons: the company either prefers not to issue actual stock or options to directors or wants to tie the amount of compensation to which the director is entitled (and thus the director incentive) to a specific business unit rather than the company's business as a whole.

The tax and accounting implications of SARs are similar to that of phantom stock. Directors generally recognize taxable income only when the SARs are exercised and payment of the appreciation is made to them, and the company receives a deduction in the same amount at that time. Like phantom stock, SARs have the disadvantage from a financial accounting standpoint that the company must record compensation expense over the time period the SARs are outstanding based on their changes in value as of the end of each accounting period.

Cash Bonuses

A third form of cash incentive compensation is director bonuses tied to achievement of specified performance goals. In contrast to phantom stock or SARs, in which the amount of cash compensation received by the director is based on the value of the company's equity or business (or a portion thereof), bonuses typically provide for payment of fixed dollar amounts to the extent that specified performance goals for the company are a particular business unit of the company are met. For example, a bonus program might provide that each participant will receive a bonus of 20% of such participant's base salary but only if revenues or net income increase by a target percentage for the year.

Bonuses are taxed to directors at the time of receipt as ordinary compensation income, subject to applicable income tax withholding, and provide a deduction for the company in the amount of the income recognized by the director. The company recognizes compensation expense for financial accounting purposes in the amount of such payments.

Other Compensation

Directors in smaller, private companies may not receive cash payments for service on the board or for board and committee meeting attendance. These types of payments, however, are quite common in public companies. Despite this, some outside directors of private companies will require these payments in exchange for their participation.

Board Retainer

For small, public companies, an annual board retainer will traditionally range from $10,000 to $20,000 annually. Oftentimes, committee chairpersons receive an additional premium of $5,000 to $10,000. In some cases, board members will receive annual retainers that exceed $50,000. These fees have increased in recent years because of the increased workload, a heightened perception of risk, competition for directors, and separate retainers that may be paid to board chairs of board committees.

Meeting Fees

Public company boards often compensate board members for meeting and committee attendance. In some cases, private companies will also compensate board members for meeting participation. Generally, the fees for public companies range between $1,000 and $2,000 for a regular board meeting and may be higher in the event that a meeting requires a longer time period or involves a separate committee meeting. For nonpublic entities that do not compensate board members for meeting attendance, some will provide reimbursement for meeting attendance.

Director and Officer Insurance

As the risk of liability has increased for directors, many directors today will require that the company purchase directors' and officers' insurance policies.

Working with the Board of Directors

Many new entrepreneurs have questions regarding the number of board meetings, the amount of information that should be provided, and the level of involvement the board should have on tactical or operational issues.

FORGETTING WHO FOOTS THE BILL

What could happen

You've gotten funding and are devoting 100% of your time to getting the product launched. Your board members and investors are full of suggestions and ways to improve the company, but you are so focused on the product development that you don't have time to deal with it. They'll be fine, right?

What to expect

One thing you'll notice about venture capital investors is that they do not view themselves as a bank: you don't take out a loan and only hear from them when the loan comes due. If things are not going well, you should provide this information to your investors before they first learn about it at a board meeting. As one successful venture capital investor said, "I never want to be surprised by the words coming out of my CEO's mouth at a board meeting."

If things are going well, you should still remember that your investors may view their investment as a right to give you their opinion and have it taken seriously. Steve Jobs was fired as CEO at Apple because of his early squabbles with his investors. The founders at Google struggled to mesh their vision with that of their early investors in the early days of the company.

TIP: Don't forget to involve and include your investors in the business.

Regularity of Meetings

The board of directors of a significant number of emerging companies will meet on a monthly basis, which will oftentimes correspond with the company's monthly accounting cycle. This allows the board to provide guidance for a rapidly growing, changing, or developing business. Although monthly meetings provide a great deal of hands-on involvement, there are also the risks of involving the board into tactical and operating issues that may be outside of their ideal scope. Additionally, such regularity of meetings may create difficulty in meeting attendance or maintaining adequate interest levels of busy board members and limiting the ability of the management to develop individual relationships with each director.

Other emerging companies will decide to meet on a quarterly or semiquarterly basis to prevent some of these issues, or companies will hold certain meetings via teleconference or Web conference for alternating or non-quarter-ending meetings, relying on a shortened agenda for these meetings. Once a company has become established or is a more mature publicly traded entity, the board will do most of its business in committees and will hold full board meetings four to five times annually.

Depending on the geographic diversity of the board members, a company may plan to hold meetings for half- or full-day time periods, with more thorough strategic planning meetings held over several day periods for certain companies. Companies may decide to hold most board meetings at their offices with other meetings held in locations more convenient for particular board members. It is important to build in time for social gatherings to engage with the members of the board and develop relationships between the board members themselves and with members of the management team.

For the sake of all members, not only those that require planning for traveling, the company should schedule meetings as far in advance as possible, perhaps setting a schedule on an annual basis. To create a cohesive entity and develop complete, shared understanding among the directors, the company should attempt to have all directors attend every meeting or attend via teleconference or Web conference. Meetings should be efficient and effective, beginning and ending on time, and following a schedule to provide information before the meetings and circulate minutes afterward.

FREQUENCY OF MEETINGS

According to research by Noam Wasserman, the frequency of board meetings decreased based on certain events.

- As the number of rounds of financing increase, a startup company has *fewer board meetings*.

- As revenues in the current year increase over the previous year, a startup company has *fewer board meetings*.
- As the number of years since incorporation increase, a startup company has *fewer board meetings*.

Meeting Content

As with any organization, the meeting agenda will serve as an important tool for a successful meeting. A company should strive to structure the meetings in such a way as to first address substantive items before moving onto administrative matters. It is a delicate balance to attempt to deal with all items on an agenda and give each its full due.

BOARD MEETINGS: WHAT TO INCLUDE AND HOW MUCH TIME TO SPEND

The structure, content, and time of a board meeting will usually be determined by what is going on with the company. However, because boards of directors are often made up of busy individuals, try to set an agenda to keep people on track. This should include the following:

- **Operating and financial reports:** These reports are an opportunity for the directors to meet and interact with the members of the management team. Approximately 25% of the time of meetings.
- **Tactical and strategic issues:** This portion of the meeting should generally not include the management team and should be discussion among the board members themselves. Approximately 50% of the time of meetings.
- **Administrative matters:** This is the formal approvals and more routine legal matters. Approximately 15% of the time of meetings. Oftentimes, the board may handle administrative matters as the first order of business because these "routine" matters can frequently be forgotten or pushed off but are actually very important for effective corporate governance.
- **Review of meeting:** This time will be used to discuss the meeting itself and set the agenda for the following meetings or highlight items that the directors need in response to questions raised at the meeting. Approximately 10% of the time of meetings.

Although the meetings themselves are important, meetings will ultimately be driven by interactions with individual directors in the time between meetings. This individual interaction is important to identify issues before the meetings and avoid surprises at a meeting. The management team should address issues and questions raised by the board after the meeting and should attempt to provide this feedback well in advance of the subsequent meetings.

One of the most important sources of information for your board and for board meetings will be the development of a board meeting package. You should attempt to develop a package in collaboration with the board members that will provide information useful to the board. This package should be simplified and should provide decision-useful insights in summary formats. Oftentimes, you will develop a "dashboard" that identifies the key business drivers in a simple and easy-to-read format that the board can review and use for decision-making.

At least one board meeting each year should include a report from an outside auditor. The company should use an outside auditor as soon as feasible (likely in the first year of operations) to instill good discipline and avoid any issues in subsequent audits. Ensuring that information is accurate early on is important to assist with future decision-making as the company matures. Having audited information readily available will assist the company in its ability to make strategic decisions going forward.

Fiduciary Duties of Directors

Corporate law defines certain duties of directors. This area of law does change, however, so a director should be certain that he continues to act in compliance with the duties of a director. The duties discussed in this section are the fiduciary duties under Delaware law, but many of the duties will be applicable for any director regardless of the state of incorporation for the corporation.

Duty of Care

A director shall discharge the director's duties, including duties as a member of a committee (1) in good faith, (2) with the care an ordinarily prudent person in a like position would exercise under similar circumstances, and (3) in a manner he reasonably believes to be in the best interests of the corporation.

COMPLIANCE WITH THE DUTY OF CARE DOCTRINE

Compliance with the duty of care is based on diligence applied to the ordinary and extraordinary needs of the corporation, including the following:

- **Regular attendance:** Directors are expected to attend and participate, either in person or by telephone (to the extent authorized by law), in board and committee meetings. Generally, directors cannot vote or participate by proxy; a director's personal participation is required.
- **Agendas:** Although agendas for both board and committee meetings are generally initiated by management, a director is entitled to place matters the director reasonably considers to be important on the agenda.
- **Adequate information:** Management should supply directors with sufficient information to keep them properly informed about the business and affairs of the corporation. When specific actions are contemplated, directors should receive appropriate information sufficiently in advance of the board or committee meeting to allow study of and reflection on the issues raised. Important time-sensitive materials that become available between meetings should be distributed to board members. On their part, directors are

expected to review the materials supplied. If sufficient information is not made available in a timely manner, the director should request that action be delayed until the desired information is made available and studied. If a director believes the board is not regularly provided with enough information to enable the director to vote or act in an informed manner and is unsuccessful in efforts to remedy the situation, the director should consider changing management or resigning.

- **The right to rely on others and the need to keep informed:** A director is entitled to rely on reports, opinions, information, and statements (including financial statements and other financial data) presented by (1) the corporation's officers or employees whom the director reasonably believes to be reliable and competent in the matters presented, (2) legal counsel, public accountants, or other persons as to matters that the director reasonably believes to be within their professional or expert competence, and (3) duly authorized committees of the board on which the director does not serve, unless in any such cases the director has knowledge that would make such reliance unwarranted. However, a director relying on others has a responsibility to keep informed of the efforts of those to whom the work has been delegated. The extent of this review function will vary depending on the nature and importance of the matter in question.
- **Inquiry:** A director should make inquiry when alerted by the circumstances.

The duty of care is qualified by the business judgment rule. This rule, well established in case law, protects a disinterested director from personal liability to the corporation and its shareholders, even if a corporate decision the director has approved turns out to be unwise or unsuccessful. In reviewing a disinterested director's conduct, a court will not substitute its judgment (particularly in hindsight) for that of the director, provided the director (1) acted in good faith, (2) was reasonably informed, and (3) rationally believed the action taken was in the best interests of the corporation.

Duty of Loyalty

The duty of loyalty requires directors to exercise their powers in the interests of the corporation and not in the directors' own interest or in the interest of another person (including a family member) or organization. Simply put, directors should not use their corporate position to make a personal profit or gain or for other personal advantage. In themselves, conflicts of interest are not inherently improper. It is the manner in which an interested director and the board deal with a conflict that determines the propriety of the transaction and of the director's conduct. The duty of loyalty has a number of specific applications.

Conflicts of Interest

Directors should be alert and sensitive to any interest they may have that might be considered to conflict with the best interests of the corporation. When a director, directly or indirectly, has a financial or personal interest in a contract or transaction to which the corporation is to be a party or is contemplating entering into a transaction that involves use of corporate assets or competition against the corporation, the director is considered to be "interested" in the matter. An interested director should seek approval by disinterested directors of interested transactions or conduct and should disclose that interest and describe to the board members all material facts concerning the matter known to the director. The board members should then act on the matter with complete candor, accuracy, and inclusiveness before the action is taken. An interested director should abstain from voting on the matter and, in most situations, leave the meeting while the disinterested directors discuss and vote.

A corporation often will enter into transactions with other corporations that share a common director. When possible, the common directors, after having disclosed all pertinent information known to them, should avoid personal participation in approving the transaction and leave review and action to disinterested directors. State statutes usually include procedures that may be used to authorize or ratify transactions with interested directors and should be followed.

Corporate Opportunity

In some circumstances, the duty of loyalty requires that a director make a business opportunity available to the corporation before the director may pursue the opportunity for the director's own or another's account.

If a director believes that a contemplated transaction might be found to be a corporate opportunity, the director should make full disclosure to the board and seek its authorization to pursue the opportunity.

COMPLIANCE WITH THE CORPORATE OPPORTUNITY DOCTRINE

Whether such an opportunity must first be offered to the corporation will often depend on one or more of the following factors:

- The circumstances in which the director became aware of the opportunity
- The significance of the opportunity to the corporation and the degree of interest of the corporation in the opportunity
- Whether the opportunity relates to the corporation's existing or contemplated business
- Whether there is a reasonable basis for the corporation to expect that the director should make the opportunity available to the corporation

Duty of Good Faith

The duty of good faith is simply the duty to act in a reasonable and deliberate manner and in the best interest of the corporation.

Confidentiality

A director should deal in confidence with all matters involving the corporation until such time as there has been general public disclosure. A director of a publicly held corporation is often asked by investors and investment advisers to comment on sensitive issues, particularly financial information; however, an individual director is not usually authorized to be a spokesperson for the corporation and, particularly when market-sensitive information is involved, should avoid responding to such inquiries. A director normally should refer investors, market professionals, and the media to the CEO or other individual designated by the corporation.

Advisory Boards

Many entrepreneurs want to recruit an advisory board (also known as a strategic or technical advisory board) for their new business. An advisory board can serve to assist the business, provide credibility when reaching out to investors, and be a source of board members as the company continues to grow. In certain sectors, particularly medical, high-tech, and software, the use of an advisory board is important for companies as a networking tool. Even so, many startups have found maintaining an effective and efficient advisory board to be a difficult proposition.

Responsibilities

A new company may struggle with multiple constituent groups to which it is responsible for working with and ensuring satisfaction. The advisory board sometimes becomes a list of references compensated by stock options, without much stake or role in the ongoing development of the company itself. Therefore, before a company decides to create an advisory board, the role of these individuals should be clearly defined and the expectations for the members themselves should be clear. As much as with a successful board of directors, the advisory board must be engaged and involved by the management to reap actual business benefits. It is important to balance the relationship: engagement of the advisory board members without creating too much of a burden.

Advisory board members can include both members of your industry as well as others from very different fields. Much like selecting your board of directors, it is important to find a balance of individuals in the advisory board and to recruit each member to provide a specific need for your business and for the advisory board itself.

The role of these advisors includes availability for quick questions or, on occasion, introductions by phone or e-mail, participation in periodic conference calls, and attendance at in-person meetings. It is important to develop an agenda and ensure that any calls are limited in time (perhaps an hour) and meetings are also limited in time (perhaps two to three hours) and should sometimes include a social component. Additionally, the calls and meetings should not only be a time for the company to present to the advisory board members but also an opportunity for members of the board to present to the company and other members (i.e., a presentation by a member of the advisory board on an emerging trend in the marketplace).

TYPES OF ADVISORY BOARD MEMBERS

Bernard Moon identifies three different types of advisory board members for technology startups. Your advisors should include the following:

- **"Prestige" advisors:** These advisors are likely to be well-known public figures that give your company some public relations "buzz" and added visibility.
- **"Credibility" advisors:** These are advisors that are not necessarily as well known as a "prestige" advisors but will be well-respected academics or professions, usually who are within the personal networks of the management team.
- **"Practical" advisors:** These advisors can provide assistance and guidance in areas in which your management team may be lacking (accounting, finance, technical) or may have startup or other industry experience that will be valuable.

Source: Bernard Moon (http://bernardmoon.blogspot.com).

Advisory Board Agreements

It is important to develop agreements for advisory board members to cover issues such as compensation, duration of relationships, and confidentiality. The company should attempt to develop agreements that are consistent. Any new company should keep in mind that the advisory board members will not have the same fiduciary responsibilities as your board members. These individuals, even if paid, will not have the same level of responsibilities generally. As such, the company should insist on signing nondisclosure agreements and ensure that the advisory board members that are providing any type of publicity do not release confidential information.

Generally, the term of an advisory board member will be shorter in duration than will a traditional board member. Members of the advisory board are often available to assist with short-term business objectives. Therefore, agreements should include term limits, such as 12 or 24 months. Term limits will help ensure that the company may not need to remove a nonperforming advisor and can simply terminate the relationship at the end of the term to avoid creating an uncomfortable situation. This will also allow you to enter into agreements to continue the relationships with those advisory board members that have developed a strong relationship with the company.

Generally, there is not a standard compensation scheme for advisors. The compensation most likely will depend on the number of advisors, the time requirements, and the type of company you have. A general strategy for fast-growing entities is that approximately 1% to 2.5% of share capitalization is reserved for all advisors, whereas 10-20% will be reserved for senior executives and key employees. Not all advisory relationships will require compensation, and it may be the strategy of a company to test the advisory relationship before issuing shares. This may allow the company to reward a particularly helpful advisor with some equity in the company over time. Other companies will only provide for travel expenses or provide some small level of nonequity compensation.

WHAT KIND OF THINGS DOES THE BOARD OF DIRECTORS NEED TO APPROVE?

Typical actions that may be necessary or desirable for the board to approve include the following:

- Issuing securities and granting options, warrants, or other rights to purchase securities
- Adopting a stock option plan
- Amending the articles of incorporation
- Amending the bylaws
- Entering into major contracts, leases, or other obligations
- Declaring distributions, dividends, or stock splits
- Borrowing significant sums and the giving of security in connection with such borrowings
- Entering into employment agreements with key employees
- Electing officers of the company and setting or changing their compensation and terms of employment
- Adopting or amending employee benefit plans
- Forming subsidiaries
- Designating committees of the board and the powers of the committees
- Calling shareholders' meetings
- Buying or selling significant assets
- Electing directors to fill vacancies on the board
- Adopting company policies
- Merging or reorganizing the company
- Entering into a new line of business
- Changing the principal place of business
- Commencing or settling material litigation
- Taking any other actions material to the business

Sample Board Minutes

HIGH-TECH STARTUP INC. MINUTES OF A [TELEPHONIC] MEETING OF THE BOARD OF DIRECTORS

[Insert Date]

TIME AND PLACE

A **[telephonic]** meeting of the Board of Directors (the "Board") of High-Tech Startup Inc., a Delaware corporation (the "Company"), was held at **[insert time of meeting]** on **[insert date]**, pursuant to notice duly given.

PRESENT

The following directors were in attendance: **[Sally Founder]**, **[Thomas VeeCee] [and Joe Industry]**. Also in attend were [________], [________] and [________]. **[Mr./Ms.]** ________ served as Secretary of the meeting.

CALLED TO ORDER

[Sally Founder] called the meeting to order at **[insert time of meeting]** and confirmed that each person attending the meeting could hear and be heard by each other person attending the meeting.

APPROVAL OF PRIOR MINUTES

The Board reviewed the minutes of the previous Board meeting. After review, upon motion duly made, seconded and unanimously carried, the following resolution was adopted:

RESOLVED: That the minutes from the Meeting of the Board held on ______________, 200___ be, and they hereby are, approved in the form presented to the Board.

CEO OVERVIEW

________ presented the Board with an update on the Company's business since the last Board meeting, including ________________________. Discussion ensued.

BUSINESS DEVELOPMENT UPDATE

________________________ then reviewed with the Board the Company's business development opportunities. Discussion ensued.

ADJOURNMENT

There being no further business to come before the Board, upon motion duly made and seconded, the meeting was adjourned at **[insert time meeting ended]**.

____

____

____________, Secretary of the Meeting

MORE INFORMATION

To download a copy of the sample board minutes, visit the book's website at http://www.myhightechstartup.com.

13

Protecting Your Intellectual Property

Intellectual property is the oil of the 21st century.

Mark Getty
Founder of Getty Images

Intellectual property has the shelf life of a banana.

Bill Gates
Founder of Microsoft

What to Watch For

Mark Getty calls intellectual property the oil of the twenty-first century, but Bill Gates says intellectual property has the shelf life of a banana. So how should we try to reconcile the quotes above from Getty, founder of Getty Images, and Gates, founder of Microsoft?

It's quite simple really (I'm sure both Bill and Mark would agree): to be a success in high technology, your company must continually be creating intellectual property, because today's invention quickly becomes last year's model.

In the high-tech industry, development cycles and product lifetimes continue to shrink. Building a successful company over the long haul requires an integrated strategy to develop and protect your intellectual property. Just gathering baskets full of patents won't be enough.

WHY INTELLECTUAL PROPERTY MATTERS

VisiCalc was the first spreadsheet program available for personal computers, and some experts credit the application for transforming the personal computer from a hobby into a tool for increased productivity.

VisiCalc was conceived by Dan Bricklin and Software Arts in the late 1970s, and it was the only product of its kind into the early 1980s. However, an attorney for Bricklin advised that it was unlikely he would be able to patent VisiCalc (at the time, patents on software were in their infancy and many were concerned they would not be upheld).

As a result of this decision, Brinklin and Software Arts only decided to copyright the code. However, this only prevents literal copying of the code but did not prevent competitor companies from writing their own code (thereby avoiding copyright infringement) to develop nearly identical application such as Lotus 1-2-3, QuattroPro, and Excel.

Some surmise that, if VisiCalc had been patented, the most popular spreadsheet tool, Excel, would not have been developed and VisiCalc would have remained the dominant (or perhaps only) player until its patent would have expired sometime nearly 2010.

Intellectual Property and High-Tech Startups

It is without question that founders of high-technology companies think about intellectual property, usually around the same time as they decide they want to start a company.

A technology startup faces unique challenges. Protecting your intellectual property costs money (oftentimes more than you have handy), but failing to protect it could cost money down the road.

The obvious next question is, what should I do to protect my intellectual property? The answer, unfortunately, is not as intuitive as the question. Although the need to protect the crown jewels of your organization (your ideas, concepts, developments, and advances) is a given, many of these protections cost money, and unfortunately, these costs can add up quickly. Filing a patent can cost $25,000 and take five years, a substantial sum at a time when many bootstrapped startups just don't have it. This conflict between protecting the most valuable assets of the company and finding a way to pay for it in the early stages of your company are a difficult tug-of-war.

The practical answer for high-tech startups is you must develop a strategy that balances the substantial upfront costs of a broad intellectual property approach with a modest budget. You need to stretch the almighty dollar to the maximum by using the least expensive protections when possible and spending whatever is necessary to protect the key intellectual property for your startup. The remainder of this chapter provides information on the various protections for your company's intellectual property and offers some insights into developing a comprehensive strategy to use these tools to provide maximum protection on a startup budget.

DOING INTELLECTUAL PROPERTY RESEARCH ONLINE

Delphion (http://www.delphio.com) provides a fee-based Web service for searching, viewing, and analyzing patent documents. Delphion provides access to the following sources:

- United States patents and patent applications (full text searching capable)
- European patents and patent applications
- PCT application data from the World Intellectual Property Office
- Patent abstracts of Japan
- Business research about patent applications

The U.S. Patent and Trademark Office website (http://www.uspto.gov/patft) provides free search functionality of its "issued patents" (full image search from 1790 to present; full text search 1976 to the present) and its "published applications" (2001 to the present) databases.

The U.S. Copyright Office website (http://www.loc.gov/copyright) provides free search of its records since 1978.

The Canadian Intellectual Property Office website (http://www.strategis.ic.gc.ca/sc_mrksv/cipo) provides free search functionality of its patent database for records since 1979.

The European Patent Office website (http://www.epo.org) provides access to documents in any official language of the European Patent Office member states.

Most countries have some Web-based services to do country-specific research on intellectual property in that specific country. This may involve some searching around (some countries do not have an easily identified website or a way to find information on the country's online databases), but, with a bit of searching, you can oftentimes find a way to access the information you want.

Understanding Intellectual Property

Intellectual property is a term used to cover a variety of intangible exclusionary rights in inventions, trade secrets, and creative works. These rights may be protected by federal law (such as patents) or by state law (such as trade secrets). For purposes of this section, the four most important forms of intellectual property rights are those associated with patents, copyrights, trademarks, and trade secrets. Certain other systems of rights exist, such as those governing design patents, but they are relevant only in very limited circumstances.

Of these four principal rights, two are based solely on federal law (patents and copyrights), one is based on both state and federal law (trademarks), and the last is based primarily on state law (trade secrets).

Both the founders of a company and its investors are very interested in ensuring that the company protects its own intellectual property and either avoids infringing on the intellectual property of third parties or obtains the necessary licenses to use such intellectual property. The company must take affirmative steps to achieve these goals. For example, the new company should ensure that the founders assign (or license) any existing intellectual property that is relevant to the products or services of the new company. The company should also ensure that all of its employees and independent contractors sign agreements assigning to the company their rights in intellectual property relating to the company's business (but note that California places limits on such assignments).

Patents

A patent gives its inventor the right to prevent others from making, using, or selling the patented subject matter described in words in the patent's claims. To be eligible for a patent, the subject matter in question must fall within the statutory subject matter of the U.S. patent statute and additionally meet the stringent tests of novelty, utility, and nonobviousness set out in the statute. Patents may protect inventions ranging from electronic circuitry, to new drugs, to living organisms, to software. Patents are initially owned by the individual inventors. Although state law implies an assignment for certain employees (employees who are "hired to invent"), the company should ensure that all employees (and appropriate independent contractors) execute appropriate assignments immediately after being hired.

Patents in the United States are obtained by application to the U.S. Patent and Trademark Office. This process is an adversarial one and takes several years. The applications must be filed by patent lawyers (or patent agents) who are licensed to practice before the Patent and Trademark Office. Patents are granted on a national basis. Consequently, a patent issued in the United States will not provide any rights in Canada or other countries. Patent laws in other countries differ significantly from those in the United States, and companies should be careful to employ an experienced attorney for these matters.

Obtaining a patent is traditionally a relatively expensive and time-consuming process, generally costing between $10,000 and $25,000 per patent and taking three to five years. However, many technology entrepreneurs will tell you that obtaining patent rights to your intellectual property is a crucial step. These rights provide small startups with a protected marketplace and are a key factor that investors and potential members of your management team or board will consider. As such, many startups will invest significantly in protecting their inventions by patents.

Patents for Software

Historically, software and related technology from databases, to computer displays, to system operating tools had been protected under copyright laws. Because securing rights under copyright laws is inexpensive and easy, this remains an important tool for protecting these assets.

PATENTS AND SOFTWARE

Relying solely on copyright laws opens software up to reverse engineering of the technology. This was no more evident than for Lotus Development Inc. and the industry-leading spreadsheet program Lotus 1-2-3.

Lotus did not acquire patents for the technology and lost exclusivity in the market when its competitors developed nearly identical software but did not literally copy Lotus's code, costing Lotus potentially billions in lost revenues. As a result of these potential issues, patents for software are now becoming another important tool for software companies with well over 100,000 patents issued for software by the mid-2000s.

In fact, research done by Josh Lerner at Harvard University found that, after this decision, software companies disproportionately accelerated patent filings. Today, more than ever, patents are a key piece of a full intellectual property strategy even for software firms.

Patents and the Internet

Because of the relative ease to which a successful and profitable Internet company can have competitors pop up, Internet technology companies have been forced to find unique ways to protect their technologies. In the late 1990s, patents began to be issued for technologies for certain online auction services, for placing orders online, for payment of bills over the Internet, and for certain online databases. Although some debate is outstanding about the relative breadth of these patents that have been issued, patent rights for Internet and Web companies have been able to serve as important differentiators.

Copyright

The owner of the copyright in an original work of authorship has the right to prevent others from reproducing, distributing, modifying (creating a derivative work), publicly performing, or publicly displaying the work or one that is substantially similar. Copyright law protects the expression of an idea but not the idea itself. Copyright protection is available for works ranging from books, to computer software, to films.

In the United States, the employer is automatically the owner of the copyright in works created by employees "within the scope of their employment." Generally, this rule applies only to full-time employees. However, this rule does not exist in many foreign countries, and the company should ensure that all employees (and appropriate independent contractors) execute assignment agreements immediately after being hired. Once again, the existence and scope of protection under copyright law varies from country to country.

RETAINING A RIGHT TO AN EMPLOYEE'S COPY

What could happen

One of your soon-to-be employees has written the research reports highlighting key aspects of your products. Would your company own this copyright?

What to expect

It depends. Ownership of a work, under copyright laws, is initially designated to be the author of the work. However, under the work-for-hire doctrine under copyright law, the employer may be held to be the author under certain circumstances for copyright purposes. For this doctrine to apply, the author of the work must be an employee. Therefore, independent contractors would not be covered, which means that a separate agreement assigning any copyrights to the company must be made. If your soon-to-be employee created the work before his employment, this doctrine will likely not apply.

Although it may seem obvious who is and is not an employee, that may not always be the case. Just because an individual has a title, works in your offices, or the parties assume the individual is an employee, a court may see this differently. A court will oftentimes examine facts such as control over the individual's time and responsibilities or ownership of equipment used by the individual. In today's employment environment, companies must be aware of these issues with employees telecommuting or working from home on company laptops. If you have doubts about the ownership of certain works produced by an employee, the company should consider entering into an assignment agreement in exchange for some form of adequate consideration.

TIP: Ensure that works made by employees or independent contractors are properly assigned or owned by the company.

Trademarks

A trademark right protects the inherent or acquired symbolic value of a word, name, symbol, or device (or a combination of the foregoing) that the trademark owner uses to identify or distinguish her goods (or those that she sponsors or endorses) from those of others. Service marks resemble trademarks but are used to identify services. The owner of a trademark (or service mark) can prevent the use of a "confusingly similar" mark or trade name. Confusing similarity is based on a comparison of both the appearance of the two marks and the goods (or services) and channels of distribution for the goods (or services).

For a high-technology startup company, many companies choose to trademark their brand name to ensure a consistent identity of their product and to provide a consumer brand. Genentech, Google, Microsoft, Yahoo!, and Facebook all began as names of the company but have since become synonymous with the products the company produces. Protecting the "next" Yahoo! or Google is why most high-tech startups will consider trademark protection for their emerging brand.

HOW CAN I FIND WHETHER MY MARK IS AVAILABLE?

Begin with a preliminary search of your mark within the United States with the U.S. Patent and Trademark Office website http://www.uspto.gov. Once on the site, click on "Trademark" and then "Search." It may be helpful to obtain a full search from a company that provides a full-service trademark search (this will prevent you from spending the time and money if your mark is unable to be protected). You may want to share this information with your attorney.

Also, check out international conflicts. For searching trademarks in other countries, you can use the following free public databases:

- **Australia:** http://www.ipaustralia.gov.au/trademarks/search_index.shtml
- **Benelux:** http://www.internetmarken.de/bxmarks.htm
- **Canada:** http://strategis.ic.gc.ca/cipo/trademarks/search/tmSearch.do
- **European Union:** http://oami.eu.int/search/trademark/la/en_tm_search.cfm
- **Hong Kong:** http://ipsearch.ipd.gov.hk/tmlr/jsp/index.html
- **India** (status only using application number): http://www.skorydov.com/tmr/Status.asp
- **Ireland:** http://www.patentsoffice.ie/eRegister/Query/TMQuery.asp
- **Japan:** http://www3.ipdl.jpo.go.jp/cgi-bin/ET/ep_main.cgi?992599551313
- **New Zealand:** http://www.iponz.govt.nz/search/cad/dbssiten.main
- **United Kingdom:** http://www.patent.gov.uk/tm/dbasc

Trademark protection in the United States arises under common law based on the use of the mark in commerce, but many of the most important means of enforcing trademark rights are granted under the federal Lanham Act and specific state laws. Marks issued since 1989 under federal registration are registered and protected for 10 years and can be renewed indefinitely. Generally, registrations of your mark under state-specific laws are for 10 year periods.

The Lanham Act permits the registration of a trademark (or service mark) based not on use but on "intent to use." These applications permit a company to "reserve" a mark for up to four years. The registration of a trademark under the Lanham Act provides nationwide rights instead of rights limited to the geographic area of actual use provided under the common law. Registration under the Lanham Act requires application to the U.S. Patent and Trademark Office. The process is an adversarial one in which the application is reviewed by trademark examiners. In addition, a number of states have separate trademark statutes patterned on the Lanham Act that create separate causes of action.

The protection of trademarks (and service marks) in foreign jurisdictions is very complex and varies from country to country. Many countries do not provide any "common law" rights and all rights are based on registration.

WHAT ABOUT INTERNATIONAL TRADEMARKS?

What could happen

You've focused on brand building in the United States and now are ready to take your efforts international. Could you have a problem?

Watch out for

Trademarks must be filed in each country where you would like protection. A number of successful U.S. brands have attempted to penetrate foreign markets only to have found that a local company is now using their trademark or trade name in that foreign market. Your sales could infringe on those foreign marks (or others that were set up without aims to leverage your brand) if you attempted to sell the product in that market. Companies should consider likely international markets when developing trademarks and trade names. As with a patent strategy, consider the markets you want to tackle in the early stages of intellectual property protection.

TIP: Ensure your product won't violate local country trademarks or trade names before selling into a market.

Trade Secrets

Trade secrets are an underappreciated tool in protection of intellectual assets of high-technology startups, and it is surprisingly easy: just keep it a secret. Of course, there is a bit more to it, but that's the concept in a nutshell. For a new company without the financial resources to build a substantial patent portfolio, trade secrets may be an important tool for your company as it grows.

The commonly understood definition of a trade secret is any information, including a formula, technique, pattern, physical device, program, idea, process, compilation of information, or other information (1) that provides a business with a competitive advantage (that is generally unknown and not readily discoverable), and (2) in which the individual or company takes reasonable steps to protect the secret and maintain these protections, absent improper acquisition or theft. A trade secret right permits the owner of the right to act against persons who breach an agreement or a confidential relationship or who otherwise use improper means to misappropriate secret information. This allows you to retain the right to the secrets that give you a competitive advantage.

In most cases, the owner of the trade secret has expended some costs to develop or exploit this trade secret. The scope of what qualifies for a trade secret is quite broad, even including negative information (in which your efforts show that something is not possible and shouldn't be researched or exploited further). The key for most startup companies is the information that provides you a distinct competitive advantage, and therefore you take steps to protect this information.

EXAMPLES OF TRADE SECRETS

- Customer lists
- Software code

- Supplier lists
- Blueprints
- Maps
- Design drawings
- Business plans
- Company records (e.g., personnel, financial, sales)
- Chemical compounds
- Business processes
- Survey results
- Prototypes
- Research results
- Sales and marketing plans

Trade secret rights are, for the most part, governed by state law. California courts have generally considered the following factors in deciding whether information constitutes protectable trade secrets of a company: (1) whether the information has economic value because of its relative anonymity in the industry; (2) the company's efforts to keep the information secret, both outside the company and within the company; (3) the time and money spent by the company in developing the information; (4) the relative commercial value of the information; and (5) the ease or difficulty with which the information could be independently obtained by outsiders. The nature of the "reasonable efforts" necessary to protect a trade secret varies depending on the nature of the trade secret. They include non-disclosure agreements with employees (and other companies to whom the trade secrets are disclosed), marking any laboratory books and other materials as confidential, and restricting access to trade secrets on a need-to-know basis. Trade secrets can range from computer programs, to customer lists, to the formula for Coca-Cola.

Trade secret law protects owners from the wrongful appropriation of their trade secrets but, unlike the patent law, not from independent development of the same information by other parties. Trade secrets are protected in most foreign countries, but the statutory protection is generally much weaker than in the United States. This weakness of the statutory scheme makes the use of contracts much more important in foreign countries.

WHY USE TRADE SECRETS LAWS AND RIGHTS?

- You are considering applying for a patent or have already applied for a patent but have not yet received the patent.
- You have a trade secret that can be kept confidential over an extended period of time without unusual effort (which can remain a trade secret longer than the information can be protected by a patent).
- You have information that can't be patented.
- You have a unique process, procedure, operating manner, etc. that differentiates the way you produce a product.
- You have valuable information, but it is not the "crown jewel" of the company.

In addition to protecting its own trade secrets, a company should be careful not to misappropriate the secrets of others. This is particularly important when a company hires a competitor's employee.

NOT KEEPING SECRETS

What could happen

Your management team has agreed that while you are considering obtaining other formal intellectual property protections, you'll keep your production process a secret. Is that enough? No.

What to expect

High-tech companies' most valuable assets may be their technical knowledge. To protect this knowledge and know how, companies can rely on protections offered by trade secrets. To receive protection for your trade secrets, (1) your confidential information must have restricted availability, (2) the information receives economic value because of the limitations on its availability, and (3) you must make "reasonable precautions" to keep the information secret and confidential.

It isn't enough to decide to keep your technical process a secret; you also need to take "reasonable precautions" to keep the information a secret. Although there isn't any absolute understanding of what would be seen as "reasonable precautions," high-tech startup companies should create and operate under a formal trade secret protection policy. Companies should follow proper protocol set forth in their policies and should provide these policies to each employee, independent contractor, or consultant.

TIP: Just because information is "confidential" doesn't mean it is a legally protected trade secret.

Domain Names

Nearly every company now has a website; it is a standard tool to reach potential customers, employees, and partners through the Web. Today's technology companies use features such as blogs, comment boards, and e-commerce to create interactive forums with their customers. Many technology companies rely on the Web as their first (and sometimes only) sales tool.

Your first step in developing a Web presence is selecting your domain name, and this choice does raise certain intellectual property issues. For some companies, their domain names represent one of their most valued assets. Domain names are unique designations used to identify a particular computer on the Internet. To communicate with each other on networks, computers must have individual identifications. A domain name is a way to identify and access a computer to a unique site on the Internet. Domain names typically consist of letters, numbers, and hyphens. An example is www.google.com or www.yahoo.com.

Domain names can be registered with any one of the over 150 registrars accredited by the Internet Corporation for Assigned Names and Numbers, a nonprofit corporation that manages the Internet domain name system. Anyone can register a domain name in the .com, .info, .net, or .org scheme. Registering a domain name does not, however, give one trademark protection over the domain. Domain name registrars are not liable for trademark infringement, dilution, or contributory infringement merely because they issue a domain name that is claimed to infringe.

Because a domain name on the Internet is a unique address designating an Internet site, it may be a trademark or service mark and have legal protection. However, the Patent and Trademark Office has stated that, if a domain name is used solely as an Internet address and not to identify a source of goods and services, it is not protectable as a trademark. A business may be wise to include its domain name in advertising its goods or services to obtain trademark protection.

YOUR DOMAIN NAME AND TRADEMARK LAW

What could happen

You've found an available domain name for your website and are prepared to launch the site to begin drawing customers to the site. Because the domain name wasn't taken, that means we are free to operate it, right? Maybe not.

What to expect

Don't assume that just because the domain name is available that you are free to use the domain name. Domain names and the use of metadata on your website have been the source of serious headaches for today's technology companies. Companies have been found to be liable for infringement if they use a domain name that is close to or similar to an existing trademark. If your site sells services or products that would confuse or mislead a consumer with another trademarked product, your company may be held to be liable. Domain names are also protected under the Anticybersquatting Consumer Protection Act and could lead to criminal penalties for your business. In certain circumstances, the registrar company may even be able to have your domain name transferred over to the owner of the similar mark (even if your mark is stronger in the marketplace). Before using a domain name for your website, be sure to conduct a trademark search (or work with an attorney to have this done), and remember to review both state and federal databases.

TIP: Before you use any available domain name, be sure to check for potential trademark infringements.

Various state and federal laws, including the Anticybersquatting Consumer Protection Act (ACPA), help form a body of legal protection of domain names. ACPA in particular prohibits the registration of, trafficking in, or use of a domain name with a bad faith intent to profit from the goodwill of another's trademark. ACPA also allows a business to protect their domain names from infringement, tarnishment, and dilution and to bring actions against cyberpirates and cybersquatters. Dilution and tarnishment are the weakening and sullying of the brand value of a domain name.

Cyberpirates appropriate and register well-known and distinctive trademarks as their own Internet domain names without permission from the trademark owners. Cyberpirates seek to profit from the goodwill and recognition associated with those marks to direct or misdirect consumers to the cyberpirates' websites. Cybersquatters register trademarks in Internet domains with no intention of developing a website with that domain but instead seek to resell the domain name to the trademark owner or a third party. A variant of cybersquatting is typosquatting, in which a domain name registrant will register a variant of a famous trademark. Courts vigilantly forbid cyberpiracy and cybersquatting because both acts sap brand strength, sully trademarks, foster consumer fraud, and incite consumer confusion.

Numerous cases have also dealt with competitors registering their adversary's trademark as a domain name. This sort of behavior is almost always illegal. When noncompetitors register a trademark as a domain name for a legitimate use, however, the law is more complex. Because trademark use is territorially and industry based, many times the same trademark may be simultaneously used with legal issues. In cases of noncompetitors suing over another business' use of a trademark as a domain name, courts have emphasized that mere registration of a domain name without more (such as an offer to sell or the intent to block the trademark holder from using the name) is perfectly legal.

As with other forms of intellectual property, trademark rights are subject to fair use. First Amendment liberties also allow a person to use another entity's trademark for purposes of commentary or parody. Fair use and the First Amendment do not protect all uses of another's trademark. Although most free speech and parody cases have been settled outside of the court system, the law that has developed suggests that registration of the exact trademark as a domain name is generally disallowed by courts, whereas registration of a name that incorporates the trademark such as www.business.X.is.terrible.com may be upheld.

Managing Your Intellectual Property as a High-Tech Startup

Effectively managing your intellectual property assets is a difficult proposition for a new technology startup given the fiscal constraints nearly every new company will face. While trying to raise money, high-tech companies are quizzed about their patent protections. Ironically, you are probably trying to raise money to file patents, right?

STRATEGIES TO MANAGE YOUR INTELLECTUAL PROPERTY

Here are a few key steps a high-technology startup should consider in managing their intellectual property:

- Develop a company-wide strategy for managing your intellectual property protection needs and incorporate this into your overall business plan.
- Document your technology development efforts and keep clear records of all documents related to the company intellectual property and employee records related to invention assignments.

- Take advantage of "free" or "low-cost" intellectual property protections.
- Focus limited resources on key intellectual assets.
- As you continue to grow, strategically expand your intellectual property through additional filings in broader markets and use of licenses for technologies you cannot or do not want to develop.

How should your high-tech startup approach intellectual property in the early stages and as the company continues to mature? First, you should come up with a strategy and integrate that strategy into your business plan. This is an opportunity to work with competent intellectual property counsel who can help you to identify a strategy that fits within your budget yet maximizes the scope of your protections. Second, you should use low-cost intellectual property protections to their fullest. Next, you should focus your resources on protecting key intellectual property assets through patents and trademarks. As your company continues to mature, consider expanding the scope of your use of the more expensive protection methodologies, but always keep in the back of your mind that, with respect to some classes of intellectual assets, you cannot "unring the bell." Therefore, be mindful of the choices you make to file or not file for certain protections, given that, if you choose not to file for certain protections, you may be unable to do so at a later point.

At the early stages of your company, potential investors may still want you to have your cupboards full of patents, trademarks, and copyrights. However, if you can present a well-thought out strategy for your company's intellectual property, you will most likely be able to offer a story that resonates with investors.

LETTING THE CAT OUT OF THE BAG (WITHOUT PROPER PROTECTIONS)

What could happen

You are planning to file for a patent (if you can) or keep a technology secret. One of your employees has written an article detailing the key proprietary aspects of your technology to be published in an industry trade journal. What will happen if this is published?

Watch out for

Many very successful companies have never filed a single patent but have relied heavily on trade secrets (think Coke's secret formula). The key to a successful trade secret is taking reasonable steps to prevent disclosure.

If you want to apply for a patent or maintain the proprietary information as a trade secret, you will need to take swift action to prevent disclosure. Establish corporate policies early in your existence to ensure disclosure is limited. Take precautions, especially with proprietary technologies that are important to the future success and health of your business venture, and remember, if you are looking to license technology from a third party, you should be wary of disclosure by the licensor.

TIP: Don't be caught off guard by improper disclosures of key technologies or you may lose protection avenues.

Taking Advantage of "Free" and "Low-Cost" Intellectual Property Protections

When new entrepreneurs are told about the average price tag for a patent (between $15,000 and $25,000), they may become concerned that their bootstrapped startup just won't ever be able to make it. However, what you must remember is that patents only represent one piece of an intellectual property strategy for a high-tech company, and several key aspects of that strategy do not involve expensive filings or fees.

There are three key areas that all high-technology companies should incorporate into their business from the beginning: (1) trade secrets, (2) copyrights, and (3) agreements to retain the rights to your company's inventions.

Trade Secrets

As discussed previously, trade secrets include any information (formula, technique, pattern, physical device, program, idea, process, compilation of information, or other information) (1) that provides a business with a competitive advantage (that is generally known and not readily discoverable), and (2) in which the individual or company takes reasonable steps to protect the secret and maintain these protections, absent improper acquisition or theft. The scope of trade secret protection can be incredibly broad and, therefore, can be a key tool for startups.

From day one, the founders of the company should put a trade secret protection policy in place. This policy will limit the disclosure of key information internally and externally to prevent information from being spread inappropriately.

ASPECTS OF A TRADE SECRET PROTECTION POLICY

Company Confidential Documents and Materials

- Label any proprietary or confidential documents and materials as "confidential" or "proprietary."
- Use password protections on key files and programs.
- Require all confidential or proprietary documents to be locked in file cabinets, desks, or company safes.

Employees

- Provide written copies of your trade secret policy and offer regular reminders of the importance of the policy.
- Limit disclosure of proprietary or confidential matters only to employees that will need such information.
- Use exit interviews for departing employees to remind of ongoing confidentiality restrictions and responsibilities.

Hiring Employees or Consultants

- For hiring procedures that will involve the disclosure of confidential information, use nondisclosure agreements before commencement of hiring process.

- Include confidentiality restrictions in all new hire and consultant agreements (include language that the new hire will not use confidential or proprietary information in subsequent jobs and will not reveal information to future employers).

Third-Party Correspondence and Meetings

- Use nondisclosure agreements in all meetings in which confidential or proprietary information may be exchanged.
- Have visitors sign nondisclosure agreements when visiting or touring your facilities.

Many startups forget that some of their employees will depart the company, so be certain that when this happens you are diligent in conducting an exit interview and reminding the employee of his ongoing obligations. With that in mind, be sure that your employees have signed confidentiality agreements that include their time as employees and their obligations after they depart. Not all employees or consultants need to know all matters; be selective in who knows what within your company (although this is difficult when your startup is only a few people with everyone pulling various oars!). Also, with the extensive use of e-mail and e-mail attachments, you should be careful about inadvertently forwarding confidential documents.

WHO DO I NEED TO SIGN A NONDISCLOSURE AGREEMENT?

Your attorney (or potential attorney)? Usually not, because your attorney is bound by certain ethical standards. However, when in doubt, you can always ask.

Potential business partner? Most definitely. You should have mutual nondisclosure agreements in place before discussions begin.

Potential hire? Yes.

New employee? You should have a policy in new hire documentation whereby new employees will sign confidentiality restrictions (among other restrictions).

Venture capital firm where I am sending my business plan? Ideally, you would want the firm to sign a nondisclosure agreement. However, most will not because they see so many similar business plans and presentations that could create problems for the firm. You can try, but know that industry practice is that VCs don't and won't.

Your accountant or auditor? Possibly, particularly in the case in which they will come into contact with proprietary information or data.

Nondisclosure agreements are an important tool for technology companies. You should prepare a mutual and one-way nondisclosure agreement early in your development (or ask your attorney for a simple form of these agreements). As discussed previously, not all individuals will sign a nondisclosure agreement (your attorney will usually not need to sign a nondisclosure agreement nor will most VCs sign one). However, you should put in place a policy within your organization that all discussions with outsiders that involve discussions or observations of proprietary information will require a nondisclosure agreement to be in place. Also note (as discussed below) that signing a nondisclosure agreement is not, in itself, enough to provide trade secret protection. Be sure that you consider other ways your trade secrets may be released including at technical presentations and conferences, or through published papers and articles. When in doubt, keep your key secrets a secret.

NONDISCLOSURE AGREEMENT AND TRADE SECRET PROTECTION POLICY

What could happen

Your company requires every employee to sign a nondisclosure agreement with the company. Is that sufficient to protect our trade secrets?

What to expect

Maybe not. A nondisclosure agreement is designed to protect the company from unauthorized disclosure of company confidential or proprietary information. However, if the company attempts to enforce the nondisclosure agreement signed with the employee, a court may find that the company did not take "reasonable precautions" required under trade secret laws. Remember, a nondisclosure agreement alone might not be sufficient to protect your trade secrets if the company fails to take other precautions to protect its confidential information.

TIP: A nondisclosure agreement itself does not necessarily provide trade secret protection.

Copyrights

Copyright protection represents another inexpensive tool that technology startups can use to protect certain aspects of their technologies. You should use the copyright notice for all documents produced by the company (© with the year the information or document was created or published, as well as the name of owner of the copyright that will typically be the name of your company). These copyright protections exist at the time your information is created.

Additionally, high-tech companies should use these protections for their websites and for any software they create or design. For companies that have a website (for both sites created solely for informational purposes and sites created by e-commerce or other Web technology companies), you should also place copyright notes on each webpage. Additionally, copyright protections should be used for any software or programs the company produces, including source code and full software packages. In the case of software code or programs, you may also want to file your code with the U.S. Copyright Office to assist with enforcement actions in which others use your code without permissions. These filings do have a small fee associated with them.

Agreements to Retain the Rights to Your Company's Inventions

Although it may seem counterintuitive, your company may not automatically own something invented by your employees or consultants while employed or retained by your company. Therefore, it is very important that you put agreements in place that will ensure your company keeps rights to the inventions your company and its employees or consultants make.

THE IMPORTANCE OF INVENTION ASSIGNMENT

What could happen

Your technical guru just invented a new patentable product in your laboratories. Can your company file for the patent?

What to expect

No. The patent rights for an invention developed by an employee of the company are held by the employee. In this case, although your employee conceived, developed, and produced this new invention while employed, did so in your laboratory during working hours while using company equipment, patent law is still clear that the employee is the holder of the patent rights. To retain rights over this new invention, the employee must assign these patent rights to the company, and to be considered a valid assignment of these rights by the employee, the assignment will need to be in writing and must be made in exchange for valid consideration.

For this reason, high-tech startup companies should be certain that proper invention assignment agreements are entered into with all employees. Typically, these agreements can be effected on employment (and by including it as a standard part of your new employee documentation, you can avoid the oversight). If such assignment is not done before the creation of an invention, valid consideration must be exchanged (such as a cash bonus, equity payment, or other similar types of remuneration). Also, be aware that certain states (including California, Washington, Illinois, and others) require that certain disclosures be made by the employer in the assignment agreement for the assignment of the patent rights to be valid. If such disclosures are not made, the assignment may be held to be invalid.

If you did not secure a valid assignment and the employee has refused to sign an invention assignment, the company still may hold "shop rights," which grants a nonexclusive license to use the invention developed by the employee, but this right may not be further assigned and would only be nonexclusive, which may scare away certain potential investors.

TIP: Be sure to receive a proper invention assignment from employees to retain exclusive patent rights for the company.

From day one, your company should be certain that each founder has executed an invention assignment for future inventions as well as properly assigns rights to any intellectual property that is being contributed to the company. Don't delay on making these assignments: the worst thing to happen to a promising startup will be a primary founder departing before assigning key intellectual property to the company and the remaining team left without key technologies.

Additionally, all new high-tech startups should put standard employment and consulting agreements in place for all employees and consultants. Be sure to properly follow state laws for proper assignment of your intellectual property. Some states have specific language that must be included in any agreement assigning intellectual property rights. Remember, each state has differing laws, so if you have employees located in different jurisdictions (at virtual offices or working from home), be certain that you've carefully drafted assignment language to avoid any issues down the road. Software companies should ensure that copyright is included in the assignment provisions to keep rights to any software code written by your employees or consultants.

Finally, be aware of unique problems that sometimes result from developing intellectual property with a third party. Startups should be careful when collaborating with a university, a government agency, or other companies or organizations. If you work with a university or university personnel, be certain that the agreement does not permit the university to hold any rights to the current inventions or any future improvements or enhancements. Likewise, research conducted with funds from government grants oftentimes permit the government to have rights to the intellectual property, permit additional research to be done on the developed technologies, or may even forfeit your rights to the intellectual property if not properly disclosed. Be certain to review any grant agreements closely for the intellectual property provisions and insist on narrowly tailored language with respect to ownership of work done at government facilities. Finally, remember that any work done with a partner company or organization could subject you to joint ownership of the developed technology. Without a specific agreement, intellectual property will likely be coowned by the partner organizations or companies.

The key to preventing ownership problems is be certain that all agreements are drafted very carefully and have clearly identified provisions of assignment. Early on, your startup should attempt to create standard agreements that have carefully crafted assignment language for the particular state where the work will be done. Be certain to closely review all agreements for any provisions that may limit any rights you hold to your current and future intellectual property.

Protecting Key Intellectual Property

Many successful startups will tell you that one of the keys to their success was steadfastly protecting the core intellectual assets of the company. For many companies, this will mean leveraging patents and trademarks.

Early-Stage Patent Strategies

Patents are a key differentiator for high-tech startups. In many cases, investors, customers, and potential competitors will look more favorably on a company with a strong patent portfolio in the key technologies of the field. Therefore, it is important to consider key steps to take as an early-stage company with respect to protecting patent rights domestically and abroad.

As discussed, the expense of filing a full patent application can be upward of $25,000. Therefore, it is important to find ways to maximize the value for patents. Listed are some key steps a startup company should consider in developing their intellectual property strategy.

Provisional Patent Applications

A provisional application permits a company or individual to make a less complicated patent filing for $75 (up to $150 for certain inventors and companies). From the date of this filing, you have one year to submit a full application. This timing may allow your company to protect itself while it raises sufficient funds to create its patent portfolio.

With the relative ease and lower fees of a provisional patent, a company may consider filing a provisional patent on both the initial invention and then on subsequent improvements on the invention. Once the one year period nears expiration, the company can make a determination as to a full patent filing.

To file a provisional patent application, you must (1) pay the required fee; (2) attach a cover sheet (one page in length); (3) submit a description of the invention with detailed information on the invention, the process for its creation, and details on its use (the description is an identical requirement as with a full patent filing); and (4) provide illustrations (if necessary).

Filing a provisional patent allows the holder to claim the date of filing of the provisional patent application as the original filing for purposes of priority but permits the holder to wait up to one year for a full filing. However, just like in the case of a full U.S. patent application, your company will need to file an international patent application within one year of the provisional filing.

Complying with Filing Deadlines

Although not all companies are prepared to file for patents early in their formation, you should be aware of certain actions that will prevent you from filing in the United States or abroad.

In the United States, you are unable to file a patent if your invention (1) has been "offered for sale" for greater than one year, (2) has been used publicly for greater than one year, or (3) has been published. Some companies have been barred from filing where they have offered the product, although they have not made any sales until before the one year deadline.

Outside the United States, most countries are not "first to invent" but operate on a "first to file" basis. Therefore, public use or offers for sale will not bar others from filing a patent on your product. As such, you should consider filing as soon as possible if your first filing will be international. Once you have filed a U.S. patent application, you will typically have one year from that filing date to file for an international patent.

"Make Special" Application

Although the patenting process is such that your patent application will not be examined by the U.S. Patent and Trademark Office for more than two or three years (with issuance dates ranging from two and a half to four or more years), in certain cases, you can apply to have the patent examined on a shortened timeline.

To make such an application, you will need to have a full-scale previous art search done that yields positive results for your claims and then you will be required to provide the results to the U.S. Patent and Trademark Office to confirm such claims.

The "make special" process is attractive, for example, in the case that you discover a potential infringer that could derail your business plans. By filing such an application, you can speed up the process and forbid the potential infringer from harming your business.

NOT THINKING INTERNATIONALLY ABOUT YOUR INTELLECTUAL PROPERTY

What could happen

You showed a prototype to a potential business partner from Japan but hadn't filed for patent there. Could you lose the right to apply for a patent in Japan now? Yes.

Watch out for

You must apply for a patent in each country where you want protection (except that you are able to file a single application for protection throughout the European Union). For the United States, you have a one-year period after an invention is sold or made public to file for a patent application. Elsewhere around the world, once your invention is made public or sold before you have filed a patent application, your invention can no longer be patented in that country.

Be careful with sales, presentations, and other disclosures to international parties if you are planning to apply for protection elsewhere in the world. Additionally, make sure you have a patent and intellectual property strategy that considers international implications. It is expensive to protect your intellectual property worldwide. So plan accordingly and consider which markets are the most likely and what costs the company can spend to apply for the necessary protections in those markets. These may be tough choices for any early-stage company, but it is better to make a choice than have a choice made for you. Discuss with an intellectual property attorney to see what makes sense for you.

TIP: Don't forget to consider international implications of your intellectual property.

Trademarks

Trademarks are an important part of developing your brand and brand awareness among potential customers. Selecting the right mark is important because, the day you begin to use that mark, you've begun the process of brand identity.

Where many companies get into trouble is being forced to change their mark down the road. Imagine a company that spent money on branding, has established a website, and begun making sales, only to receive a letter from another company claiming infringement. You have invested time and money in establishing your brand only to see it slip away as a result of incomplete diligence at the start. Companies get into trouble because they select a mark, begin building a brand, and then find out that mark violates another company's intellectual property.

FREQUENTLY ASKED QUESTIONS

Trademarks

What are the benefits of registering a trademark?

- You reserve a mark you intend to use in the future.
- You make your mark appear in searches others do before they adopt a mark, thereby preventing others from adopting your mark inadvertently.

- You provide yourself additional remedies when someone else uses your mark or a mark likely to be confused with it.
- You make your rights in a mark federal and nationwide.
- You enable your use of ® next to your registered mark. ® is the symbol for showing that the U.S. Patent and Trademark Office has confirmed that you have the exclusive right to use this mark in connection with your goods or services. Using the ® gives notice to others that you have this right.

When should I apply to register my mark?

For a high-tech company, in which product adoption is the key, you should consider protecting your mark as soon as you select it, particularly if there is a lengthy research and development process before you begin selling the product. Although using a mark without registering it confers rights in the United States, you need to register it to enforce those rights if someone else adopts your mark.

May I use ™ next to my mark?

You may use ™ as soon as you have ascertained that no one else in the United States has rights in the mark. It may be good to check with your attorney before making this determination to avoid potential infringement. However, it is usually not necessary to file an application to register your mark before you use the ™ symbol.

Which country should I begin with?

Most U.S. companies will first register in the United States. The Paris Convention rights will allow you to apply to register in most countries of the world for six months and still obtain the U.S. filing date.

What does it cost to file marks in the U.S.?

To register a mark in the U.S., the registration process (including legal fees) generally costs between $2,500 and $5,000 (but remember, it can cost more if there are objections from the Trademark Office). Also remember that, in many cases, these costs are spread out over the entire registration process, which oftentimes takes between one and two years.

How can you prevent your mark from violating another trademark? The best course of action is to hire a third party to do a search once you've selected a mark; oftentimes, your attorney can recommend a service to assist you. Before you present your mark to this service, be sure to select a mark that is inherently distinctive. Start with a thorough search on the Internet, searching your word choice and similar derivations of those words. Once you've come up with a list of possible names, begin looking at trademark databases maintained by the various government agencies (a list of databases is available in the "Trademark" section of this chapter). Once you've selected a mark or series of marks, consider filing.

You should also consider the international implications of selecting a particular mark, especially if you consider international markets to be a key marketplace for the sale of your products.

FREQUENTLY ASKED QUESTIONS

International Trademarks

Which countries should I register in?

- Most U.S. companies will first register in the United Sates. The Paris Convention rights will allow you to apply to register in most countries of the world for six months and still obtain the U.S. filing date.
- You should also consider filing a European Community Trademark (CTM). Most countries, other than the United States, are "first to file" countries (meaning that the first to file an application for a mark, as opposed to the first to use a mark in commerce, has superior rights). The CTM provides coverage in 25 countries (Austria, Belgium, Cyprus, Czech Republic, Denmark, Estonia, Finland, France, Germany, Greece, Hungary, Ireland, Italy, Latvia, Lithuania, Luxembourg, Malta, The Netherlands, Poland, Portugal, Slovak Republic, Slovenia, Spain, Sweden, and United Kingdom). As of the writing of this book, Switzerland and Norway are not in the community and require separate applications.
- If you are a U.S. company and envision your product being sold into other countries in the Western Hemisphere, you may also file in Canada, Mexico, and in a number of countries within South America.

With so many countries throughout Asia, where should I begin?

- You may consider a modest filing program in Asian countries including Japan, China, Taiwan, South Korea, Singapore, and Hong Kong.
- Many companies will also look at filing in Australia and New Zealand because of their use of English.

What does it cost to file marks internationally?

- Foreign applications cost $1,500 to $3,000, depending on the country.
- A CTM application costs from $4,000 to $6,000 through registration.
- As with U.S. filings, in many cases and countries, these costs are spread out over the entire registration process, which oftentimes takes between one and two years.

Later-Stage Patent Strategies

As your company continues to grow and raise investments to fund a growing patent issuance and protective strategy, you'll want to continue to modify your patent strategies. In many cases, you'll expand your filings outside of a small, subset of key intellectual assets into ancillary inventions. As such, you'll need to establish a process for protecting these invention rights.

As your business continues to expand, you'll begin to think more broadly about acquiring technology rights in a particular area of your company. These patent rights will come from in-house technologies developed by your researchers as well as licensing technologies from third parties (this is discussed additionally below).

You will likely begin to undertake discussions of offensive and defensive strategies for protecting your technologies. As it sounds, a defensive strategy will involve building a set of patents and licenses to protect your core technologies from potential competitors. Generally, this is a more focused approach and concentrates on patents that are directly related to your technology. Conversely, other companies may consider an offensive approach to patenting if they are able to raise the necessary funds to accomplish this strategy. Offensive tactics involve such actions as filing a series of patents for likely improvements to a competitor's invention, designing solutions around a competitor's patents, or making broad claims with respect to future product development to create additional barriers to future development.

In either case, as you continue to develop your patent portfolio through some combination of offensive and defensive measures, you will begin to see the value in monitoring filed patents in your industry and near your technology. As you find ways to enhance your own position through your own actions, be aware that other industry players (in some cases, with more financial resources) may be considering undertaking offensive or defensive patent strategies to limit your options as you grow.

Licensing of Intellectual Property

Startup companies are often unable to develop every component needed for their products. Therefore, many high-tech companies will need to license certain technologies and inventions from third parties. Additionally, some startup companies may find that a particular invention can be licensed to a third party as an additional stream of revenue. Because of these issues, it is important for high-technology companies to be familiar with key aspects of licensing intellectual property.

Intellectual property owners may license all or part of their intellectual property to others. An intellectual property licensing is simply a promise by the licensor, the entity issuing the license, not to sue the licensee, the entity obtaining the license, for the licensee's use of the intellectual property. License terms can vary widely given the particular needs of the parties in each licensing agreement. Licenses are written contracts and must therefore adhere to relevant contract and intellectual property law principles.

University Licenses

In particular, many key technologies of today have been licensed from universities and colleges around the country. For example, the algorithms that went on to become Yahoo!'s and Google's proprietary search engines began as part of the research of doctoral candidates in Stanford University's computer science program. The benefits of licensing technology include the obvious such as exclusive rights to the technology, but oftentimes entrepreneurs fail to recognize some of the ancillary benefits. These benefits may include ongoing involvement of a professor or research department (perhaps as an advisory board member or another role) and the added prestige of licensing technology from a prestigious institution, department, and professor.

In the event you are considering licensing technology from a university, you should be aware that different universities have different policies and practices with respect to licenses. Consider involving an attorney in the licensing process that is familiar with negotiating licenses with the university. The university's technology transfer office is generally the party with the bargaining leverage, so involving an experienced intellectual technology licensing attorney may allow you to address certain obstacles with a creative and practical solution.

Key Terms for Licenses

To be effective, an intellectual property license must meet four requirements. First, the licensor must legally own the intellectual property or have authority to license the intellectual property. Second, the intellectual property must have or be eligible for legal protection. Third, the license must specify what intellectual property rights it grants to the licensee. Fourth, the license must specify what the payment for the intellectual property rights granted will be.

Intellectual property licenses may play an important part in a company's overall business strategy. Licensors can profit from a licensee's use of the licensor's intellectual property in ways not available to the licensor, such as with overseas markets, sublicensing, or application of the intellectual property to other technological fields. Licensees may have better manufacturing capabilities, wider distribution outlets, and greater local knowledge and management expertise than the licensor. Licensees may profit from a license by obtaining intellectual property from a licensor by reducing research and development costs, reducing market entry time, obtaining superior technology, using branding power, or obtaining protected technology or information not otherwise available.

In many industries, standard licensing agreements and terms have been developed. Other times, licensing agreements can be negotiated and drafted from scratch. Negotiations for an intellectual property license can be long, difficult, and drawn out. Frequently, the licensor and the licensee are competitors, and potential licensees often refuse to sign confidentiality agreements during the early stages of negotiation.

KEY PROVISIONS OF LICENSING AGREEMENTS

Although the scope of a licensing agreement can vary widely, the following terms are some of the most important and most common:

- **Grant clause:** This is perhaps the most important clause in a technology licensing agreement. The grant clause identifies the licensed technology and the rights being licensed. The clause defines the technology, the scope of rights licensed, the field of use limitations, any territorial limitations, and any temporal limitations. Technology that is patented may also be limited to the country or region where patent protection exists or where it is expected to exist.

- **Exclusivity:** Licenses may be exclusive or nonexclusive. In an exclusive license, only the licensee can exploit the licensed property, either directly or through sublicenses to others. A nonexclusive license permits the licensor to issue equivalent licenses to many licensees.
- **Transfer/sublicense:** A license may not be limited to the named licensee. It may also allow the licensee to transfer or sublicense its rights to other entities. Although transfer and sublicense rights are often resisted by transferors, sublicense rights are more typical when the parties have agreed on a marketing plan.
- **Compensation:** Compensation to the licensor from the licensee is usually in the form of a lump-sum or periodic royalty payment, or a combination of the two. The form and amount of royalty payments are usually dependent on the nature of the marketplace and the demand for the licensed intellectual property. Scaled royalties that provide for changes in royalty payments on the occurrence of defined events are often used. Cross-licensing, wherein each party licenses to the other, usually for related technologies and often to settle potential disputes, is also common.
- **Termination:** Termination provisions define the specific conditions that trigger the early termination of the agreement. These conditions can range from material breaches (e.g., failure to pay royalties or other amounts due, failure to maintain quality controls in trademark licenses, or failure to maintain confidentiality of trade secrets) to failure of the licensee to meet defined sales levels or produce a product using licensed intellectual property within a specified time period. The termination provision also sets forth the obligations that remain after termination.
- **Territory:** The territory clause should provide a clear statement of the geographic locations in which the license is to apply. Statutory and regulatory restrictions exist on the countries in which and the foreign nationals to which information and products may be used. These regulations are complicated and subject to changes in international policy and should always be consulted in updated form.
- **Representations/warranties:** This clause states that each party has the right, power, and authority to enter into the agreement, that the agreement has been duly executed and delivered, that the agreement is valid, legal, and binding, and that the agreement does not contravene any other agreement to which the licensor or licensee is a party or its certificate of incorporation or bylaws.

  The licensor of intellectual property often will seek to minimize or eliminate the representations and warranties. Licensors are often wary about stating, in absolute terms, that the property does not infringe rights of a third party. This is not necessarily because of any incriminating knowledge that the licensor may have; it may be a combination of the fact that claims of infringement and plagiarism may be very common in the particular industry and that a licensor is unwilling to assert that absolutely no other party may have intellectual property rights conflicting with or superseding the licensor's rights.

In addition to the explicit representations and warranties included in the text of the license agreement, certain implied warranties may be deemed applicable to the license if the license is considered to involve goods. These include the implied warranties of merchantability and fitness for a particular purpose.

- **Confidentiality:** Often during licensing agreement negotiations, confidential information is shared between parties. Although some negotiations may carry an implied requirement of confidentiality, it is good practice to include a confidentiality agreement in license agreements restricting use and disclosure of the confidential information by the receiving party absent written consent from the disclosing party.
- **Indemnification:** This clause allows one party to bear or reimburse all the costs, expenses (including associated legal fees, which are often limited to a reasonable amount), and related damages owed or paid by another party. Often indemnification is limited to situations in which the damages are related to the first party's (the indemnitor) negligent action or inaction or to that party's breach of one or more terms of the license. Often an obligation to indemnify will be coupled with an obligation to defend wherein the indemnitor agrees to take on the legal defense of the indemnitee in those situations involving indemnification.

Licensing overseas may present problems not encountered domestically. Accurate language translations in written documents and negotiations are a problem. New and evolving technology may not have foreign translations. The legal and business systems of different countries may also conflict. In some countries, certain technologies may not be licensable, may be subject to forced or "compulsory licensing" to local third parties, or may be deemed to be in the public domain after a short period.

Licensing may also occur in the context of a joint venture. A joint venture is simply a business entity involving two or more enterprises pooling their resources with a common business objective. In intellectual property-based joint ventures, often one party will contribute intellectual property and the other will contribute financial resources.

International Intellectual Property

In today's global economy, the importance of thinking internationally about your intellectual property cannot be understated. Throughout this section of the book, we've identified some key considerations a technology company should entertain with respect to your intellectual property portfolio.

International efforts oftentimes entail individual efforts in many foreign jurisdictions. In some cases, these efforts are substantial and ongoing compliance is significant. For instance, some countries will require you to have a local attorney to file your application; others may require regular filing updates and reports. Before undertaking a strategy to acquire international protections, be certain you have a process in place to manage your filings and international compliance; many times your attorney may be able to assist you with these ongoing efforts.

There are some entities that can assist with efforts by providing a streamlined process for submitting multiple applications. These include the Patent Cooperation Treaty, which provides a simplified process for filing for protections in more than 100 countries. In Europe, the European Patent Office (under the European Patent Organisation), which was established by the European Patent Convention of 1977, offers a single patent filing process for its member countries, whereas the Office for Harmonization in the Internal Market provides trademark registration in European Union member states.

INTERNATIONAL CONSIDERATIONS FOR AN INTELLECTUAL PROPERTY STRATEGY

Listed are some key steps a startup company should consider in developing their intellectual property strategy:

- **Using the Paris Convention:** Most of the key nations that an investor would be likely to consider for patent filings are members of the Paris Convention. The convention provides entrepreneurs with one year from the date of a patent application filing in any of its member countries to effectively file on that first date in other member countries. For a U.S. company filing an application with the U.S. Patent and Trademark Office (without public disclosures before your filing), you'll have a year to file in other countries even if you subsequently make a public disclosure. Nevertheless, the costs of a broad international patenting strategy may be a hurdle to a filing in many countries. Be aware of the costs of the strategy and consider the costs to protect your key technologies in important markets.
- **Selecting a global mark:** A startup can save itself a great deal of effort in international mark registration by researching its mark internationally early. Although you may not choose to register your mark in other jurisdictions at formation, by selecting a mark that allows for international expansion, you may prevent future issues from arising.
- **Disclosure risks:** One of the key differences between U.S. and international patent regimens is that most other countries give priority to the first-to-file. In the United States, the "first-to-invent" is given priority. Therefore, be aware of the risks of disclosing your invention before filing in international markets. Although you would not be barred after disclosure within the United States, you may be barred internationally. Putting your product on sale or marketing your product for sale could immediately bar your product from patent protections abroad, so be certain to file in the United States before sales efforts if you intend to file for patents abroad. One area that startups may not realize is that an issued patent will qualify as "publication" in international jurisdictions. Therefore, if you intend to file for patents abroad, be certain to make those filings before the publication of your patent in the United States.

14

Product Development

In years past, product development in the technology industry was largely driven by engineers, who usually designed products based on their own interests and left the marketing department with the task of finding consumers to buy them. But with the accelerating pace of technological innovation, product cycles are measured in weeks rather than years. Many companies are finding that it makes better business sense to design products based on what consumers want rather than what engineers think they want.

Peter Kupfer
Reporter for the San Francisco Chronicle

What to Watch For

For a high-technology entrepreneur, development of a product or service may seem to be one of the most familiar or straightforward aspects of the startup company. However, developing a good product simply isn't enough. According to Robert Cooper, for every seven new product ideas that companies put forth, approximately four enter the product development phase, approximately one and a half are launched, and only one will succeed (meaning only 15%–20% of new product ideas will ever be a success with consumers). To produce a product that consumers want to buy, collaboration between the engineers, scientists, and technologies developing the technology and the marketing and sales personnel reaching consumers is crucial.

IMPORTANCE OF NEW PRODUCT INNOVATION

According to Robert Cooper, new product innovation drives sales and profits for new and existing businesses:

- New products (those products that have been on the market for less than five years) account for 33% of company sales on average.
- For companies recognized for their innovation:
  - New products account for 49.2% of sales and 49.2% of profits (compared with 25.2% and 22% for all other companies, respectively).
  - It takes only 3.5 ideas to achieve one successful new product (compared with 8.4 ideas for all other companies on average).
- The average ROI for successful new products averages 96.9%, with a payback period of 2.49 years and an average market share in their defined target market of 47.3%.
- The median results show similar trends, with an ROI of 33%, a payback period of less than two years, and market share above 35%.

It is clear that new product innovation drives success in a business, particularly a startup company. But what determines whether a product launch is successful or a failure?

Raj Karamchedu, in his book *It's Not About the Technology: Developing the Craft of Thinking for a High Technology Corporation*, discusses the link between successful innovation and the collaboration between the marketing and engineering departments of a high technology company. Says Karamchedu, "How is it that, in spite of making remarkable strides in high technology product design, development and deployment of these products in markets, we are still struggling to create a harmony between marketing and engineering professionals?"

Karamchedu posits that the best technology doesn't win the day: it ultimately is won by the product that best satisfies a customer's needs.

As a result, it is wholly appropriate that the chapters discussing product development and sales and marketing are next to one another. The importance of each cannot be understated, and, in fact, these concepts are intimately linked for the success of an entrepreneurial venture.

According to a survey of high-technology revenue-producing companies conducted by Launch Pad, the companies in the study spent an average of $6.3 million annually on the areas of marketing, sales, and engineering (in which the average revenues for the company were $7.4 million). These three areas, besides being some of the most significant expenses for high-technology organizations, are also crucial to the success of product development and launch in a high-technology business.

Startup Product Development

Numerous experts note that the key to developing a successful product begins at the end: deciding what your target market will buy. Sounds simple, right?

Unfortunately, for high technology, the window of opportunity may only open for a short time or may change significantly over the time it takes to develop a new product. As a result, the product development cycle requires much greater flexibility within the high-technology arena. For an additional discussion of opportunity identification, go to Chapter 2.

A Market-Centric Product Development Approach

A formal approach to product development may not be ideal for entrepreneurs or new companies. Rather, many of the key stages of product development could be skipped or combined. The steps set below tend to act like a funnel because numerous ideas could be developed in the idea generation phase and the business and market forces will point the company toward a single new product or product suite. The product development phases are as follows:

- Strategy/market selection
- Idea generation
- Idea screening
- Concept development and testing
- Business analysis
- Product development and testing
- Market testing
- Commercialization

The process to generate new ideas is fairly inexpensive, but the actual development of a new product is anything but. As a result, successful companies encourage idea generation but apply substantial resources in the idea screening stage to determine where best to apply necessary resources or development efforts. Ultimately, the decision to pursue a new product from an initial idea will be answered by looking into the customers, competitive positioning, target markets, product benefits, product production process, and thorough market research.

ACCORDING TO VENTURE CAPITALISTS

According to William Davidow, when it comes to product development, being "slightly better is dangerous."

The next stage in the process is concept development and market testing. The purpose of concept testing is to provide a fuller description of the product to understand how the product would be seen in the marketplace, primarily focused on external consumer reaction. In essence, this stage attempts to mimic the sales and marketing efforts of a new product to determine what features are attractive to various potential target markets and how the product would be perceived in the marketplace. Oftentimes, concurrently or shortly after the concept development and testing stage, the business will have its marketing and sales professionals build a business analysis for the product concept, leading to a set of financial projections that management can use in its decision-making. In addition, the business analysis will include estimates of returns on investment, break-even points, and profitability.

At this point in the product development cycle, little actual product development has even occurred. Instead, the emphasis has been on determining the right opportunity to pursue. However, once management is satisfied from the concept development and testing and business analysis, the company will move onto product development and testing. At this stage, the business will build or create product prototypes and test the prototype for performance internally. Once satisfactory results are reached internally, many companies will provide prototypes to test groups or directly to end users in a beta testing situation. The ultimate aim of this stage is to determine the weaknesses of the product that can or should be improved as well as identify the likelihood of target customers to purchase the product. After this stage, a company may decide to begin market testing the product with a selected geographic, industry, or customer market. Depending on the resources, a startup company may forgo this stage as a result of the substantial investments and limited resources of the company. Additionally, the downside of a selected product launch is that it could alert potential competitors to the market strategy.

Finally, the company will decide to make a commercial launch, also referred to as a full-scale launch. A commercial launch tends to be preceded by a technical implementation strategy setting out issues from sourcing, to production, to distribution, to what-if scenario building. In the high-technology space, a full commercial launch is often met with technical challenges and product failures, so a well-executed contingency plan is crucial to successful customer relationships.

THE PERFECTIONIST

What could happen

You have instilled the highest standard in your organization. This means you do things right, every time. You want everything to be perfect, but this also means that you are struggling to meet deadlines and your employees are afraid to make mistakes.

Watch out for

Don't make your business become stuck because you are unable to reach perfection. Small businesses will inevitably have struggles, missteps, and challenges. Some of the most successful products and brands have, at one time, had flaws that required recalls, changes, and re-dos. Embrace calculated risk-taking in a high-tech environment. Build an enterprise that strives for perfection but isn't paralyzed by that quest.

TIP: Don't get stuck in neutral because of a failure to take risks.

Product Development Teams

The development strategy requires substantial marketing and engineering resources and expertise. As a result, building cross-functional teams has become a very typical way for high-technology organizations to manage innovative product development. In practice, the use of this type of cross-functional team allows management to task the team with nearly all aspects of the development effort with the addition of other company resources for certain tasks or responsibilities. Generally, such a team will be given substantial leeway in the early stages of the idea-generation stage and will report to a senior management position within the organization such as a vice president or a program manager.

For some organizations in the high-technology industry, development of new products is extremely time and resource consuming. As a result, companies may look to build partnerships in various stages of the product development cycle.

Product Development Resources

For new companies, initial product development may consume nearly all of the human and financial resources of the company. As the company continues to grow, additional innovation is likely to continue coupled with ongoing sales and marketing efforts for existing products. Determining the proper resources allocation is a difficult challenge for new organizations creating high-tech products.

Between 2000 and 2003, Launch Pad, a marketing consulting organization, surveyed a group of high-technology companies on their expenses related to sales, marketing, and engineering. The companies in the survey were focused in the information technology field, with the most companies involved in sales of business-to-business products. As a result, some of the information may not be as relevant for industries such as life sciences, energy technology, or telecommunications. However, the information does provide some generic insights into the activities of any high-tech startup company.

Engineering expenses represented 54.2% of the amount spent collectively in the areas of engineering, marketing, and sales during 2000 through 2002 by the technology companies surveyed by Launch Pad. For companies that are prerevenue, it is likely that nearly all spending will be in the areas of administrative and research and development. However, for these revenue-generating companies, engineering expenditures still represented the most consistent expense item: seeing a much smaller cutback by the technology companies during overall declines in spending by the surveyed companies between 2000 and 2002. This represents an average budget for the sales department of the surveyed organizations of approximately $3.4 million annually between 2000 and 2002. Among the surveyed companies, in which the average revenue for the companies was approximately $7.4 million, engineering expenses were approximately 46.4% of total revenues.

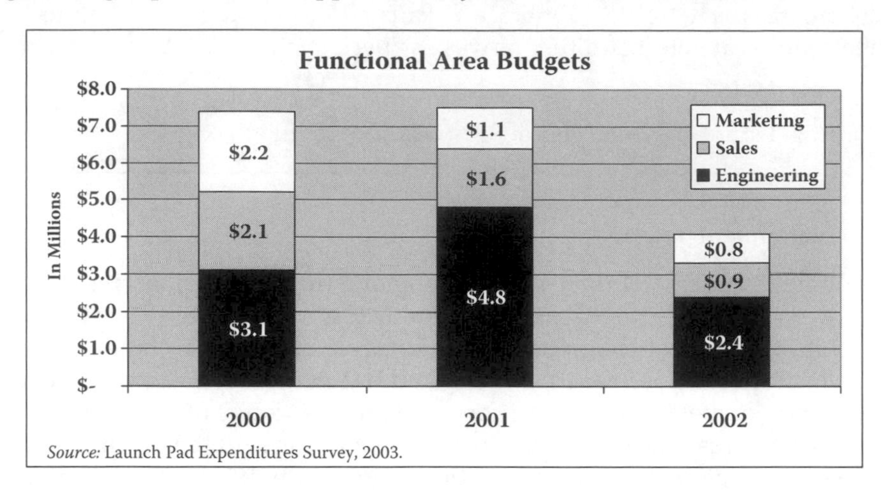

Source: Launch Pad Expenditures Survey, 2003.

Changing Consumer Needs

Product development of high-technology products involves greater flexibility when compared with other industries. According to research by Launch Pad, 19% of high-technology companies surveyed said they had changed their product line in the past year, and an equal number said they had changed their target market.

When building a product development strategy, emerging companies should be aware of the need to react quickly to consumer demands to capitalize on the changing marketplace.

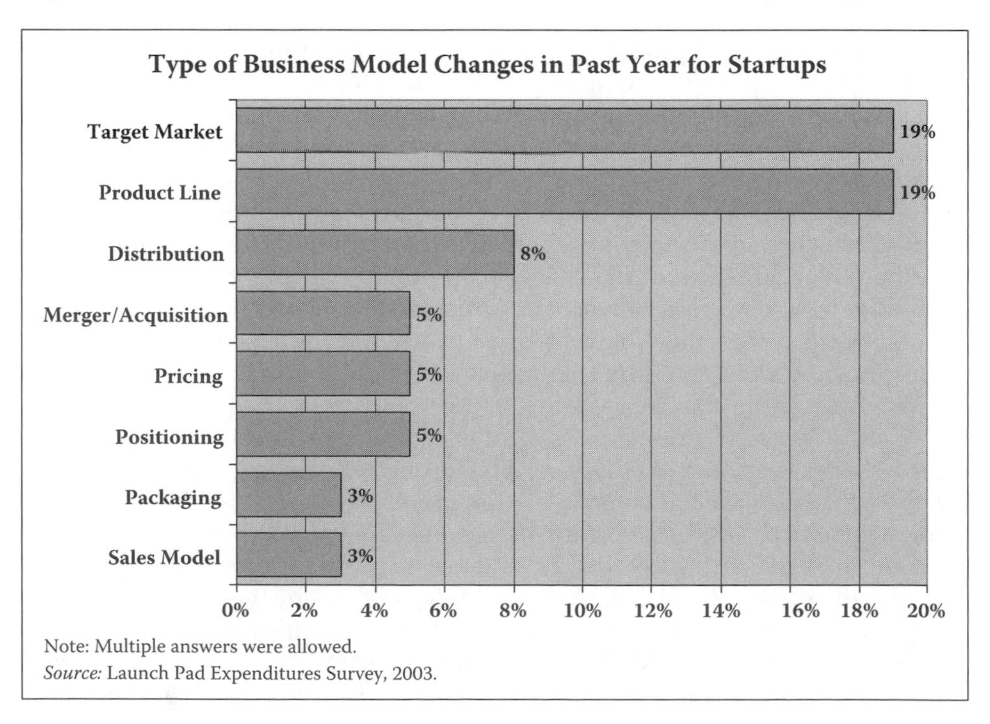

Note: Multiple answers were allowed.

Source: Launch Pad Expenditures Survey, 2003.

Product Development for Specific Sectors

Biotechnology

Biotechnology, which includes pharmaceuticals, biosensors, and biotechnology equipment, has historically been one of the top three largest investment categories for venture capital, taking in 3.71% of the total investment dollars in 2007. One of the reasons for its size is the time and money it takes to get a drug compound to market: more than 10 years and $64 million according to Peter Kolchinsky, PhD. In the biotechnology industry, the regulatory landscape represents one of the key steps in product development.

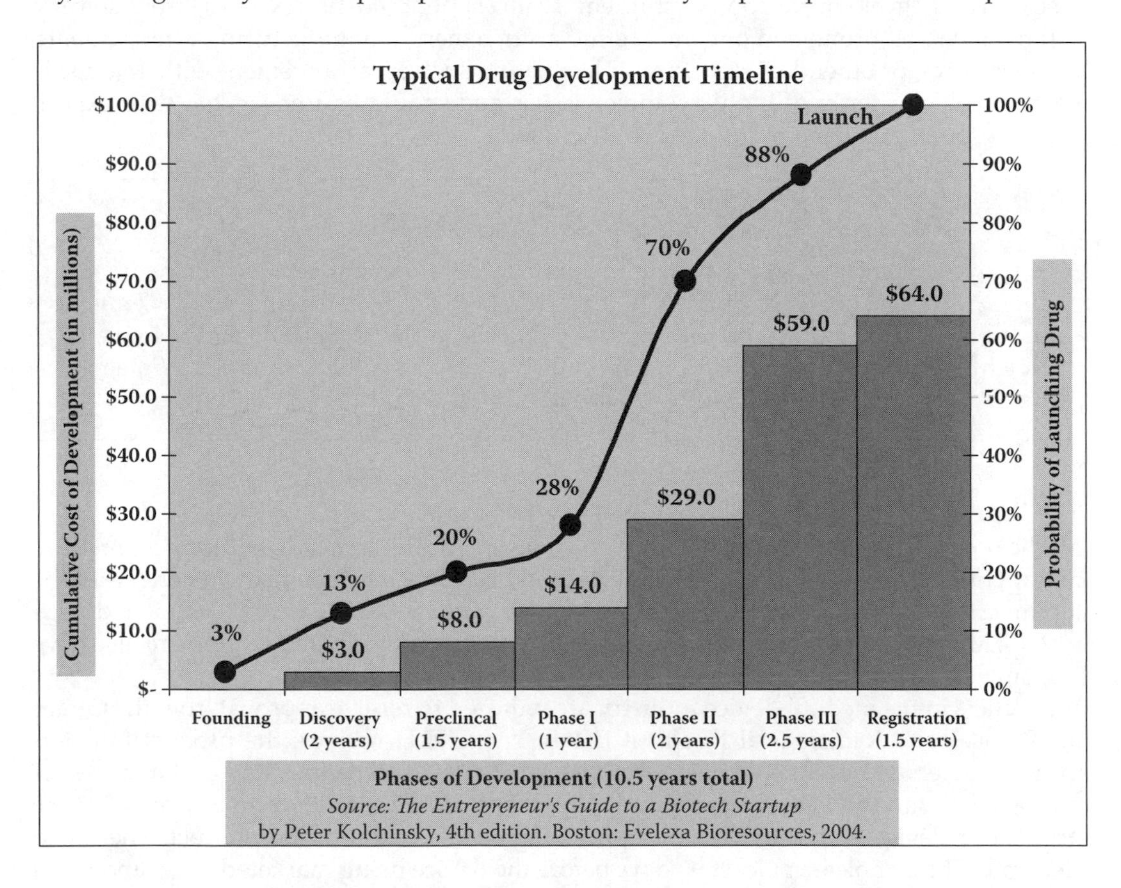

Source: The Entrepreneur's Guide to a Biotech Startup by Peter Kolchinsky, 4th edition. Boston: Evelexa Bioresources, 2004.

The clinical development process for a new entity or a biological within the United States begins with submission of the Investigational New Drug Application (IND) to the Food and Drug Administration (FDA) to authorize administration of an experimental drug to human subjects. The process to file an IND is an extensive one and tends to require the assistance of outside parties familiar with the regulatory scheme of the FDA. Following a 30-day period after submission, unless the FDA acts to prohibit the study, the company will then begin the Phase I study.

Phase I of the clinical process is usually completed in a one-year time window through a series of smaller studies that address areas such as basic safety of the compound. Phase II, also referred to as the "Proof of Concept" stage, is accomplished over a two-year period with a patient sample size of between 50 and 250 individuals. These trials are done on afflicted patients to answer questions of efficacy, side effects, and dosage requirements. After a successful Phase II, the trials will be expanded to 300–1,000 afflicted patients to closely mimic actual clinical practice of administration of the compound.

Many emerging biotechnology companies will not intend to develop a product and take it through a complete clinical development process. Rather, many of these companies will only plan to license development rights to a large pharmaceutical company after Phase I or Phase II of the clinical development cycle.

With the extensive time and money costs of the clinical development process, many startup companies find that they must partner with other organizations to share certain costs. These relationships can be structured around upfront fees, milestone payments, research and development sponsorship, equity investments, loans, royalties, profit-sharing arrangements, copromotion or comarketing arrangements, manufacturing arrangements, joint ventures, or other deal structures. Negotiating such an arrangement with an outside party involves a series of highly complex issues and should be undertaken with experienced resources on hand to guide such discussions.

FOR MORE INFORMATION

For an excellent resource on the clinical drug development process for new ventures, read the chapter entitled "Clinical Drug Development," authored by Kenneth B. Klein, MD, found in *The Entrepreneur's Guide to a Biotech Startup*, edited by Peter Kolchinsky, PhD. To download a copy, go to http://www.ibioinstitute.org/images/resource_guidekolchinsky.pdf.

Medical Devices and Equipment

Medical devices and equipment, which includes medical diagnostic equipment, medical therapeutic equipment, and other health-related products, has historically been one of the top investment categories for venture capital, taking in 2.8% of the total investment dollars in 2007. Much like in the biotechnology industry, medical devices and equipment also have a complex and time-consuming regulatory landscape.

Products in the medical device industry are required to receive approval from the Center for Devices and Radiological Health (CDRH). The CDRH reviews the product to determine whether the benefits outweigh the costs of the device. The first step in the process is registration with the CDRH as a medical device business. Next, the company must list the product or device with the CDRH by providing a description of the product. These two steps must be completed at least 90 days before the device being marketed. New and novel products will usually file a premarket approval application, whereas other products will file for an exemption (on the basis of a variety of exemptions based on the category or use of the product) or for recognition as substantially equivalent to a previously marketed device or product.

FOR MORE INFORMATION

For an excellent resource on the approval process for medical devices, read the chapter entitled "Medical Device Approval," authored by Christopher P. Pimentel, Esq., found in *The Entrepreneur's Guide to a Biotech Startup*, edited by Peter Kolchinsky, PhD. To download a copy, go to http://www.ibioinstitute.org/images/resource_guidekolchinsky.pdf.

Software

Software has historically been one of the top three investment categories for venture capital, taking in 3.76% of the total investment dollars in 2007, the highest investment amount for any single industry category. Unlike biotechnology or medical devices and equipment, software development rarely involves the same level of regulatory approval before launch.

Software development methods tend to be quite varied depending on numerous factors. The development process tends to follow fairly closely with general product development schedules. Roger S. Pressman, in his book *Software Engineering (A Practitioner's Approach)*, lays out some of the specific stages of software development:

- Domain analysis
- Software elements analysis
- Scope analysis
- Specification
- Software architecture
- Implementation
- Testing
- Deployment
- Documentation
- Software training and support
- Maintenance

CATCHING AN OPEN-SOURCE VIRUS

What could happen

One of your software developers has embedded a string of open-source code into your product's software code. Could this be a problem?

What to expect

Perhaps. There is some debate about what could occur if proprietary software incorporates some portion of open-source code. Open-source software has been referred to as "potentially viral software" because of the end-user licenses. Although most experts note that open-source license terms cannot simply move from one software program to any other program it touches, there is still some lingering concerns about the use of open-source software. Be sure to review the license agreements for any open-source code that is used in your software products and keep copies on file to allow for ease of review by third parties. If you have questions about the use of open-source software or software code, check with an experience intellectual property attorney.

TIP: Review any open-source software licenses and check with your attorney if you have any questions or concerns.

Some software development methods include the waterfall model, the spiral model, model-driven development, user experience, top-down and bottom-up design, the chaos model, evolutionary prototyping, standard prototyping, ICONIX process (UML-based object modeling with use cases), unified process, V-model, extreme programming, software development rhythms, and incremental funding methodology.

FOR MORE INFORMATION

For an excellent resource on selecting the appropriate software method for your development, read "Choose the Right Software Method for the Job" by Scott W. Ambler. The article contains several tables comparing the methods, the project categories, and the relative strengths of certain methods. To read the entire article, visit http://www.agiledata.org/essays/differentstrategies.html.

15

Marketing Efforts

Great devices are invented in the laboratory; great products are invented in the Marketing Department.

William H. Davidow
General Partner of Mohr, Davidow Ventures

What to Watch For

Research has indicated that one of the most important factors in the success of a startup company is its reputation, which has a strong bearing on the ability to attract and retain customers, obtain funding, recruit employees, and countless other necessary activities of a startup company. Particularly for a startup company, the marketing and sales departments play integral roles in building that reputation of a new entrant into the marketplace.

Author Geoffrey Moore describes the marketing process for startup companies in the technology industry as two distinct phases: the first involves marketing to early adopters, whereas the second involves a mainstream effort to develop customers out of a pragmatic, mainstream audience. Moore postures that many companies incorrectly believe that, by securing the early adopters through an extensive marketing and sales effort, mainstream customers will simply "follow the lead." Instead, Moore's research indicates that an entirely different approach is required for each distinct customer base, and targeting the right message to each party is crucial.

There are literally hundreds of excellent books available on sales and marketing strategies, techniques, opportunities, and tips for companies, many of which are targeted directly at startups and emerging businesses. This chapter is not designed to be the definitive answer on marketing for high-technology startup businesses. Instead, this chapter aims to identify some of the strategies and approaches for high-tech startups in marketing and selling their products.

DOES A MARKETING STRATEGY MATTER?

A look at the video cassette wars between Betamax and VHS tells the story of how a product with a better marketing and sales strategy (VHS) beat out the product that many of the experts claimed was a superior technology (Betamax).

In his book *Fast Forward: Hollywood, the Japanese, and the Onslaught of the VCR*, author James Lardner details how Sony's decision to market its product to the high-end market led to its downfall. The developer of the Betamax, Sony, attempted to dictate an industry standard and stand alone as the single supplier of Betamax video players. However, JVC decided to develop its own technology, the VHS, based on much simpler technology made more widely available. In the end, VHS became the industry standard primarily because of a wider availability of its products and a lower cost to consumers, although many experts noted that the Betamax was the superior product.

NOTE: Some experts have claimed that the Blu-ray versus HD-DVD war for high-definition DVD players represented a similar battle with the winner once again coming from a superior marketing strategy.

Marketing Basics

The ultimate aim of a successful business (be it a startup or a large multinational company) is to attract and retain an increasing group of satisfied customers. The marketing process tends to follow a fairly standard process, beginning with an analysis of the company and the external factors facing the markets of the company. After this analysis, the company will set out its marketing strategy and a more specific approach to allocate resources appropriate to transition potential customers into satisfied customers.

Market Analysis

Companies tend to use a variety of different techniques or tools to summarize the key market factors that will influence and drive the business. Generally, market analysis is a two-stage review of internal and external factors facing a company.

One of the traditional tools used in market analysis is the SWOT analysis. This analysis was first described in the late 1960s by a team of researchers including Edmund Learned. SWOT (sometimes referred to as TOWS) stands for strengths, weaknesses, opportunities, and threats, with the first two items a part of the internal analysis and the latter categories part of external factors. Under a SWOT analysis, the company will identify internal strengths and weaknesses in areas such as exclusive contracts, intellectual property, market share, key staff members, financial resources, operational efficiencies, and other similar factors. When analyzing external forces, the analysis will focus on customers, competition, partners, suppliers, macro-level trends, political factors, and other similar external-focused factors.

The chart below lays out a common approach for companies to organize the items from a SWOT analysis for strategic planning purposes. This approach tends to allow a company to better use information generated from this type of an analysis. The purpose of the chart is to identify the interplay between the company's internal strengths and weaknesses with the external trends that may be opportunities or threats, depending on how the company is situated.

Strengths	**Weaknesses**
1. 2. 3. …	1. 2. 3. …
Opportunities	**Threats**
1. 2. 3. …	1. 2. 3. …

Similar to the SWOT analysis, the five-C analysis separates its categories into internal and external factors. The five Cs are (1) company, (2) collaborators, (3) customers, (4) competitors, and (5) climate (or context). The climate category often uses information from a PEST analysis discussed in greater detail below. The purpose of this analytical tool is to lay out key areas applicable to marketing decisions and identify factors that match the company's strengths.

Organizations will use a PEST analysis to identify and analyze external macro-environment factors, usually on a country-by-country basis. The factors identified in this analysis tend to be out of the control of the company but are crucial for a company to recognize and understand. In many cases, a company will use this analysis before entering a new foreign market. A PEST analysis includes (1) a political analysis, which addresses factors such as political stability, corporate, employer and intellectual property laws, taxes, and relationship with other nations, (2) an economic analysis, which analyzes the economic system of the country or region, skill level of potential employees, and key economic indicators such as unemployment, inflation, and interest rates, (3) a social analysis, understanding educational structures, demographic data, entrepreneurial activity, health, and cultural differences, and (4) a technological analysis, looking at the state of technology, recent developments, costs structure, and diffusions of technologies.

More information on market research and sources of market data can be found in Chapters 2 and 8.

Market Strategy

Once an analysis of the appropriate markets has been completed, the next step in the progression is development of the company's marketing strategy. The steps in development of a market strategy are as follows:

- Market segmentation
- Selection of a target market or markets
- Product positioning
- Development of the value proposition

This process is not necessary a single action, isolated decision or a static process, but the process in itself involves a progression that starts with a broad market (i.e., restaurants, construction companies, parents), moves to specific market segments (i.e., restaurants in Baltimore and Washington DC, single-family home construction companies, affluent parents), then targets specific customers and adopters (i.e., Asian-cuisine restaurants in the DC Metropolitan area, jumbo single-family home construction contractors, affluent parents with children aged 4–10), and finally works to position the product and identify the specific value proposition for these targeted customers.

Segmentation

When analyzing market segments, it is important to evaluate whether a market segment is reasonable to target. You should ask yourself whether the market has the following characteristics:

- Is identifiable?
- Is accessible?
- Is substantial?
- Has unique needs?
- Is durable?

Segmentation is often different in consumer and industry markets. For consumers, segmentation will oftentimes be based on geographic (regions, size of city, country, climates, etc.), demographic (age, gender, income, occupation, ethnicity, social class, etc.), psychographic (activities, opinions, values, beliefs, attitudes, etc.), or behavioralistic (usage, readiness to purchase, status of a user, timing of purchases) segmentations. Conversely, industrial markets tend to focus on locations, types of companies, and behavioral characteristics.

Target Market or Markets

Selection of a target market dovetails with proper segmentation. The aim of segmentation should be to develop a list of likely market candidates that additional research and analysis will allow a company to select which markets to initially target, which markets to target in future periods, and which markets to avoid.

The most valuable market segments will be ones in which the participants are most similar within the segment but can be easily distinguished from other segments. The most substantial risks to targeting any particular market segment is that you are unable to identify and reach these targeted customers, which oftentimes occurs if the segment participants are not properly analyzed initially. This involves additional analysis into the segment participants to identify its true size, growth, sales processes, and competitive landscape. This is a place in which market research may be crucial. Some key factors to examine in a potential target market are as follows:

- Value of the market (number of customers, number of annual products purchased or to be purchased)
- Maturity of market (growing, stable, shrinking, no market yet)
- Competitors (are customers satisfied or unhappy with current products?)
- Opportunity for new competition and new products
- Purchasing process and buying decisions
- Margins in the market

Companies will also need to determine whether to focus on a single target market, several key markets, suite of products within selected similar markets, or full market penetration. In the chart below, high-tech companies surveyed identified the functional areas they had targeted for their sales and marketing efforts.

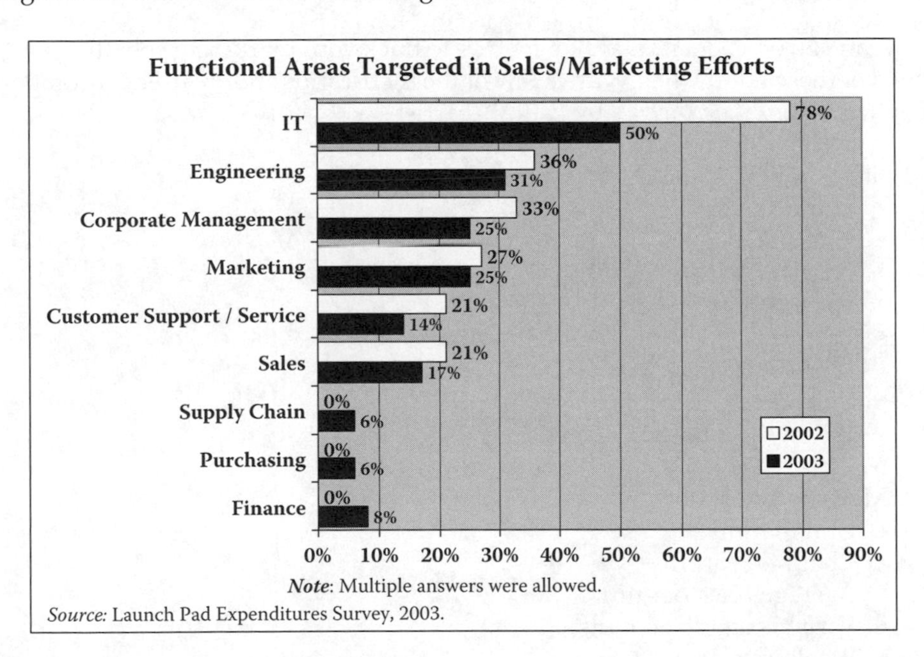

Note: Multiple answers were allowed.

Source: Launch Pad Expenditures Survey, 2003.

Positioning and Value Proposition

One of the most consistent struggles for startup companies is positioning their products in a market and providing a value proposition to customers. These decisions will likewise involve market research and a thorough understanding of the target market and the customers for the product.

Marketing Mix and Implementation

E. Jerome McCarthy identified four categories of marketing decisions made by companies, which have been referred to as the 4 Ps of marketing:

- Product
- Price
- Place (distribution)
- Promotion

Each of these areas represents a particular constraint of the company in its marketing strategy. As a company makes decisions to modify the product by adding additional features, warranties, or design aspects, this will affect decisions in pricing and promotions.

TRACKING MARKETING EFFECTIVENESS

A marketing program represents a significant portion of a startup's budget and oftentimes is a key component to the overall health and success of the business. At the same time, determining the effectiveness of your marketing dollars and efforts oftentimes proves difficult.

What can you do about it? Develop metrics to track your marketing spending and the results of these efforts. Below are a few of the metrics commonly used by companies taken from *Marketing Metrics* by Paul W. Farris:

1. Share of hearts, minds, and markets
   - Market share
   - Unit share
   - Penetration share
   - Top of mind
   - Loyalty
   - Likeability
   - Customer satisfaction
   - Willingness to recommend

2. Margins and profits
   - Unit margin
   - Channel margins
   - Average price per unit
   - Variable and fixed costs
   - Break-even sales
   - Target volume
   - Target revenues

3. Product and portfolio management
    - Trial
    - Repeat volume
    - Penetration
    - Cannibalization rate
    - Growth (percentage)
4. Customer profitability
    - Customers
    - Retention rate
    - Customer lifetime value
    - Average acquisition cost
    - Average retention cost
5. Sales force and channel management
    - Workload
    - Sales potential forecast
    - Sales total
    - Inventories
    - Direct product profitability
6. Pricing strategy
    - Price premium
    - Reservation rate
    - Optimal price
7. Promotion
    - Baseline sales
    - Redemption rates
    - Price waterfall
8. Advertising media and Web metrics
    - Impressions
    - Net reach
    - Effective reach
    - Page views
    - Clickthrough rate
    - Cost per click
    - Visits
    - Visitors
9. Marketing and finance
    - Net profit
    - Return on sales
    - ROI
    - Payback

TRYING TO BECOME PROFITABLE TOO QUICKLY

What could happen

We've done our beta testing and think we are ready to launch our product. We believe it is a good product and can still sell it at a high price to pay off our investors very quickly. Is that the right approach?

What to expect

The goal of a successful startup business should be making something that people want. This concept sounds obvious, but many startups fail because they focus more on making a profitable product rather than a desirable product. With the later, most often comes the former. This is not to suggest that you shouldn't consider how to become a profitable business, but remember that your goal should be to "grow" your business. Be certain to balance your desire for profits early on with the ability to retain key customers. Build long-term success through short-term emphasis on meeting customer wants.

Google represents a great example of a company that focused on creating a successful product and then found a way to make a profit from that product. Some experts had suggested early on that Google should charge consumer for its search tools, but Google decided to keep its bread-and-butter search product a free-consumer product and found alternative ways to create profit from its use.

TIP: Be wary early on of sacrificing customers to raise profits.

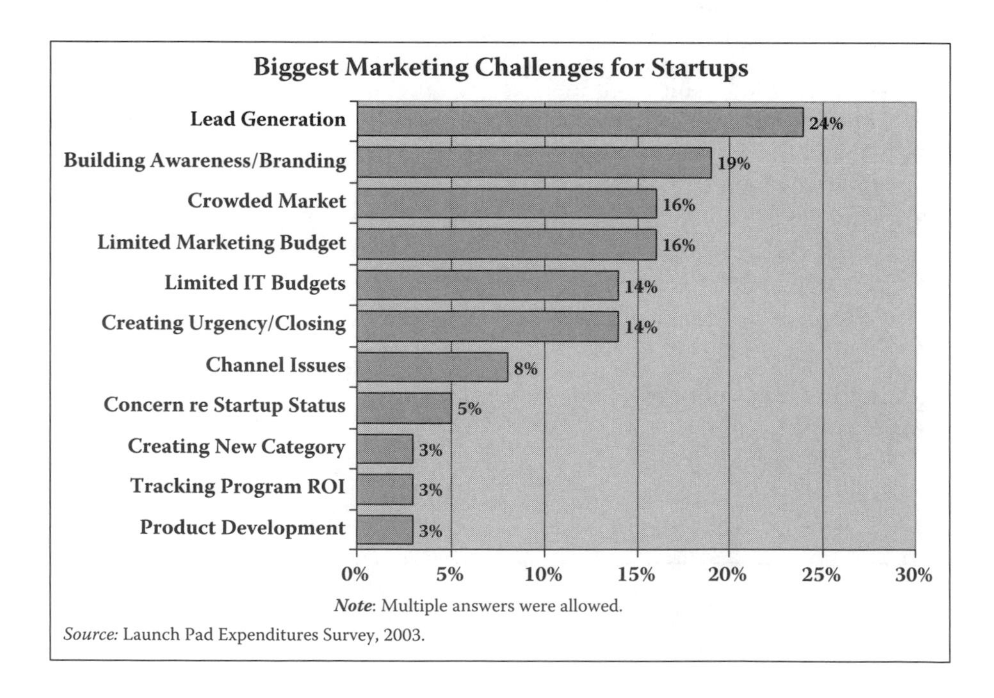

Note: Multiple answers were allowed.

Source: Launch Pad Expenditures Survey, 2003.

Startup Marketing

Marketing of a startup company may be constrained more than a well-established business because of limited marketing budgets and, oftentimes, more limited windows of opportunity. As a result, startups cannot afford to "miss" in their efforts to segment their markets, target a key customer segment, and position their products. In addition, startups must focus their spending on key areas that produce sufficient returns for the startup company.

Generally, for technology startups, segmentation is based on either company size (i.e., midsized businesses, Fortune 1000 businesses), by vertical industry (i.e., insurance industry, hospitality industry), or a combination of both (i.e., midsized insurance companies). It is less likely to see technology companies focus on a specific geography because of the necessity to provide a more focused target market that is not conducive to a limited geography. In addition, because of difficulties reaching high-level executives and some concerns about purchasing certain technology applications or services from startups, many businesses decide to focus their marketing efforts on individuals at the vice president level and below of their target companies, in many cases avoiding sales directly to an information technology department to approach a segment of the business such as research and development or finance.

Startup Marketing Spending

In the 2003 Launch Pad survey, marketing expenses represented 21.6% of the amount spent collectively in the areas of engineering, marketing, and sales during 2000 through 2002 by the technology companies surveyed by Launch Pad. This represents an average budget for the marketing department of the surveyed organizations of approximately $1.4 million annually. Among the surveyed companies, in which the average revenue for the companies was approximately $7.4 million, marketing expenses were approximately 18% of total revenues.

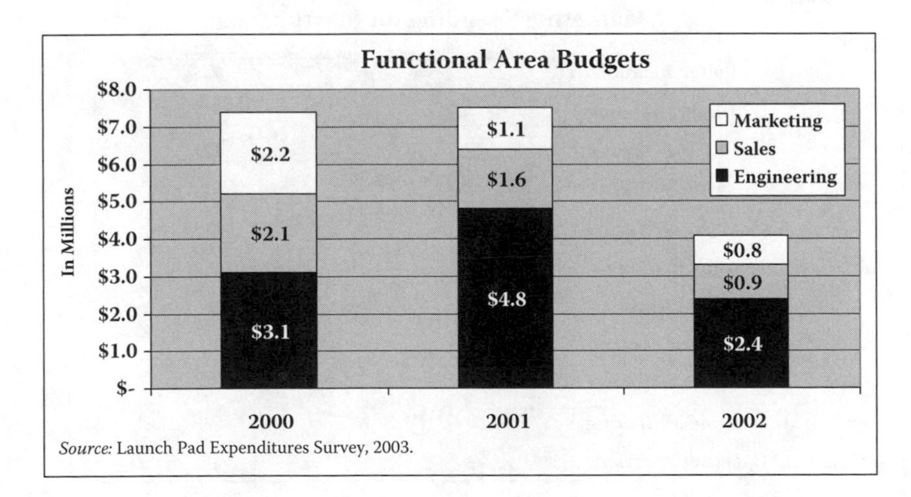

Source: Launch Pad Expenditures Survey, 2003.

In 2000 and 2001, the average size of the marketing staff for the technology startups surveyed was 7.6 and 6.0, respectively, with an average of 1.0 vice president-level marketing positions, 1.4 director-level marketing positions, 2.0 manager-level positions, and 2.4 line staff-level positions. In 2000 and 2001, the average marketing budget was $2.243 million and $1.143 million, respectively, with $698,000 (31%) spent on staff and $1.545 million (69%) spent on marketing programs in 2000 and $518,000 (45.3%) spent on staff and $625,000 (54.7%) spent on marketing programs in 2001.

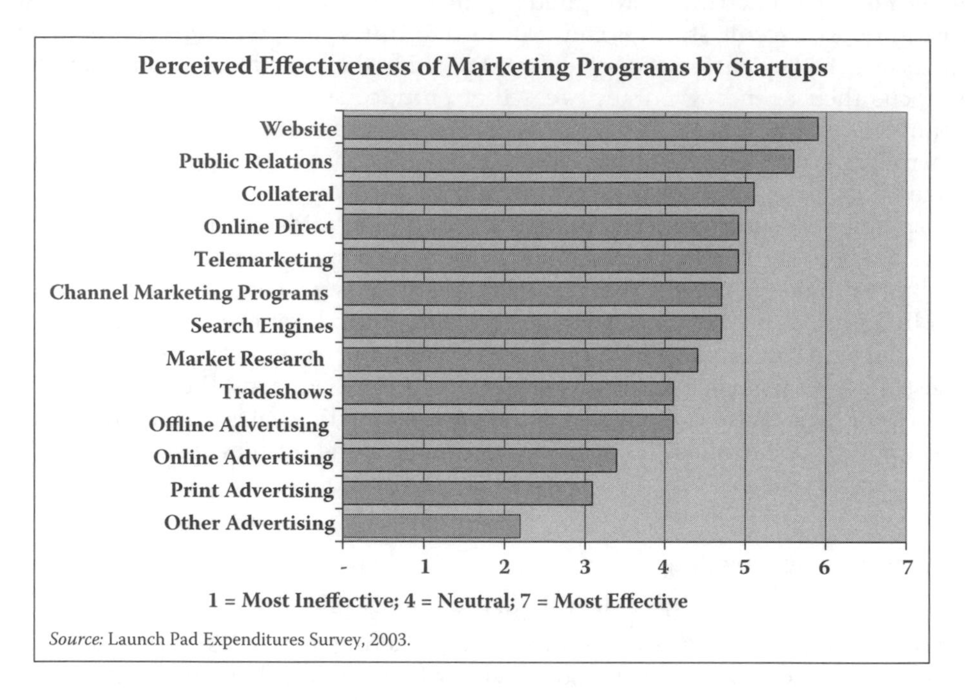

Source: Launch Pad Expenditures Survey, 2003.

Most companies, 51% in a 1997 study, currently outsource one or more of the marketing functions such as advertising or public relations or telemarketing.

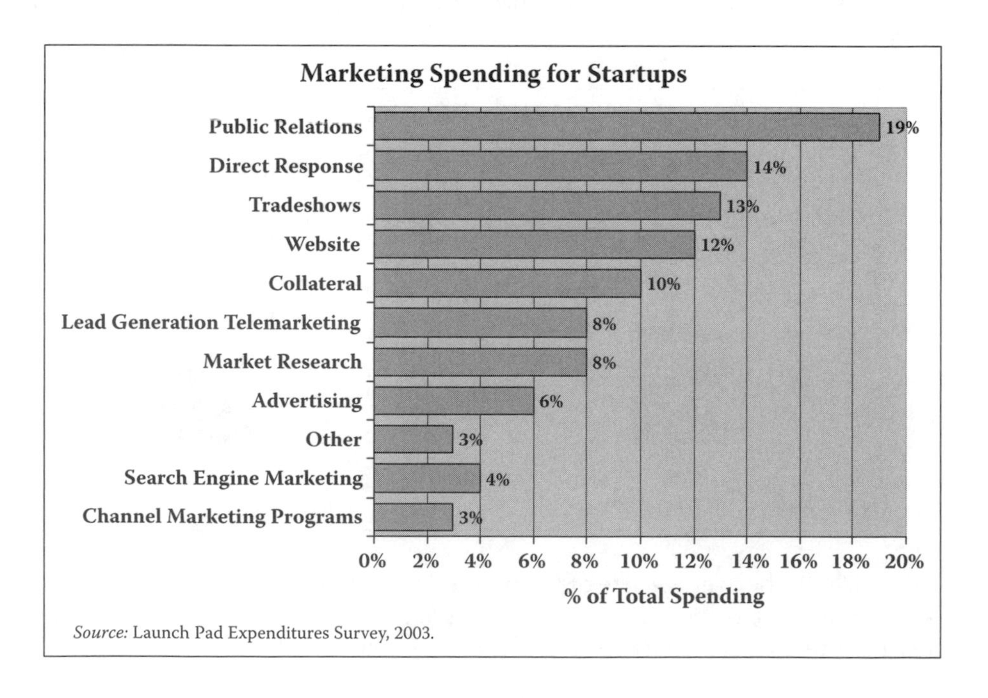

Source: Launch Pad Expenditures Survey, 2003.

EFFECTIVE USE OF PUBLIC RELATIONS

According to companies surveyed, spending on public relations represents one of the largest as well as most effective tools used in marketing efforts. Public relations can be a crucial tool to increase awareness of the product as well as to create additional interest from potential investors, business partners, and employees.

How do you develop an effective public relations strategy? Be specific in your goal of a public relations strategy, the target audience, and the message you want conveyed. For example, an article in a technical trade publication may be good press, but it might not reach the decision-makers who will purchase your products, or a general press release may fall on deaf ears if you can't get it into the hands of individuals who can produce press on the issue.

- **Doing it yourself:** Any public relations strategy will involve some amount of "doing-it-yourself."
- **Hiring public relations professionals:** A professional public relations firm will offer a suite of services, including drafting press releases, soliciting journalists to write stories on your company or interview you for expert opinions, and various other services from website and search customization to mass mailing and advertising.

  A public relations firm will generally charge either a per-hour fee or a monthly retainer fee for their services. The hourly rates will usually be based on the experience level or skill set of the individual. Fees tend to range between $5,000 and $6,000 per month for services and up to $10,000 to $15,000 monthly for a top-flight public relations firm. The retainer approach tends to provide for a set suite of services to be provided monthly with additional expenses billed separately. In most cases, a firm will ask to enter into a six-month or one-year contract, with the option to extend the relationship. Usually most contracts will have a 30- or 60-day notice period before termination.

  In some cases, certain public relations (or similarly situated marketing services) firms will take a reduced fee in exchange for stock or stock options. This agreement may involve certain milestones for the stock or options to vest.

16

Sales Organizations

People get caught up in wonderful, eye-catching pitches, but they don't do enough to close the deal. It's no good if you don't make the sale. Even if your foot is in the door or you bring someone into a conference room, you don't win the deal unless you actually get them to sign on the dotted line.

Donald Trump

What to Watch For

Think having the right sales people doesn't matter? Think again. According to the HR Chally 2006–2007 Survey of more than 50,000 North American corporate buyers, salesperson effectiveness counted for more than 39% of customer choice, more than price, quality, or solution. Developing an effective sales organization is crucial to expanding your product and reaching key customer markets.

Startup companies face a unique set of challenges in building and maintaining a sales team and a sales strategy. In addition, management must decide when to begin sales efforts and how to grow the sales staff as the organization matures.

Sales

In the 2003 Launch Pad survey, sales expenses represented approximately 24.2% of the amount spent collectively in the areas of engineering, marketing, and sales during 2000 through 2002 by the technology companies surveyed. This represents an average budget for the sales department of the surveyed organizations of more than $1.5 million annually between 2000 and 2002. Among the surveyed companies, in which the average revenue for the companies was approximately $7.4 million, marketing expenses were just under 20% of total revenues.

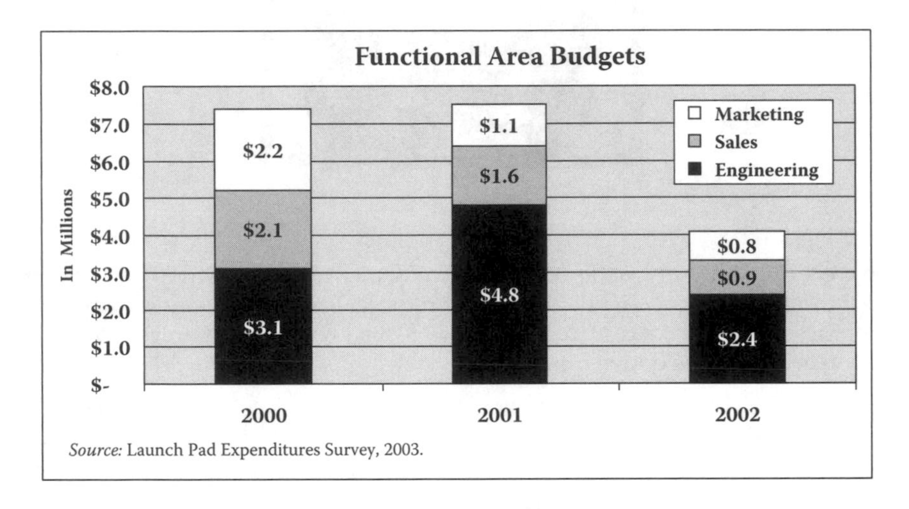

Source: Launch Pad Expenditures Survey, 2003.

The importance of the sales organization and its structure to a startup business is substantial. As one VC said, "I view the entire startup organization as basically customer support for the sales team."

Similar to the marketing strategy, an effective sales strategy for high-tech startups will change with the growth of the organization. Researchers Andris Zoltners, Prabhakant Sinha, and Sally Lorimer have identified four key factors in a company's sales strategy that a company will modify during different phases of its growth. The factors are as follows:

- The roles that the sales force and selling partners play
- The size of the sales force
- The sales force's degree of specialization
- How salespeople apportion their efforts among different customers, products, and activities

The factors affect the ability of the company to quickly respond to market opportunities, to reach potential customers, and to change the revenues, costs, and profitability of the organization. Although companies will weigh each of these factors differently, these researchers note that the importance of certain factors tends to correlate to the companies' growth stage: the relationship between an internal sales force and external selling partners is much more important for a startup company, whereas the degree of specialization of the sales force tends to be of greater importance during the growth phase.

	Business Life Cycle Stage			
	Startup	**Growth**	**Maturity**	**Decline**
	Emphasis			
Role of sales force and selling partners	♦♦♦♦	♦♦	♦	♦♦♦
Size of sales force	♦♦♦	♦♦♦♦	♦♦	♦♦♦♦
Degree of specialization	♦	♦♦♦♦	♦♦♦	♦♦
Sales force resource allocation	♦♦	♦	♦♦♦♦	♦
	Underlying Customer Strategy			
	Create awareness and generate quick product uptake	Penetrate deeper into existing segments and develop new ones	Focus on efficiently serving and retaining existing customers	Emphasize efficiency, protect critical customer relationships, and exit unprofitable segments

Source: "Matching Your Sales Force Structure to Your Business Life Cycle," by Zoltners, Sinha, and Lorimer, *Harvard Business Review*, July-August 2006.

In each of the business life cycle stages of a company, the organization will generally focus on two of the sales force factors. For companies in the startup phase, the organization will need to identify the size of their internal sales staff as well as identify the proper mix of selling partners and internal sales resources. Companies in the growth phase will tend to place a greater emphasis on the size of their sales organization and the degree of specialization as the company looks to grow past early adopters into a broader market.

Sales Force Structures

The structure of the sales force for an early-stage company is generally fairly straightforward because of the size of the organization. However, as the company continues to grow, offer additional products, and reach out to new markets, more complexity will be required.

Sales forces tend to be organized in the early periods as flat sales organizations, with all sales people reporting to a single sales director or manager. However, as an organization grows and depending on the level of support required for the sales force, additional layers of management may be required, creating tall sales organizations. The decision between these structures will typically depend on the decision-making autonomy by the sales force, the management and operational support required by the sales force, and the size of the selling organization. Many early-stage businesses will also choose to organize the business around a series of major accounts as well as specialization of the various sales people, perhaps based on the previous industries or selling markets of a sales person.

For many companies, the sales organization will be structured depending on the needs of the customers and the complexity of the product offering. Companies with a fairly basic product portfolio or a suite of products that are not highly technical that are consistently sold across customers will tend to organize themselves around geographic boundaries. Conversely, a company with a complex product suite and unique customers needs tend to organize their sales efforts in a product- or market-driven sales organization. Organizations may also choose to create a hybrid organization, such as a market-driven organization focused by geography.

Market-Driven Sales Organization

The market-driven sales organization is structured around the types of customers or end users (i.e., government versus commercial customers). Benefits to the organization are a better understanding of the customer needs and greater management control over specific markets. Drawbacks are an increased cost to the organization as well as a duplication of geographic efforts.

Product-Driven Sales Organization

The product-driven sales organization is structured around a product line or product category (i.e., hardware versus software, drug A versus drug B). Benefits to the organization are a more customer/product-focused sales staff and greater management control over selling efforts. Drawbacks are an increased cost to the organization as well as duplication of customer and geographic efforts.

Geography-Driven Sales Organization

The geography-driven sales organization tends to be based on regions, states, countries, or other defined areas that are usually grouped together to limit travel (i.e., Western, Eastern, Southern, and Midwest regions). Benefits to the organization are lower costs, limited duplication of customers and geographies, and less management. Drawbacks include less sales person specialization and lesser customer/product focus.

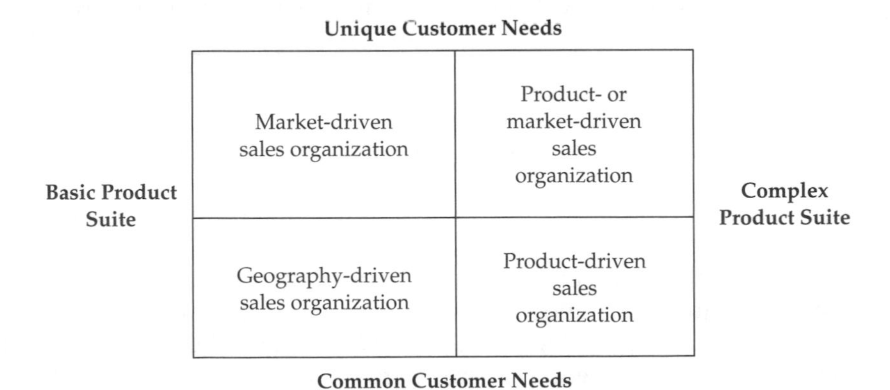

What Is the Right Size for Your Sales Staff?

Determining the size of your sales staff is dependent on a number of factors, including the use of selling partners, the ROI for new sales people, productivity of sales people, and turnover of the sales staff. For an early-stage business, the addition of sales people will usually increase sales much more than the selling costs. However, as the organization grows, there usually is a point at which the addition of a salesperson will not produce new sales to offset the additional selling costs.

DETERMINING THE PROPER SIZE OF YOUR SALES STAFF

Companies tend to determine the appropriate sales force size by using a number of analytical tools. These tools include the following:

- **Breakdown approach:** Determine the average sales per salesperson and then, using the forecasted sales, determine the appropriate sales force to reach those sales figures.
- **Selling expenses approach:** Determine the forecasted sales for the following year and set selling expenditures as a percentage of forecasted sales to determine the total selling expenses and the per salesperson selling expenses for the sales force.
- **Workload approach:** Determine the total selling effort needed (based on number of sales calls, prospect meetings, conferences, etc.) and then determine the average selling effort per salesperson to calculate the appropriate sales force to accomplish the desired selling effort.
- **Incremental approach:** Determine the marginal contribution of each additional salesperson against the marginal cost and only add new salespeople when the contribution exceeds the costs.
- **ROI approach:** Calculate the three-year ROI of your sales force (calculations below) to determine whether an appropriate return is being made on the sales force.

The breakdown, selling expenses, and workload approaches tend to be very rough instruments but can be useful in determining some basic information about the sales force. The incremental and ROI approaches tend to require more detailed analysis and may be more applicable for a more established organization.

Researchers Zoltners, Sinha, and Lorimer identified three common scenarios early-stage companies use in establishing the size of a sales force:

- **Earn your way:** The company only hires additional salespeople as additional profits are generated
- **Play it safe:** The company initially hires its sales staff for the current year based on the expectations or projections for next year's sales staffing needs (i.e., hiring one year ahead)
- **Quick build:** The company fully staffed the sales force based on an estimate for the size of the sales force three or more years in the future

One of the most common models for startup sales force staffing is this earn-your-way approach that hires additional sales people as additional profits are generated, usually establishing a barebones sales force initially and ultimately ramping up the size of the sales group. In the study by Zoltners, Sinha, and Lorimer, they examined each of the three different staffing models to project the impact on contributions (gross margin less selling expenses) over a three-year period. According to the modeling, when adopting an earn-your-way model in which sales people are added systematically as sales grow, the total contribution by the sales force over the three-year period was 17% lower than the total contribution over the same three-year period in which the company adopted a quick-build model in which a fully staffed sales force was put in place in year one and remained consistent over the three-year period. Likewise, a middle model, the play it safe that involved a high initial sales force that only increased after year one, had a 7% increase over the total contribution from the earn-your-way model.

IS YOUR SALES FORCE TOO SMALL?

Data suggest that many startup companies continue to maintain smaller sales organizations, although empirical evidence may suggest that companies are leaving money on the table by failing to deploy sufficient sales people.

ZS Associates, a global business consulting firm, provided an optimal sales force size analysis to 11 different startups in the healthcare industry, based on models that forecast the sales and profit implications of various sales force sizes. In 10 of those 11 companies, the sales leaders created teams that were smaller than the optimal size communicated to them by ZS Associates, an average of 64% of the optimal.

The researchers in this study developed a model to assist companies in determining an appropriate staffing model for their sales organization. The model determines a three-year ROI for the sales force based on the calculations of the break-even ratio (the ratio of the incremental sales revenue per additional salesperson to the break-even sales) and the carryover sales rates (the percentage of sales that the company will retain without any sales force efforts). According to these models, the optimal sales force will have an ROI between 50% and 150%. A sales force with an ROI under 50% is likely to be too large, whereas a sales force with an ROI above 150% is likely to be too small and additional sales resources could be deployed.

SIZING THE SALES FORCE BY THE NUMBERS

Calculating the three-year ROI on the sales staff

$$[(M_R \times I) + (M_R \times I \times K_2) + (M_R \times I \times K_3) - C] \div C$$

C = Estimated annual cost of a salesperson
I = Estimated incremental annual sales revenue for an additional salesperson
M = Gross margin
M_R = Gross margin rate
K = Annual carryover sale percentage

Results

Less than 50% ROI = sales force is likely to be too large
50%–150% ROI = optimal size of sales force
More than 150% ROI = sales force is likely to be too small

Source: "Matching Your Sales Force Structure to Your Business Life Cycle," by Zoltners, Sinha, and Lorimer, *Harvard Business Review*, July-August 2006.

Do It Yourself or Outsource?

Early-stage companies are often unable, depending on their resources, to solely rely on internal sales resources. As a result, many companies must choose how to allocate resources for their sales efforts.

Companies will generally make a determination to outsource the sales function by comparing the fixed costs of a direct sales force with the variable costs of an outsourced sales organization. At a certain level of sales, the direct sales force will exceed the value of the variable costs of an outsourced sales force. However, firms should also consider additional factors such as coverage efficiency and selling effectiveness. Additionally, companies may consider a hybrid approach in which certain markets or functions are outsourced to third-party providers and key markets and customers are the focus of an internal sales force.

Companies that initially use third-party partnerships must also determine when and whether to transition from a heavy reliance on partnerships. Some companies are hesitant to move from sales through partners for fear of alienating customers or eliminating a successful market.

Targeted Approach

Startup companies face a unique challenge in their sales strategy: convincing decision-makers that they are going to be around for the long haul. According to a survey of chief information officers (CIOs) by Launch Pad, only 48% of the individual surveyed were willing to purchase products from a startup company. In fact, only 5% of those surveyed said they were very willing to purchase from a startup company. In the survey, several CIOs raised the point that they were less willing to purchase a "mission critical" product from a startup company because of concerns of long-term sustainability and service.

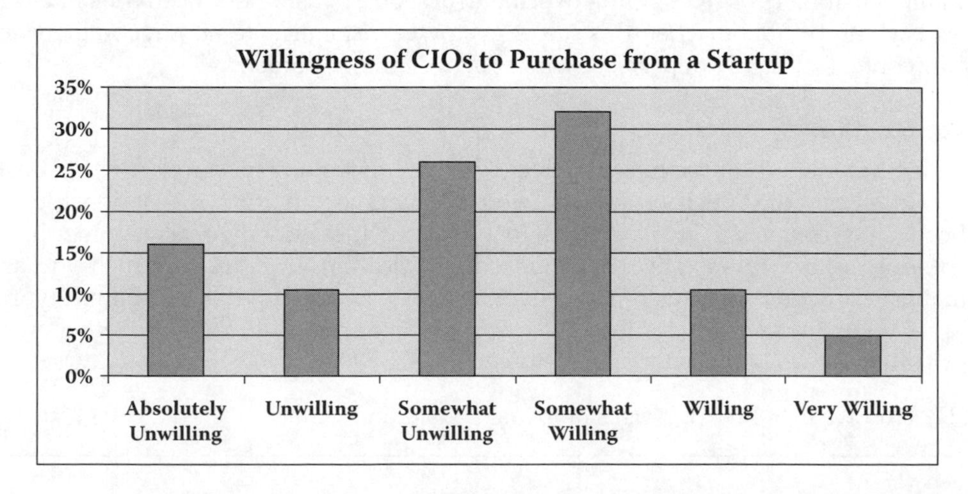

As a result of the challenges faced by startup companies selling products to customers who are concerned about relying on a startup company, companies have expanded their sales channels to target individuals in positions outside of the CIO or similar role or in other functions altogether.

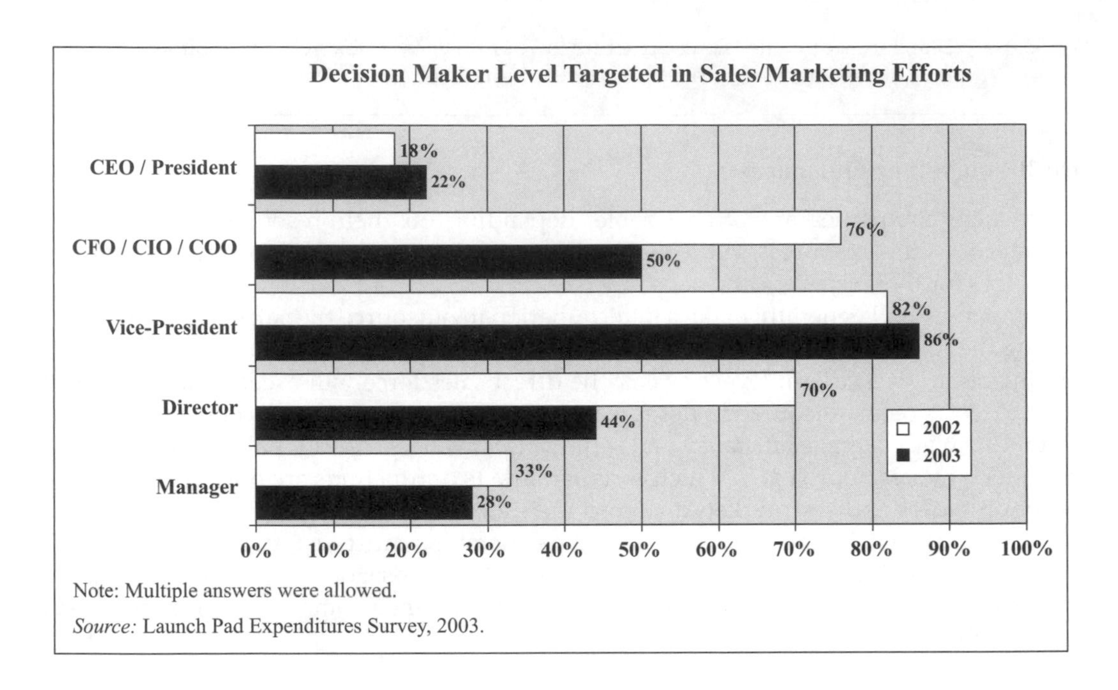

Note: Multiple answers were allowed.
Source: Launch Pad Expenditures Survey, 2003.

LOVING A CUSTOMER (TO DEATH)

What could happen

Over the first year of operations, you've built a strong relationship with one major customer (who is a good family friend). You are continuing to build a number of smaller customers but struggling to bring in any other customers. Scenarios like this can become troubling when this major customer becomes late on payments, when your contact suddenly departs, or when you lose the customer.

Watch out for

Remember that one of the major sources of problems for any new and growing business is relying too much on one or two customers, or one or two relationships at those major contacts. A sign of health for a new business is a diverse customer base. Attempt to limit this exposure through diversification. In cases in which you are unable to broaden your customer base, extend your reach into that company beyond one or two points of contacts; broaden into finance, purchasing, sales, and other areas for your contacts with that major customer.

TIP: Find ways to limit the risks of losing a major customer.

HARD TO SAY GOODBYE . . .

What could happen

You've got a customer who has becoming increasingly difficult to work with and yields a low profit margin for your company. However, you've worked with this customer from the beginning and are hesitant to lose any relationship.

Watch out for

You should not view a client relationship as indispensable, and, in the event that the relationship does not work (from a resource demand, profit margin, or other metric), you should consider finding a way to restructure or end the relationship. New companies are sometimes hesitant to end any customer relationship, but, in certain cases, this may be the desired approach.

TIP: Evaluate unprofitable or difficult customer relationships regularly and ensure they are in the best interest of your company.

17

Operational Matters of the Business

Go for a business that any idiot can run—because sooner or later, any idiot probably is going to run it.

Peter Lynch
Research consultant at Fidelity Investments

What to Watch For

One of the surprises for new entrepreneurs are the number and amount of administrative items that suddenly become their responsibility. The business requires office space, will sign leases for buildings, offices, and equipment, must sort through business, directors and officers, and health insurance, and must look into setting up a website, just to name a few of the administrative issues any new business will face.

In particular, health insurance programs for employees represent a major challenge for startups and small businesses. According to a Discover study, three of four small businesses surveyed did not offer health insurance programs for their employees, and two of three that did offer health insurance had considered cutting back on the program because of increasing costs.

Choosing an Office Location

For some companies, they may be looking for their first official office as their business receives funding to move out a founders' basement or garage. Other companies may be looking to expand from an initial office to a larger space as the company looks to scale its efforts. Still other companies, as discussed previously, will choose to relocate the business at some point in their early stages. This may be for a variety of reasons. In the event a company wants to potentially partner with Google and Yahoo!, they may relocate to Silicon Valley to find ex-employees of those companies and to be closer to these potential partners. Other companies looking for financing may move to a state such as California, Massachusetts, New York, Texas, or Washington, states in which more than 70% of investment dollars are invested.

Ultimately, each of these companies will be faced with a decision regarding selecting where to locate their offices and facilities. Generally, you can determine a great deal about appropriate requirements for your office space by investigating your competitors to find the type of locations most appropriate for similarly situated companies. Once you've done your homework, you may look to hire a commercial realtor to assist your efforts in finding the proper location.

Key Considerations in Choosing an Office Location

What is the most important consideration for high-technology companies in selecting their office location? According to research by Paul Dettwiler, Peter Lindelof, and Hans Lofsten, high-technology startups felt the most important criteria in selecting an office location was the access to staff. For some firms, this means locating an office near other companies in their industry or near a university. For others, it means locating the business in an area that is convenient for staff members (with an easy commute or public transportation) that may consider working for the firm. In fact, some businesses have had great success in attracting top talent frustrated by a long commute, simply by locating their facilities outside of a downtown area and nearer to residential concentrations.

This same research found that the infrastructure provided by the office space or facility was critically important in the selection of the space. Firms were even willing to pay a premium to be certain that the facility would provide access to technology infrastructure such as network technology or research equipment. Many entrepreneurs have found that a well-managed science or technology park will offer a higher level of support for technology infrastructure, especially given the high concentration of technology companies located in a given park.

As with all things real estate, selecting a location truly comes down to location, location, location, but the importance of location may vary depending on the structure of the business. In an organization involved in technology development such as biotechnology or Web software application, customer foot traffic may not have the same importance as a retail establishment. Therefore, it is most important to select an office that can provide the proper space for your needs.

Science Parks

One option for many technology-based companies is selecting a science park, also referred to as a research park, technology park, technopolis, and biomedical park. These science parks are facilities and offices developed to attract high-tech-, science-, or research-related businesses. In some cases, these parks may be associated with a university or college. The primary difference a science park offers over a more traditional high-tech business district is that science parks would be more structured, organized, planned, and managed.

The research by Dettwiler, Lindelof, and Lofsten looked at the implications of facility locations for new technology-based firms. According to this research, technology companies located in science parks saw higher levels of growth of the companies, which the researchers felt was attributable to the quality of the facilities, the access to talent, and the ability to collaborate with other similar companies.

Leases

Once you've secured an office location, the next step will be securing the facility through a lease. A lease document may be more than 100 pages with numerous provisions that seem unnecessary for obtaining space for a few desks. The problem is that, for many people, when they find a space to rent that has the right mixture of the locations, price, and amenities, they're ready to just sign the lease and move on. However, these 100 pages of text can be very important down the road.

As long as the lease document has the agreed-upon price, square feet, and address, what more is there, right? Not so. A lease has the potential to be a source of a major headache if problems arise down the road or if situations change in your business. Here are a few key items to consider that may or may not be in your lease:

- Personal guarantees by any of the founders
- Future expansion or contraction of office space
- Other costs for which you are responsible (janitorial, renovations, etc.)
- Who will be responsible for preparing the space for your use
- Activities that require landlord consent (such as sale of additional stock or purchase of assets)
- Assigning or subleasing the space in the event of a merger, sale, or restructuring.

Particularly for a startup company, renting office space is a fairly substantial step, regardless of whether it is the company's first office or the fourth in five years. New office space serves as an important way to focus the efforts of the business, provides an added degree of professionalism, and offers a greater ability for collaboration and teamwork within the business.

Few people (other than a couple of real estate attorneys I know) enjoy reading a lease. Leases are oftentimes lengthy documents with poorly drafted legalese and contain provisions upon provisions buried within a few more addendums and attachments. Oftentimes, it isn't what is in the lease, but what isn't in the lease that is really important.

FAILING TO PLAN FOR OFFICE (SPACE) EXPANSION

What could happen

You know you are going to expand, but how fast and how quickly? Should you start out with a small space and move once you've outgrown it, or should you lease space large enough to support your team if all goes according to plans in two years?

Watch out for

Multiple moves of offices probably won't be the death knell of your new company, but it does represent a significant distraction for a growing company. If substantial growth is part of your plans over the first two years, discuss possible ways to expand your initial space with your initial landlords. Oftentimes, you can find ways to lease or sublease extra space rather than require a complete move.

TIP: Discuss with a potential landlord expansion opportunities before signing an initial lease.

Because office space represents an important item for a startup business, we should really pay more attention to the lease and the terms that it subjects us to. What will happen if the space won't work for the company and you need to break the lease? What happens if you need to expand the space you need? What happens if you rent too much space and are stuck with a lease payment that is going to drive your business out of business?

So your lease does matter and you should be certain that the lease agreement provides appropriately for your company over the term of the lease, not just on the day you sign your lease.

OFTEN OVERLOOKED IN NEGOTIATING A LEASE

What are a few of the terms that provisions that oftentimes get buried within a lease and overlooked?

- **Acts included in force majeure:** In an area prone to a particular event of Mother Nature, such as earthquakes in California, hurricanes in Florida, or flooding in low-lying areas, be certain to carefully review these clauses so such acts that are likely to occur given the location are not included to excuse the landlord.
- **Terminating the lease for convenience:** Will the company be reimbursed for improvements to the property if it is terminated?
- **Indemnification of the other party:** Can you have a cap on the damages to indemnify or can you limit the indemnification to the acts of the landlord and not third parties?

- **Adding the landlord as an additional insured party:** Does the landlord need to be an additional insured under the commercial general liability policy or can you obtain a separate policy in which the landlord would be a named insured? Can you make the company insurance secondary to the landlord's coverage?
- **Responsibility for consequential and liquidated damages:** Can you strike liquidated damages or limit the timeframe? Can you have a mutual waiver of consequential damages?

Get out a red pen, sit down with the list below, and answer each of the questions as you read through the document. If you get through and still have questions unanswered, check with the landlord or your attorney.

For high-tech companies that are likely to grow quickly, raise future financing rounds, and may have an exit event in the short term, it is usually a best practice to have you lease reviewed by your attorney. To minimize the costs of such a review, some entrepreneurs will undertake initial review and negotiations but will provide the lease to the attorney for a review to identify any other issues or items that were missed.

A LEASE CHECKLIST

As you read through any lease document, be certain that each of the questions below are answered in the agreement. If they are not addressed or you are not happy with the provision, you should discuss with the landlord.

1. How much space will the business need, and where should the business be located? Is there a precise plan (written and drawn) of the space included in the lease documents?
2. Are there any restrictions on how the premises of the building may be used? If so, you should ensure that the uses are broad enough to include current, planned, and incidental uses of the space by your business are covered, as well as uses for a potential sublessor that may have not yet be contemplated. Do the uses comply with local zoning laws?
3. How long will the lease term be, and are there any options to renew?
4. How much will rent be, and are any increases scheduled? Must the tenant pay a security deposit?
5. Will the tenant be responsible for any other costs, such as the maintenance of common areas or operating costs? Do the operating costs include costs that would be more appropriate for the landlord to bear, such as costs for renovations, legal fees incurred to enforce lease obligations, the cost of structural repairs, and the cost of removing asbestos or other hazardous materials? Will the landlord cap the fee increases for any of the costs to be borne by the tenant?

6. Does the tenant have the right to inspect the landlord's books and records? Does the tenant have the right to receive a detailed list of expenses in the event such expenses increase?
7. Who is responsible for improvements to the premises, and may the tenant make such improvements if necessary or desired? For initial improvements, will the landlord be responsible for performing certain renovations to prepare the space for your use? If the lease is terminated before the end of the term and improvements were made, will you be reimbursed?
8. Who is responsible for repair and replacement costs, and will the tenant receive a reduction in rent if the landlord fails to provide them in a timely manner? If repairs, alterations, or improvements are required by law, is the landlord solely responsible for such costs?
9. How much are utilities, and how are they apportioned among other tenants in the building?
10. Who must provide services for the office space, including janitorial services, washing windows, cleaning the sidewalks, etc.? If services are the responsibility of the landlord and are interrupted, does the tenant have any form of remedy (such as money damages, abatement of rent, or cancellation of the lease)?
11. May the tenant assign or sublet the space? If the company is acquired, can the lease be automatically assigned? Does the tenant have the right of first refusal or first offer on additional space in the building or to purchase the building in total?
12. What events will lead to a default on the lease? Does the lease provide for advance written notice of default for nonpayment of rent? Does the lease provide for at least five days for the tenant to remedy any breach for nonpayment and at least one month for nonmonetary defaults?
13. Are there assigned parking spots? Are a particular number of parking spaces included in the lease? Are there additional storage areas in the facility? Are those spaces currently allotted to other tenants? Are any storage spaces available and at what cost?
14. Can the lease be entered into by the corporation or LLC? Does the lease include nonrecourse language? Avoid, when possible, from entering into a lease in the name of an individual founder or founders, or when a personal guarantee is required.
15. Is the landlord obligated to insure the building for full replacement cost? Has the landlord agreed to waive claims against the tenant for any claims covered by insurance?
16. Are there any terms or provisions included in the lease that should probably be covered elsewhere (confidentiality, nondisclosure, etc.)?
17. What is included in force majeure? Does the lease require the landlord to be an additional named insured? What does the lease say about consequential and liquidated damages? What does the lease say about indemnification?

Consider taking the first step at reviewing the lease and getting a better understanding. Once you've settled on all the key terms and have given the document a fairly detailed review, ask your attorney to review the lease for any items that may trip you up. In particular, a startup attorney will look at items that could trip up a future financing event or cause a delay in a potential sale or merger of the company.

Bank Accounts

Obtaining a bank account for your business is one of the first steps you'll take for your new business. To open an account, most banks will require that you provide certain information, such as the date of formation, the type of business, and names and addresses of the business owners. In some cases, a bank may require formation documentation or a resolution of the board of directors for a corporation or the managers of an LLC authorizing setting up a bank account. In many cases, you may find it advantageous to call the bank before trying to set up your account, to avoid having to make multiple trips if you don't bring certain paperwork.

WHICH BANK?

Does it matter which bank you select? Yes and no.

Yes, it does matter because banks have various fees, programs, services, and features that will affect your banking experience. Much like selecting a personal bank, find a bank that will give you the lowest fees for the maximum level of service. So shop around and let a bank know what its competition is offering because, oftentimes, they will match fees, rates, or services.

But no, don't worry that by going with one bank over another that investors may stay away from you, customers won't buy your products, or potential employees won't sign on to join you. Pick the bank that has a good reputation and ask other entrepreneurs or service providers who they recommend. Sometimes your bank will also provide services such as a line of credit or payroll services.

Remember, don't wait too long before opening a business account. First, it will get you in the habit of properly tracking business-related expenses. One of the things potential investors look at is how the business handles its finances. Also, separating these expenses is important for tax purposes. So there is a risk of commingling of business and personal expenses.

Websites

Setting up a website has become standard fare for nearly any new business. For some businesses, the website is the business. For others, the website is a way to communicate more about your product or company.

BASICS OF SETTING UP YOUR WEBSITE

There are three primary steps involved in setting up a new website:

- Selecting and registering a domain name
- Finding a hosting company
- Designing your website

Although you may decide to undertake each of these steps separately, some online services and websites will offer a complete service to assist with registering a domain name, designing your site, and hosting the site.

Domain Names

Selecting a domain name raises both business and certain intellectual property issues. On one hand, you want to select a domain name that will generate hits to your website (particularly if you are selling your product or service online). On the other hand, you don't want to select a name that infringes on another company's trademark or brand name.

> **FINDING A DOMAIN NAME**
>
> Selecting a domain name and extension is a difficult process, particularly because many names are already taken, especially shorter names and any names with the .com extension. Visit http://www.nameboy.com, which provides a free service that will search for and identify not-yet-taken domain names based on one or two words you input that describe your business or your site.

A general rule of thumb for a startup company looking to select a domain name for their company is to select the name YourBrandName.com. The next best option is YourBrandName.net. In the event neither of these alternatives is available, you may consider adding additional language to the brand name such as "home" or "web" or consider abbreviation of some portion of your name.

Because a domain name on the Internet is a unique address designating an Internet site, it may be a trademark or service mark and have legal protection. However, the Patent and Trademark Office has stated that, if a domain name is used solely as an Internet address and not to identify a source of goods and services, it is not protectable as a trademark. A business may be wise to include its domain name in advertising its goods or services to obtain trademark protection. Additional information on intellectual property issues related to domain names can be found in Chapter 13.

> **YOUR DOMAIN NAME AND TRADEMARK LAW**
>
> **What could happen**
>
> Although this was covered in Chapter 13, here it is again to remind you of potential issues when selecting your domain name. You've found an available domain name for your website and are prepared to launch the site to begin drawing customers to the site. Because the domain name wasn't taken, that means we are free to operate it, right? Maybe not.
>
> **What to expect**
>
> Don't assume that just because the domain name is available that you are free to use the domain name. Domain names and the use of metadata on your website have been

the source of serious headaches for today's technology companies. Companies have been found to be liable for infringement if they use a domain name that is close to or similar to an existing trademark. If your site sells services or products that would confuse or mislead a consumer with another trademarked product, your company may be held to be liable.

Before using a domain name for your website, be sure to conduct a trademark search (or work with an attorney to have this done), and remember to review both state and federal databases.

TIP: Before you use any available domain name, be sure to check for potential trademark infringements.

Once you have selected a name, you will need to register that domain name on the Internet. Numerous services provide domain name registration, and you can find numerous options by running a simple query through a Web search engine for "domain name registration." Before choosing a provider, ensure they offer and support the extension you want (some registrars do not offer the .cc extension) and compare prices among the various service providers because prices can vary a great degree. The registration will often offer you the ability to register the name for a single year or over a multiyear period.

Hosting Your Website

After selecting and purchasing a domain name, your business will need to select a hosting company to host the files for your website. The domain name will be set to "point" to the hosting company so that visitors to your site will be directed to view the files on the host's servers. As with selecting a domain register, there are numerous hosting companies, and you can find a number of options simply by a query of "Web hosting." Information from the hosting company and domain name registrar will need to be matched to ensure that the website is configured correctly.

The article "Choosing the Right Web Host for Your Site" (located at http://www.hostsearch.com/article21.asp) can provide you important information to match your hosting service provider to the type of website and level of activity that will be generated by your website.

Designing a Website

For entrepreneurs planning to operate a Web-based business, this section will be unnecessary. However, in the event that a business wants to configure a simple website, the entrepreneur will need to generate the website.

The language that informs a Web browser how to display a page is known as HTML, or hypertext markup language. To create or edit a website, you will need to edit the HTML that will tell the browser to display text, graphics, or run various other applications in other software such as Java applets. Editing the HTML language can be done through an HTML editor software or an online site builder/HTML generator.

An HTML editor will provide a designer greater functionality but may require a basic understanding of HTML for some of the functions available. Numerous editor software packages are available, including Microsoft FrontPage, HotDog Professional, and Macromedia Dreamweaver. Additionally, you may find a variety of shareware and freeware HTML editors available for download.

For parties that are less comfortable with HTML editors or prefer a simpler system to design a very basic website, you may instead use an online site builder or an HTML generator. Online site builders do not require the website designer to know anything about HTML. Instead, these Web tools will use the information you provide through a series of questions to build the website. This site will likely have less functionality than a site built using an HTML editor, but you should be able to modify the appearance of the site, write text, add graphics, and include basic application scripts (for example, a script can be included to allow a visitor to search all the information on your site).

FREE ONLINE SITE BUILDERS

You can find numerous free online site builders including the following:

- http://www.uweb.ucsb.edu/generator/basic.php
- http://www.3zweb.com/free/index2.htm
- http://www.searchbliss.com/free_web_tools_html.htm
- http://www.applelinks.com/tools/webpage.shtml

For more advanced users and for future designs, companies should consider various new technologies that can be integrated into websites allowing greater interaction with users. These tools, ranging from the ability to tag information, to survey tools, to comment boxes, to blogs, are often referred to collectively as Web 2.0.

EMBRACING THE WEB

What could happen

Your company isn't a Web or Internet company, so why should you worry about Web 2.0, Internet marketing, blogs, or text messaging?

Watch out for

If your company isn't thinking about the Web and advanced communications strategies, your competitor is. In Gartner's New 2006 Tech Report, it listed Web 2.0 as the key technology theme for business in the next 10 years. Investors and partners are looking at how all companies (not just Web companies) will use the Internet and the Web to distinguish their products.

TIP: Web 2.0 isn't just for Web companies.

Insurance for the Business

In most cases, new corporations will need to obtain insurance coverage for all anticipated contingencies to protect the individual participants from personal liability as well as to protect the assets and future retained earnings of the business. The coverages that may be appropriate include the following:

- Directors' and officers' liability insurance
- Key-man insurance on the lives of important employees with benefits payable to the company
- Business interruption insurance to pay for expenses and loss of earnings while the business is interrupted as a result of natural disasters or other designated events
- Liability insurance for death or bodily injury of employees, customers, and others on the company's premises
- Insurance to cover damage to corporate property from fire, windstorms, explosions, and accidents

The key provisions to consider in analyzing an insurance package include (1) scope of losses covered, (2) periods covered, (3) property covered, (4) persons covered, (5) deductible amounts, (6) maximum dollar amount of coverage, (7) time period of coverage, and (8) enumerated exclusions. The company should seek the services of a competent insurance agent or broker to determine the policies that are appropriate given the company's operations and financial situation. In many cases, other entrepreneurs, board members, your accountant, or your attorney can recommend an individual to assist in these efforts.

INSURANCE ISN'T A LUXURY ITEM

What could happen

You are a new company and money is tight. One area where you can cut costs (just until you get the business rolling, raise additional money, and establish a smooth running enterprise) is insurance.

Watch out for

Some potential customers, potential investors, or potential business partners will insist on your business having certain levels of insurance before entering into a contract with you. Although the costs may seem high, you should look at these as a cost of doing business. Remember, a catastrophic loss can derail your business, permanently. There are ways to protect your business in a cost-effective manner. Discuss those options with an insurance provider.

TIP: Certain insurance coverage is necessary for doing business.

Health Insurance

Providing health benefits for employees is a substantial challenge for startup companies. According to a survey by the National Association for the Self-Employed, nearly all respondents agreed it was important or even necessary to offer a health insurance benefit to find and hire qualified people. However, the options available for smaller-sized companies to provide such a benefit are extremely limited.

In addition, health insurance plans continue to increase in price annually. In 2006, premiums for small group plans (plans with less than 50 people) averaged $3,732 for a single coverage and $9,768 for family coverage, according to the research of America's Health Insurance Plans. Although many startups or small businesses will only cover a portion of the costs, these could still be in excess of $3,000 for single coverage and $8,000 for family coverage.

For startup companies, the costs of providing health insurance in the early days of the organization are particularly challenging. As a result, many entrepreneurs are forced to postpone offering any health insurance programs for employees until the company is able to produce operating capital or receives outside financing.

ACCORDING TO ENTREPRENEURS

Healthcare

Do you provide healthcare to your employees?
No — Approximately 75%
Yes — Approximately 25%

If you offer healthcare, have you considered curtailing coverage for yourself or your employees due to costs?

No — Approximately 33%
Yes — Approximately 66%

NOTE: This includes all small business owners, not only startup companies discussed throughout this book.

Source: Discover Small Business Watch.

What are some practical steps you can take to address the challenges of providing health insurance for a startup company?

- Participate in your spouse's health insurance plan
- Have your insurance broker see whether health insurance may be available through a traditional provider
- Join with other self-employed individuals, an industry, or trade association or similar group that provides group rates on healthcare to members
- Combine a high-deductible health plan with a health savings account

ADDITIONAL INFORMATION ON HEALTH INSURANCE FOR STARTUPS

Visit http://www.startupnation.com/articles/1392/1/deciding-health-insurance.asp for a series of articles by Christopher Cameron of Main Street Insurance on dealing health insurance issues for startups and small businesses.

18

Contracts and Product Warranties

A verbal contract isn't worth the paper it's written on.

Samuel Goldwyn
Academy- and Golden Globe Award-winning producer

What to Watch For

For any startup business, you'll find that there are countless contracts you'll need to deal with. There are contracts for employees, for office supplies, for Internet and telephone access, for server space, for travel, and for countless other needs of a new business. In every contract, you'll find pages of legalese and language that may seem harmless but could cause problems farther down the road.

So why bother reading this chapter? Well, for one thing, you should have some familiarity with contracts. You'll run across numerous contracts in the process of creating and growing a business. For some key contracts, you'll have your attorney intimately involved in the drafting process. For others, you'll be responsible for managing the contract and ensuring it meets certain key business and legal standards you have adopted for the business. For still others, you may have a standard form contract you've prepared, perhaps for customers or for employees, and you'll need to ensure that any modifications won't harm your interests.

Most startup and small businesses are unable to afford to have their attorney read every contract, warranty, or agreement, so you should be certain you are aware of some key aspects of contracting, including key terms, provisions that may be included in contracts, and when to involve your attorney in the negotiation process. Remember, when in doubt, involve your attorney—they can save you from substantial issues with a quick review.

Basics of Contracts

In the United States, contract law may be governed by three different sources: common law, statutory law (such as the Uniform Commercial Code [UCC]), and international commercial law (such as the Convention on Contracts for the International Sale of Goods).

Common Law

Many of the laws governing contracts in the United States do not exist in statutes but in the "common law" developed by judges and the court system. When judges decide cases, their rulings on particular issues form a body of precedent that will determine the outcome for similar cases in the future. Whether this body of precedent is "binding" on a court depends on which court issued the decision. Decisions from courts in other jurisdictions or lower courts in the same jurisdiction create "persuasive" precedent. Thus, a judge may look to the precedent for guidance but is free to disregard it. Decisions that come from a higher court in the same jurisdiction are binding and the judge must apply the same rule. Nevertheless, courts have some leeway even in these circumstances because they can distinguish their case from the precedent to reach a new conclusion. In very rare circumstances, a court may decide to overrule earlier precedent. To do so, the court deciding the issue must be either the court that created the precedent or a higher court in the same jurisdiction. Courts typically overrule cases only when they determine that the previous decision was wrongly decided and resulted in significant injustice.

Statutory Law

Two of the most important bodies of statutory law governing contracts are the "statute of frauds" and the UCC. The statute of frauds was initially passed by the English Parliament in the seventeenth century but has subsequently been adopted in almost every American state. Under the statute of frauds, certain contracts must be in writing and signed by the parties for a court to enforce them. (The section on "Written Contracts" later in this chapter provides examples of contracts governed by the statute of frauds.)

The UCC is a body of uniform laws that govern commercial contracts. It has been adopted in one form or another by every state in the United States, which is important in a country in which commercial transactions frequently occur across state borders. Because similar rules will apply to a transaction whether it is conducted in Wyoming or New York, the UCC facilitates uniformity and predictability for businessmen. Additionally, courts attempt to interpret the UCC consistently across jurisdictions.

The UCC is broken down into several different articles that govern different commercial subjects. For example, Article 9 governs secured transactions, Article 3 addresses negotiable instruments, and Article 5 deals with letters of credit. This chapter focuses primarily on Article 2 of the UCC, which governs the sale of goods. Article 2 is notable because a number of provisions in this section of the UCC modify the common law rules as they would apply to contracts for the sale of goods. Some of these differences are illustrated in the chart below.

International Commercial Law

The Convention on Contracts for the International Sale of Goods (CISG) is a multilateral treaty governing international sales law. It generally applies to international sales contracts between parties that have their principal place of business in different contracting states. (Countries that have ratified the treaty are termed "contracting states.") As a treaty, the CISG is a source of federal law and therefore preempts state common law and the UCC on any issue to which it applies. Consequently, parties must expressly exclude the CISG from their contract if they do not want to be governed by it.

The CISG contains several provisions that differ from American common law and the UCC. Some of these differences are explored in the chart below. Businesses that engage in international sales with foreign entities should consult an attorney to determine the applicability of the CISG to their transactions.

Differences in Contract Law

When entering into commercial transactions, it is important to consider which body of law will govern the substance and the form of the contract. American common law, UCC Article 2, and the CISG differ in a number of ways that can have important consequences for contracting parties. A few of these differences are highlighted in the chart below.

	U.S. common law	UCC Article 2	CISG
Scope	Applies to all contracts between parties, including contracts for the provision of services and the sale of real estate.	Only applies to contracts for the sale of goods; thus, it will not apply to real estate transactions or the provision of services.	Applies to sales of goods between merchants in different countries that are a party to the CISG, unless the merchants expressly opt out.
Battle of the forms	An acceptance must be on the same terms as the offer (i.e., the "mirror image" of the offer)	An acceptance that contains additional or different terms is treated as a proposal for an addition to the contract; such terms may become part of the contract if they are not material and the original offeror does not object.	An offer that purports to be an acceptance but actually modifies the original offer is a rejection and counteroffer.

(continued)

(continued)

	U.S. common law	UCC Article 2	CISG
Warranties	Express warranties may become part of the contract.	Three types of warranties: (1) express warranties, (2) implied warranty of merchantability, and (3) implied warranty of fitness for a particular purpose.	Covers express warranties, implied warranty of merchantability, and implied warranty of fitness for a particular use.
Writing requirement	Oral contracts are enforceable. However, the statute of frauds requires certain contracts to be in writing and signed (i.e., transfers of real estate, contracts that cannot be performed within one year, marriage contracts, etc.).	Sales of goods worth $500 or more must be in writing.	Sales contracts need not be reduced to writing to be enforceable.

Written Contracts

Although contracts may be formed through oral agreement or the performance of a specific act, the most advisable method for forming a contract is to write it down and to have it signed by all the parties involved. Adhering to the formal process of writing down the agreement serves two important functions:

1. It encourages parties to deliberate over the terms of their contract and to negotiate over important terms and issues.
2. It provides evidence of the existence of a contract and its terms if a dispute arises.

Additionally, some types of contracts are governed by a state's "statute of frauds" and must be written down and signed to be enforceable. Although the rules vary by jurisdiction, examples of the types of contracts that must be evidenced in writing are as follows: sales of interests in land, agreements to pay another person's debt, contracts that take longer than one year to complete, contracts for the sale of goods for $500 or more, and contracts for the sale of securities. For lists of the types of contracts that must be evidenced in writing, contracting parties should review their state statute of frauds and the UCC.

Written contracts should be drafted in clear, simple, and specific language. Sentences should be kept short to avoid ambiguity. The grammatical structure, tone, and usage of words in the document should be consistent. If a contract is lengthy, the contract should be broken down into subsections with headings so that parties can refer to important clauses quickly and easily. The document should be signed by all of the parties.

WHEN TO INVOLVE YOUR LAWYER

Here are a few key times when you should look to involve your attorney:

- **License agreements:** License agreements oftentimes involve complicated scenarios over ownership. Be careful in cases in which you are licensing technology from a third party or are licensing your technology to a third party.

- **Nonstandard employment or consultant agreements:** Many startups will work with their attorneys to create standard form agreements for employees and consultants. In the event the employee or consultant has proposed changes to legal terms (ownership of intellectual property, nondisclosure, nonsolicitation, or noncompetition, just to name a few), you may want to involve your attorney.
- **Investment agreements:** Compliance with securities law exemptions is very important for any privately held business. You should involve your attorney when you negotiate any nonstandard investment contracts.
- **Joint development agreements:** Be careful of intellectual property issues that may arise when you negotiate a contract that involves two parties making contributions.
- **A "material contract":** Any time you are entering into a contract that you would consider to be material to your business, you should get your attorney to review it. A material contract is any contract that is not entered into in the ordinary course of business. If it isn't a contract you'll enter into with regularity in your business or involves a major amount of expense or revenues in your business, you should consider outside review as well.
- **When the other party involves their attorney:** In the event the other party has elected to have their attorney participate in the negotiations and the contract preparation, you may consider involving your attorney.

When in doubt, you should check with your attorney on provisions that raise issues. You may find that your attorney can quickly review select provisions of a contract even if they do not review the entire contract or may be able to provide a final review before signing to flag any potential issues.

Forming a Contract under Common Law

At a very basic level, a contract is a legally enforceable agreement between two or more persons to do something in the future. Although exceptions exist, contracts generally consist of the following three elements.

Offer

An offer is an expression by one party of his intent to enter into a contract and to be bound by the terms proposed. The offer must be definite and certain. Unless stated otherwise, an offer remains open for a "reasonable" amount of time. What amount of time is "reasonable" under the circumstances, however, is subject to interpretation and may vary depending on the nature of the business and the circumstances in which the offer was made. To avoid confusion, an offer should include an expiration date. The person making the offer (the offeror) has the power to revoke his offer, but the revocation must occur before the other party has accepted. Additionally, a party can give up his power to revoke an offer by agreeing to leave the offer open for a specified amount of time. This type of offer is called an "option" and usually is made in exchange for some type of payment.

Acceptance

The person to whom the offer is addressed (often referred to as the offeree) has the "power of acceptance." If the person chooses to accept the offer, his acceptance must be a clear expression of his agreement to the terms of the offer. If the person chooses to propose new or different terms, that person's proposal becomes a "counteroffer" and the initial offeror has the power to accept or reject the new counteroffer. A contract may not come into existence at all if the person with the power of acceptance chooses to reject the offer or delays too long in expressing his acceptance.

Consideration

Consideration is the legal term for the bargained-for exchange between the parties, in which each party gains some benefit from the transaction and suffers some detriment or obligation in exchange for that benefit.

If all of these elements are present, a binding contract is formed between the parties and may be enforced by the courts. For an illustration of how these common law rules may differ from contracts governed by the UCC or CISG, see the table above.

Typical Contract Provisions and Clauses

Contracts will vary in form and content depending on the substance of the transaction between the parties. The following is provided as a general description of some common provisions and clauses in business contracts; parties should consult general legal forms for additional ideas and consult with an attorney on technical issues.

- **Date:** A contract should set forth the date on which it will be effective. The agreement may be effective as of the date it is signed or may specify a particular date in the future when it will enter into force.
- **Identity of the parties:** The beginning of the contract should list the names of the parties and should state whether they are individuals or business entities. If a business is a party to the contract, the contract should specify the name and type of the business (partnership, corporation, limited liability company, etc.). Also, the official title of the person signing on behalf of the business should be included in the document, and the parties should make sure that the signer has authority to bind the business to the contract.
- **Purposes of the contract:** Setting forth the purpose of a contract in the preamble to a contract may help to clarify the contract if the meaning and intent of certain clauses is later disputed. In addition, the preamble will serve as strong evidence of the parties' intent to enter into, and be bound by, a contract.
- **Rights and duties of the parties:** The contract should specify the rights of the parties as well as their duties and obligations. Some examples of a party's rights may include the right to access certain information, the right to vote on certain matters, or the right to receive a specified quantity of a good at a particular place. A party's duties may include the obligation to pay on a certain date or to deliver a quantity of goods to a particular location.

- **Representations and warranties:** Representations and warranties are statements of fact, conditions, or circumstances that are material to the contract and express particular assumptions on which the contract rests. They are generally included in contracts as inducements to encourage another party to enter into the contract. If they later turn out to be inaccurate, misleading, or false, the party who relied on the representation or warranty may have a claim for misrepresentation. Parties who want to limit their liability for misrepresentation should include a statement that no party to the contract has relied on any representation or warranty besides those expressly contained in the contract.
- **Conditions:** Obligations contained in a contract may be modified, extinguished, or made contingent on the occurrence of certain events (called "conditions"). Parties may use simple and straightforward language to include conditions in contracts, such as "if," "only if," "unless," or "until."
- **Disclaimers and exclusions:** One party may want to disavow responsibility for particular events or occurrences (a "disclaimer") or may want to remove obligations from the contract that could otherwise be implied (an "exclusion").
- **Arbitration clauses, governing law, and venue for bringing suit:** Parties should decide what laws will govern their contract (for example, the laws of the State of California), whether issues will be submitted to final and binding arbitration, and which court will hear any disputes or litigation arising out of the contract. If you have a choice, choose governing law and venue that is convenient to you. If the other party is arguing with you over the law and venue, choose an alternative venue and governing law that does not favor either party.
- **Attorney's fees:** If the contract does not contain a provision for awarding attorney's fees in the event a dispute arises, then each party is responsible for their own costs. However, contracts may vary this rule by specifying that attorney's fees will be awarded to the prevailing party.
- **Merger clause:** Because the terms contained in a contract may be worked out through a series of negotiations, a "merger clause" provides definiteness and finality to the process by stating that the contract constitutes the parties' entire agreement and that no other representations, warranties, or agreements are made. It limits the possibility that one party will assert that other promises were exchanged and not included in the agreement.
- **Signatures:** The agreement should be signed by all the parties to the contract. If individuals are signing on behalf of an LLC, partnership, or corporation, the signature lines in the contract should include the name of the company or organization, the name of the person signing the document, and the person's title.

GOTCHA! CONTRACT PROVISIONS THAT YOU SHOULD PAY ATTENTION TO

There are a few provisions that may trip up a company that doesn't read the language closely or fails to understand the impact of certain provisions. Here are a few provisions to pay close attention to:

- **Ownership of intellectual property:** In the event you are looking to enter into a contract with a consultant, an employee, or any other third party, you should be certain to consider how any newly created intellectual property will be owned. For many companies, they will want to be certain that they retain ownership rights to the intellectual property. However, for more complex joint development projects, there may be cases in which you will assign all or some portion of rights to the intellectual property to another party.

- **Assignment provisions:** Assignment provisions will require a company to get permission from the other company in the event that they want to assign or transfer the contract to another party. This is particularly important in the event a company is acquired or sold because it may require the company to obtain consents from all third parties before the acquisition can be completed (which is sometimes difficult).
- **Attorneys' fees:** Be aware of any provision that requires either party to be responsible for attorneys' fees associated with negotiating or finalizing the agreement.
- **Use of trademarks or customer names:** In the event you want to be able to publicize the names of your customers, use customer logos on your website, or announce an important contract or agreement, be certain there are no restrictions on use.
- **Termination and renewal provisions:** Be aware of any provisions that allow the parties to terminate the contract early, particularly provisions that allow for termination in the event of even a minor breach without the opportunity to rectify the breach. In addition, be aware of any provisions that allow for the automatic renewal of the contract.
- **Compliance with reporting provisions:** Some contracts will include language obligating the other party to provide certain ongoing reports or information to various parties such as federal or state agencies.
- **Obligations on employees:** Contracts will sometimes require all employees that interact with the other party to be screened or drug tested. Ensure that complying with these requirements will not breach any contracts you have with your employees.
- **Access to premises:** Be aware that contracts may permit the other party to enter your facilities.
- **Payment terms:** Look for the terms of payments due as well as whether any payments due will be accelerated in the event that either party fails to perform certain activities.

What Happens if a Contract Is Breached?

A party may allege breach of contract when the other person fails to perform a material condition of the contract in a timely or satisfactory manner and does not have a proper excuse for nonperformance. To seek remedies for breach of contract, the plaintiff will have to establish that (1) a contract was formed between the parties, (2) the plaintiff performed his obligations under the contract or was excused from such performance by the other party's breach, (3) the defendant failed to perform a material element of the contract in a timely manner, and (4) the plaintiff was injured by such failure.

If the plaintiff wins his claim for breach of contract, he is ordinarily entitled only to monetary compensation for his damages. The amount of damages is generally limited to the amount necessary to put the plaintiff in the position in which he would have been had the contract been fully performed, plus compensation for any foreseeable losses that were caused by the breach. Very seldom would a court force the other party to perform what they had failed to perform, so don't expect the breaching party to be forced to build you the widgets or perform the consulting they agreed to do.

Electronic Contracts and E-Commerce

With the rise of computers and the Internet, people are increasingly using electronic communications to negotiate and enter into contracts. New state laws and federal legislation recognize that electronic records and signatures may be given legal effect and be enforceable. This section addresses two of these new laws: the Uniform Electronic Transactions Act (UETA) and the Federal Electronic Signatures in Global and National Commerce Act (ESIGN Act).

UETA

UETA is a model law that state legislatures may choose to adopt at their discretion. As of 2004, 46 states had enacted UETA. For UETA to apply, the parties must agree to conduct their business with each other electronically. Such consent may be determined from the context and circumstances of the parties' conduct. Once the parties have consented to do business electronically, UETA sets forth several rules applicable to the parties' transaction: (1) the parties' records and signatures will not be denied legal effect or enforceability simply because they are in electronic form; (2) a contract between the parties will not be denied legal effect or enforceability simply because electronic records were used to form the contract; and (3) electronic records and signatures will satisfy laws requiring records to be in writing and signed (such as state statutes of fraud). For purposes of UETA, almost any electronic process or mark may constitute a "signature" that has legal effect. The only key requirements are that the sound, symbol, or process be associated with an electronic record and that such elements be undertaken with an intent to sign the electronic record. Thus, typing one's name at the bottom of an e-mail, signing a fax, clicking an "I agree" button on a webpage, or pushing a number on one's phone to accept certain terms may constitute signatures under UETA.

ESIGN Act

The ESIGN Act was adopted by Congress and applies to interstate and foreign commerce. Like UETA, the ESIGN Act provides that signatures, contracts, and other records pertaining to a transaction will not be denied validity simply because they are in an electronic form. The ESIGN Act also defines an electronic signature very broadly to include electronic sounds, symbols, or processes, as long as the signature can be attached to an electronic record and can be attributed to the person against whom enforcement of the transaction is sought. Finally, the ESIGN Act provides that contracts and other records will not be denied legal effect simply because an electronic agent was used in the process (such as a computer program or other automated equipment).

As a body of federal law, the ESIGN Act preempts all state laws that are inconsistent with its provisions. Although the ESIGN Act allows state laws to limit or modify its provisions to the extent that such state laws are consistent with ESIGN, it still is not clear what state laws will be determined "inconsistent."

Exclusions from UETA and the ESIGN Act

Certain types of documents are excluded from the rules of UETA and the ESIGN Act and therefore may not be enforceable if they are formed electronically. For example, both UETA and the ESIGN Act exclude the following documents from their scope:

- Wills and trusts
- Contracts relating to family law matters, such as divorces and adoptions
- Contracts governed by certain provisions of the UCC

The ESIGN Act also excludes certain documents that are not excluded from UETA:

- Court orders and official court documents
- Notices pertaining to mortgage or rental agreements on a person's primary residence, such as notices of default, foreclosure, or eviction
- Notices relating to the cancellation of life or health insurance benefits
- Documents required for transporting hazardous materials or notices announcing the recall of a product that poses health or safety risks

Although UETA and the ESIGN Act provide for the enforceability of electronic contracts, parties still need to remember that all of the other elements governing the formation of a contract must still be met. Thus, if parties are contracting in the United States and the common law governs their contract, they must still establish that a contract was formed through offer, acceptance, and the exchange of consideration.

Product Warranties

UCC: Article 2 Warranties

Article 2 of the UCC provides for three different types of warranties in transactions for the sale of goods: an express warranty, an implied warranty of merchantability, and an implied warranty of fitness for a particular purpose. Each of these warranties is discussed below, as well as possibilities for limiting one's liability under the UCC.

Express Warranty

An express warranty is an explicit statement by a seller or manufacturer about the particular quality or nature of a good. For example, a statement that "this car gets at least 26 miles per gallon" or that "this shirt is 100% silk" is an express warranty. To be an express warranty, two elements must be met: (1) the seller must provide an affirmation of fact, a promise, a description, a sample, or a model relating to a good; and (2) the buyer must rely on the statement, promise, sample, or model of the good as a basis for his bargain with the seller. Mere "puffery," or statements about the seller's opinion of a particular product, are not considered express warranties. However, the dividing line between express warranties and puffery is difficult to draw, especially because express warranties may be formed even if the seller does not use the words "I warrant" or "I guarantee."

Implied Warranty of Merchantability

For the implied warranty of merchantability to apply, the seller must be a merchant who regularly deals in goods of the kind the buyer purchased or who holds himself out as having particular knowledge of that type of good. If the seller is a merchant, then the implied warranty requires at least that (1) the goods pass without objection in the trade according to their contract description, (2) the goods be fit for the ordinary purposes for which they are used, (3) the goods be of even kind, quality, and quantity, (4) the goods be adequately packaged and labeled as required, and (5) the goods conform to any promises or affirmations of fact made on their containers or labels.

Implied Warranty of Fitness for a Particular Purpose

Unlike the implied warranty of merchantability, the implied warranty of fitness for a particular purpose is not limited to merchants. Instead, liability under this warranty is established when (1) the buyer desires goods fit for a particular purpose, (2) the seller uses his skills and expertise to select such goods, (3) the buyer relies on the seller's judgment, and (4) the seller has reason to know of the buyer's reliance. A seller will be liable for breach of this implied warranty if the goods do not meet the buyer's particular purpose, even if they are not defective.

How to Limit Liability

It may be difficult for a seller to avoid liability under the UCC's express warranty, because even simple descriptions of goods may constitute such a warranty. Although courts traditionally would not look beyond the terms of the written contract for evidence of an express warranty, some courts have found grounds to admit evidence of oral warranties that were not contained in the parties' final document. Furthermore, the UCC invalidates any disclaimers that are inconsistent with express warranties. Thus, if a manufacturer specifies that a computer will operate at a certain speed, the courts likely will not uphold the manufacturer's subsequent attempt to disclaim such a warranty by stating in the contract that it "makes no warranty, express or implied."

Despite its rigidity with respect to express warranties, the UCC does permit parties to disclaim the implied warranties of merchantability and fitness for a particular purpose. To disclaim the implied warranty of merchantability, the contract must specifically mention the word merchantability in the disclaimer. To avoid the implied warranty of fitness for a particular purpose, a seller may state that there are no warranties that extend beyond the description written in the contract. Other implied warranties may be disclaimed by indicating that the goods are sold "as is," thereby putting the buyer on notice that he assumes risk for the goods.

Whenever a seller wants to disclaim a warranty, the disclaimer should be obvious and noticeable. Techniques for drawing the buyer's attention to the disclaimer include using capital letters, contrasting the color of words, and placing the disclaimer on the front of an agreement where it is more likely to be read.

Magnuson–Moss Warranty Act

Businesses that choose to offer written warranties on consumer products that are sold in interstate or foreign commerce must comply with the Magnuson–Moss Warranty Act. This act requires the following: (1) the warrantor must designate the written warranty as "full" or "limited"; (2) the warrantor must specify the details of the warranty in a single, clear, and easy-to-read document; and (3) the warranty must be available when the product is sold so that the customer can read it before buying the product. In general, a "full" warranty is one that does not limit the duration of implied warranties, offers broad coverage to all purchasers of the product, and guarantees free warranty service. "Limited" warranties typically offer more restricted warranty coverage and are valid for a shorter period of time.

Under the Warranty Act, sellers are forbidden from disclaiming implied warranties. However, businesses that offer limited warranties may restrict the duration of the implied warranty to the term of the limited warranty. Additionally, sellers are not permitted to include any deceptive or misleading terms in their warranty.

If a seller or a manufacturer breaches a warranty, the Warranty Act makes such a breach a violation of federal law. Although the act makes it easier for consumers to bring breach of warranty claims, the purpose of the act is not to encourage litigation. Instead, the act's goals are to enable customers to compare warranties before they buy, to guarantee customers at least the minimum implied warranty of merchantability, and to protect consumers from misleading and deceptive practices. Furthermore, the act encourages warrantors to settle their warranty disputes with customers through informal dispute resolution systems (such as an arbitrator). If a company requires that customers submit a warranty claim to an informal dispute resolution mechanism before suing in court, the company must review and comply with the Federal Trade Commission's Rule on Informal Dispute Settlement Procedures.

19

Startup Accounting

There's no business like show business, but there are several businesses like accounting.

David Letterman

What to Watch For

It's easy for entrepreneurs and small business owners to be completely focused on everything but accounting when they are starting their business. As an entrepreneur, you're responsible for developing your product, finding the perfect location, meeting with potential clients, ironing out the production and distribution details, hiring staff, etc., etc., etc. If you didn't devote your time and energy to these things, you wouldn't have a business to account for, right? The to-do list when starting a business goes on and on, and accounting is inevitably very low on the list.

Perhaps, however, accounting is on the bottom of this list because small business owners don't understand the power of accounting. Yes, that's right, the *power* of accounting. The whole point of accounting is to provide you, as a business owner, with useful information for decision-making purposes about your business. Accounting helps you manage your cash flow so you know how much of your line of credit you need to use each month. Accounting allows you to identify trends in sales revenue and expenditures and identifies which products and services contribute the most to your bottom line. Accounting information tells you whether you can expand your operations to that second office, how much money you'll need to borrow to invest in that new product line you've been working on, and, perhaps most importantly, how much you can afford to pay yourself each month. Understanding accounting information gives you the power to understand what drives the profitability of your business and to make the best decisions for your company.

Language of Accounting

The first step to harnessing this power is to become familiar with accounting lingo so you can better understand your company's financial information. Here's a brief introduction to some of the most frequently used terms so you too can speak the "language of business."

Accounting refers to the process of identifying, recording, and summarizing your business' economic transactions (i.e., revenues and expenses) so the information can be used for decision-making purposes.

The term **bookkeeping** is commonly used to refer to the acts of identifying and recording all information relevant to your business' economic transactions. This information includes the date, amount, and source of all business revenues, expenses, assets, and liabilities. Bookkeeping is the beginning of the accounting process.

Relevant information relating to your business' economic transactions is recorded in a **ledger**. Back in "the old days" companies recorded their financial information in books preprinted with lined ledger paper. Hence the origin of referring to a company's "books" when discussing its financial information. Today companies generally house their ledgers electronically on company computers.

The term **operating cycle** refers to the flow of transactions necessary for your business to make a sale and collect the sale proceeds. The operating cycle begins when your company acquires goods from its suppliers. The next step involves your company selling its manufactured/purchased merchandise or services to customers. The operating cycle ends when you finally collect the cash associated with these sales.

A company can choose to use either accrual accounting or cash accounting when recording its economic transactions. **Accrual accounting** accounts for revenues when they are earned and expenses when they are incurred even if the revenues are not received and the expenses are not paid during the current operating cycle. At the heart of accrual accounting is the **matching principle**. Companies match expenses to revenues to ensure that the proper expenses are recorded in the same time period as the associated revenues are generated.

Cash accounting accounts for revenues when cash is received and expenses when cash is paid even if the revenues and expenses do not relate to the current operating cycle. Because cash accounting is fairly simple (you just record cash inflows and outflows), most of our discussion will focus on the less familiar accrual accounting.

Assets are the economic resources your company owns that will help generate its future cash flows.

EXAMPLES OF ASSETS

- **Cash**
- **Accounts receivable:** Amounts you expect to receive from customers related to sales made on credit.
- **Inventory:** Merchandise to be sold to your customers.
- **Prepaid assets:** A good or service you have paid cash for in advance that has not yet been consumed. If you prepay your 2008 rent at the end of 2007, this prepayment would be considered a prepaid asset in 2007.
- **Buildings, equipment, and land**
- **Intangible assets:** Patents, copyrights, etc. purchased by your company from a third party. Note that internally developed intangibles are not considered assets for accounting purposes because it is too difficult to determine the fair market value of intangibles while they are being internally developed.

Liabilities are the economic obligations of your company to third parties and claims against your assets by third parties.

EXAMPLES OF LIABILITIES

- **Accounts payable:** Amounts you owe your vendors and suppliers for purchases you made on credit.
- **Wages payable:** Amounts you owe your employees for services rendered.
- **Unearned revenue:** Cash received from customers for goods you have not yet provided or services you have not yet performed.
- **Loans payable:** Amounts you borrowed from and owe to your creditors.

Equity is an owner's residual interest in the company's assets once its liabilities have been satisfied. This means that equity is equal to assets less liabilities.

EXAMPLES OF EQUITY

- **Stock:** The total capital investment in your company by its owners.
- **Retained earnings:** Cumulative owner's equity from company profits retained within the company.

When a company makes a purchase it can capitalize or expense the purchase. **Capitalization** refers to recording an asset when something is purchased because you expect the purchase to provide future benefit to your company. When you purchase inventory from a third party, you capitalize the goods because the goods are expected to be sold to your customers, thereby providing your company with future economic benefit. When a purchase is capitalized, an asset account is increased. Capitalized purchases appear on a company's balance sheet (discussed below). In contrast, **expensing** refers to recording an expense when something is purchased because you don't expect the purchase to provide future benefits. When you purchase electricity from the electric company, you expense this purchase because the electricity purchased and consumed today provides no future economic benefit for your company. Expensed purchases appear on a company's income statement (discussed below).

Accounting relies on the **double-entry system** of recording transactions. This means that there are at least two parts to every accounting transaction. When you pay a $500 electric bill with cash, that transaction is recorded as a $500 expense (part one) and a $500 reduction in cash (part two). Because of the double-entry nature of accounting, a **journal entry** is used to reflect both parts of this one transaction. Journal entries have three pieces: (1) accounts, (2) amounts, and (3) directions. Accountants use debits and credits to indicate directions.

Debits and **credits** are the foundation of all accounting systems. Understand how debits and credits work and you'll understand pretty much all there is to know about basic accounting. The word debit refers to the left part of the journal entry, and the word credit refers to the right part of the journal entry. Here's the most difficult part of accounting for beginners, so pay attention:

- Debits *increase* assets and expenses and *decrease* liabilities, equity, and revenues.
- Credits *increase* liabilities, equity, and revenues and *decrease* assets and expenses.
- Debits *always* equal credits so every journal entry *always* balances.

SAMPLE JOURNAL ENTRY

If you can remember the three bullet points listed previously, you can account for almost any business transaction you encounter. To keep track of the $500 electric bill paid with cash, we recorded an expense and a reduction in cash. We just learned that debits increase expenses and credits decrease assets. Therefore, the journal entry associated with this transaction would be a debit to an expense account for $500 and a credit to an asset account (cash) for $500. Note that this last sentence talked about all three pieces of the journal entry: an expense <u>account</u> in the <u>amount</u> of $500 <u>increased</u> (debiting an expense account increases expenses). Similarly, a cash <u>account</u> in the <u>amount</u> of $500 <u>decreased</u> (crediting an asset account decreases assets). The journal entry for this transaction would be written like this:

Date	Debit Rent Expense	$500
	Credit Cash	$500

As discussed earlier, each journal entry must balance. See how the left part of the journal entry (the debit) is equal in amount to the right part of the journal entry (the credit)? This is what was meant by the statement that debits always equal credits so every journal entry always balances. If one journal entry doesn't balance, your balance sheet (discussed below) won't balance. No investor wants to see a balance sheet that doesn't balance. Additional examples of using journal entries to account for business transactions are provided at the end of this chapter.

So far we've been talking about how to identify and record economic transactions. Now let's talk about the part of the accounting process that deals with summarizing economic information so that it can be used for decision-making purposes.

Financial Statements

All accounting information is summarized into four financial statements:

- Income statement (or profit and loss statement)
- Balance sheet (or statement of financial condition)
- Cash flow statement
- Statement of equity

The term **operating cycle** was defined previously as the flow of transactions necessary for your business to make a sale to a customer and collect the proceeds from the sale. This section will discuss operating cycles in more detail because your operating cycle determines the dates over which your financial statements provide information. Because the exact number of days in a company's operating cycle is difficult to determine and is constantly changing, many businesses adopt the calendar year as their operating cycle. A company that has adopted a calendar year operating cycle generally assumes that its operating cycle begins on the first day of the calendar year (January 1st) and ends on the last day of the calendar year (December 31st). An income statement (defined below) generally includes revenues and expenses from the first day through the last day of a company's operating cycle. A balance sheet (also defined below) generally includes asset, liability, and equity information as of the last day of a company's operating cycle. Although it makes sense for most companies to use a calendar year as their operating cycle, a calendar-year operating cycle might not make sense for industries with the bulk of their operations occurring during the winter months. For example, a ski resort owner probably wants to know how much net profit he made during the 2007–2008 ski season, not how much net profit he made during the 2007 calendar year. A ski resort might adopt a July 1st to June 30th operating cycle so that its income statement reflects net profit from the 2007–2008 ski season.

An **income statement** summarizes your business' revenues and expenses during the current operating cycle. **Net income**, which is equal to total revenues less total expenses incurred during the operating cycle, is the bottom-line number presented on the income statement. This number summarizes your business' profit during the current period (generally the current operating cycle). The term **net profit** is often used interchangeably with the term net income.

A **balance sheet** summarizes your business' assets, liabilities, and equity at a specific point in time (generally the last day of the business' operating cycle).

A **cash flow statement** shows all cash inflows and outflows that occurred during the current operating cycle and illustrates the liquidity of your business' operations. Cash inflows include cash payments from customers and cash from external financing, and cash outflows include cash payments made to suppliers and debt repayments. There are three types of cash flows: operating, investing, and financing.

The **statement of equity** tracks your cumulative equity in your company. Net income increases equity, and dividends paid to shareholders decreases equity.

Basics of Bookkeeping

OK, enough with the glossary definitions. Let's talk about how to set up your accounting system so you can start using your accounting information for decision-making purposes. Accounting information can be stored in hardcopy format or electronically through the use of accounting software. You can keep your records in any way that makes sense to you as long as your records accurately reflect your business transactions. For thousands of years, before the advent of computers, businesses kept track of all their accounting information by hand through the use of ledger books. Although manually keeping track of your transactions in ledger books is completely acceptable, using electronic accounting software is much more efficient. Today's accounting software is very user-friendly, very reasonably priced, and will literally save you hundreds of hours of time when compared with recording and summarizing all of your transactions by hand. See the table below for more information about some of today's most popular accounting software.

Accounting Software Review's 2008 Software Report

Rank	Software	Price	Overall rating (max of 4)[1]	Ease of use (max of 4)[2]
1	PeachTree Complete	$265	4.0	4.0
2	MYOB Business Essentials Pro	$300	3.5	4.0
3	QuickBooks Pro	$200	3.0	4.0
4	NetSuite Small Business Accounting	$1,188[3]	3.0	2.5
5	Cougar Mountain	$1,499	3.0	3.0
6	Small Business Manager	$1,175	2.5	3.0
7	Simply Accounting Pro	$300	2.5	4.0
8	CYMA IV Accounting	$595[4]	2.5	3.0
9	DacEasy	$450	2.5	2.5
10	Microsoft Money	$80	2.0	2.5

Source: http://accounting-software-review.toptenreviews.com.
[1] Summary measure of 11 categories.
[2] One of eleven categories.
[3] Price per year.
[4] Price per module.

Keeping Track of Records

Make sure to keep good records associated with every sale, purchase, payment, and receipt your company is involved in. A good record includes the date, amount, and important details associated with the transaction (i.e., who you are paying or who is paying you, what the payment is for, etc.). Make sure to keep multiple records associated with a single transaction together. If we go back to our $500 utility bill example, the important records associated with this transaction include the bill you receive from the electric company and a copy of the check you write to the utility company. Staple these two records together if you need to.

In the very beginning of your business operations, it might seem like you don't really have a business to account for. Don't let the absence of revenues fool you into not keeping track of your startup expenses. These initial expenses are part of your operating costs. If these expenses are unaccounted for, your financial records won't reflect the true costs of your operations. Relying on incomplete financial information leads to poor business decisions. In addition, many of these expenses are deductible for tax purposes. Failing to keep track of these expenditures means that you won't remember to deduct them on your tax return, which results in you paying more income tax than you are legally obligated to pay. I'm sure you don't want to pay more tax than you have to, so keep track of all expenses you incur, even before you start generating revenues. More tax tips are provided in the following chapter.

TIPS FOR EFFECTIVELY HANDLING EXPENSE REPORTS

- Develop a policy on expense reports and publish it to employees.
- Use a standard expense form for ease of use and consistency of information.
- Require all expense reports to be completed and submitted in a timely fashion.
- Attach receipts for all out-of-pocket expenses. (Note that the IRS may deny certain deductions if copies of receipts are not available.)
- On the expense report, the date, time, and business relationship should be included for each expense.
- All expense reports should be signed and dated by the submitting employee and the individual approving the transaction.
- Track all expenses related to meal and entertainment in separate line items from travel costs (only a portion of meals and entertainment expenses are allowed to be deducted).
- Have an approval process in place and process expense reports in a manner that is consistent with processing of other approved invoices.
- Retain copies of all paid expense reports consistent with internal record retention policies (that should match applicable tax recordkeeping requirements).

Inputting Information into Your Accounting System

At the end of every day or every week, set aside time to enter all of your company's transactions by date and event into your accounting system. It's generally a bad idea to wait until the end of the month to record your transactions: thinking about entering an entire month's worth of information can be overwhelming. Doing a little bit each day or each week is more reasonable and realistic. The more transactions you have, the more often you should record your transactions into your accounting software. After you enter each individual transaction into your accounting software, the software will "post" the information to your **general ledger**. As mentioned previously, your general ledger keeps track of every business transaction you enter into your software.

Once all of your transactions have been recorded (and subsequently posted to your general ledger), you can use the information in your ledger to create financial reports with a few clicks of your mouse. These financial statements summarize your business' activities for whatever period of time you specify. This aggregated and summarized information is what you use to determine how profitable your company was this week (or month, quarter, year, etc.) and to make future business decisions. Financial reports allow you to see your company's operations at the aggregate level (am I making a profit?) and at a detailed level (how much did I spend on office supplies last month?). This detailed level will provide you with very specific information that you can use to improve your company's operations. For example, you need to be able to generate and understand your financial statements to be able to tell how fast your customers are paying you for purchases made on credit, how much money is being tied up in inventory sitting in your warehouse or on your store shelves, which individual product or service contributes the most to your bottom line, and whether you'll need to use $5,000 or $50,000 of your line of credit next month.

Internal Controls

If you don't want to deal with keeping track of all your business transactions yourself, you can hire someone to do your accounting work for you. Many startup companies find they need someone to help with their bookkeeping on a part-time basis. Keep in mind that, if you do decide to employ an internal bookkeeper, delegating some of the responsibilities of bookkeeping also means delegating some of the control you have over your business operations. You need to have adequate **internal controls** in place any time you delegate responsibilities to other people within your company. The phrase internal controls refers to a system of internal checks and balances that help ensure that all internal actions and decisions are in the company's best interest.

Adequate internal controls most often deal with the **segregation of duties**. Duties are properly segregated when no single individual has control over two or more phases of a transaction. This makes it more difficult for any one individual to do something that is not in the company's best interest (i.e., commit fraud). For example, if your bookkeeper is responsible for paying a company bill (one transaction), he shouldn't be able to determine the amount due from the billing statement (phase one of the transaction) and sign the company check authorizing the payment (phase two of the transaction). These two phases should be segregated to help ensure your bookkeeper doesn't write a check for an amount greater than is due and pocket the extra money when the excess payment is refunded to your company. Alternatively, your bookkeeper could simply write checks to himself if the duties of determining the amount of a liability and paying the liability were not segregated. One way to segregate bill-paying responsibilities is to have your bookkeeper compile all of the bills your business needs to pay and provide you (or another employee) with both the invoices and the populated checks (phase one) so that you can review his work and sign the checks (phase two). You should try to separate all duties with financial implications to the best of your ability.

OUTSOURCING YOUR BOOKKEEPING

If you don't want to have to worry about overseeing the bookkeeping work of your part-time employee, consider outsourcing your bookkeeping to a small accounting firm. Bookkeeping fees at smaller firms can be pretty reasonable, ranging from a few thousand dollars to tens of thousands of dollars per year depending on the level of service and frequency of reports you require. Search online or visit websites such as the following:

- http://www.buyerzone.com
- http://www.osibusinessservices.com
- http://www.sbsuite.com

Most of these websites will provide you with quotes from various bookkeepers in your area.

External Uses for Financial Statements

So far, we've reviewed the importance of financial statements for internal decision-making purposes. Financial statements are also used by external parties for decision-making purposes. Publicly traded companies in the United States are required to provide investors with audited annual financial statements. Although your company might be a few steps away from being publicly traded, your investors (be it lending institutions, angel investors, VCs, or your family members) will also want to receive financial statements showing your company's financial performance.

Although we can assume that any reader of this book is a very trustworthy and honest business owner who properly keeps track of all business transactions, there are many business owners out there that have proven to be less than trustworthy. Financial statement **audits**, or examinations of business' financial transactions and financial statements by a knowledgeable and independent third party, are performed so investors can be confident that companies' financial statements are free from material error. **Material** refers to something that might change an investor's decision about a company if the item was misstated in or omitted from a company's financial statements. For example, if a company forgets to record a $500,000 expense, its annual income statement will understate expenses by $500,000. This item is likely to be considered material for most small companies because its omission could change an investor's decision about the company. Alternatively, if a company forgets to record $100 of expense and its annual income statement understates expenses by $100, this item is considered **immaterial** because its omission is unlikely to change an investor's decision about the company. Financial statements omitting this $100 of expense would be free from material misstatement although the financial statements are not 100% correct. Audits provide investors with reasonable assurance that companies' financial statements are free from material misstatement. Any small business that receives investment money from external parties will likely have to provide audited financial statements to its investors.

As previously discussed, your audits are performed by knowledgeable and independent third parties, but who are these third parties, where you can find them, and how much is a financial statement audit going to cost your company? **Auditors** are the knowledgeable and independent third parties who perform financial statement audits. Some auditors run their own accounting practices, and some auditors are employed by accounting firms that range in size from two people to tens of thousands of people. Auditors can charge by the hour or by the job. For small companies, audits can range from a few thousand dollars to tens of thousands of dollars depending on who you employ as your auditor and the complexity of your company's financial transactions. Your first audit is generally more expensive than subsequent audits. This is because, during your first audit, your auditor must spend time understanding and documenting your unique business operations and internal controls to develop an appropriate audit plan. Because this audit plan is relied on during subsequent audits, the process requires less up-front work, which results in lower audit fees. Audit fees are generally positively correlated with accounting firm size: the larger the firm, the larger the audit fees.

You might be able to reduce your compliance costs by determining whether your investors will accept reviewed financial statements in place of audited financial statements. Whereas **reviews** do not involve auditor data verification and therefore provide less assurance than audits, financial statement reviews are generally less expensive and less time consuming.

Because auditors must be independent of their clients in fact and appearance, the audit firm you employ to do your bookkeeping is not permitted to also perform your company's financial statement audit. (This is because the auditors would be auditing the transactions they had themselves recorded, which doesn't make much sense.)

Paying Uncle Sam

So far, we've learned the basics of accounting lingo, how to record transactions, how to use individual transaction data to create financial statements, and how to use these financial statements for internal and external financial decision-making purposes. We haven't yet discussed another important reason for keeping orderly financial records: for tax return purposes.

Your tax liability is significantly impacted by your choice of business structure. As discussed in Chapter 20, you can choose to structure your business as a sole proprietorship, a general partnership, a limited partnership, an LLC, a S-corporation, or a regular corporation (referred to in the tax world as a C-corporation). Chapter 7 covers the tax advantages and disadvantages associated with each business structure.

Taxes are discussed extensively in Chapter 20. This section only briefly discusses taxes because you need good accounting records to prepare (or have someone else prepare) your business' tax returns. Businesses are subject to a variety of taxes: federal income tax, state income tax, payroll tax, sales and use tax, property tax, etc. You want to minimize all of your company's tax liabilities to the lowest legal amounts possible. This is only attainable with solid accounting records. Keeping organized accounting information is also very helpful in case you are audited by the IRS or your state taxing authorities. When small business owners are audited by the IRS, the businesses with proper accounting records saved themselves hundreds of hours of time compared with the businesses with poor accounting records. Imagine sorting through thousands of poorly organized documents in search of anything supporting an expense your business incurred three years ago. Spending your time tracking down receipts and invoices takes you away from what you do best: running your business. Don't let poor accounting records take you away from the very activities that make your business so successful.

MORE INFORMATION

For more information on accounting issues and examples of accounting for startup transactions, visit the book's website at http://www.myhightechstartup.com.

EXAMPLES OF ACCOUNTING FOR BUSINESS TRANSACTIONS USING JOURNAL ENTRIES

The following examples will provide you the basics of some common transactions almost every company encounters.

Most companies purchase supplies or inventory from vendors. Some purchases are made with cash and some purchases are made on credit. Companies pay their employees for services provided and pay their operating bills (i.e., rent, utilities, etc.). Companies sell their inventory and/or services to their customers with some purchases made with cash and some made on credit. Now let's walk through how to account for these types of transactions with journal entries so you know how to record your business transactions with your accounting software. The transactions are listed below along with some dollar amounts for illustrative purposes:

1. Purchase $10,000 of inventory from vendors on credit.
2. Pay cash to vendors for previous purchases made on credit.
3. Owe $2,000 to employees for weekly services rendered.
4. Pay $2,000 to employees for services previously rendered.
5. Sell $3,000 of your inventory to customers for $4,500 on credit.
6. Collect cash payments from customers for previous sales made on credit.

1. Purchase $10,000 of inventory from vendors on credit. Remember, we need three things to record a journal entry: (a) accounts, (b) amounts, and (c) directions. When this transaction occurs, your inventory account increases by $10,000 and your vendor liability account decreases by $10,000. We know that debits are used to show that assets are increasing and credits are used to show that liabilities are increasing. Therefore, the journal entry associated with transaction 1 would look like this:

Date	Debit Inventory	$10,000
	Credit Accounts Payable	$10,000

Let's try a few more journal entries.

2. Pay cash to vendors for previous purchases made on credit. Your cash is decreasing (credits decrease assets) and your liability to your vendors is also decreasing (debits decrease liabilities). Therefore:

Date	Debit Accounts Payable	$10,000
	Credit Cash	$10,000

3. Owe $2,000 to employees for weekly services rendered. Employee wages are an expense to your company, and debits are used to show that expenses are increasing. At the end of the week, you owe your employees for the services they provided that week regardless of how often you actually pay your employees. Remember, accrual accounting accounts for an expense when it is incurred regardless of when the expense is actually paid. Because you haven't paid your employees yet, a credit is used to record this increase in liability to your employees.

Date	Debit Wages Expense	$2,000
	Credit Wages Payable	$2,000

4. Pay $2,000 to employees for services previously rendered. When you actually pay your employees for their previously rendered services, you reduce your cash and reduce the wages liability you previously recorded because you no longer owe them anything. Note that you don't record more expense just because you are paying your employees with cash. The expense associated with their service was recorded in transaction 3. This journal entry just reduces cash and reduces a liability.

Date	Debit Wages Payable	$2,000
	Credit Cash	$2,000

5. Sell $3,000 of your inventory to customers for $4,500 on credit. There are two separate journal entries for this transaction. The first journal entry relates to the revenue piece, and the second journal entry relates to the expense piece. Making a sale on credit increases revenues and increases accounts receivable. We know that credits increase revenues and debits increase assets (i.e., accounts receivable). Therefore:

Date	Debit Accounts Receivable	$4,500
	Credit Sales Revenue	$4,500

Selling inventory also results in reducing inventory and recording inventory expense. Inventory expense is commonly referred to as "cost of goods sold" in the accounting world. We know that debits increase expenses and credits decreases assets (i.e., inventory). Therefore:

Date	Debit Cost of Goods Sold	$3,000
	Credit Inventory	$3,000

6. Collect cash payments from customers for previous sales made on credit. This transaction results in cash increasing and accounts receivable decreasing. If debits increase assets and credits decrease assets, the journal entry associated with this transaction is:

Date	Debit Cash	$3,000
	Credit Accounts Receivable	$3,000

20

Tax Considerations for a Startup

The hardest thing in the world to understand is the income tax.

Albert Einstein

I shall never use profanity except in discussing house rent and taxes.

Mark Twain

What to Watch For

Taxes. Most people shudder at the word. You find yourself writing what seems like endless checks every month to pay for your taxes. For this reason, many people would prefer to avoid anything further on this dreadful topic, but there is hope. Learning more about taxes can ultimately help save you money on taxes down the road. Although you may think that taxes are just for your accountant, truthfully many of the key decisions that will affect your taxes are made long before an accountant begins to prepare your tax returns.

Many founders and nonaccounting types may find this section to be wholly overwhelming and too much to handle. I'll even understand if your eyes begin to glaze over now and again.

This chapter discusses the tax implications of forming a corporation, issuing stock to investors, and distributing dividends to stockholders. It also describes the different layers of tax that a corporation might owe, such as corporate income tax, AMT, employment tax, and accumulated earnings taxes. Additionally, there is information provided for certain state taxes sometimes overlooked by startups.

Key Tax Considerations for a Startup Company

Taxes are an important consideration for a startup company. Yet, because the perception is that taxes are best left to your tax accountant, oftentimes a small company will fail to understand the effect of its actions on the taxes it will pay.

According to the IRS, the vast majority of issues for small businesses and startup come from poor recordkeeping and oversight rather than any intentional actions of deceit or fraud, and many of these problems could be prevented (and money saved) simply by focusing on a few key areas:

- Proper recordkeeping
- Tax planning
- Understanding key "small business" tax considerations

Founders should pay close attention to these initial three sections that follow and may find additional information throughout this chapter helpful as well. At a minimum, you should have a record retention policy in place, meet with your tax accountant before your taxes are due to engage in some level of tax planning, and gain a broad understanding of key tax provisions that are designed to benefit small businesses.

A FEW FEDERAL TAX FORMS

Any new business will need to be aware of a few tax forms and publications from the IRS. These forms will provide you information on what needs to be filed and when the filing must be made. Prepare to meet federal employment tax requirements by reading to following IRS publications:

- Publication 15, *Employer's Tax Guide*
- Publication 15a, *Employer's Supplemental Tax Guide*
- Publication 51, *Agricultural Employer's Tax Guide*

Ensure that your employees have completed the proper withholding certificates:

- Form W-4, Employee's Withholding Allowance Certificate
- Form W-5, Earned Income Credit Advance Payment

Estimate any federal tax payments and make any required payments:

- Form 1040-ES, Estimated Tax for Individuals
- Form 1120-W, Estimated Tax for Corporations

Proper Recordkeeping

Although it may seem fairly obvious, proper recordkeeping is one of the most important considerations for any startup company, especially in the early stages when you do not have an established set of policies, an experienced accounting staff, or a CFO. Recordkeeping for tax purposes should be integrated into an entire document retention policy for the company. Most of the items discussed below will be key documents for accounting purposes.

Receipts of Income

Retain all documentation that provides evidence of income. Gross receipts are the income you receive from your business. You should keep supporting documents that show the amounts and sources of your gross receipts. Documents that support your receipt of income should include the following:

- Receipts of sales
- Customer invoices
- Shipping records
- Bank deposit records
- Credit card payment records (particularly in the event you are primarily selling goods or services via the Internet)

Payroll and Employee Information

Early on, startup companies should create and use standard documents for all new employees that join the company. These records should be kept for a number of purposes, including for tax purposes. In addition, as you begin to provide compensation to your employees and make withholdings from the employee's compensation, you should carefully track and record these amounts. Documents that show payroll and employee information should include the following:

- Records evidencing hiring
- Records evidencing changes in title, job status, or salary

- Termination documentation
- Cancelled checks
- Paystubs or payroll run reports
- Withholding calculations

Records of Purchases

The purchases described in this section reference items purchased and resold to customers or goods purchased for manufacturing into finished products to be sold to customers. Because these purchases will likely represent an important portion of your costs of goods and your inventory, the company should keep records that identify the costs of purchase and payment information. Documents that show purchases should include the following:

- Cancelled checks
- Credit card records
- Invoices
- Shipping and receiving documentation and reports

Business Expenses

These business expenses are those expenses that are used for aspects of the business other than those directly related to items to be sold to customers. These include research and development costs, meals and entertainment, electricity bills, and similar costs. These expenses may be in the form of expenses that are submitted by your sales people for reimbursement or could involve more traditional purchases of office supplies. Documents that show purchases should include the following:

- Cancelled checks
- Credit card records
- Receipts
- Invoices

Records of Company Assets

Assets for your company may range from machinery, vehicles, or technical research equipment to buildings, software, or furniture. The key with company assets is to retain documentation on the purchase, ongoing improvements and maintenance, and disposal of the assets. Documents that show purchases should include the following:

- Acquisition records (invoices, shipping/receiving information, payment information)
- Maintenance records and costs
- Depreciation reporting
- Disposal records
- Real estate records
- Leases and rental records

TAXES: THE IMPORTANCE OF RECORDKEEPING

You need good records of all business revenues and expenses when you prepare your tax return. You need even better records if you are selected for audit by your state taxing authority or the IRS. Here are some tax-related recordkeeping tips to remember:

- **Identify the source of every receipt:** As a business owner, you will receive cash or property from various sources. Keeping track of what the receipt relates to will help you identify whether the receipt is taxable or not. For example, receipts from the sale of inventory or services are taxable, but receipts from business contributions made by owners are not taxable.
- **Identify the business purpose behind every expenditure:** Although you will be keeping track of your business expenditures for accounting purposes anyway, it's a good idea to keep track of the business purpose behind each expenditure. A business expenditure must be both "ordinary" and "necessary" to be deductible for tax purposes. An ordinary expense is one that is considered to be common in your trade or business, and a necessary expense is one that is considered to be appropriate for your trade or business. Note that some expenditures are deductible for accounting purposes but not for tax purposes (and vice versa).
- **Keep the records you use to prepare your tax returns:** If you are audited by the IRS or your state taxing authority, you may be asked to provide documentation supporting the revenues, expenses, and tax credits you report on your tax return. As a result, it is important to keep your tax records and all supporting documentation until the relevant statute of limitation expires. The IRS has three years to give you a tax refund or audit your tax return and 10 years to collect any tax due. This time period begins on the date that your tax return is due or the date you file your tax return (whichever is later). Therefore, always keep at least three years of tax return information on hand in case you are audited. A complete set of records makes an audit examination much less painful.

Tax Planning

Like visiting the doctor's office, many startup companies will avoid interacting with their tax accountant until there is an immediate need. This means that tax accountants will usually receive a telephone call when taxes are due to be paid, oftentimes once a year or periodically for interim tax bills.

However, visiting with a tax accountant on a regular basis for tax planning purposes may provide some tangible results when the annual tax bills are to be mailed. In particular, you may find that your tax accountant can provide you with important information with respect to timing of large capital-intensive purchases, interstate activities that can be affected by locating company assets within or outside state laws, or particular deductions that may benefit the company.

Consider meeting regularly with your tax accountant to understand the impact certain business decisions will have on your pending tax liabilities. In addition, if your company is profitable and will be paying taxes, you may need the assistance of your tax accountant to calculate the current year estimated tax payments to avoid underpayment penalties.

THE TAXMAN COMETH

What could happen

In its first year, your business had a profitable year (surprise to you and everyone!) and now has some cash in the bank. You've decided to take all the profits and invest in paying cash bonuses out to the team. But wait. . . . Four months later, you find out you have not one but two tax bills!

In April of your second year of operations (possibly later if you file proper extensions), you will have to pay taxes on the profits from the first year. That's not the end of it. Based on the profit numbers from the first year, you'll also be asked to prepay taxes for your second year based on the numbers from the first year. That's right, a double whammy of tax bills!

Watch out for

You should properly plan for cash outlays in your second year (if profitable) for the entire first year, as well as for quarterly estimated payments of taxes on the second year. Don't be unprepared in this case.

Proper planning is important, so check with your accountants and make sure to inform them of any changes in the business that could lead to increased taxes due at the end of year two (a rapid increase in revenues or shrinking costs) to avoid paying penalties.

TIP: Remember that, in your second year of profitability, the previous year and estimated current year taxes become due.

Understanding Key "Small Business" Tax Considerations

Although there are numerous tax considerations at issue for a company, seven key areas are of particular importance for small businesses and startup companies:

1. Depreciating capital assets
2. Section 179 expensing
3. Startup and organizational expenses
4. Research expenditures
5. Meals and entertainment expenditures
6. Employment taxes
7. Qualifying domestic production activities deduction

Depreciating Capital Assets: Follow the IRS's Depreciation Schedules

Capital assets are assets that have a useful life of more than one operating cycle, which is generally defined as more than one year. Examples of capital assets include machinery, office equipment, buildings, etc. Because a capital asset benefits your company for more than one period, the cost of the asset is "capitalized" when it is purchased and a percentage of the asset's cost is deducted (i.e., depreciated or amortized) each year over the number of years that the asset will be used.

The IRS provides taxpayers with specific depreciation schedules for every type of capital asset imaginable. Assets with shorter lives and accelerated depreciation methods result in more tax deductions now. For example, office equipment is considered seven-year property and is depreciated using an accelerated depreciation schedule. A nonresidential building is considered 39-year property and is depreciated using a nonaccelerated depreciation schedule. *IRS Publication 946: How to Depreciate Property* provides detailed instructions on how to depreciate each type of capital asset you own.

Section 179 Expensing: Deduct up to $128,000 of Capital Assets in 2008

To help stimulate the economy and support small businesses, a specific dollar amount of certain capital assets acquired for business use and placed in service during the year can be immediately deducted for tax purposes. The ability to immediately deduct a capital asset helps you reduce your taxable income and therefore your tax liability. Examples of these types of assets include off-the-shelf computer software and tangible personal property such as machinery and office equipment. The maximum dollar amount you can elect to deduct under Section 179 for property placed in service in 2008 is limited to $128,000. This $128,000 limitation is reduced dollar for dollar by the amount by which the cost of all Section 179 property placed in service during 2008 exceeds $510,000. If the total cost of your Section 179 property placed in service during 2008 is $638,000, you cannot take a Section 179 deduction.

Startup and Organizational Expenditures: Deduct up to $5,000 in 2008

A maximum of $5,000 of expenditures related to starting/organizing your business may be deducted in the year incurred. This limitation is reduced dollar for dollar by the amount by which all startup expenditures exceed $50,000. Startup costs relate to expenditures incurred to create an active trade or business or to investigate the creation or acquisition of an active trade or business. To qualify as a startup cost, the expense must be one that you could deduct if you were already in business. Examples include travel to suppliers, training for your new employees, advertising, utilities, and other preopening expenses. Organizational costs are the direct costs associated with creating your business. All costs not deducted in the current year may be amortized over a minimum of 180 months.

Research Expenditures: 20% Tax Credit

Businesses incurring qualified research expenditures may be eligible to claim a tax credit of up to 20% of all qualified expenditures in excess of a certain base amount. Per IRS Form 6765, "This research must be undertaken for discovering information that is technological in nature, and its application must be intended for use in developing a new or improved business component of the taxpayer. In addition, substantially all of the activities of the research must be elements of a process of experimentation relating to a new or improved function, performance, reliability, or quality."

Meals and Entertainment Expenditures: Only 50% Is Deductible

Small business owners generally find themselves in the position of entertaining customers, vendors, and other business-related parties. Fifty percent of qualifying business meals and entertainment expenditures are deductible for tax purposes. Examples of qualifying expenditures incurred while entertaining customers, vendors, and business associates include the following:

- The cost of meals (including food, beverages, taxes, and tips) provided to customers or clients
- Nightclub, social, athletic, and sporting club expenditures
- Theater tickets

Employment Taxes: Pay Attention to Employees and Your Contributions

Employment taxes include Social Security and Medicare tax, federal unemployment tax, and federal income tax withholding. Social Security and Medicare taxes pay for the benefits that workers and their families receive under FICA. Social Security tax is 12.4% of an employee's taxable wages, and the maximum amount of employee income subject to Social Security tax in 2008 is $102,000. Medicare tax is 2.9% of an employee's taxable wages, and 100% of employee taxable wages are subject to Medicare tax (there is no maximum income limitation). Employers pay half of these two taxes and employees pay the other half. This means that employers and employees each pay Social Security tax of 6.2% and Medicare tax of 1.45% of an employee's wages each year. Employers are responsible for withholding and remitting to the federal government employees' portion of Social Security and Medicare taxes. Self-employed individuals must pay both halves of FICA (a total of 15.3%). Use IRS Form 941 to remit FICA tax on a quarterly basis.

The Federal Unemployment Tax Act (FUTA) authorizes the IRS to collect federal employer tax to fund state workforce agencies and unemployment compensation programs. Although FUTA is technically 6.2% of the first $7,000 of annual taxable wages paid to each employee, employers who pay their state unemployment tax on a timely basis receive a federal credit offsetting up to 5.4% of their FUTA tax. This credit is awarded regardless of the tax rate paid to their resident state. Therefore, the net FUTA tax rate is generally around 0.8% (6.2% – 5.4%), with a maximum annual tax of $56 per employee (0.8% × $7,000 = $56). Each state determines its individual state unemployment insurance tax rates. Only employers (and not employees) pay FUTA tax. Employers must report and pay FUTA tax on IRS Form 940. Employers must also withhold and remit to the federal and state taxing authorities income tax related to employees' taxable wages.

Qualifying Domestic Production Activities Deduction: 6% in 2008

A business engaged in a "qualifying domestic production activity" (QDPA) is eligible for a tax deduction of 6% in 2007–2009 and 9% after 2009. Examples of "qualified production activities" include U.S.-based manufacturing, producing, growing, and extracting tangible personal property, construction and substantial renovation of real property (including infrastructure), computer software and video game development, sound recordings, and engineering and architectural services relating to U.S.-based construction projects.

Under a safe harbor rule, businesses are eligible to take the tax deduction if at least 20% of their total costs are the result of direct labor and overhead costs from operations based in the United States. Note that "cosmetic" construction services (i.e. painting) and selling food or beverages prepared on-site at retail establishments are both explicitly excluded from being considered eligible QDPA expenditures.

HOW TO CALCULATE THE QDPA DEDUCTION

QDPA Income
– QDPA Expenses

QDPA Net Income
× 6%

Tentative QDPA Deduction

QDPA Income is all taxable income arising from qualified production activities; QDPA Expenses is all expenses directly related to qualified production activities. The QDPA deduction cannot exceed taxable income for C-corporations (adjusted gross income for sole proprietors and partnerships) or 50% of current year W-2 wages paid to employees. An entity in a single line of business can simplify its calculation by comparing 6% of its net income with its annual W-2 wages paid to employees. See IRS Form 8903 and IRS Instructions for Form 8903 for additional information.

Corporations and Taxes

If you've gone the route of forming a corporation, this decision will impact the taxes your corporation will owe.

DIFFERENCES BETWEEN A CORPORATION AND PARTNERSHIP (OR OTHER PASS-THROUGH ENTITY)

A quick refresher on the key differences of a corporation from a partnership (or another similar type of pass-through entity):

1. A corporation is an entity separate and distinct from its owners. It has many rights and privileges of a person and is liable for its own taxes.
2. Stockholders generally have limited liability in a corporation. In the event of a bankruptcy, the losses of stockholders are generally limited to any capital investment they have in the corporation.

3. Ownership rights in a corporation are transferable. Ownership rights are signified by shares of capital stock, which may be disposed of by sale. In contrast to a partnership, the decision to transfer ownership is made by the individual stockholder and generally does not require the consent of each owner in the corporation.
4. A corporation may acquire capital by issuing stock. Investors generally are more attracted to buying stock than investing in a partnership, because they receive the benefits of limited liability and ready transferability of their shares. When a corporation issues dividends of its net earnings to stockholders, these are taxable to the shareholders. Thus, many people say that corporations are double-taxed: once at the corporate level and once at the individual level.

Tax-Free Incorporation

A corporation must have assets in order to operate. To obtain this capital, investors will transfer property into the corporation in exchange for stock. Although exchanges of property are generally taxable events, §351 of the I.R.C. provides nonrecognition treatment for transfers of property into a C-corporation. To achieve nonrecognition, three elements must be met: (1) one or more persons must transfer property to the corporation; (2) the transferor may only receive stock of the corporation in exchange; and (3) immediately after the exchange, the transferor(s) must control the corporation. These elements can be broken down as follows:

- **Property:** The word "property" is not defined in I.R.C. §351 but is interpreted rather broadly and may include cash, accounts receivable, inventory of cash-basis taxpayers, and the transfer of substantially all the rights to intangible property, such as patents, copyrights, and trademarks. Taxpayers seeking to transfer intangibles that cannot be evidenced by documentation, such as trade secrets, know-how, and other secret processes, must meet stringent requirements to qualify these items as property. Property does *not* include services rendered to the corporation. If a person receives stock from a corporation in exchange for services, the stock is treated as income to the recipient and he must pay taxes on it.
- **Transfers:** To qualify as a "transfer" of property, there must be a complete surrender of rights to the property. This issue may become problematic in the context of licenses and limited use transfers of intangible property.
- **Receipt solely of stock:** The requirement that the transferor receive "solely stock" of the corporation is not strictly applied. Transferors may receive money or property other than corporate stock and still qualify under §351 as long as they receive enough stock to meet the control requirements (see below). However, transferors may have to recognize gain on part of the exchange. When receiving stock in exchange for property, transferors do not all have to receive the same type of stock.
- **Control after the exchange:** To be in control of the corporation after the exchange, the transferor or group of transferors must own at least 80% of the combined voting power of all classes of stock entitled to vote and at least 80% of the outstanding shares of each class of nonvoting stock immediately after the exchange.

Understanding Tax "Basis"

The word "basis" refers to the costs of acquiring property, subject to certain adjustments. A taxpayer determines his taxable gain or loss in a transaction by taking the fair market value of property he received in the exchange, less his basis in the property given up. Thus, for corporations and transferors to know how much gain (or loss) they must recognize on transfers of property, they must know their basis in that property.

Section 358 of the I.R.C. governs the transferor's basis. It says that the transferor generally takes a substituted basis in the stock he receives equal to the basis of the property exchanged. For example, if a taxpayer exchanges $1,000 of cash for stock in a corporation, the taxpayer's basis in the stock is $1,000. The transferor must *increase* this amount by any gain he was required to recognize on the exchange of property. The transferor must *decrease* his basis by any cash received in the exchange, the fair market value of any other property received, any loss recognized on the exchange, and the amount of any liability the corporation assumed for the transferor in the exchange (depending on the circumstances). When liabilities are involved in the exchange, a transferor should consult a tax professional.

When an investor exchanges cash or other assets for the stock of a corporation, the assets the corporation acquires are called "paid-in capital." Contributions of paid-in capital are not taxable to the corporation. The corporation's basis in its assets is determined under I.R.C. §362. This provision states that the corporation's basis in the property contributed to capital by a shareholder is the same as the basis the shareholder had in the property, increased by any gain the shareholder was required to recognize in the exchange. Increases for any gain recognized may be limited by tax laws, however.

C-Corporation Items of Income and Loss

Although a corporation will not pay tax on its accumulation of paid-in capital (see above), it will owe tax on its earnings throughout the year. A corporation's taxable income is determined by taking gross income plus or minus various items.

CALCULATING TAXABLE INCOME

The following sections set forth some of the rules applicable to corporations when determining taxable income.

- **Gross income:** Under I.R.C. §61(a), a corporation must report income from "whatever source derived," including gains from property transactions, rents, royalties, and dividends.
- **Costs of going into business:** Until a taxpayer is actually carrying on a trade or business, the expenses incurred are merely preparatory costs that must be capitalized and recovered through depreciation or amortization. However, for startup costs incurred after October 22, 2004, I.R.C. §195 permits a corporation to elect to deduct a limited amount of startup or organizational costs. For purposes of that section, startup costs are costs incurred to create an active trade or business or to investigate the creation or acquisition of an active trade or business. Organizational costs are the direct costs of creating the corporation. Any cost that is not deductible in the current year can be amortized over a period of 180 months.

- **Ordinary and necessary business expenses:** Generally, a corporation may deduct any "ordinary and necessary" expenses incurred in carrying on a trade or business. These deductions include reasonable allowances for employee salaries and rents.
- **Research and experimentation expenses:** Expenditures for research and experimentation generally are required to be capitalized under I.R.C. §263. Nevertheless, under I.R.C. §174, taxpayers may elect to deduct research or experimental expenditures that are paid or incurred during the taxable year in connection with a present or future trade or business. The taxpayer has the option to treat such expenditures as current deductions or deferred expenses that are amortizable over a period of not less than 60 months, beginning with the first month the taxpayer realizes the benefits of such expenditures. The term "research and experimental expenditures" means expenditures incurred in connection with the taxpayer's trade or business that represent research and development costs in the experimental or laboratory sense. This term generally includes the costs of obtaining patents, including attorneys' fees expended in making the patent application. It does not include expenditures for (1) ordinary testing for quality control, (2) efficiency surveys, (3) management studies, (4) consumer surveys, (5) advertising or promotions, (6) the acquisition of another's patent, model, production, or process, or (7) research in connection with literary, historical, or similar projects. Any deductions for research expenditures must be reasonable under the circumstances.
- **Capital gains and losses:** All property held by a corporation is a capital asset, except for (1) inventory or other property held mainly for sale to customers, (2) depreciable or real property used in a trade or business, (3) certain copyrights, literary, musical, or artistic compositions, (4) accounts or notes receivable acquired in the ordinary course of business for services or from the sale of inventory, and (5) U.S. government publications. Corporations may deduct capital losses only to the extent of capital gains, but excess capital losses may be carried to other taxable years. The corporation must first carry the loss back for three years before the loss year and then may carry forward any remaining loss for five years. When the loss is carried backward, it is treated as a short-term loss and does not retain its original character as long term or short term. Special rules apply for ordering the capital loss carryovers and carrybacks. Under the alternative tax for corporations, I.R.C. §1201, a corporation may pay a tax of 35% on net capital gains.
- **Net operating losses:** A corporation calculates its net operating loss (NOL) by subtracting any allowable deductions from gross income, subject to certain limitations and rules. If the deductions exceed gross income, the corporation has an NOL for the taxable year. Generally, the corporation must carry the NOL back two years before the year in which the NOL is incurred. If the NOL is not used up during that time, it may be carried forward up to 20 years after the year in which it was generated. A corporation may also elect to use only the 20-year carryforward period. A corporation that expects to have an NOL in its current year is entitled to an automatic extension of the time for paying all or a portion of its income tax for the immediately preceding year. To gain the benefits of

this extension, the corporation must file Form 1138 and must explain why it expects the loss. The extension lasts until the end of the month in which the return for the NOL year is due, including any extensions.

- **Charitable contributions:** Corporations are entitled to limited deductions for charitable contributions made in cash or other property.
- **Related persons:** A corporation that elects to use the accrual method of accounting is not permitted to deduct business expenses and interest owed to a related person who uses the cash method of accounting until the payment is made and included in the related person's gross income. The same rule applies to the deduction of losses on the sale or exchange of property between related persons. For purposes of this rule, related persons are as follows: (1) another corporation that is a member of the same controlled group (see I.R.C. §267(f)); (2) an individual who owns, directly or indirectly, more than 50% of the value of the outstanding stock of the corporation; (3) a trust fiduciary when the trust or the grantor of the trust owns, directly or indirectly, more than 50% in value of the outstanding stock of the corporation; (4) an S-corporation if the same persons own more than 50% in value of the outstanding stock of each corporation; (5) a partnership if the same persons own more than 50% in value of the outstanding stock of the corporation and more than 50% of the capital or profits interest in the partnership; and (6) any employee-owner if the corporation is a personal service corporation, regardless of the amount of stock owned by the employee-owner.
- **At-risk limits and passive activity limits:** Closely held corporations may be subject to at-risk rules that limit their losses to the amounts "at risk" in the activity. For purposes of the at-risk rules, a corporation is closely held if, at any time during the last half of the taxable year, more than 50% of the value of its outstanding stock is owned directly or indirectly by five or fewer individuals. The amount at risk typically equals any money and the adjusted basis of property contributed by the taxpayer to the activity, plus money borrowed for the activity on a recourse basis. The passive activity rules also apply to closely held corporations (other than S-corporations). These rules limit passive activity losses to the amount of passive activity income. Generally, passive activities are found when a taxpayer has a trade or business activity in which he does not materially participate in the year or has a rental activity.

To calculate items of income and loss for the taxable year, a corporation must choose an accounting method and follow appropriate recordkeeping procedures. Permissible methods for accounting include cash, accrual, and any other method authorized by the I.R.C. If a corporation has average annual gross receipts that exceed $5 million, it must use the accrual method of accounting.

A corporation that has never filed an income tax return before must adopt a tax year for filing its returns. A corporation may choose between the calendar tax year (12 consecutive months ending December 31) and the fiscal tax year (12 consecutive months ending on the last day of any month except December). Once a corporation has adopted a tax year, it may have to get IRS approval to change it.

For recordkeeping purposes, any records that substantiate items of income, deductions, or credits on a corporation's return must be kept for three years from the date the return is due or filed, whichever is later. Corporations should maintain records that verify the corporation's basis in its property for as long as necessary to calculate the basis of original or replacement property. Corporations should also keep copies of all filed returns, because they are helpful in preparing future and amended returns.

Corporate Distributions

A holder of common stock in a corporation is entitled to share in the corporate earnings of a corporation. The stockholder shares in such earnings through "dividends," which are distributions to shareholders on a pro rata basis. A dividend may take place in four forms: cash, property, scrips (promissory notes to pay cash), and capital stock. Below is information on cash dividends and stock dividends.

ISSUING DIVIDENDS: CASH OR STOCK

- **Cash dividends:** For a corporation to issue a cash dividend, the corporation must first meet three requirements:
  1. Depending on state laws, the corporation must have retained earnings from which to distribute cash or must be allowed to distribute dividends from paid-in capital. A dividend that is distributed out of a corporation's paid-in capital is termed a "liquidating dividend" because the amount of the stockholders' initial contribution is being reduced.
  2. The corporation must have adequate cash. The corporation's board of directors must carefully consider current and future cash flows before issuing a dividend.
  3. The board of directors must determine how much of the corporation's income will be distributed and how much will be retained.

Once a cash dividend has been declared, the corporation will not have to recognize gain or loss on the distribution. However, if the dividend comes out of the corporation's net earnings, the shareholder will have to report dividend income for the year and will owe tax on that income. If the dividend comes out of paid-in capital, the liquidating dividend is treated as a return to capital for the shareholder. The shareholder must reduce his basis in his stock by the amount of the dividend. If the amount of the dividend exceeds his basis in the stock, according to any adjustments, then he must pay capital gains on the dividend. Until 2010, the difference between dividend income and capital gain income will essentially be immaterial for shareholders because corporate dividends are taxed at capital gains rates.

- **Stock dividends:** A stock dividend results in a decrease in the corporation's retained earnings and an increase in the corporation's paid-in capital. Stock dividends serve several purposes: (1) satisfying shareholders' dividend expectations without expending cash; (2) increasing the number of shares in the market, which consequently reduces the share price and makes the stock more marketable to small investors; and (3) reinvesting stockholders' equity in the corporation so that it is not available for cash distributions. Stock dividends are generally tax free to shareholders until the stock is sold, at which time the sale is taxed at capital gains rates. The corporation also will not be taxed on the stock dividend. A corporation is not permitted to deduct the expenses of issuing a stock dividend.

> If a corporation issues a dividend, it must report the distribution on a Form 1099-DIV. For dividends that are nontaxable because they come out of paid-in capital, the corporation must file a Form 5452, Corporate Report of Non-Dividend Distributions.

Taxes of C-Corporations

C-Corporations are liable for income taxes, estimated taxes, and employment taxes. They may also be liable for the AMT, accumulated earnings tax, and excise taxes. These taxes are summarized below:

- **Income taxes:** A corporation's first federal income tax return on Form 1120 must be filed on or before the 15th day of the third month following the close of the taxable year. Every corporation must file an income tax return regardless of whether or not it actually has taxable income. Corporate profits are generally subject to two levels of taxation: one at the corporate level and another at the shareholder level when profits and earnings are distributed (see the discussion under "Corporate Distributions" above). Corporate income taxes are imposed progressively based on taxable income, up to 35%.
- **Estimated taxes:** Corporations that expect to owe tax of $500 or more when they file their return must make estimated tax payments. These taxes arc due in four quarterly installments, and each installment must equal at least 25% of the required annual amount. Failure to pay these estimated taxes when required may subject the corporation to an additional, nondeductible penalty tax.
- **Employment taxes:** Employment taxes include Social Security and Medicare taxes, federal income tax withholding, and FUTA tax. Social Security and Medicare taxes pay for benefits that workers and their families receive under FICA. These taxes must be deducted from the employee's wages and withheld by the corporation. The corporation also must pay a matching amount. FUTA pays unemployment compensation to workers who lose their jobs. Corporations must report and pay FUTA tax separately from Social Security, Medicare, and withheld income tax. The FUTA tax is paid only from the corporation's funds; employees do not pay the tax and do not have it withheld from their pay. The willful failure to collect and pay over FICA and withholding taxes is a felony under I.R.C. §7202 and may result in up to five years of imprisonment and fines.
- **AMT:** The AMT is designed to reduce the possibility that some corporations with substantial economic income can significantly reduce their regular tax. A corporation owes AMT if its tentative minimum tax is more than its regular tax.
- **Accumulated earnings tax:** Corporations are allowed to accumulate earnings for potential expansions and other bona fide business motivations. However, if a corporation permits its earnings to accumulate beyond the reasonable needs of its business, it may be subject to an accumulated earnings tax of 15%.
- **Excise taxes:** Businesses that manufacture or sell certain products, operate certain kinds of businesses, use various kinds of equipment, facilities, or products, or receive payments for certain services may be subject to excise taxes. The general categories of excise taxes include the following: environmental taxes, communications and air transportation taxes, fuel taxes, tax on the first retail sale of heavy trucks and tractors, and manufacturers taxes.

Tax Advantages of Small Business Corporations

Equity investments in small business corporations receive preferential treatment under three provisions of the Code: I.R.C. §§1045, 1202, and 1244. Sections 1045 and 1202 (the "gain" provisions) permit favorable treatment of gains from the stock of "qualified small business corporations." Section 1244 (the "loss" provision) provides for favorable treatment of losses attributable to the stock of small business corporations.

The definition of small business corporation differs between the gain provisions and the loss provisions. Under the gain provisions of I.R.C. §§1045 and 1202, a qualified small business corporation cannot have more than $50,000,000 of gross assets up through the time they issue stock. Under the loss provisions of I.R.C. §1244, the determination of small business corporation depends on the corporation's receipt of capital equity investments up through the time of the stock's issuance. I.R.C. §1244 only applies to corporations that have received less than $1,000,000 in equity capital.

Section 1045 permits the tax-deferred rollover of gain from the sale of stock of a qualified small business corporation into the stock of another qualified small business corporation.

Under §1202 of the Code (adopted in 1993), noncorporate shareholders of qualified small business corporations may qualify for a 50% exclusion from taxable gain on the sale of such stock if it was acquired at original issue and held for at least five years before sale. Section 1202 imposes a number of restrictions on eligibility for the exclusion, including requirements that (1) 80% of the corporation's assets be used in an active business, (2) the corporation be a qualified small business, (3) no repurchases of the shareholder's stock, other than certain de minimis repurchases and certain repurchases in connection with the death, divorce, or bona fide termination of employment of the shareholder, occurred within two years before or after the stock was issued (and no significant repurchases of any shareholder's stock occurred within one year before or after the stock was issued, other than the excepted repurchases noted above), and (4) the corporation agrees to file reports with the IRS to the extent required by those authorities. In addition, there is a $10 million per issuer limitation on gain qualifying for the exclusion for joint filers and a requirement that 42% of the excluded gain be taken into account as an item of preference for AMT tax purposes.

The tax rate benefit of §1202 qualifying stock has been reduced somewhat as a result of the enactment of the Taxpayer Relief Act of 1997. The act established a maximum federal tax rate of 20% on capital gains attributable to assets held for more than one year (18% in the case of assets acquired after December 31, 2001 and held for more than five years).

Together, Sections 1045 and 1202 reduce the effective shareholder-level tax rates and thereby reduce the double-tax burden on C-corporation businesses. However, taxpayers should carefully weigh whether retaining earnings and availing themselves of these provisions makes the "double tax" regimen of C-corporation status more favorable than pass-through regimens. Businesses that predict they will distribute almost all of their earnings on a current basis will achieve fewer benefits from these provisions and may choose to organize themselves as pass-through entities.

Section 1244 allows for ordinary loss treatment on the stock of small business corporations. The benefits of this section only apply to individuals, whether directly or through a partnership, who recognized a loss on §1244 stock that was originally issued to them in exchange for money or property by a domestic small business corporation. As a consequence, S-corporation shareholders are not permitted to report ordinary losses as a result of the sale of §1244 stock that they hold.

Section 1244 eliminates some of the discrepant treatment of losses sustained by unincorporated businesses and losses sustained by C-corporations. Unincorporated businesses have the advantage of passing losses through to the individual equity participants, so the losses are deductible in full against ordinary income. In contrast, operating losses of corporations are only deductible against corporate income and are not deductible against the income of shareholders. Thus, for new ventures that may prove to be unsuccessful, the losses may go to waste. By allowing shareholders of small business corporations to take ordinary losses, §1244 reduces these discrepancies.

Individuals deciding between various types of entities should also remember that §§1202 and 1244 of the Code are not elective. For taxpayers in the 15% ordinary income tax bracket, §1202 may actually result in a tax rate increase rather than a tax savings.

Because the provisions of §§1045, 1202, and 1244 are complex, please contact a tax professional for additional information if you are interested.

Taxes on S-Corporations

An eligible corporation may avoid the double taxation of C-corporations by making an election to be treated as an S-corporation. S-Corporations are generally exempt from federal income taxes. Items of income, deduction, loss, and credit are passed through to the S-corporation's shareholders.

Among a few other requirements, a corporation is eligible for S-corporation status if (1) it is a domestic corporation created under the laws of the United States or any state, (2) it does not have more than 100 shareholders (75 shareholders in tax years before 2005), (3) all shareholders are individuals, estates, trusts, or tax-exempt charitable organizations, and (4) the corporation has only one class of stock.

If it qualifies, the S-corporation's income, loss, deduction, and credit are allocated pro rata among each shareholder for each day in the tax year. Certain items of income, loss, deduction, and credit must be separately allocated to each shareholder. The following items are examples of the items that need to be accounted for separately:

- The net amount of gains and losses from sales or exchanges of capital assets
- Charitable contributions
- Tax-exempt income of the corporation
- The corporation's AMT adjustments

Taxes on Partnerships

A partnership is a relationship between two or more people to carry on a trade or business. The partners contribute money, property, or labor with an expectation of sharing in the profits and losses of the business.

Like S-corporations, partnerships are pass-through entities and therefore are only subject to tax at the partner level. The partnership is not a taxable entity itself; instead, all items of income, loss, deduction, and credit flow through to the partners and must be recorded on their individual tax returns.

LLCs may elect to be treated as either a partnership or a corporation. Some advantages of choosing the partnership classification over S-corporation classification are as follows: (1) there are no limitations on who, or how many, may be a partner; (2) there is great flexibility in allocating the partnership's profits, losses, and credits among the partners; and (3) the partner's basis in his interest in the partnership includes his share of liabilities (a shareholder's basis in the shares of an S-corporation does not include liabilities).

State and Local Taxes

Businesses will oftentimes be subject to the state and local taxes of multiple jurisdictions, including the state of incorporation, the states where you have officers or employees, and the states where you make sales. The first step you will need to undertake is determining whether your business is subject to these taxes. Once it is established that you have nexus with the state (usually through some combination of sales, payroll, and property within the state), you'll need to qualify to do business in the state and remit the necessary taxes required.

At the end of this chapter, you will find information on certain states and some of the likely taxes a business may be subject to if they operate in that state. These represent only a summary and any business should check with each jurisdiction to find the applicable taxes, fees, and other filings necessary.

What Is Nexus?

One of the first steps to determine whether or not you will be subject to certain state taxes is to determine whether you have nexus with the state. Nexus is the amount or degree of business activity that the business taxpayer needs to have with a particular state before the state is able to impose taxes. You can think of this as a basic threshold test, and, when your activity with the state exceeds a certain amount, you will be responsible for filing and paying taxes in the state. (Note that, even if you don't have nexus with a particular state, you will still typically be responsible for payment of state sales or use taxes on sales you make in the state, based on the rates the state or local jurisdiction imposes.)

Each state has its own laws that govern how that state will calculate nexus, and so you will need to make a determination based on the state rules (oftentimes with the assistance of your accounting department or outside auditors). Generally, the state will look at a combination of three factors to determine whether you have nexus and to calculate the taxes due. These factors are (1) sales, (2) payroll, and (3) property. Sales will usually include sales in the ordinary course of business as well as sales of fixed assets located in the state. Additionally, some states may factor in other revenue items, including interest, royalties, etc. Payroll will include all employees, including officers located in the state and should tie directly to the payroll numbers on your W-3 forms. Property is oftentimes a combination of items, including fixed assets, equipment, building, land, and items such as leases and rental equipment.

In each case, you will generally be asked by the state to calculate ratios for these three factors. The ratio will usually look at the portion of sales, payroll, or property attributable to the state over the totals for the entire company. Unfortunately, what constitutes each of these factors may well be different in each state. In general, if the sales factor is the only factor that applies in the state, you may be eligible for an exemption under PL 86-272 and liable for an annual license fee or franchise tax.

CALIFORNIA STATE TAXES

Corporation Franchise and Income Tax

All corporations that derive income from sources within California must pay corporate income tax. Corporate income tax in California is imposed on net income, which is equal to gross income less specified deductions under California law. The tax rate on general corporations is 8.84% of net income for the preceding year or, if greater, the minimum tax. Because California provides for special deductions and credits, taxpayers who receive favored treatment under these provisions are still required to pay at least an AMT. The AMT rate is 6.56% of alternative minimum taxable income for the year that exceeds the exemption amount.

Corporations that are organized in California, but do not conduct business there, are subject to a minimum franchise tax of $800. The minimum franchise tax also applies to any corporation if the tax they would owe under the general corporate income tax rate is less than the $800 minimum.

S-corporations must pay California minimum tax, the regular income tax for S-corporations, or the franchise tax rate for S-corporations. S-corporations are generally taxed at 1.5%. LLCs that elect to be treated as corporations for federal income tax purposes will be treated the same way for California tax purposes. Taxes on LLCs that elect to be treated as partnerships have been repealed since 1996. Although partnerships are not subject to the corporation franchise or income taxes, the partners remain liable for individual taxes in California.

The first California corporate income tax return is due on or before the 15th day of the third month after the close of its first taxable year.

Tax Treatment of Qualified Small Business Stock

California permits special tax treatment for rollovers of qualified small business stock. To qualify for this treatment, the stock must meet the federal rules for qualified small business stock (I.R.C. §1045). One of the primary restrictions under §1045 is that the company cannot have more than $50 million of gross assets.

In addition to the federal requirements, California has its own restrictions on qualified small business stock. First, during the period in which the taxpayer holds such stock, at least 80% of the issuer's assets must have been used in the active conduct of a qualified business in California. Second, certain stock redemptions by the holders and related persons will destroy qualified small business stock status. Finally, at least 80% of the issuer's payroll must have been attributable to California employment when the stock was issued, as well as during the period in which the stock was held.

If all of these elements are met, a stockholder is entitled to "rollover" benefits. The stockholder will not have to recognize gain on the sale of stock as long as he reinvests those earnings in replacement California qualified small business stock.

Additionally, stockholders who hold their qualified small business stock for more than five years will not have to include 50% of the gain from the sale or exchange of qualified small business stock in their gross income (subject to some limitations).

Please be aware that, if a corporation's stock qualifies as California small business stock, a report must be filed with the California Franchise Tax Board on Form 3565 for each year in which such stock is issued (and a copy must be provided to each shareholder who acquired such stock during such year) no later than the due date, including extensions, for the corporation's income tax return. You should consult with your accountants regarding this filing requirement.

Sales and Use Tax

California usually imposes either a sales or a use tax on transfers of tangible personal property, but not both.

Retailers must pay sales tax, unless exempted, for the privilege of selling tangible personal property at retail in California. Typical exemptions include goods sold out of state, destined for export, intended for resale, or sold to the U.S. government. The tax, when levied, is a percentage of the gross receipts earned by the seller from the sale (which percentage varies depending on the county of sale). The retailer may collect the tax from the purchaser, provided the amount of the tax is separately stated or he posts notice that the purchase price includes the tax.

The use tax complements the sales tax. It generally does not apply in situations in which the sales tax does. In general, the use tax acts as a mechanism for collecting tax on personal property purchased outside of California from a retailer and brought into California for storage, use, or consumption in the state. The use tax is measured as a percentage of the sales price of the property. Any retailer engaged in business in California and making sales of tangible personal property for taxable use in California is required to collect the tax and remit it to the state.

California Property Taxes

Real property and tangible personal property located in California are generally subject to property tax imposed by local governmental units. Intangible personal property includes items such as money, shares of capital stock, and computer programs (other than "basic operational programs"). Intangible property is exempt from tax in California. Tangible personal property is property that may be seen, touched, or is perceptible to the senses in some other way. Certain real and tangible personal property is exempt from the tax, such as property owned by the state of California and inventories used for business.

Each business should file a business property affidavit with the county assessor setting forth information on the value of the business property that it owns. The affidavit must be filed between March 1 and the last Friday in May. However, the county assessor may require filing by an earlier date after April 1. Late filings can be penalized.

DELAWARE STATE TAXES

Corporate Income Tax

Delaware's corporate income tax is imposed on all corporations that conduct business or have property within Delaware, unless exempted. If a pass-through entity does business or has tangible real or personal property in Delaware, then corporate members of that entity must pay corporate income tax on their share of the pass-through entity's income. Banks, trust companies, and building and loan associations are subject to a franchise tax instead of the corporate income tax.

The corporate income tax is imposed annually and is based on net income derived from business conducted in Delaware or property located in Delaware. Net income is based on the entity's federal taxable income, with certain adjustments according to state law.

Entities that derive income from activities in multiple states must allocate and apportion their income to determine Delaware taxable income. The corporate income tax rate is 8.7%. Calendar year corporate returns are due and payable on April 1; fiscal year corporate tax returns are due on the first day of the fourth month after the close of the corporation's fiscal year.

Stock-Based Franchise Tax

Delaware also levies an annual stock-based franchise tax on domestic corporations that choose to incorporate in Delaware. Foreign corporations that choose not to be incorporated in Delaware are not subject to the franchise tax; however, they may still be subject to the corporate income tax.

Nearly all corporations incorporated in Delaware are subject to the franchise tax. A few exemptions do exist, including banks (subject to a separate franchise tax), religious corporations, certain nonprofits, etc. Also, because the stock-based franchise tax applies to stock, this tax does not apply to noncorporate business entities such as partnerships and sole proprietorships.

Corporations subject to the stock-based franchise tax may determine their tax liability by computing their tax base according to two methods: (1) authorized shares method, which is based on the number of authorized (but not necessarily issued) shares of capital stock; and (2) assumed par-value capital method, determined according to the number of issued shares, par values, and gross assets. The corporation must pay tax according to whichever method yields the lesser tax liability; however, corporations that have stock with no par value may only use the authorized shares method. Every corporation in existence under Delaware's General Corporation Law must make an annual franchise tax report to the Secretary of State. These reports are due on March 1, and there is no exception for corporations that use a fiscal year.

Business and Occupational License Fees

Delaware does not impose a state sales tax; however, much like Washington state, it imposes a business and occupational license fee (tax) on the gross receipts of most businesses. Some examples of the types of business activities subject to such a tax include contractors, manufacturers, retailers, wholesalers, and lessees of tangible personal property intended for use within Delaware. Various tax rates, exemptions, deductions, and credits may be applicable to the corporation depending on its activities.

Property Taxes

All real property located in Delaware is taxable; however, tangible and intangible personal property is not taxed. Shares of stock of domestic corporations owned by out-of-state persons or corporations are also exempted from property taxes. Property tax rates vary by county and locale.

NEW YORK STATE AND LOCAL TAXES

Franchise Tax

New York imposes a franchise tax on every domestic or foreign corporation doing business, employing capital, exercising its franchise, or owning or leasing property in New York. LLCs and partnerships that are classified as corporations for federal income tax purposes are subject to this tax. Taxable income is determined under four alternative tax bases:

- Entire net income base
- Capital base
- Minimum taxable income base
- Fixed dollar minimum base

Entire net income is the primary base for calculating the franchise tax and is determined by making additions, subtractions, and modifications to federal taxable income.

The applicable tax rate depends on the taxpayer's base. The rate for entire net income taxpayers is 7.5%, subject to variations. For example, small business corporations pay the franchise tax at progressive rates ranging from 6.5 to 7.5%. S-Corporations must pay tax on the greater of the following: (1) tax computed based on entire net income, or (2) the fixed dollar minimum tax, which is based on the taxpayer's gross payroll. Sole proprietorships are not subject to the franchise tax; instead, they must pay tax on their distributive share of LLC and limited liability partnership income according to personal income tax rates.

New York City Taxes

In addition to state corporate income taxes, businesses operating in New York City should also be aware that the city imposes a general corporation tax and an unincorporated business tax.

Sales and Use Taxes

New York also has a statewide sales tax of 4% that is imposed on sales of tangible personal property and certain services. In addition, the state imposes a 4% use tax on purchases made outside the state that will be used within the state.

Property Taxes

New York property taxes apply to all real property in the state, regardless of whether the property is owned by individuals, corporations, residents, or nonresidents. Neither tangible nor intangible personal property is taxed. The property tax rates vary according to county/locality.

WASHINGTON STATE TAXES

Business and Occupation Tax

Washington does not have a corporate income tax. Instead, the state levies a gross receipts tax called the Business and Occupation (B&O) tax (RCW 82.04.220). Because this excise is imposed at every level of the economic process (from manufacturing to retailing), the B&O tax may result in duplicative taxation.

Almost every entity engaging in business in Washington is subject to this tax, including joint ventures, partnerships, and other entities that may not be subject to an entity-level tax for federal income tax purposes. Importantly, Washington taxes affiliated entities separately. Thus, transfers, allocations, and services provided to wholly owned subsidiaries and other affiliates will be subject to the B&O tax.

The base on which the B&O tax is levied depends on the taxpayer's business activities. There are three general ways in which to measure the tax. Manufacturers and extractors are taxed on the value of their products. Retailers, wholesalers, and other sellers are taxed on their gross proceeds of their sales. All other taxpayers besides manufacturers, extractors, and sellers are subject to a tax on their gross incomes. The taxpayers who typically fall into this category are businesses that provide legal, medical, consulting, management, or other types of services. Because the tax is levied on gross receipts, taxpayers generally are not allowed a deduction for the costs of doing business.

The B&O tax is applied at varying rates, depending on the nature of the taxpayer's business and the activities engaged in by that taxpayer. Calculating the tax may become very complicated, because taxpayers engaging in multiple activities may be subject to multiple tax rates. Examples of the different activities in which a taxpayer may be involved include extracting, manufacturing, distribution, processing, publishing, providing financial services, etc. Taxpayers must determine the revenue they receive from each activity and apply the relevant tax rate.

Some exemptions, deductions, and credits are applicable to the B&O tax. A few of the relevant tax credits available are as follows:

- Multiple Activities Tax Credit
- Small Business B&O Tax Credit
- High Technology B&O Tax Credit
- Distressed Area B&O Tax Credit

In addition to the statewide B&O tax, a local B&O tax may also be imposed. B&O taxes are paid on an excise tax return or by electronic filing. They are paid monthly, quarterly, or annually. Monthly returns are due on the 25th of the following month, quarterly returns are due at the end of the month after the close of the quarter, and annual returns are due on January 31.

Sales and Use Taxes

Washington also has sales and use taxes. Sales taxes are imposed on purchases made inside the state, primarily of tangible personal property. Use taxes are imposed on goods that were purchased outside Washington but will be used inside the state. The sales and use taxes are mutually exclusive.

Property Taxes

All real and personal property in Washington, whether tangible or intangible, is subject to property taxes, unless exempted by law. Real property includes land, buildings, structures, improvements, and other fixtures on the land. Personal property includes goods, stocks, leases of real property and leasehold interests, etc. Certain things are exempted from the definition of personal property, including mortgages, notes, accounts, bonds, and business inventory. Property taxes are arranged in a two-year cycle such that property taxes are assessed in the first year and are due for payment the following year.

21

Issuing Securities

Don't gamble; take all your savings and buy some good stock and hold it till it goes up, then sell it. If it don't go up, don't buy it.

Will Rogers
Social commentator and humorist

What to Watch For

Although issuing securities in your company may seem like a very simple concept, even successful companies have found themselves forced to clean up problems as a result of improper stock issuances. These problems can range from minor ones that cause time and energy to clean up to major problems that bring about fines from the SEC.

Proper stock and option issuances are crucial for companies as they raise funds from sophisticated investors, consider mergers and acquisitions with other organizations, and contemplate doing an IPO. For this reason, it is important that companies exercise due care with their securities issuances.

Key Considerations for Issuing Securities

In general, companies should be aware that both federal and state laws regulate the offer and sale of securities, including stock, options, and warrants. Within the company, it is generally the board of directors that will approve all offers and issuances of securities, oftentimes requiring additional approval from a percentage of the stockholders.

Before you issue any company securities, you should ask the following questions:

- Has this issuance been approved by the board of directors, if necessary?
- Has this issuance been approved by the stockholders, if necessary?
- Is this issuance valid under federal law?
- Is this issuance valid under state law, including the state of incorporation, the state where our headquarters is located, and the state where the stockholder will reside?

Be sure that you have properly considered these key questions before issuing stock, options, warrants, or other securities to avoid costly cleanup down the road.

Securities and the Startup

There are a number of securities that can be used with a startup company. At various points, a company may choose to issue equity or debt securities. A security is simply a fungible financial instrument that represents value. The entity issuing a security is called the issuer. What qualifies as a security depends on the regulatory structure of each country. Securities laws govern the raising of capital for business purposes.

There are many different types of securities to aid in the financing of an emerging business. Securities are broadly categorized as debt and equity instruments, such as bonds and common stocks, respectively. Debt instruments generally are fixed obligations to repay a specific amount at a specific date in the future, with interest. Equity instruments generally are ownership interests entitled to dividend payments but carry the specific right to a return on capital. For debt and equity securities, the issuer may establish a wide variety of rights, privileges, and limitations.

When deciding how to finance a startup, various strategies can be used to obtain a favorable debt and equity combination. Aside from business strategy, the laws regulating securities is complex, with a mix of federal and state regulations. Competent securities counsel is crucial to the successful financing of a startup.

The equity securities may involve ownership of a class of stock, typically either common or preferred stock. Debt securities may involve promissory notes, loans, or other debt instruments. Other securities operate as more of a hybrid between securities: convertible promissory notes and convertible preferred stock, to name a few. Finally, other securities such as stock options and warrants provide a mechanism to provide for a future interest in ownership of stock of the company. Although much of this discussion will focus on the securities of a corporation, LLCs do offer companies the opportunity to issue membership units in the company.

COMMON SECURITIES ISSUED BY STARTUPS

- **Common stock:** Common stock is an ownership interest in a corporation, entitling one to a portion of the corporation's earnings and assets after the corporation's dissolution. If a company only chooses to issue a single class of equity, common stock will be the class. Generally, the founders, management, and employees will be issued common stock of the corporation (it is more typical for investors to be issued shares of preferred stock). After liquidation of the company, the common stock will typically be the lowest claim priority, after secured and unsecured creditors, bondholders, and preferred stockholders.

  Common stock is the primary form of equity financing. Common stock holders may or may not have associated voting rights in the corporation's management decisions. Common stock holders are typically given voting rights for the election of directors of the corporation's board. The benefit of common stock is closely aligned with the corporation's success and failure. When a corporation is profitable, common stock holders are entitled to unlimited appreciation in the value of their common stock. Alternatively, common stock holders may obtain no assets after a corporation's dissolution after all other creditors and preferred stock holders are paid with priority.

  Common stock is advantageous to a corporation because there is no obligation to repay the principle equity invested in the company for the stock, there is no obligation to pay dividends, investors obtain a right to share in the growth of the corporation, and investors may influence the management of the corporation through voting for directors. Common stock may be disadvantageous because it dilutes management's interest in the corporation and imposes great risk on investors.
- **Preferred stock:** Preferred stock represents a particular class of stock of the corporation with certain rights that common stock does not have. These rights may include liquidity preference, decision-making management control, dividends, antidilution protection, participation, veto provisions, and others.

  Preferred stock is an alternative equity instrument to common stock and is typically used for VC investments. Preferred stock will usually have priority in dividend payment and liquidating proceeds over common stock. Preferred stock may be voting or nonvoting. Preferred stocks may be beneficial because they do not directly dilute the ownership stake of the management of the corporation in the common stock (until a future event) and have predictable dividend payments and priority in liquidation. However, issuing preferred stock may be disfavored because it lessens the amount of dividends that can be paid to common stock and used by the corporation generally.
- **Convertible preferred stock:** Convertible preferred stock represents a subset of preferred stock that provides the owner the right to convert the preferred shares into common shares of stock. This subset of preferred stock is the typical equity security issued to venture capital firms and other private equity investors. In most cases, convertible preferred stock will convert automatically into common stock when the company undertakes an IPO.

- **Warrant:** Warrants provide a holder with the right to purchase shares in a company at a predetermined price, generally at a price above the current price or value of the stock. Warrants represent long-term purchase options that will typically be valid over a period of several years or indefinitely. Oftentimes, warrants will be issued in conjunction with preferred stock or bonds to increase the attractiveness of the stocks or bonds.
- **Stock option:** Stock options represent the right of the holder to purchase or sell a share of stock at a specific price within a specific period of time. Stock options to purchase common stock of the company are often used for employees and management of high-technology or startup companies as incentive compensation. Many companies will offer these options as a part of an employee compensation plan, usually in the form of a stock option plan.
- **Debt instruments:** Generally, debt instruments, such as notes, bonds, and debentures, are entitled to receive payment before preferred or common stockholders. Debt instruments may be secured by assets of the issuer or may be unsecured. Debt instrument holders generally cannot participate in the management decisions of a business, although they may impose certain affirmative or negative obligations on the business. Debt instruments can be long or short term and carry fixed or variable interest rates. Debt instruments are sometimes preferred because they create a predictable payment schedule to investors, do not lessen management's interest in the growth and voting power in a business, and involve less risk for investors than equity investments. They may be disfavored, however, because they potentially restrict operations, limit the use of working capital, and tie up assets through pledges of collateral on the debt.

Corporate Approvals

Issuing securities will oftentimes require certain approvals of the board of directors or the stockholders. Determining what approvals are necessary will usually be found in the corporate laws of the state of incorporation and the company's articles of incorporation and bylaws. Many companies will prepare a summary sheet for the board of directors that detail the authorized securities outstanding and the required approvals for the issuance of any additional securities.

Typical actions that may be necessary or desirable for the board to approve include the following:

- Issuing securities and granting options, warrants, or other rights to purchase securities
- Adopting a stock option plan
- Declaring distributions, dividends, or stock splits

Some of the actions for which shareholder approval may be required or desirable include the following:

- Amendment of the articles of incorporation (including the authorization of additional shares of stock)
- Issuance of securities
- Adoption of stock option plans

Before the issuance of any securities, the company should first be certain that the securities are authorized under the articles or certificate of incorporation. After such a determination (or authorizing the shares by filing a proper amendment to the articles with the state), the company will then need to obtain the proper approvals by the board of directors and/or the shareholders of the company.

Federal Regulation of Securities

Issuing securities in a new startup company is a way to allocate ownership of the company among various stakeholders. For private companies, there are unique considerations to issuing securities.

SECURITIES LAWS AND THE ENTREPRENEUR

For a new or experienced entrepreneur, much of this information may seem like complex legalese, too technical, highly irrelevant, or even unnecessary. Most startup companies will hire an experienced attorney who can address many of these issues; however, it is still important to familiarize yourself with the basics of these rules. Remember to avoid issuing any securities without discussing the securities laws with your attorney. Securities laws are quite complex and can be implicated through the simple act of giving your brother 100 shares of stock in the company.

For most early-stage, private companies, you will be primarily concerned with ensuring that any security issuance is exempt from registration with the SEC or state agency regulating securities laws. Before you make a sale of stock to a friend, family member, relative, or another individual or company, you should be certain that you remain compliant with a federal exemption. This is particularly important as the company continues to grow. For startups considering more advanced funding sources (for example, raising money from the public markets via an IPO), you'll start to be concerned with different federal securities provisions. However, at this point, work with your attorney to ensure each issuance qualifies for some exemption (and usually any properly structured issuance will be).

The Securities Act of 1933 (which will oftentimes be referred to as simply the Securities Act) and the Securities Exchange Act of 1934 (often referred to simply as the Exchange Act) are the two primary sources of federal law regulating securities transactions. The Securities Act regulates all issuances of securities by a corporation. The Exchange Act regulates the resale of securities and generally is not applicable until the corporation has made its initial offering of stock to the public. The principal federal securities regulatory agency is the SEC.

The Securities Act prohibits all sales of securities unless a registration statement containing certain required information has been filed with the SEC and has been declared effective or unless either the security or the transaction is exempt. Registration with the SEC is expensive and time consuming, and, as a practical matter, startup companies are usually unable to use the registration process. Accordingly, startup companies must structure their securities offerings so that one of the exemptions from the registration requirements of the Securities Act will be available.

ONLINE SALES OF SECURITIES

What could happen

You have decided to use the Internet to post an online advertisement offering to sell securities in your company in exchange for cash.

Watch out for

Federal and state laws limit "general solicitations" for private securities offerings. In most cases, posting an advertisement for securities on a general website, a listserv, an online blog, in newspaper ads, or other similar locations will violate these provisions. There are some new means whereby sales of securities through websites advertising to "prequalified" individuals have been allowed. However, when in doubt, ask a professional before soliciting investors online. Some individuals or websites may provide you with assurances about using their site. Don't risk a securities law exemption without consulting your own expert first.

TIP: General solicitation of investors (even online advertisements) is not allowable. When in doubt, check with a professional.

Common Exemptions to Registration for Company Sales and Issuances

For a startup company, your sales of securities will tend to be exempt from registration under three key provisions:

- Private placement exemption
- Regulation D exemption
- Intrastate offering exemption

ALL ACCREDITED INVESTORS ARE *NOT* CREATED EQUAL

Practical Pointer

Securities laws provide certain exemptions that companies can use to avoid registering sales of their securities. For private companies, a key exemption is based on investments made by "accredited investors." Accredited investors are defined as persons with a net worth of $1 million or annual income of $200,000 in each of the previous two years with a reasonable expectation of making $200,000 in the current year.[1]

Critics note that these limits, meant to provide a substantial threshold for investors eligible to invest under this exemption, no longer do so, because these limits have not been changed since 1982! Now, it is not uncommon for individuals to have $1 million dollars in equity just in their homes. Today, that same $1 million in assets would be more than $2.25 million adjusted for inflation.

As a result, simply because an individual is accredited, does not make them suitable as an investor in a high-tech startup venture and should not be your only threshold. If possible, look to investors that have previously invested in startup or emerging companies. When that is not possible, be sure that any investor truly has the means to invest (and isn't liquidating their child's college fund) and fully appreciates the risks a new company faces.

Avoid a potential scenario in which an investor who is not familiar with the limited liquidity, higher risk, and greater uncertainty of an investment into a startup decides to hire a lawyer to file a complaint to a state agency or to bring suit against your company. Sadly, these suits do happen and, many times, on the basis of relatively small investment amounts. Even if the suit is without merit, your new company does not need to spend time and energy on an investigation by the state's securities investigators.

[1] See Rule 501 of Regulation D of the Securities Act of 1933 for the full description.

Private placements of securities involve no advertising in any media and no widespread solicitation of prospective investors. Regulation D is a set of more specific rules and guidelines related to sales or securities that will be exempt from federal registration requirements. These rules set out specific limitations on issuance including limitations based on the amount of securities sold and the number of unaccredited investors. The intrastate offering exemption includes any security that is offered or sold solely to persons residing within a single state, as long as the issuing company is a corporation incorporated within such state and conducts substantially all of its business from that state.

The following textboxes lay out the specifics of each exemption and should be reviewed before issuing securities.

PRIVATE PLACEMENT EXEMPTION

Summary

A private placement of securities involves no advertising in any media and no widespread solicitation of prospective investors. In a private placement, the issuer usually has a preexisting relationship with the purchasers. The issuer or its agent will personally contact a limited number of prospective investors.

Section 4(2) of the Securities Act

Section 4(2) of the Securities Act exempts an offer and sale of securities in transactions that do not involve a public offering. The exemption applies to transactions with a limited number of sophisticated investors who possess sufficient financial bargaining power and sophistication to compel the issuer to provide them with all information concerning the issuer critical to their investment decision.

Although the number of offerees must be small, there is no bright line number of offerees above which the transaction will no longer qualify under Section 4(2). The issuer must be able to prove that the number is small. If an offering package is used, limiting its distribution and recording all persons who receive it is one typical means of doing this. In addition, each investor will be asked to sign an investment representation statement, representing that he or she is purchasing the securities for investment purposes only and not with a view to reselling or distributing the shares.

REGULATION D EXEMPTION

Summary

Regulation D is a series of substantive and procedural rules, numbered 501 through 508, that provide three main transaction exemptions (Rules 504, 505, and 506) for private offerings. Regulation D is the SEC's attempt to provide precise guidelines that, if followed, will ensure issuers of the availability of an exemption from the federal registration requirements. Securities issued in reliance on one of the Regulation D exemptions are subject to certain restrictions on resale.

Form D Filing

Regulation D requires that the issuer of securities file with the SEC five copies of a notice on Form D no later than 15 days after the first sale of securities in a Regulation D offering. One copy of every notice must be manually signed by an authorized person of the issuer.

- **Rule 504:** Rule 504 of Regulation D provides a safe harbor for offerings of up to $1,000,000 over any 12-month period. No specific sophistication level is required of the purchasers, and Rule 504 contains no specific disclosure requirement. However, all issuances of securities are subject to Rule 10b-5 promulgated under the Exchange Act, which makes it unlawful for the issuer to make any untrue statement of a material fact or to omit to state a material fact necessary to make any statements actually made not misleading. In general, subject to certain very limited exceptions, no advertising or general solicitation of investors is permitted.
- **Rule 505:** Rule 505 of Regulation D permits the sale of up to $5,000,000 of securities during any 12-month period to any number of accredited investors and up to 35 other purchasers. If any unaccredited investors will be purchasing securities, Rule 505 requires that a disclosure statement be prepared that contains substantially the information that would be contained in a formal registration statement. No advertising or general solicitation of investors is permitted under Rule 505.
- **Rule 506:** Rule 506 of Regulation D relates to transactions that are deemed to be exempt from registration under Section 4(2) of the Securities Act. It permits the sale of securities (without regard to dollar amount) to any number of accredited investors and up to 35 other purchasers. Before the sale of the securities, the issuer must reasonably determine that each nonaccredited

investor (or his or her purchaser representative) has sufficient knowledge or experience in finance and business and that he or she is capable of evaluating the merits and risks of the investment. The securities must have limitations on resale, and the issuer must exercise reasonable care to ensure that the purchasers are not statutory "underwriters" because of their intent to redistribute the securities acquired. If any nonaccredited investors will be purchasing securities, a disclosure statement must be provided containing substantially the information that would be required in a formal registration statement. No advertising or general solicitation of investors is permitted.

INTRASTATE OFFERING EXEMPTION

Section 3(a)(11) of the Securities Act exempts from its registration requirements any security that is offered or sold solely to persons residing within a single state, provided that the issuer is a corporation incorporated within such state and conducts substantially all of its business from that state.

Each purchaser and each offeree of the securities must be a resident of the state. In addition, the ownership of the securities must be restricted to residents of the state for a period extending nine months after the last offer has been made. Because an option or a warrant is considered to be a continuing offer, if options or warrants are sold in the transaction, the restriction on transfer must be maintained until nine months after the expiration of the last option or warrant.

In addition, Rule 701 under the Securities Act provides an exemption from registration to certain issuers for issuances of securities to employees if the offers and sales of the securities meet various conditions. Issuers must not be reporting companies (i.e., subject to the reporting requirements under Sections 13 or 15(d) of the Securities Exchange Act of 1934, as amended) or investment companies registered or required to be registered under the Investment Company Act of 1940. The exemption only applies to issuances of securities in connection with employee compensation and not in connection with capital raising transactions. Sales and issuances must be pursuant to a written plan or agreement, and the aggregate annual dollar amount of exempt securities is limited by the rule. Securities sold in reliance on Rule 701 are subject to certain resale restrictions under the rule.

HIRING AN AGENT TO SELL YOUR STOCK TO INVESTORS

What could happen

To motivate your team to help raise capital, you offer anyone on your management team or your sales force a commission if they can make a sale of your securities to an investor or investors.

Watch out for

Securities laws limit your ability to pay commissions (or similar types of compensation) for sales of your company's securities. Only a licensed broker-dealer may receive commissions under securities laws. In addition, be sure that, if you engage a third party to assist with your efforts to obtain investment funding and if this individual or organization will receive a commission or other fees based on finding you investors, that they are licensed as a broker-dealer. Some individuals will attempt to work around these rules by using a title of "finder" or "funding consultant" but remember that, despite their assurances or previous dealings, only registered securities brokers can earn commissions on sales or placements of your securities.

TIP: You may only pay commissions on sales of your securities to licensed broker-dealers.

Sales and Transfers of Securities by Other Parties

After you have issued securities and complied with federal securities laws for the issuance, what happens when stockholders want to sell their shares? These resale transactions must also comply with securities laws.

In the event that any of the stockholders of your company want to sell their securities, those transactions will need to comply with certain federal laws. The offer and sale of shares by persons other than the company issuing the securities are also regulated by federal and state securities laws. These sales and transfers will most often be exempt under two exemptions: Section 4(1½) Exemption and Rule 144 Exemption.

Under federal securities laws a person wishing to sell restricted stock must register the shares under the Securities Act or sell in a transaction which is exempt from the registration requirements.

Section 4(1½) Exemption

The basic idea behind the Section 4(1½) exemption is to provide some leeway from the registration requirements for offers and sales by a securities holder of the issuer under circumstances in which the Section 4(2) exemption would have been available for a sale by the actual issuer of the securities. Although the so-called Section 4(1½) exemption is the exemption most widely used in the resale of securities of a nonpublic company, it is not specifically recognized under the Securities Act or by the SEC. This exemption has evolved through securities law practice, SEC "no action" letters, and the commentaries of securities law scholars.

Accordingly, the factors that apply to private placements under Section 4(2), such as purchaser sophistication, ability to bear the risk, economic bargaining power, and investment intent, must be present for the transaction to fall under the Section 4(1½) exemption. The issuing company must regulate the resale of restricted stock by its securities holders and may require either an opinion of counsel satisfactory to the company to the effect that the transfer may be made without registration or a "no action" letter to the same effect issued by the SEC, in which the staff of the SEC provides written assurance that it will not recommend that the SEC take enforcement action against the parties to the transaction.

Rule 144 Exemption

Rule 144 allows the resale of restricted securities to the public if the following applies:

1. The securities have been held for more at least six months.
2. The issuing company is a "reporting" (i.e., public) company that meets Rule 144's requirement that current public information about the company be available through the filing of all required reports with the SEC.
3. The sale is made in an unsolicited "broker's transaction" or directly with a market maker.
4. The number of shares sold by the seller (and certain affiliated sellers) does not in any 90-day period exceed the greater of 1% of the total outstanding shares of the issuer or the average weekly trading volume of the issuer's securities for the four weeks preceding the sale.
5. A notice on Form 144 is filed with the SEC before or simultaneous with the sale.

Nonaffiliates of the issuer (i.e., persons who are not officers, directors, major shareholders of the issuer, and certain family members and other persons, controlling or controlled by any of the foregoing) who have held their shares for more than three years may sell their securities without complying with the requirements listed above, pursuant to Rule 144(k). Under Rule 144, the holding period commences when the shares are fully paid. In the case of securities purchased by promissory note, such securities are not deemed fully paid for purposes of Rule 144 unless the note is secured by adequate collateral other than the securities purchased or until the note is paid off.

The SEC has adopted the position that, unless a resale is effected under Rule 144, the parties to an unregistered sale of restricted stock bear a substantial burden of proof in establishing that an exemption from the registration requirements of the Securities Act applies. Because many uncertainties exist in this area of the securities law, it may be wise to seek guidance from the SEC by requesting a no action letter in which the facts are not clear cut.

State Regulation of Securities

In addition to the federal securities laws, issuers and sellers of securities must comply with state securities regulations also referred to as "blue sky laws." Exemptions are specific to the state. Many blue sky laws provide a statutory transactional exemption modeled in some manner on Section 4(2) or portions of Regulation D under the Securities Act. Variations in these state statutes include the following:

1. Limitations on the number of offerees within the state and/or total offerees
2. Similar limitations on the number of purchasers within a 12-month period
3. Affirmative filing requirements with the state securities administrator before and/ or after the offering
4. Possible limitations on commissions paid
5. Limitations on the manner of the offering and prohibitions on general solicitation and advertising

6. Requirements concerning minimum amount of investment
7. Use of state-registered broker-dealers in connection with the offering

Many states also provide an exemption for "isolated" transactions, involving very few sales within a prescribed period of time meeting the statutory, regulatory, or case law interpretation of the relevant state.

You may be responsible for compliance in states where the holders are to reside, the state of your company's incorporation, as well as the state where your headquarters is located in certain cases. In most cases, your attorney will be able to assist with the necessary filings and researching the applicable exemptions.

California

California's securities regulatory system is perhaps the most highly developed system among the states. Generally, all offers and sales of securities in California (including sales and transfers among shareholders and third parties in which the issuer is not involved) must be qualified with the Department of Corporations unless either the security being offered or sold or the transaction is exempt under the California Securities Law or the regulations. California securities regulations also reach certain other transactions, such as modifications of the rights of existing security holders, which are exempt under federal law and the laws of other states. California regulations, like federal regulations, apply both to the offer and the sale of securities, and qualification must be obtained or an exemption identified before an offer commences.

To qualify an offer and sale of securities with the California Commissioner of Corporations, a permit application must be filed with the commissioner. In granting a permit, the commissioner may restrict the class of persons to whom the securities may be offered and sold and may impose certain requirements on the rights, privileges, preferences, and restrictions of the securities themselves. Several exemptions from the qualification requirement are available.

CALIFORNIA: EXEMPTION UNDER SECTION 25102(F)

Startup companies most often rely on the Section 25102(f) exemption, which exempts offers and sales of securities as long as the following criteria are met:

1. The security is sold to no more than 35 "counted" persons.
2. Each purchaser has a preexisting business or personal relationship with the issuer involving contacts of a nature and duration to enable a reasonably prudent purchaser to be aware of the character, business acumen, and general business and financial circumstances of the issuer or seller, or is a "sophisticated investor."
3. There is no general advertising of the offering.
4. Each purchaser represents that he or she is acquiring the shares for his or her own account for investment purposes only and not for distribution to the public.
5. A filing is made with the Department of Corporations along with payment of a fee within 15 days after the first sale of a security in California under Section 25102(f).

In addition, an exemption contained in Section 25102(o) applies to the issuance of securities issued pursuant to stock purchase and stock option plans and agreements, provided that such securities are exempt from federal registration pursuant to Rule 701 under the Securities Act and that the terms of the plan or agreement conforms to certain Department of Corporations regulations. The filing and fee requirements of Section 25102(o) are similar to those under Section 25102(f), except that the notice is due within 30 days of the first issuance of securities under the plan.

California also regulates the resale of securities that have been issued pursuant to a permit. Under these regulations, the commissioner's consent must be obtained before any transfer of securities in California, unless an exemption from the consent requirement is available. California law provides for a number of exemptions to the consent requirement, including transfers by gift, transfers back to the issuer, transfers to another holder of the same class of securities of the issuer, and transfers in which neither the buyer nor the seller is a California resident.

Part IV

Next Steps and Stages

Money is only used for two things. One, it's to make you comfortable, and the more comfortable you are the more creative you will become. And the other purpose is it enables you to extend the service you provide far beyond your own presence.

Bob Proctor
Author of The Science of Getting Rich

22

International Expansion

Globalization has changed us into a company that searches the world, not just to sell or to source, but to find intellectual capital—the world's best talents and greatest ideas.

Jack Welch
Former CEO of General Electric

What to Watch For

So you have had some entrepreneurial success—you had a great idea, made a plan, built a team, formed a company, received funding, and, most of all, marketed your service or product profitably. Or perhaps your business plan contemplated an international component from the very beginning. In either case, although you may still have room to grow domestically, you might be ready to start thinking about expanding abroad.

This chapter covers some of the common issues faced when doing business abroad, including simple sales of products internationally, opening offices abroad, and engaging in more complex international transactions. In addition, a couple helpful case studies can be found in this chapter, including one on forming a joint venture in China, opening a Japanese subsidiary, and, in the following chapter on IPOs, listing on the London Alternative Investments Market. Obviously, any in-depth discussion of these topics could fill its own book, so keep in mind that this chapter aims only at providing a starting point for considering these issues.

If you believe you are ready to expand your footprint across the globe, your next step will be to begin discussing these plans with your managers, investors, accountants, and legal counsel.

Expanding into Global Markets

Will your service or product be successful overseas? It might. Many foreign markets are eager to have access to what U.S. consumers enjoy, but foreign markets are just that: foreign. You will need to revisit your business plan and make sure that all of the ingredients of your domestic success either apply in the foreign market or can be feasibly substituted with elements more appropriate to that market. What works in the United States might work abroad as is, although this type of globalization is becoming increasingly rare. More likely, you will need to adjust your product or service and think again about the various facets of your business. How will you do this?

First, you must anticipate the shape of your business activities in this new market. Many startup companies will first look to sell their product internationally, perhaps in similar markets such as Canada or the United Kingdom. As you continue to grow, you may explore other geographies and may even look to expand your footprint to have an office in a growing market. At this point, you will have to think again about your choice of business entity. Foreign jurisdictions have business forms that differ from American business forms in important ways. Once you have selected what type of an overseas presence you want to form, you will need to address some details of doing business there. You will almost certainly need to contract or partner with local businesses for a variety of support. If you are starting from scratch in the new country, you will need office space, telecommunications services, employees, and, if you sell a product, production, marketing, and distribution as well. This list resembles the list of issues you faced when you first started your business but with the added complications that you must now execute your business in a foreign market and at the same time integrate this effort with your existing U.S. operations.

Beyond requiring a large amount of business adaptation, your overseas ambitions will present a set of legal issues not relevant to solely domestic concerns. Of course, you will have to comport with the laws of the target country, but you will also be subject to additional U.S. laws (mostly federal, but some state as well) that govern U.S. companies operating abroad. You will have to navigate new, often complex, tax issues. Foreign jurisdictions differ in important areas of substantive law as well, such as intellectual property rights.

OUTSOURCING TO INDIA

A growing number of startup companies in the information technology industry have begun to give outsourcing a long, hard look. According to some experts, Indian information technology providers offer gross savings of 40% to 50% and 25% to 35% net savings to U.S. businesses. Remember to do your due diligence on any foreign service provider. For more information on Indian firms providing outsourcing services, visit the following websites:

- National Association of Software and Service Companies (http://www.nass com.in)
- Dataquest (http://www.dqindia.com)
- Global Services (http://www.globalservicesmedia.com)

Selling Your Products Abroad

One of the first steps of international expansion a company will take is selling products into an international market. Today, sales into global markets represent an increasing part of any company's long-term growth strategy, but such a strategy is not without its share of challenges and obstacles to overcome. Determining where to sell the products, how to reformulate or reconfigure the product, how to ship the product, and understanding the rules, domestically and internationally, represent a few issues you'll face.

Selecting Markets

After you have made the choice to pursue international sales opportunities (or perhaps you've seen demand from those markets on their own), you'll need to determine which markets to pursue and understand the feasibility of introducing your products abroad. For most new market introductions, you will need to consider a feasibility study looking at the market demand and the opportunities for your product to fill market needs.

The feasibility study will be based on an understanding of individual country or regional markets, including local demand, pricing structures, distribution mechanisms, and delivery means. Experts will recommend a company focus their efforts on the top two or three best prospect markets rather than attempt a broad platform of international growth up front. Determining which markets should be based on a larger review of a number of markets, perhaps beginning research to understand a particular region or regions such as Europe, Southeast Asia, or South America, for example, before selecting individual countries or localities.

U.S. DEPARTMENT OF COMMERCE RESOURCES

One of the first places to begin your study is from information compiled by the U.S. Department of Commerce, specifically on their Export.gov website (found at http://export.gov). Visit the U.S. Commercial Service Market Research Library, which houses more than 100,000 industry and country-specific market reports that contain local information useful in understanding numerous markets. More specifically, the U.S. Commercial Service Market Research Library contains the following:

- Country commercial guides
- Industry overviews
- Market updates
- Multilateral development bank reports
- Best markets
- Industry/regional reports

One of the first steps in identifying potential global markets is to research trading information to identify countries currently purchasing products in your market. One of the mistakes some companies make in identifying a country for growth is selecting a market that is doing little importation in a particular market of interest. Therefore, it is helpful to research legal considerations, taxes, tariffs, market openness, common practices, and distribution channels to identify targets.

Once you have narrowed the search down to a few key markets of interest, you will want to further research macro trends in the country, including economic and demographic trends. At this stage, you should also examine the price and cost implications of a marketplace. In addition, be cognizant of barriers to entry in any new marketplace, including tariffs, or the potential existence of U.S. or foreign incentives to exportation of your product or service.

Finally, once you've selected a market or markets, you may want to perform actual testing of demand within an international market. The U.S. Commercial Service offers various services to assist with these efforts, including catalog exhibitions, access to the magazine *Commercial News USA*, and services such as foreign partner matching and trade lead services.

EXPORT ASSISTANCE CENTER

For companies that are new to exporting and are considering exporting into certain target markets, consider scheduling an appointment with your local Export Assistance Center, a service of the U.S. Department of Commerce. These local offices are located in more than 100 U.S. cities and 80 international cities. To find a local office in your city or state, visit: http://www.export.gov/eac/. For general questions about exporting such as tariff rates or U.S. Federal Government export assistance programs, call the Trade Information Center at 1-800-USATRAD(E).

Understanding International Issues

After you have selected a market, you will need to research specific issues affecting the country where you will be selling.

Shipping and Distribution Issues

Shipping internationally involves a number of issues, including protecting the products as they are shipped, international tariffs, taxes, and fees, distribution routes, documentation, insurance, and language barriers. To overcome these challenges, many companies will hire a freight forwarder to assist in the process. A freight forwarder is generally aware of the majority of potential problems, can provide you information about the process before you ship the product, will be able to give you packaging and distribution guidance, and can assist with distribution once the product passes through the local port. The costs of a freight forwarder can usually be included with the overall costs to ship your products to an international destination.

Legal Issues

Many countries have very specific rules and regulations governing imports from another country. In addition, you may find that sales will trigger certain licensing or filing requirements. Some products such as technology, medicines or biotechnology, and agricultural products have stringent restrictions on international transactions. Likewise, you should be cautious with items that could be used for potential terrorism acts that are subject to both export and import restrictions. In most cases, it is helpful to retain local counsel to address some of the likely issues that will arise entering into a new country.

Product Modifications

Entering into some regions or countries may require substantial changes to your products. In particular, companies will face obstacles with respect to the exclusive use of the metric system, differing electrical components, varied mobile and wireless infrastructure, and requirements for use of different languages for certain products. For example, software companies have been required to change the language displayed for users. Before you consider entering a new geography, be certain to research the product modifications necessary to determine whether these changes represent a major overhaul or simply a minor modification and the associated costs.

Packaging and Labeling

Language will make labeling a product abroad different from the packaging used within the United States. Even countries that speak English may not associate certain phrases, markings, or graphics in the same manner as Americans. Be aware of local restrictions on the use of certain international trademarks or confusion that may result from the use of a name associated with another product. Some countries have rules requiring packaging to be readable in numerous languages, as well.

Instructions and Warranties

As with packaging or labeling, product instructions or warranties will require translation into the local language. With technical products, it is even more crucial to find a local service that can provide translation of technical terminology to avoid confusion. (Remember the time you tried to read instructions to a product manufactured abroad with poorly translated English?) Likewise, be certain to provide clear and understandable instructions related to any product warranties as well as contact information.

Avoiding Pitfalls

Selling your products internationally is a significant step (perhaps not as significant as opening an office abroad, as will be discussed later in this chapter). Here are a few of the key items you should consider when selling internationally:

- **Failing to secure a local partner (or to secure a good one):** Much the same way as selling domestically can be driven by the talent or lack of talent you employ in your sales and marketing forces, so too can you struggle in your selection of local partners. Selling products in international locations involves a series of challenges, including import-export rules and tariffs, local distribution, finding local retailers, and transactions sometimes without face-to-face interactions. In some cases, international operations can involve different ethical considerations that may not come to light until after the transaction has taken place. Be sure to find a partner to work with in the local jurisdiction that is trustworthy and capable of assisting with your business efforts.
- **Failing to adequately commit to international operations:** International markets aren't the place to enter quickly and hope for the best. International companies may be viewed with a suspicious eye at first and will require a sustained commitment to secure significant penetration. Before you consider entering a market, be certain that you are prepared to commit for the time necessary.
- **Failing to understand how to sell and market your product:** If you invested a great deal of time and energy into determining how to sell your product domestically and, after a bit of trial and error, found success, you may be tempted to apply those same lessons to international markets. Remember that international markets may have some similarities but will also require customizations. Starbucks found that it needed to market to younger women in Japan to get a foothold. So remember that just getting your product onto shelves won't be enough. Be sure to develop a specific marketing and sales plan for the locality.
- **Failing to modify products to meeting local needs or demands:** Selling a product into a foreign market will oftentimes involve a number of modifications from sizes (oftentimes metric), to languages, to electrical conversions, to packaging. Have you ever traveled to Europe or Asia and tasted the differences in the flavor of sodas such as Coke or Pepsi? If you did, you'd know that the beverages were reformulated to match the tastes of the local consumers. For some products, the packaging will need to be completely changed to make your product appealing to a consumer from another country. Likewise, you may need to modify the product or packaging to comply with local laws, rules, or even customs. Be certain to research effectively before making a single sale into an international jurisdiction.

FAR FROM HOME . . .

What could happen

You are approached by a distributor who is ready to take your product global. You'd originally planned to hold off on international expansion, but this may allow expansion without a sizeable investment in international sales staff.

Watch out for

Different rules of business and business ethics apply in other countries, even countries we would consider to have well-established business regimens. In particular, watch out for different revenue recognition policies (when is a sale really a sale?). Remember that your reputation is tied to the reputation of your distributors.

Corporate culture is difficult to instill in markets far away from your home, but if you try to go global, do your best to instill and maintain your culture. Schedule regular visits and organize trips back to the home office for training and regular communications.

TIP: Ensure compliance with corporate policies, rules, and laws for all international personnel and employees.

Tax Planning

You should consider tax issues at the outset of your overseas business expansion planning. Errors in the structure of overseas business formation can result in tax liability at home or in the foreign jurisdiction that can wipe out the gains of a successful new business, and net operating losses in one jurisdiction will not offset steep corporate taxes in another. In fact, the range of tax exposure varies so dramatically by jurisdiction and type of business form used that tax issues should be considered when deciding whether or not to expand overseas in the first place.

A few examples will serve to make the point. Many countries look to whether a new business has established a permanent establishment (PE) for tax purposes. Those companies that do will fall under the corporate tax regimen of that country, and even small scale business activities, such as an employee working from home or a hotel room, may be considered a PE, require registration, and trigger tax liability. Keep in mind that local officials may look to the type of business activity being undertaken to determine whether a PE exists, so the phrasing of business name or title may not be enough to avoid being labeled as a PE. Once under a corporate tax regimen, your business may be subject to corporate taxes that reach as high as 50%. At the same time, bilateral tax treaties may allow a business to claim a tax credit in the United States for corporate taxes paid overseas, and many corporate tax regimens tax smaller businesses at a lower rate.

Your foreign employees will be subject to the tax regimens of their home countries. In many jurisdictions, employers must deduct employee income taxes and social security (or its foreign equivalent) from employees' salary. These can often be tricky regulations with which to comply, and it is often easier for you to hire a payroll or accounting service to manage these payments and make certain that the proper amounts are withheld from your employees' salaries and paid to the proper government agencies.

DOING RESEARCH ON INTERNATIONAL TAXES

Where to go if you are looking for more information:

- Need information on the relevant international tax policies of more than 80 countries and foreign jurisdictions? Visit Tax Sites at http://www.taxsites.com/international.html.
- Need information to assist foreign nationals to comply with U.S. tax laws? Visit the Nonresident Tax Guide at http://www.thetaxguy.com/nrtaxguide.htm for a summary of the filing requirements.
- Need information to assist U.S. expatriates overseas to comply with U.S. tax laws? Visit the Expat Network at http://www.expatnetwork.com/?ID=65 or Global Tax Help at http://www.globaltaxhelp.com.
- Need software to assist you with planning for international taxes or for payment of various international taxes such as VAT or GST? Visit Tax Sites at http://www.taxsites.com/International_2.html for a list of and links to various software packages.

Some jurisdictions have additional tax laws that are rare or unknown in the United States. In Europe, subsidiaries of foreign companies must register for national, regional, and local taxes much as you would expect but must also register for a value-added tax (VAT). A VAT is a tax paid on each transaction in a chain of transactions, provided that the transaction adds some value to the product being transacted. In some cases, your foreign subsidiary may benefit from early VAT registration even if it will not be liable for VAT payments for some months. In Japan, corporations must pay certain local taxes, in addition to the national corporate income taxes, such as the corporate inhabitant tax. The corporate inhabitant tax is assessed not only on income but also on a per capita basis using the corporation's capital and the number of its employees as the tax base. All of these examples are offered to encourage you to seek professional tax advice at the earliest stages of your overseas business planning.

Direct Overseas Activity: How Big a Footprint?

As you are choosing service providers, managers, accountants, and counsel to help you in your overseas expansion and beginning to consider important tax issues, you can start to think about what type of overseas presence you want to establish. This decision will be informed by your business aims, an appraisal of the value of your goods or services in the new market, your familiarity with the target country's language and business customs, and what capital and infrastructure you have available to support overseas efforts.

This section introduces four ways of establishing a direct overseas presence:

1. Establishing a representative office
2. Establishing a branch office
3. Establishing a foreign subsidiary
4. Forming a joint venture

These options represent a sampling of ways to expand overseas with varying degrees of risk and commitment. In general, a smaller overseas footprint entails less risk but can limit your overseas growth, whereas a larger overseas footprint puts more of your capital at risk but may increase the likelihood and size of your success. Keep in mind that these are only representative examples. The names and details of these business forms will vary from country to country.

Also remember that an increasing amount of overseas activity can be accomplished indirectly, with no overseas footprint at all. Selling your products worldwide via the Internet is one example. Such indirect overseas activities offer a very inexpensive way to test foreign markets. They also present a different set of issues beyond the scope of this introduction to direct overseas activity.

Representative Office

Sometimes called a liaison office, a representative office is a minimal overseas presence and is usually relatively easy to establish. A typical representative office consists of a handful of people, sometimes even just one person, who generate leads to refer to the parent company. The representative office itself does no more than this, because many jurisdictions limit the business activities of representative offices to marketing and lead generation. Business transactions beyond these, such as sales, must be made between the client and the parent company directly. A representative office has the advantages of being easily set up and inexpensive to maintain.

Branch Office

A branch office is an overseas office of a U.S. parent company. Because the branch office is an office of the parent company, the branch office's profits and losses are reported on the parent company's books. Similarly, when forming a branch office, the parent company's corporate information is usually included in the application to the foreign authorities. Many jurisdictions also require regular disclosure of the parent company's financial statements, and some companies may choose to open a subsidiary rather than a branch to avoid making the parent company's financial information publicly available in the target country.

Subsidiary Company

A subsidiary is a separate legal entity incorporated in the target country. This option represents a greater commitment to the foreign market than does opening a branch or representative office and is generally more expensive to establish and operate. At the same time, because of its status as a separate legal entity, a subsidiary will only have to disclose its own financial information, and not the parent company's financial information, in the course of its registration and reporting. Furthermore, a subsidiary's assets are insulated from its parent company's, and this structure limits the parent company's exposure in the case of unexpected liabilities, such as lawsuits.

Despite their additional cost, many companies choose the subsidiary form over the simpler representative and branch offices when expanding overseas. These companies seek to avail themselves of the subsidiary's separate legal structure and greater capacity for growth. In addition, the business communities in some countries tend to shy away from transacting with foreign businesses that have opened a representative or branch office and not formed a subsidiary.

Joint Venture

A joint venture is a partnership with an existing company in the target country. The advantages of forming a joint venture lie in gaining access to the partner company's experience and expertise. Your business can benefit from the partner's reputation, understanding of the local market, connections to clients and other business partners, and infrastructure, saving you the cost and time required to develop these business assets on your own.

The potential difficulties of a joint venture derive from the fact that the structure requires you to add a separate legal entity—with different corporate resources, outlook, and goals from your own—into your business plans. It is important to consult counsel for this structure. Should you decide to pursue a joint venture, be careful to take time to articulate the goals of the venture, the scope of its activities, the conditions of termination, and procedures for dispute resolution and dissolution in the beginning. This may seem obvious advice, but, during the early stages of negotiation, many businesspeople shy away from discussing unhappy matters such as resolution of disputes and disposition of joint assets in the event of dissolution. It is usually too late to address these issues once problems have arisen, because goodwill between the companies will have already been spent. Establishing these procedures in advance are key for an orderly resolution of problems should they arise. Furthermore, having these procedures in place can go a long way to providing a sense of security in the relationship that helps avoid the problems in the first place.

Special Concerns Relating to Overseas Subsidiaries

As noted previously, you have a range of options from which to choose when deciding to form a direct presence in a foreign country. In practice, many entrepreneurs have firm plans to expand overseas over a long time frame, and this certainty (among other factors) leads them to choose to open a foreign subsidiary. Subsidiaries are complex entities, and the next section discusses a few of the additional issues involved in establishing an overseas subsidiary.

Capitalization Requirements

Minimum capital requirements vary significantly by jurisdiction. U.S. entrepreneurs are accustomed to an incorporation process in most states that does not require any specific amount of minimum capital. Only a few states have minimum capitalization requirement for new companies, and the most frequently used corporate jurisdictions, such as Delaware and California, have no minimum capitalization requirement.

In contrast, foreign jurisdictions can have quite high capitalization requirements. In Europe, The Netherlands, Switzerland, and Germany all require significant minimum capital. Until May 1, 2006, the effective date of Japan's new Corporation Law, Japan required a minimum capitalization of ¥10 million, or approximately $80,000 at current exchange rates. With the new law, Japan has fallen in line with most common law jurisdictions and reduced the capitalization minimum to ¥1, or less than a cent. (However, a Japanese subsidiary with less than ¥3 million, about $24,000, in net assets will not be allowed to pay dividends.)

Regardless of what the jurisdiction's requirements are, you should be cautious about leaving your foreign subsidiary with little or no capitalization. Having a low capitalization may discourage potential clients and business partners, who may look to such factors to assess your firm's financial health and commitment to the new market.

Board Structure and Meetings

Some jurisdictions place nationality or residence requirements on board members. For example, a majority of the board might be required to be local directors, either citizens or permanent residents of the country. In such case, the parent corporation will have to carefully select the local directors to ensure that they will work smoothly with the U.S. parent company and management.

Additionally, some jurisdictions require all board meetings to be held in the foreign country and may not allow directors to participate by written consent, telephone, or videoconference. If you expand in such a jurisdiction, you must consider carefully whether to ask a nonresident to serve on the board, because the inability of that person to attend board meetings in the foreign location may make it difficult for the foreign subsidiary to meet its governance obligations.

Other Regulatory and Reporting Requirements

Foreign jurisdictions will also place regulatory and reporting requirements on corporations with which the foreign subsidiary must comply. Many of these requirements have analogs in the U.S. corporate system and will be familiar to you, but some will not. Unless you have extensive experience operating a business in the foreign jurisdiction, putting in place local staff that can help navigate the foreign country's corporate regulatory system is paramount.

At the same time, you will want to take care to strike the right balance between control of the foreign subsidiary and delegation to the local staff. If you keep too much control at the parent company, you risk noncompliance with local regulations. Additionally, you may jeopardize your ability to take advantage of the experience and understanding of the local market that local staff possess. Conversely, too much delegation to local staff will increase the likelihood of independent action by the subsidiary and may lead to problems if they stray too far from the directives of the parent company. These can be costly problems to unravel, so you will want to keep as informed as possible of daily activities that the subsidiary develops.

International Issues with Intellectual Property

As an entrepreneur, you are already well aware that the world of business shrinks with each new advancement in the digital age. This ever-increasing interconnectedness has important consequences for all aspects of your business, but a few international issues relating to intellectual property—often your business's greatest asset—merit specific comment. This section briefly discusses the treatment of inventions by employees in a foreign arena, protection of your established intellectual property assets abroad, the relationship between intellectual property licensing, taxes, and corporate structure, and compliance with U.S. export regulations relating to intellectual property.

Inventions by Employees

If your business has a large research or technology component, you will likely have already required your U.S. employees to sign agreements assigning the intellectual property rights of inventions and work product to your company. Some foreign jurisdictions assign the intellectual property rights to inventions by employees to the employer as a matter of course, but you will want to confirm this. In other jurisdictions, assignment clauses of the form used in your U.S. employment agreements may not be given effect. You will want to research this issue thoroughly with your local counsel.

Protection of Established Intellectual Property Assets

You will want to prepare a complete plan for protecting your intellectual property assets in every country in which you intend to do business. The regimens for recognizing and enforcing intellectual property rights vary significantly among jurisdictions. You will want your legal counsel to consider the timing of trademark registration and patent prosecution in various jurisdictions so that your activities in one country do not prevent you from asserting your intellectual property rights later in another. The following chart briefly summarizes U.S. and international protection for various intellectual property assets.

Intellectual property asset	U.S. protection	International protection
Copyright	Copyright protection is automatic once an idea is reduced to tangible form (and other requirements are met). Registration is therefore not required, but copyright holders are advised to register to obtain added protections, such as statutory damages for infringement.	Copyright protection is automatic in countries that are signatories to the Berne Convention, although the details of that protection can vary, and, importantly for many high-technology businesses, not all countries view software as copyrightable content. *Relevant treaties: Berne Convention*
Patent	Patent protection is federal and requires registration with the U.S. Patent and Trademark Office. U.S. patents are not protected abroad.	Patent registration must be sought in each jurisdiction, but an increasing effort to harmonize laws among countries is making the process easier. Patents can be filed simultaneously in multiple countries to secure protection in all of the relevant foreign jurisdictions. *Relevant treaties: Agreement on Trade Related Aspects of Intellectual Property Rights (TRIPPS), European Patent Convention, Paris Convention*

(continued)

(continued)

Intellectual property asset	U.S. protection	International protection
Trademark	Trademarks are protected by both federal and state laws, but federal registration provides greater protection, such as national rather than simply regional rights, and is typically sought. U.S. trademark protection does not extend internationally.	Protection for the mark must be sought in each jurisdiction, although some international treaties and systems can facilitate the registration and recognition of marks in multiple countries. *Relevant treaties: TRIPPS, the Madrid System, the Community Trademark System (in the European Union)*
Trade secret	Trade secrets are governed by state law, although an increasing number of states have adopted the Uniform Trade Secrets Act, so the law governing trade secrets varies less between American jurisdictions than it did in the past.	Trade secret protection varies by jurisdiction, and foreign protection for trade secrets is often much weaker than U.S. protection. *Relevant treaties: North American Free Trade Agreement, Generally Acceptable Accounting Principles, TRIPPS*

In addition to the above categories of intellectual property assets, some additional protections may be available abroad that are not available in the U.S. Moral rights are one example. Moral rights are available in many civil law jurisdictions and are similar to but distinct from copyrights. They include an author's right of attribution, right to have a work published anonymously or pseudonymously, and right to the integrity of the work. In 1990, the United States recognized a limited set of moral rights in the Visual Artists Rights Act, but, as the name implies, this recognition extends only to visual works of art. You will want to be aware of any additional protections, such as moral rights, that are available abroad and consider what value those protections might provide as you structure your business operations overseas.

Finally, keep in mind that, in some countries with developing legal systems, enforcement mechanisms may not sufficiently protect your company's intellectual property. No global intellectual property policy for your company is complete unless you address whether you will do business at all in a location that provides inadequate mechanisms for enforcing your rights. In some cases, changes can be made to your business model that will keep your intellectual property assets from being put in jeopardy. Otherwise, you may have no choice but to forego doing business in such a country until the intellectual property protections improve.

Intellectual Property Assets and Your International Corporate Structure

Unlike real property, intellectual property assets are easily moved. If your business model focuses on licensing or otherwise exploiting these intangible assets, you should consider structuring your international operations to allow the income from exploiting these assets to be captured in low-tax jurisdictions.

One common method of doing this is to create a holding company in a low-tax jurisdiction to which you transfer the rights to license the U.S. parent company's intellectual property assets abroad. This holding company would transact with foreign customers referred to it by the subsidiary companies in each country and perhaps from the U.S. parent company as well. Profits from these sales would be captured by this holding company in the low-tax, foreign jurisdiction and would be available for reinvestment abroad. Note that repatriating the profits to the United States would normally lead to tax liability on the repatriated portion of those profits, so you typically would want to transfer only the portion of the offshore profits that you need to use in the United States back to the U.S. parent company.

These structures implicate the tax laws of several countries. Successful implementation of a distributed intellectual property licensing structure requires a large amount of accounting and legal work, which can be quite costly. However, if implemented correctly, these structures can be quite tax efficient, and, if intellectual property licensing forms a large part of your business, then the added accounting and legal costs will be offset by even larger tax savings. You will want to thoroughly discuss these alternatives with your legal counsel and tax advisers when planning the international corporate structure that is most tax efficient and best suited to your international plans.

Export Compliance and Deemed Exports

Finally, you will want to be familiar with U.S. export controls relating to your intellectual property. Export controls apply not only to physical shipments out of the United States, but also to the communication of technology or technical information in electronic form.

Furthermore, you will want to be aware that the transfer of technology to foreign nationals *within the United States* may be "deemed an export" under U.S. export regulations. These deemed exports can occur through routine business activities, such as inviting a potential customer from abroad to visit your office (if the visitor comes in contact with restricted technology during the visit) or hiring a nonnational to work in your U.S. office (permanent residents and certain other noncitizens are exempt, but H-1B visa holders are not). Your business may come in contact with foreign nationals routinely, and you may be in the habit of giving tours of your facilities or demonstrating your technology. To be safe, be sure to discuss in advance with your legal counsel whether your anticipated activities expose controlled technologies to foreign nationals.

Factors to Consider When Choosing Overseas Counsel

You will require legal advice as you navigate the issues that arise from expanding internationally. You have a spectrum of choices when it comes to foreign counsel, ranging from sole practitioners and small local firms on the one end to large, full-service international firms on the other. Each type of legal service provider has advantages and disadvantages, and you will want to keep these in mind as you select counsel to help you internationally.

If you haven't already worked with attorneys domestically, take a moment to review Chapter 5: Working with an Attorney. Chapter 5 introduces and discusses basic issues relating to selecting attorneys as you grow your business venture domestically, and the concepts discussed in that chapter underlie the more specific issues presented when choosing counsel to aid you in expanding overseas.

CHOOSING A LAW FIRM TO HANDLE INTERNATIONAL ISSUES

- **International sole practitioners and small firms:** Sole practitioners or small firms can be a cost-efficient way of creating a local business entity and getting it up and running. These legal service providers' primary advantage is price. Sole practitioners or small firms typically offer their services for comparatively low fees and, for this reason, can be an attractive option if you want to test foreign waters with little upfront monetary commitment.

However, smaller-sized local counsel can be limited in their knowledge of international issues. This lack of exposure to complex international legal and business issues may force you to seek additional counsel when such issues arise. Even if local counsel has exposure to general international issues, it is likely that they may be limited in their ability to assist with the complex tax and regulatory requirements that accompany international expansion.

- **Full-service international law firms:** In contrast, a full-service international law firm will be more familiar with the wide range of issues facing international entrepreneurs than will be a small local firm. Although such firms may be more expensive than smaller local counsel, many entrepreneurs have found that these firms provide added value. For starters, they offer the obvious advantage of having offices overseas, and they can draw on this global presence to provide unified legal advice across jurisdictions.

  Another advantage that global firms possess is their high-quality client base and strong community connections. Large international firms frequently interact with other international companies, chambers of commerce and international business organizations, and local governments and regulatory agencies. These connections can be excellent sources of information and introductions for a small entrepreneur who is entering a new market in a foreign country. If you are lucky, you may already have a relationship with a full-service domestic law firm that has a presence in the countries or regions in which you plan to expand.

In summary, it should be clear that you should give as much care to selecting a legal service provider for your overseas work as you do for your domestic work. You may want to select a small provider to keep costs low in the beginning stages of your overseas venture or you may want to rely on an established, global firm that can attend to all of the complex international issues that will arise. As in the domestic arena, a large international law firm can provide additional value by assisting with nonlegal issues, ranging from introductions to local community members and potential business partners to providing insight on business issues in the new market. Expanding overseas is an exciting endeavor but a large commitment, and selecting the right legal partner is crucial.

The Reach of U.S. Laws Abroad: The Foreign Corrupt Practices Act

Should you decide to take your new business to a foreign country, your activities there will be governed by the law of that country. In addition to these foreign laws, certain U.S. laws will also apply to your business activities. You must take care to comply with all applicable U.S. laws as well.

This section introduces the Foreign Corrupt Practices Act of 1977 (FCPA) as amended by the Convention on Combating Bribery of Foreign Public Officials in International Business

Transactions in 1998. In general, the FCPA seeks to prohibit bribes and certain other payments by American citizens or companies to foreign officials in connection with obtaining or retaining business with foreign countries. The FCPA also applies to foreign firms and persons that take any act in furtherance of such a payment while in the United States.

APPLICABLE FCPA PROVISIONS

The FCPA accomplishes its goal through two major provisions:

- Accounting and record keeping controls, which obligate public companies to maintain accurate books and records
- Prohibitions on payments (i.e., bribes) to foreign government or political party officials for the purpose of obtaining business.

The FCPA imposes severe civil and criminal penalties for violations, applicable to both the company and individual employees.

The FCPA does not limit other U.S. laws from applying to businesses acting outside the U.S. For example, conduct that violates the antibribery provisions of the FCPA may also give rise to private causes of action under the Racketeer Influenced and Corrupt Organizations Act or to actions under other federal or state laws. In addition, other statutes such as the mail and wire fraud statutes and the Travel Act, which provides for federal prosecution of violations of state commercial bribery statutes, may also apply to business activities occurring outside the United States.

Who Is Covered by the FCPA?

The FCPA applies to U.S. citizens, residents, and foreign persons who take any act in furtherance of a bribe while in the United States. Additionally, the FCPA applies to U.S. companies and foreign firms in the U.S., as well as their officers, directors, employees, and agents. The application of the FCPA to U.S. companies includes the foreign branch offices of U.S. companies. In addition, foreign subsidiaries of a U.S. company, foreign partnerships, joint ventures in which domestic concerns are a part, or foreign corporations in which domestic concerns are a majority or minority owner may be prosecuted under the FCPA if they aid, abet, or conspire in an FCPA violation.

You should proceed with the understanding that the FCPA will apply generally to all of your overseas business activities, regardless of whether those activities are undertaken by the U.S. company directly, by a branch office in a foreign country, or by a foreign subsidiary.

Penalties for Noncompliance

After conviction for corrupt payments, a company is subject to a fine of not more than $2,000,000. Any director, officer, or stockholder who willfully violates the corrupt payment provisions and is convicted may be fined up to $10,000 or imprisoned for not more than five years. Employees and agents of the company who are U.S. citizens, nationals, residents, or otherwise subject to the jurisdiction of the United States, after conviction for a willful violation, are subject to similar penalties. A company is prohibited from paying the fine of a convicted officer, director, stockholder, employee, or agent.

Key Tips to Complying with FCPA

Doing business abroad can raise issues with respect to compliance with the FCPA. As such, you should try to avoid activities that could unwittingly draw attention to the activities of your company.

WHAT RAISES RED FLAGS UNDER FCPA?

The presence of any of the following factors in a transaction may render a company more susceptible to FCPA compliance questions:

- **Latin America and Middle East activities:** Most FCPA cases have involved a number of countries in Latin America and the Middle East; there is a much greater risk of FCPA violations in countries that lack a strong tradition of governmental integrity.
- **Publicity:** Media coverage or industry knowledge of corruption in a particular country or government ministry may make FCPA problems more likely.
- **Working with former government officials:** There may be a greater risk of FCPA violations if principals of a company's agent or distributor are current or former government officials or if they emphasize their business or personal connections with government officials.
- **Unusual commissions and discounts:** Requests by an agent or distributor for unusually high commissions or discounts on government sales may indicate FCPA problems.
- **Unusual payment patterns:** Requests for payment of commissions in cash, to third country bank accounts or to a third party may indicate FCPA problems.
- **Failure to follow and/or include FCPA compliance provisions:** Objections by an agent or distributor to an FCPA compliance provision in a contract with the company also may indicate FCPA problems.

To adhere to the FCPA, companies should at a minimum (1) know the local laws regarding bribery and grease payments, (2) conduct a due diligence investigation of an agent in a country in question, (3) inform their employees of the FCPA requirements, and (4) obtain representations from a foreign joint venture partner that it will refrain from participating in conduct prohibited by the FCPA.

INTERNATIONAL EXPANSION

China: Forming a Joint Venture

For many companies, the lure of the ever-expanding marketplace in China is substantial. This section offers a brief outline of certain issues that you must consider when forming a commercial joint venture in the People's Republic of China. The discussion below assumes that your U.S. corporation is entering into a joint venture with a foreign partner for the distribution and sale of its products in China. Your company would contribute its products or distribution rights to its products, and the foreign partner would contribute capital, local contacts, and distribution expertise in China (referred to in this box as the "Joint Venture").

The information here is only a summary of a small number of issues related to forming a joint venture in the People's Republic of China. With its increasing market size, China can be an attractive place for you to expand. However, given the variety of legal and tax issues involved, it is important that you consult with legal and accounting experts experienced in forming joint ventures in China as you begin to make your plans for expansion.

CHINA JOINT VENTURE

Formation of China Joint Venture

If your Joint Venture is formed as a Chinese company, Chinese law may require the following series of organizational steps. Depending on the registered capital of the Joint Venture, the organization of the Joint Venture may be subject to Chinese Government approval by either the Ministry of Foreign Trade and Economic Cooperation (MOFTEC) or a local government agency to which evaluation and approval has been delegated by MOFTEC. This process involves (1) approval of the parties' preliminary proposal (typically based on a letter of intent or memorandum of understanding), (2) review and approval of a detailed feasibility study, the definitive Joint Venture agreements, and the articles of association of the Joint Venture, and (3) the issuance of a business license by the State Administration for Industrial Cooperation. The joint venture parties should allocate responsibility for taking such steps in the Joint Venture agreement. Because the approval process is fairly complicated and unpredictable, it is important for you to identify a foreign partner with proven business and government contacts in China.

Rules Affecting Wholly Foreign-Owned Entities

The foreign joint venture partner will not always be a Chinese entity. Instead, you might partner with an entity organized in a third jurisdiction, for example, a Singapore or Taiwanese company. In such a case, the Joint Venture would be subject to the Chinese rules governing wholly foreign-owned enterprises. Such rules materially inhibit the parties' flexibility in structuring their relationship. For example, if your company's contribution takes the form of a transfer of technology or distribution rights, Chinese law would restrict the value of intangibles contributed to the capital of a wholly foreign-owned enterprise to no more than 20% of the total equity of the Joint Venture. Correspondingly, allocation of profits and losses in a manner disproportionate to the equity interest of each of the joint venture parties in the Joint Venture would not be permitted. In addition, under Chinese law, your Joint Venture would own any intellectual property rights developed by the Joint Venture, including localized versions of your U.S. company's products. This ownership scheme typically presents a problem, because a U.S. partner normally has a strong interest in ensuring that it owns all such intellectual property rights to be free to market localized Chinese versions of its products in the event the Joint Venture is terminated by the parties for any reason.

Thus, if you expect that the Joint Venture will be a wholly foreign-owned enterprise, you should consider forming the Joint Venture in another jurisdiction, with a wholly owned subsidiary in China, to have greater flexibility in structuring the relationship, in making capital contributions, in allocating profit and loss, and in allocating intellectual property rights.

Trademarks

Many countries in Asia, including China, are "first to file" jurisdictions with regard to trademarks. A first to file jurisdiction allows a company to prospectively register its trademark, even before it has used the mark in commerce. As a result, you should take steps to register your company's trademarks in expected sales territories, even in anticipation of the formation of the Joint Venture, to protect those trademarks as widely as possible. In addition, certain former British Commonwealth jurisdictions, such as Hong Kong and Singapore, require a registered user agreement to be filed with the local trademark authorities to have a valid trademark license. In the absence of such a registered user agreement, your Joint Venture's use of the trademarks could be deemed a fraud on the public or an adverse use. You will want to research any potential trademark issues thoroughly in every jurisdiction in which your Joint Venture plans to do business.

Governing Law and Dispute Resolution

The Chinese government is not likely to approve a joint venture agreement relating to a Chinese enterprise that purports to be governed by the laws of another jurisdiction, such as California or Delaware. At the same time, the Chinese People's Courts have yet to acquire experience in resolving international commercial disputes. Consequently, you will usually want to specify an offshore dispute resolution forum, such as Singapore, for example, to ensure that any disputes will be resolved in a forum with international commercial dispute resolution experience.

CHINESE AND U.S. TAX ISSUES

Choice of Entity

A China joint venture could be structured in any of three different entity forms: a corporation (as noted above, either U.S. or foreign), a partnership, or an LLC. The U.S. tax consequences of each vary considerably.

If a corporation is used, the income or loss of the Joint Venture will not generally be included in the U.S. taxable income of the U.S. parent corporation unless the subsidiary is a "controlled foreign corporation" earning "Subpart F income" (see discussion in next section). Instead, income or loss will generally be reported by the Joint Venture (if it is a corporation formed under U.S. law) or, assuming that all of the income or loss is derived from activities of the Joint Venture outside of the United States, not at all (if it is formed as a foreign corporation). If a partnership or LLC is used, in contrast, the income or loss of the Joint Venture will pass through to and be includible in the U.S. parent company's income. Therefore, a partnership or LLC may provide U.S. tax advantages if the Joint Venture generates losses but may be less advantageous than a foreign corporation if the Joint Venture generates net income. Because a corporation has certain other nontax advantages over a partnership (e.g., the joint venture's losses may not need to be included with the U.S. parent company's results for accounting purposes and a corporation provides insulation from the liabilities of the Joint Venture), this discussion assumes that you will use a corporation as the form for the Joint Venture.

Choice of Jurisdiction for Joint Venture Corporation

The decision on where to incorporate your Joint Venture has significant tax consequences. As noted above, a corporation formed under U.S. law will be subject to U.S. income tax on all of its worldwide income, whereas a non-U.S. corporation as a general rule is taxable only on income effectively connected with a U.S. business or derived from U.S. sources. For this reason, joint ventures of this kind are often structured as corporations incorporated in a low-tax jurisdiction, such as the British Virgin Islands, so as to avoid U.S. income tax on the non-U.S. income of the joint venture.

However, such income will be subject to U.S. tax if the joint venture corporation is a controlled foreign corporation (a CFC) that earns Subpart F income. A CFC is a corporation formed under the laws of a country other than the United States, the stock of which is owned more than 50% (by either voting power or value) by U.S. shareholders. If a corporation is a CFC, then the U.S. shareholders of the CFC, including your U.S. company, are treated as receiving dividends from the CFC each year in an amount equal to the shareholders' pro rata share of the CFC's Subpart F income. Subpart F income includes, among other things, interest, dividends, royalties, and income arising from sales of property for use in a country other than the country in which the CFC is incorporated if the CFC purchased that property from a related party.

Also note that the income of your Joint Venture could potentially be considered Subpart F income if the Joint Venture's products are sold for use outside of the country in which the Joint Venture is incorporated, because the Joint Venture's products will be purchased from its U.S. joint venture partner (which is a "related person" for the purpose of discerning Subpart F income). In that event, if the Joint Venture were a CFC, your U.S. company would be subject to U.S. taxation on its pro rata share of the Joint Venture's income as though the Joint Venture distributed that income to the U.S. joint venture partner as a dividend. However, by incorporating the Joint Venture or a subsidiary of the Joint Venture in the sales territory, you may be able to avoid having the sales of the Joint Venture's products recognized as Subpart F income, because the products will be sold for use in the same jurisdiction in which the Joint Venture will be incorporated.

This deemed dividend problem, as noted, will arise only if your Joint Venture is a CFC. Therefore, if your U.S. company maintains an ownership interest of 50% or less (in terms of both voting power and value), the Joint Venture's income generally will not be subject to U.S. income taxation if the Joint Venture is formed as a foreign corporation and derives all of its income outside of the United States.

U.S. Taxation on Transfer of Property to Joint Venture

If the Joint Venture is formed as a foreign corporation, then any contribution of tangible property by your U.S. company to the Joint Venture will be taxed in the same manner as if your company had sold the property to the Joint Venture. Under U.S. tax law, a capital contribution of property to a foreign corporation is (with certain exceptions) treated as a taxable sale, unlike the tax-free treatment normally accorded contributions of property to a subsidiary. If your U.S. company transfers intangible property to the Joint Venture, then your company generally recognizes no immediate taxable income but instead must report U.S. taxable income in amounts reflecting a reasonable royalty for the use of the intangible property over its useful life. This is true even if the Joint Venture's income is earned entirely outside of the United States.

ACCOUNTING ISSUE: CONSOLIDATION OF THE JOINT VENTURE LOSSES

Ownership of more than 50% of the equity interests of a subsidiary normally will require the parent corporation to include the subsidiary's losses in the parent company's results for financial reporting purposes. As a result, in addition to a potential CFC problem, acquiring a greater than 50% stake in the Joint Venture may require consolidation of the Joint Venture losses.

INTERNATIONAL EXPANSION

The United Kingdom: Listing on the London AIM

Today, more U.S. companies are considering the AIM, the Toronto Stock Exchange (TSX), and other international stock markets as a source of funds from an international market. More information is available in Chapter 23 about listing on the AIM as one example of an international marketplace available to U.S. companies.

INTERNATIONAL EXPANSION

Japan: Establishing a Japanese Subsidiary after the New Corporation Law

Although the phenomenal recent growth of the economies of China and India have drawn increasing attention in the international business media, Japan remains the world's second largest economy and one of the U.S.'s largest trading partners. Japan has a stable, developed infrastructure, a mature business and legal environment, and a population of savvy consumers that are aware of and often seek American goods and services. Additionally, Japan has recently overhauled the legal framework in which corporations operate. Japan's new Corporations Law (*Kaisha Hō*), which took effect in May 2006, allows for much easier and flexible corporate formation and governance, making Japan even more accessible to foreign businesses. In short, Japan offers U.S. businesses a highly sophisticated, technologically advanced market in which to expand.

Despite recent attention to other Asian economies, Japan remains an attractive market for overseas entrepreneurs. In addition to its mature, stable market and savvy consumers, Japan's new Corporations Law has streamlined and modernized much of the law governing corporate formation and activity. Forming a Japanese joint stock company (*kabushiki kaisha* [KK]) subsidiary in Japan has become even easier, requiring less in the way of procedure and paid in capital than ever before. You may want to give serious thought to bringing your business to Japan.

This section discusses the method for establishing a KK subsidiary under Japan's new Corporations Law and also introduces some issues relating to Japan's legal environment. Japanese law also recognizes a number of other business entities that might be suitable for foreign investors, such as branch offices, representative offices, restricted transfer joint stock companies, and even forms analogous to limited liability partnerships. However, of all of these, the KK most resembles the U.S. corporation in its structure. It offers benefits, such as being a separate legal entity and limiting shareholders' liability to their investment that U.S. investors will be familiar with from the standard Anglo-American corporation model. Furthermore, many Japanese customers and business partners will expect a serious market entrant in Japan to form a KK. Although this may begin to change as people become familiar with the new corporate law regimen, the perception in Japan that business forms other than the KK lack commitment to the market, and are therefore less prestigious, remains strong.

GETTING LEGAL HELP: UNDERSTANDING THE DIFFERING TYPES OF LEGAL PROFESSIONALS IN JAPAN

As you would in any foreign country, you will need legal help in Japan to establish your KK. In this regard, Japan provides a good example of how other countries' legal systems can be both familiar and foreign. Japan is a civil law country, with a legal system based on the German Civil Code (Bürgerliches Gesetzbuch) that has undergone major revisions at various time in the past century. Thus, the structure of the Japanese legal system will be somewhat familiar to a Western observer. However, Japan's legal system differs in many details. For example, Japanese consumers access legal services through a number of legal professionals, and a brief introduction to the variety of legal service providers you may encounter will be useful.

Japanese lawyers (*bengoshi*) are authorized to counsel on any matter of law and are also authorized to represent clients before Japan's district and high courts and Supreme Court. These lawyers closely resemble their U.S. counterparts in that their licenses allow them to both give legal advice and represent clients in court. The number of Japanese attorneys is quite low, and it is frequently said that Japan has fewer lawyers than a country of its population requires. Although there is some truth in this observation, the scarcity of actual lawyers is ameliorated by the fact that many legal services are provided by other licensed professionals.

Two of the more common non-lawyer legal professionals include judicial scriveners (*shihō shoshi*) and administrative scriveners (*gyō sei shoshi*). Judicial scriveners are qualified to work on matters such as real estate filings, court pleadings, and company registrations and may even represent clients in summary court proceedings. Many law offices in Japan will have judicial scriveners on staff to assist the Japanese lawyers in legal matters for their clients. Some judicial scriveners maintain their own offices and can be contacted directly for help establishing a subsidiary in Japan. Administrative scriveners are somewhat similar to judicial scriveners but are empowered to make applications and filings to the government, for example, in governmental licensing matters.

Unlike in the United States, Japanese lawyers do not handle patent prosecution. Japanese patent attorneys (*benrishi*) manage this work, and this profession requires a separate certification than that of Japanese attorneys. Although most Japanese patent attorneys specialize in patent matters, they are empowered to handle other intellectual property issues as well, such as trademark, copyright, and trade secret matters.

Finally, you may encounter foreign attorneys working in Japan. Some of these attorneys will be registered as attorneys at foreign law (*gaikokuh jimu bengoshi*), although many foreign attorneys practice in Japan without this qualification. From the standpoint of a consumer of legal services, it is worth noting that, regardless of whether or not a foreign attorney is registered as an official attorney at foreign law, that attorney may not advise on Japanese law and is only qualified to give advice on the law of the attorney's home jurisdiction. However, working with a law office that employs both Japanese and foreign attorneys can provide you with an easy way to coordinate the establishment and operation of your Japanese subsidiary with your U.S. parent company.

THE PROCESS FOR ESTABLISHING A KK

Incorporating a KK in Japan generally takes about two months from start of the process to the end of registration and commencement of business activities. With a little planning, this time frame can be slightly reduced. This section describes the general procedures for incorporating a Japanese joint stock company as a subsidiary of a U.S. company. If you are giving serious consideration to establishing a KK subsidiary, you might use this description as a preliminary list of materials you will need to prepare to make the incorporation and registration process go smoothly. However, this description should not be considered a substitute for legal advice tailored to your situation, and you should discuss your plans with your legal and tax advisors before proceeding.

PROCEDURES FOR ESTABLISHING A KK

Begin preparing the necessary U.S. documents. These include documents relating to the U.S. parent company's legal status, such as notarized copies of the articles of incorporation, registration certificate, and establishment certificate. Additionally, begin preparing other required documents, such as documents certifying the profile of the U.S. parent company, the authority of the U.S. parent company's representative to act on behalf of the parent company, as well as certification of the authenticity of the representative's signature. These documents should be notarized as well.

Conduct an examination at the Legal Affairs Bureau for Japanese companies with similar corporate names and confirm whether the KK's anticipated business activities are in an industry that requires special previous notification to the Bank of Japan.

- Prepare the KK's articles of incorporation. The articles of incorporation typically include the total shares issued and their classification, the amount of paid in capital, the names of the required officers, the KK's business objectives, the location of the head office, and the method of public notice. These documents should be notarized in Japan.
- Apply to a Japanese bank for capital custody and issuance of a capital custody certificate.

- Remit initial capital to the special bank account and receive share custody certificate from bank.
- Conduct a shareholders' meeting to appoint the required officers.
- Review all KK application materials and apply to the Legal Affairs Bureau for registration of incorporation as a joint stock company and for registration of the company seal.
- Receive certification of registration of incorporation and certification of the company seal. This generally takes four to six weeks from the time of application.
- Notify necessary government agencies of the incorporation, including the Bank of Japan, national and local tax offices, the Labor Standards Inspection Office, the Public Employment Security Office, and the Social Insurance Office.

Once notification is complete, the KK may open a bank account in its name, send and receive funds, and begin its business activities.

REQUIRED OFFICERS, SHAREHOLDING REQUIREMENTS, AND RECORDKEEPING REQUIREMENTS

Your KK must have at all times at least four officers. As noted above, notice of the names of these officers must be given in the articles of incorporation, and these officers must be elected at the first shareholders' meeting. For a KK, the required officers must include at least three directors (*torishimariyaku*), of which one must be a representative director (*daihyō torishimariyaku*) and additionally a statutory auditor (*kansayaku*). The representative director must be a Japanese resident. All directors are appointed to two-year terms; the statutory auditor is appointed to a four-year term. These requirements may change for large companies (those with capital greater than ¥500 million or liabilities greater than ¥20 billion).

Shareholding requirements have been significantly eased by the new Corporations Law. However, a KK with less than ¥3 million in share capital will not be allowed to pay dividends to its shareholders.

Other requirements include that the KK hold a general meeting of shareholders annually and that the KK maintain its books and records for 10 years.

23

IPOs, Mergers, Acquisitions, and Sales

John Masher proposed the following humorous mergers for 2005:

1. Hale Business Systems, Mary Kay Cosmetics, Fuller Brush, and W. R. Grace Co. will merge and become: **Hale, Mary, Fuller, Grace**.
2. Polygram Records, Warner Brothers, and Zesta Crackers join forces and become: **Poly, Warner Cracker**.
3. 3M will merge with Goodyear and issue forth as: **MMM-Good**.
4. Zippo Manufacturing, Audi Motors, Dofasco, and Dakota Mining will merge and become: **ZipAudiDoDa**.
5. FedEx is expected to join its major competitor, UPS, and become: **FedUP**.

The rest of the list can found at http://www.alphadictionary.com/fun/mergers.html.

What to Watch For

As a startup company continues to mature, most companies will reach a stage whereby a substantial infusion of cash is necessary to continue the growth of the company. High-tech companies may reach a point at which they will begin to consider an "exit strategy" for the founders and investors: a point at which the company will move from a privately held entity or independent entity to a company whose stock will trade on public markets or will operate as a part of a larger entity.

At this stage, companies will begin to consider a number of options. In most cases, companies will look for an infusion of cash from an IPO on traditional domestic stock markets such as National Association of Securities Dealers Automated Quotations (NASDAQ) or the New York Stock Exchange (NYSE). Other companies may pursue an alternative fund-raising course from an international stock market such as the Toronto Stock Exchange (TSX) or the London Stock Exchange's Alternative Investment Market (AIM). Still other companies will entertain acquisition and merger offers, opting to leverage the size and capital of a larger entity with greater resources.

For many of these companies, this will take the form of a liquidity event: going public with an IPO or selling the company under a merger or acquisition. In each case, the company will be looking for additional capital (provided by outside investors or through the synergies of combining with another company) and the added benefit of some level of liquidity for the stockholders.

As a company begins to reach this level of maturity, new issues and considerations will arise. These efforts may require retaining an investment bank or financial advisor to assist with sourcing new opportunities. The company will probably need to hire staff with experience in public company compliance, particularly in the accounting and finance staff. The company may desire to hire additional sales and marketing staff to showcase higher growth. These new considerations will require changes within the organization and necessitates advanced efforts.

An Exit and Liquidity Event

Oftentimes, startup companies will discuss an exit event or exit strategy, or a future liquidity event. For a new company, predicting a future event such as an IPO on a U.S. or foreign market or a future merger or acquisition event is simply impossible to forecast. However, these events represent important considerations for a company, particularly for a venture-backed company. Investors understand that investing in a private company offers them the opportunity to supply necessary capital for a company they hope will produce exceptional gains in the future. However, these investors also understand that it is very difficult for the investors to recoup these gains if the company stays as a private entity without available liquidity in the company's stock.

Owning stock of a privately held company is very different from holding stock in IBM, Disney, or Ford. To avoid registration with the SEC and other state regulators, private companies must place restrictions on the transfer or sale of their stock. Although this helps private companies avoid the expensive process of registering with the SEC (as well as providing ancillary benefits of a more manageable capitalization structure of the company), these restrictions will prevent investors, executives, and other employees from selling their ownership of the company. Therefore, after a startup business reaches a certain level of sustained growth and consistent results, many key stakeholders may begin to be looking for an opportunity to gain liquidity in their shares of the company. For this reason, even an early-stage investor will want to be certain that the company has considered potential options for a future liquidity event.

AN END IN SIGHT . . .

What could happen

Your business is maturing and you think it may be time to look at various exit strategies for the business.

Watch out for

Don't wait too long to begin thinking about the future of the business. You should always be watching the horizon (before it is on you) to give yourself the widest breadth of alternatives to consider. Discussing alternatives is not meant to imply that you want to sell the business, but it is meant to help you continuously evaluate the business and its future.

TIP: Always be thinking about the future of the business and evaluating potential avenues for exits.

Not all merger and acquisition activity results in the startup company being acquired by a larger company; there are oftentimes cases in which a medium-sized startup company will acquire or merge with another medium- to small-sized competitor with the aim to better compete in the marketplace. Oftentimes, these events will not represent a liquidity event for any shareholders but instead signal a new direction and could be coupled with an additional infusion of investment dollars or new technologies and products.

Both an IPO and a sale of the company offer founders, employees, and investors liquidity in their investments and new sources of capital for the company. Determining whether such a major transaction is in the best interest of your company and which type of transaction to pursue depends on a number of short- and long-term strategic, financial, and other considerations. Once a company has decided which transaction is in its best interest, many of these same considerations will affect the timing and, in the case of a sale of the company, the structure of the transaction. There is no one-size-fits-all decision tree and things can change quickly as your financial projections, capital needs, and market conditions change as well. This is a major decision for the company, and the entire process will take a significant amount of attention from the company management, which may be a distraction from normal business operations and growth.

WHERE COMPANIES LOOK TO "GO PUBLIC"

Although your initial vision may be of brokers frantically trading your shares in the NYSE, there are actually several markets in which you can choose to "go public."

- **NYSE:** The NYSE is the largest stock exchange in the world by volume of dollars.

- **NASDAQ:** NASDAQ is the largest stock exchange in the United States by number of companies and is a popular exchange for IPOs because the listing fees are generally cheaper than the NYSE.
- **American Stock Exchange:** Generally, smaller companies are listed here attributable in part to more liberal listing requirements.
- **Foreign Stock Exchanges:** If you are an international company, you might also consider foreign stock exchanges such as the London Stock Exchange (and the its AIM), TSX, Tokyo Stock Exchange, and other more regional markets.

NOTE: More information on offerings on the London Stock Exchange AIM follows in this chapter.

Perhaps you or the other founders want to move on to your next idea and look to convert your invested "sweat equity" into cash. Your venture capital or angel investors may seek or even demand liquidity. New sources of capital may be needed for day-to-day operations, product development, or company growth, and traditional early-stage financing options are no longer sufficient to meet your company's needs. A red-hot IPO market in your industry or an attractive acquisition inquiry may also compel your decision. Often, a combination of these factors will drive the decision to pursue a liquidity event.

Pursuing an IPO or Merger or Acquisition

Obviously, not every company represents a viable candidate for pursuing an IPO or for a strategic merger or acquisition transaction. The decision to undertake an IPO or to enter into acquisition negotiations oftentimes will be done over many months, if not years. A number of key factors must be examined before any company definitively decides to enter into either of these transactions.

THE ODDS OF GOING PUBLIC

In 2007, 80 venture-backed companies went public, representing an increase of nearly 30% over the 57 venture-backed companies that went public in 2006. Although 2007 may have represented an increase over 2006, these numbers are still way down from the 260 and 264 companies that went public in 1999 and 2000, respectively.

Source: National Venture Capital Association.

Determining the "Right" Timing

There is no "right" time to enter into any substantial liquidity transaction. However, there are certain factors that will begin to drive the decision, one of which is that the company begins to receive significant interest from investors, advisors, or even potential acquirers.

Said one startup CEO recently, "I didn't realize we were a good candidate to be acquired by our acquirer until we started getting calls from companies we were emulating."

Perhaps the decision will be driven by the need for substantial capital for scaling of your technology or product, perhaps you are unable to compete in certain markets because of customer perceptions, perhaps a key investor is not willing to invest in follow-on rounds, or perhaps key members of management have expressed a desire to begin to transition out of the company. In each of these cases, the management team and the board of directors may begin early discussions into the various options the company has, including retaining advisors or consultants to begin early research. The result of this research may guide the steps the company takes or may provide insights into the outside perception of the business. Ultimately, although the company and its management will be involved in the decision to pursue a significant liquidity event, a significant factor will be legitimate interest from parties outside of the firm.

Whether your company decides to pursue an IPO or a sale of the company, the factors for deciding the proper timing of these major transactions are essentially the same. The decision ultimately hinges on (1) the needs and vision of the founders, employees, and investors, (2) the current position of the company, and (3) market conditions in your industry. In making your decision, you should consult with accountants, attorneys, and, particularly in the case of an IPO, with investment bankers. The window of opportunity for an IPO can quickly close for your industry or for the IPO market at large, virtually halting all IPO activity until the market warms again.

Likewise, you may find that advisors will recommend that the company consider different public markets at different stages. Companies that may not be mature enough for listing on NASDAQ may be better candidates for raising money through an IPO on the TSE or AIM. Fortunately or unfortunately, companies often find that outside factors will drive their choices and options for pursuing various liquidity transactions.

Strategic Considerations

A liquidity event will also present your company with new strategic opportunities and challenges. An IPO may increase the publicity and stature of your company, resulting in new customers, employees, and joint ventures. Additionally, the publicly traded shares available post-IPO provide the company with a currency of sorts, aside from cash reserves, to make acquisitions of its own. With a sale of the company, a well-positioned acquirer may provide economies of scale, new business contacts, and additional expertise for the target company. (In an acquisition, the buyer is referred to as the "acquirer" or "acquiring company," and the company being acquired is the "target" or "target company.")

Along with these new opportunities are new challenges, many stemming from the fact that you will inevitably lose some control over the decisions and direction of your company. The IPO process and life as a public company place additional pressures and restrictions on management through SEC filing and disclosure requirements (including disclosure regarding executive compensation, customer contracts, and other previously confidential information), restrictions on publicity, increased fiduciary duties and liability of the directors and officers, increased public scrutiny of management decisions, and the impact of the Sarbanes–Oxley Act (see "Public Companies and the Sarbanes–Oxley Act" at the bottom of p. 550). Failure to satisfy these many requirements may expose the company and its directors and officers to securities litigation from the shareholders.

A sale of the company may allow you to avoid the requirements of a public company (if the acquirer is private) but can also result in a total loss of control for the founders or, at a minimum, an additional layer of oversight from the acquiring company depending on the structure of the transaction and the intent of the parties. The acquiring company may implement major changes in personnel, compensation, benefits, policies, or company culture.

Financial Considerations

An IPO and a sale of the company provide shareholders with liquidity, but, with a sale of the company, liquidity will be more immediately available. In an IPO, the founders of the company are typically limited by the investment bank (referred to as the "underwriters") for a period of time (known as a "lockup" period) in which the founders are unable to sell their shares or pursue certain hedging strategies such as options. This period, typically six months, leaves the founders subject to fluctuations in the company share price (including market-wide fluctuations).

A sale of the company gives shareholders immediate liquidity, in the form of cash consideration or in shares of the acquirer under a stock-for-stock exchange, shares that are not subject to lockup provisions like in an IPO. For a cash transaction, the amount of consideration for shareholders is guaranteed; the tradeoff is that the upside is capped because consideration is fixed (unlike in an IPO in which there is no ceiling on share price and thus the shareholders' upside). If consideration is in the form of stock in the acquirer, the target shareholders do have some potential upside but that depends on the performance of the acquirer as a whole, not just the performance of the target. Founders may be willing to sacrifice the benefits of immediate liquidity for the opportunity to enjoy all the upside of the company.

Another consideration in your decision should be the transaction costs and, in the case of an IPO, the ongoing compliance costs for a public company. When taking into account the underwriters, filing, legal, and other fees, the IPO alone can cost in excess of $1.5 million even for relatively small offerings (according to *The Initial Public Offering: A Guidebook for Executives and Boards of Directors*, Second Edition, by Schultheis, Montegut, O'Connor, Lindquist, and Lewis, Wilson Sonsini Goodrich & Rosati, 2004). Once your company is public, disclosure, legal, and accounting requirements will add additional expense to the company. Studies have shown that, with the Sarbanes–Oxley Act, the average annual corporate governance compliance costs for companies with less than $1 billion in revenue is $2.8 million! A sale of the company will also incur transaction costs in the form of legal, accounting, and consulting fees, although many of these expenses will be borne by the acquirer.

PUBLIC COMPANIES AND THE SARBANES–OXLEY ACT

The Sarbanes–Oxley Act of 2002 represented the most sweeping change to federal corporate securities law in nearly 70 years. Most of the requirements of the Sarbanes–Oxley Act and the related rules recently adopted by the SEC, NASDAQ, and NYSE, and American Stock Exchange, by their terms apply only to public-reporting companies and companies that have filed a registration statement for an IPO, but many of the provisions affect private companies that are contemplating an IPO at some point in the future or are seeking to be acquired by a public company.

The Sarbanes–Oxley Act imposed a number of new requirements on public companies in three principal areas. First, the act enhanced existing reporting requirements for public companies and their insiders, by expanding disclosure requirements, requiring senior officer certifications of certain reports, and dramatically shortening reporting due dates. Second, the act directed the exchanges and listing agencies to strengthen board independence and other corporate governance requirements for listing. Finally, the act significantly amended the auditor independence requirements for public companies and enumerated specific responsibilities for the Audit Committee as part of its auditor oversight.

Private companies planning an IPO, or hoping to be acquired by a public company, should be particularly cognizant of the Sarbanes–Oxley Act's requirements, because some of the requirements have long lead times and cannot be safely postponed until an IPO or public company acquisition is imminent. These companies should begin the process of documenting and evaluating their internal controls, recruiting independent board members, adopting required board committees (such as audit and compensation committees), and considering voluntarily complying with other requirements of the Sarbanes–Oxley Act. Although not required, some of these items are relatively inexpensive to implement and help build confidence by investors, lenders, acquiring companies, and other third parties in the company's corporate governance practices.

Industry Considerations

The sector or industry of the company also affects the choice to pursue an IPO or a sale of the company. For IPO candidate companies, the business model must make sense to investors who may be unfamiliar with certain technologies, markets, or customer bases. Therefore, certain industries tend to have products that more easily translate into IPO candidate companies. In industries with active acquirers, companies may be less willing to undertake the investment required to pursue an IPO.

For example, medical device companies tend to be less likely to go public through an IPO. Rather, companies in this sector are more likely to see exit events from medical device giants such as Medtronic, Boston Scientific, or Guidant. In certain subsectors, such as neuro/spinal with 10-12 active acquirers and cardiovascular with four to five acquirers, the number of attractive acquirers leads companies to pursue an merger or acquisition exit over a potential IPO.

Other Considerations

The decision of whether to pursue an IPO or a sale of the company will also depend on the position of your company as well as current market conditions. Companies experiencing rapid growth or that are in an industry currently favored by investors (see the many dotcom IPOs in the late 1990s) are better suited for an IPO, whereas those companies experiencing constant but slower growth might find a sale of the company to be more appropriate. The market will also ebb and flow from year to year, quarter to quarter, with various levels of mergers and acquisitions and IPO activity. Recently, mergers and acquisitions have been the more popular route in the United States for venture-backed companies with

$31.9 billion in deals for 2006 compared with $3.75 billion in IPOs during that same period according to VentureSource.

Preparing the Company for a Potential Transaction

Preparing for a major transaction such as an IPO or a sale of your company is like preparing to sell a house: you'll attract more buyers and obtain a better return if you engage in some house cleaning and maintenance before the transaction. In particular, companies that have been engaged in excellent corporate governance throughout their existence will find this process much less tedious (and probably much less expensive). For your company, this means resolving any potential problems with corporate or financial matters, management structure, employment issues, or anything else that may deter potential buyers or investors.

Underwriters and investors in an IPO and potential acquirers in a sale of the company will review the following: the company's corporate organization, quality and experience in management, internal financial and accounting controls, growth trends and potential, the health of the company's industry, and the company's competitive position within that industry. Although some factors, such as industry health, are beyond the control of management, the company can be proactive in shoring up these other factors before the IPO process.

Either before or during the transaction, underwriters, investors, or the acquiring company in the case of the latter will perform a thorough legal and financial review of the company through the due diligence process. You can anticipate potential problems in due diligence by reviewing these items with the help of your legal counsel in preparation of a sale. This will allow the IPO or sale transaction to go more smoothly once underway and may provide the company a more favorable valuation by investors or potential acquirers.

WHAT TO THINK ABOUT BEFORE YOU BEGIN ANY TRANSACTION

- **Corporate matters:** Are the company's organizational documents (articles of incorporation and bylaws) up to date and reflective of the current organizational structure? Are the number and type of issued shares of the company's capital stock consistent with the authorized shares in the articles of incorporation? Have all security issuances been properly approved by the board of directors and other shareholders when appropriate? Are all stock records and minute books up to date? Is the company in good standing in all jurisdictions it conducts business? Does the company have a complex capital or organizational structure that would be difficult for an investor or potential acquirer to understand? If so, is there a compelling tax or business reason for the structure?
- **Financial matters:** Are the company's financial statements correct, up to date, and properly audited? Has the company properly filed all necessary federal, state, and foreign tax returns? Are there any liens, encumbrances, mortgages, or other charges on the personal and real property of the company?

- **Management and operations:** Are the company's business plan and financial projections accurate and up to date? Does the company have internal controls and is the company in compliance with those internal controls? Does the company have rights in all the intellectual property it is using, by patent, trademark, license, or otherwise?
- **Employee matters:** Is the company in compliance with all relevant labor and employment laws? Are all employee confidentiality, intellectual property assignment, or noncompetition agreements signed and current? Has the company stock option plan and each individual employee stock option grant been properly approved by the board?
- **Insurance:** Are all company insurance premiums, including workers compensation and directors and officers liability insurance, current and sufficient in coverage for the company's needs?
- **Litigation:** Is there any current or pending litigation (consider breached contracts, employment disputes, etc.) that can be resolved before the sale of the company?

In the case of an IPO or acquisition by a public company, the company may additionally consider the following:

- **Board of directors:** The NYSE, NASDAQ, and SEC place a number of requirements on the composition of the board of directors of public companies with respect to the number of independent directors and the composition of compensation and audit committees. Making some of these changes before being subject to the requirements may help ease the transition.
- **Changes in management:** If the company is lacking in board members or officers with public company experience, it is advisable to hire an executive with that experience to help guide the company through the IPO process. This is particularly useful for a CFO given public company financial reporting requirements.

Wherever possible, the company should take the necessary steps after its review to rectify any correctable problems you have encountered. This could mean resolutions of the board of directors ratifying previous actions or hiring outside counsel, auditors, or consultants to get the company in order. Although this may seem burdensome, the problems will be revealed during the due diligence process, and addressing as many as possible before IPO or the negotiations process is in the best interest of your company.

Merger and Acquisition Activity: Mergers, Asset Acquisitions, and Stock Acquisitions

Merger and acquisition activity may be initiated by the acquirer or the company seeking to be acquired. For companies interested in being acquired, the first step may involve hiring an investment bank or financial advisor to assist with potential acquisition partners. In these cases, an acquisition target may look to approach several potential or promising acquirers. In other cases, an acquiring company will seek out the target as your company grows in revenue and reputation. The potential acquirer may be a customer, supplier, or even a competitor. Private equity firms are another potential buyer.

How this process is initiated will primarily depend on the company and the industry. For some industries such as software or biotechnology in which growth by acquisition is common, larger companies may seek potential targets to secure their position in the marketplace. For other industries or companies, the process of finding acquirers may be less obvious. In the event your company wants to take a more active level role in selling your company, it can be advantageous to consult with your attorneys, investors, or investment bankers who may be able to make introductions to potential buyers.

Types of Transactions

If you have decided to sell your company, the next step is to familiarize yourself with the various forms of transaction. In deciding which transaction is in the best interest of your company, many of the same considerations discussed above will weigh on your decision along with the needs of the acquiring company, because the various transactions have different tax, liability, and posttransaction structural consequences.

Merger

The first kind of transaction is the statutory merger or consolidation in which two corporations combine to become one "surviving" corporation. There are typically three forms that the statutory merger can take, but all result in the acquiring company holding an equity stake in the target. The first is a direct or forward merger, in which consideration is paid to the target company shareholders and then the target company merges into the acquiring company, with the acquiring company as the surviving entity.

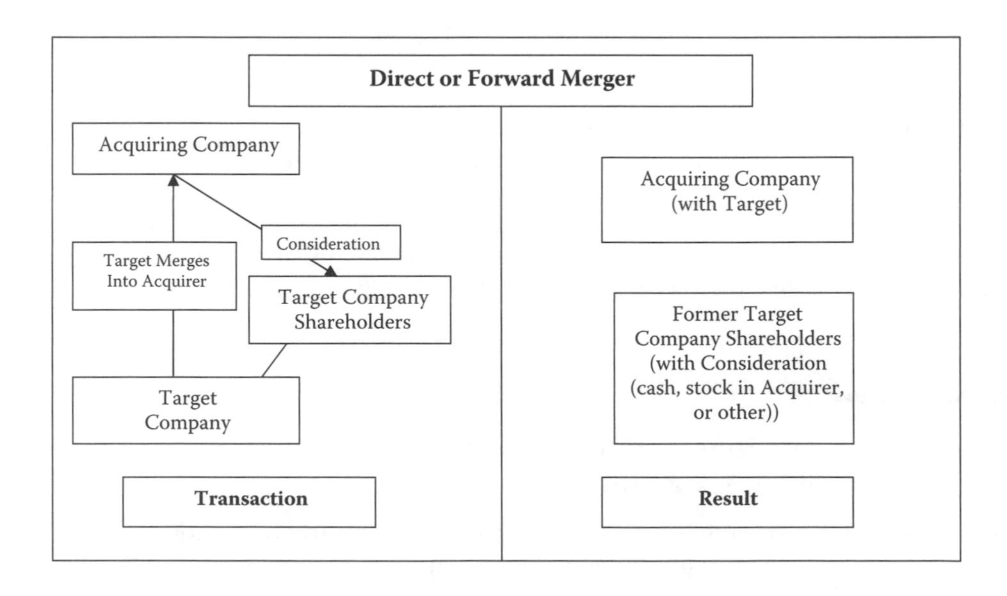

The remaining two forms are "triangular mergers," in which the acquiring company forms a subsidiary that either merges into the target (reverse triangular merger) or the target merges into the subsidiary (forward triangular merger). The result is that the target is part of a wholly owned subsidiary of the acquiring company. One advantage of a triangular merger is that the acquiring company may be able to limit any liability stemming from the merger solely to the assets of the subsidiary (the assets of the parent company cannot be touched). Additionally, there may be tax advantages in structuring the deal this way (more on this below). All three forms of statutory mergers require the approval of shareholders of both the target and the acquiring company.

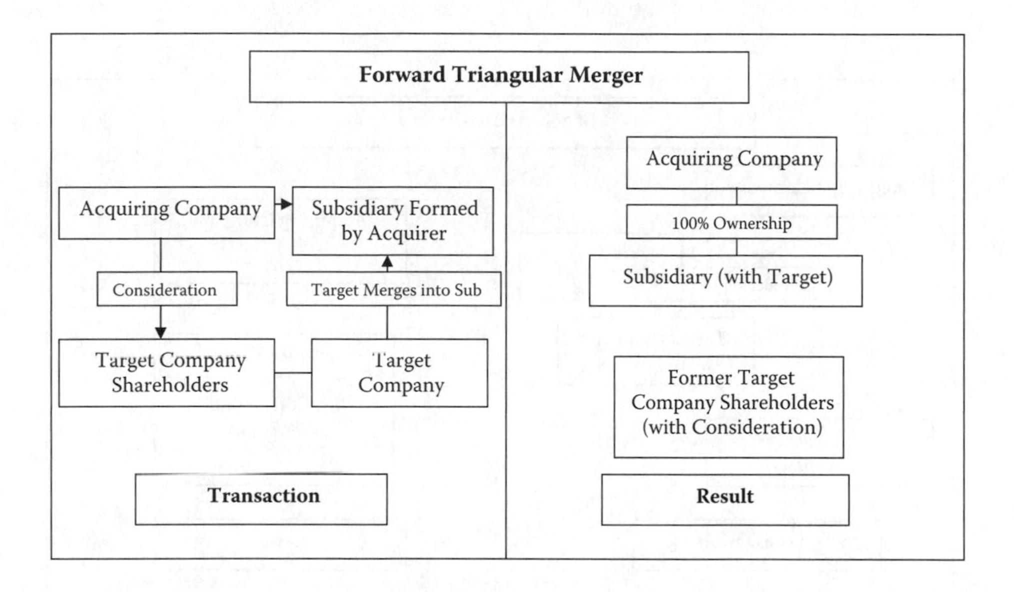

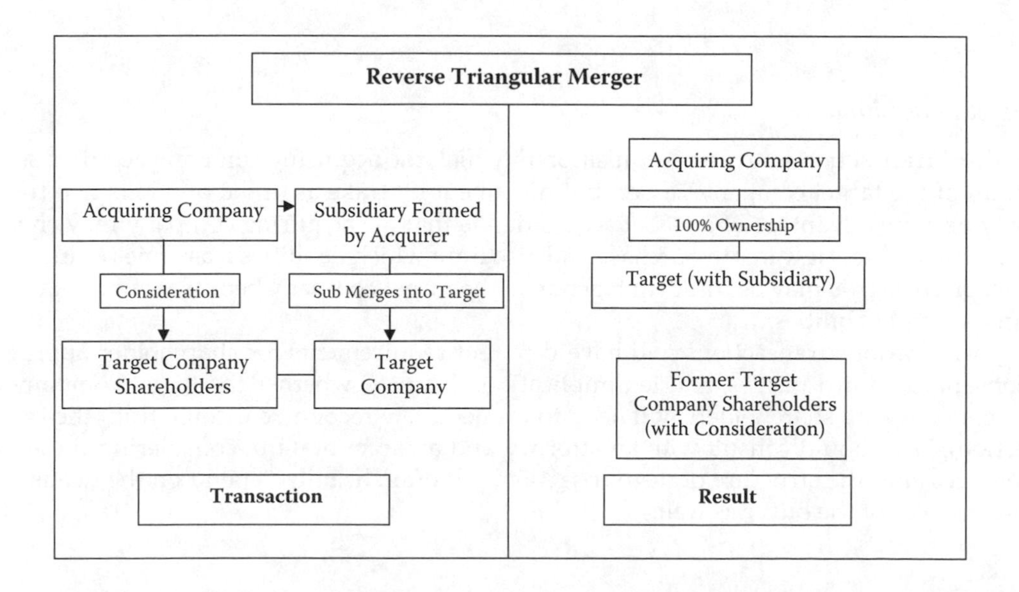

Stock Acquisition

A second transaction is known as a stock acquisition in which the acquirer agrees to purchase all outstanding shares of the target company's stock. This is actually an agreement between all the stockholders of the target company and not necessarily the company itself (although the company is often a party as well). Like in a merger, the result is that the acquiring company holds an equity stake in the target. Even if the acquirer is not able to acquire all of the target company's stock, they may be able to acquire a majority and then, depending on relevant state laws, acquire the remaining stock in what is called a "second-step merger."

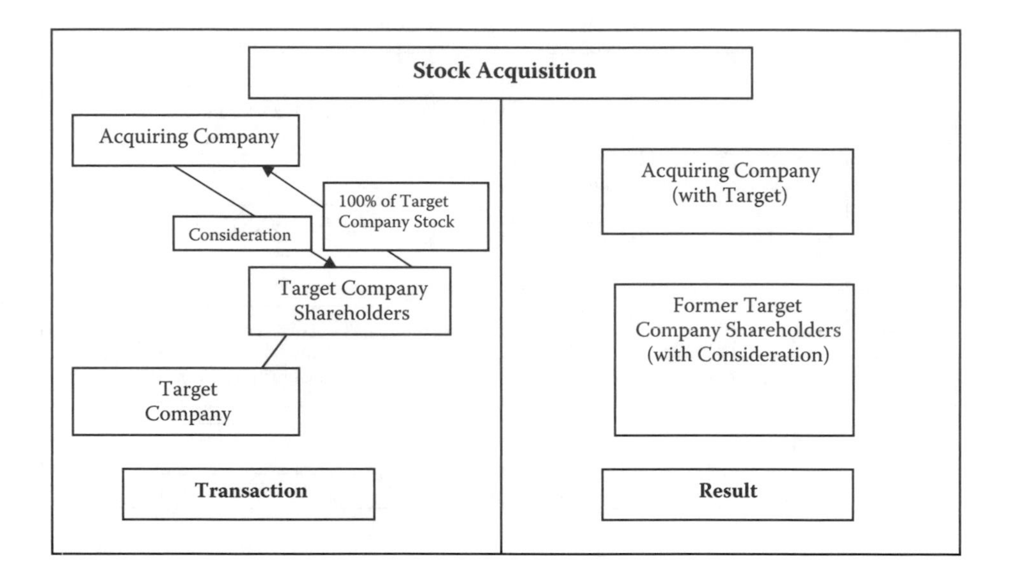

Asset Acquisition

A third transaction is an asset acquisition in which the acquiring company acquires some or all of the target company's assets but not an equity stake as in the previous two transactions. One advantage of an asset acquisition is that the acquiring company may choose only those assets it wants to purchase and also limit which liabilities it assumes (with some exceptions). This may be disadvantageous to the target company because it will retain all unassumed liabilities.

These various transactions will have different requirements for shareholder approval, consents by third parties, and tax implications (namely, whether the target company or target company shareholders will need to immediately recognize income from the transaction). You should consult with an attorney and an accountant in considering these differences, and the structure of the transaction will often heavily depend on the needs and preferences of the buyer as well.

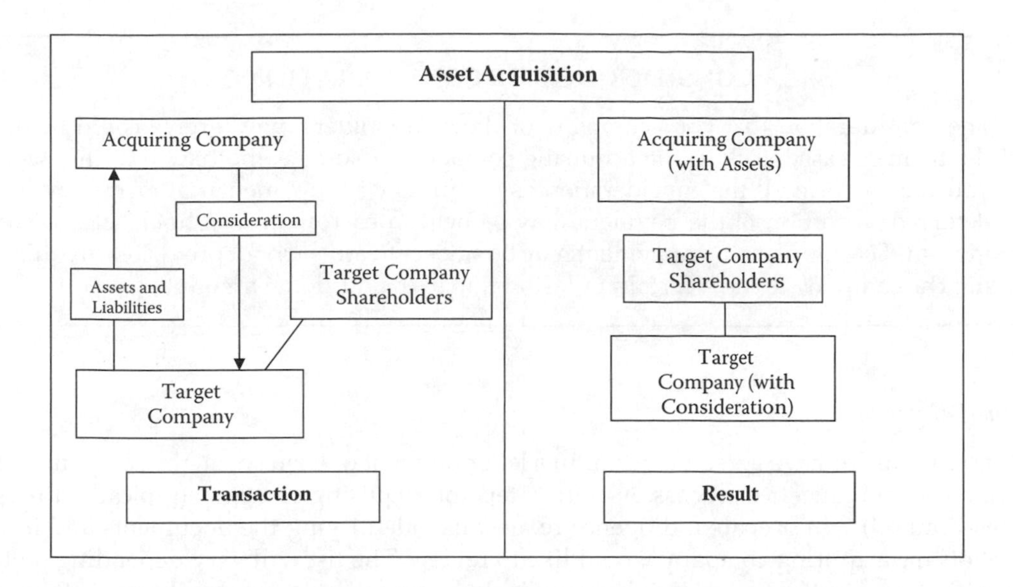

Understanding the Transaction Process

Initial Negotiations

Initial negotiations will take place between representatives of both the target and the acquirer to discuss the basic aspects of the deal (estimated purchase price or structure, assets to be acquired if an asset acquisition, postclosing corporate organization, etc.). If the parties decide to proceed with a deal, the next step is for the parties to sign confidentiality agreements and perhaps exclusivity agreements. Significant confidential information is exchanged throughout the due diligence and negotiation processes, and confidentiality agreements are necessary to protect that proprietary information, especially in case the deal fails. Your lawyer should be able to draft a confidentiality agreement. An exclusivity agreement may also be appropriate and demanded by the acquiring company given the substantial time and cost involved in the diligence and negotiations process. If an acquiring company is going to dedicate the resources to the acquisition, they want to make sure they are not going to get "scooped" on the deal at the 11th hour by another company.

Letter of Intent or Term Sheet

The next step of the process is for the parties to draft either a "letter of intent" or a "term sheet" outlining the key terms of the transaction. The parties may want to avoid signing a binding letter of intent or term sheet in case negotiations fail. The level of specificity in the term sheet or letter of intent will vary depending on the transaction but may include such terms as a description of the assets or liabilities to be transferred (if an asset acquisition), the purchase price and form of consideration, employee matters, indemnification provisions, termination provisions, and tax considerations. The more terms that are established at the outset, the more likely that the parties will be able to reach a final agreement. The final agreement, be it a merger agreement, asset purchase agreement, or stock purchase agreement, may mirror the terms of the letter of intent or term sheet but not necessarily so.

CONSIDERING THE CONSIDERATION

The consideration that your company or the shareholders may receive could be in the form of cash, stock in the acquiring company, or some combination of the two. You may receive all the consideration as an initial cash payment, payment may be deferred, a portion of the payment may be held in escrow or as a holdback for an indemnification period, or payment may be subject to an "earnout provision," requiring the company to meet certain milestones to earn additional amounts.

Due Diligence

After the parties have agreed to terms in a letter of intent or term sheet, the companies can begin the due diligence process. As a first step, the acquiring company (typically through their counsel) will prepare a diligence request list, identifying the documents and information the acquiring company would like to review. The list will vary depending on the specific terms of the transaction but may include the following: general corporate documents, customer/supplier lists, material contracts, list of real and personal property, list of intellectual property and copies of licensing agreements, corporate financing documents, evidence of insurance coverage, list of current or pending litigation, financial information, tax information, employee information (including compensation, benefit plans, etc.), and environmental issues and liabilities. The target company then either provides hard copies of the documents or places electronic copies in a data room for the acquiring company's counsel to review. The target company may also conduct due diligence of the acquirer if, for instance, the consideration is in the form of acquirer stock but that due diligence is typically much less extensive.

If problems arise during the due diligence process, they may be resolved by the company providing additional documentation, adjusting the purchase price or structure, or addressing the problems in the representations and warranties of the target company (discussed more fully below).

Transaction Documents

The merger or asset purchase agreement sets forth the terms of the agreement and the obligations of the parties. The agreement will set the mechanics of the closing, the purchase price and structure, the deliverables of the parties, and the assets and liabilities to be transferred if an asset acquisition.

A large and occasionally heavily negotiated section of the agreement contains the "representations and warranties" of the parties. For the target, representations and warranties will often address the organization and good standing of the company, capitalization, proper authority to enter the agreement, necessary consents (third party or governmental), accuracy of financial statements, compliance with laws, taxes, real and personal property, material contracts to which the company is a party, benefit plans, intellectual property, insurance, customers, personnel, litigation, environmental matters, and other necessary facts. Any exceptions to the representations and warranties are listed in a "schedule of exceptions" or "disclosure schedule" to the agreement. For the acquiring company, organization and standing and authority for the agreement may be sufficient (unless the consideration is in the form of stock in the acquiring company in which case additional reps and warranties may be appropriate). The representations and warranties not only serve as a means of disclosure of relevant information between the parties but are also critical in connection with the indemnification provisions discussed below.

Unless the merger or asset acquisition is structured as a "sign and close," in which the merger or sale closes immediately after the signing of the agreement, the agreement will contain conditions to the parties' obligations to close at a later date (known as a "delayed closing"). For example, the acquirer's obligation to close may be dependent on the obtainment of certain approvals, no material adverse events since the signing, the representations and warranties of the target remaining true as of the date of closing, performance or compliance with covenants in the agreement, or designated employees have signed employment or noncompetition agreements. Many of the same obligations may apply to the acquirer but again tend to be less extensive. If the conditions have not been satisfied, the other party can walk away from the deal. Given this possibility, the precise wording of these provisions is very important so as to avoid giving the other party too much discretion in whether or not to close the deal.

With a delayed closing, the agreement will also likely contain a termination provision allowing a party to terminate the agreement in the case of mutual agreement by the parties, failure to close before an agreed on date, a material breach by the other party, or a governmental order enjoining, restraining, or prohibiting the transaction.

Another section of the agreement with which you should be familiar contains the indemnification provisions. Often, the target company and its shareholders will indemnify the acquiring company in the case of a misrepresentation or a breach of a warranty in the agreement (highlighting the importance of the target company's representations and warranties). To cover these claims, a portion of the purchase price may be held in escrow for the period of indemnification.

IPOs

If you have decided to take your company public, this section should provide you with an overview of the timeline, key events, and issues in the IPO process.

FROM FOUNDING TO IPO

In 2007, the median time from the founding of a startup until its IPO was nearly eight years. This average time from formation to IPO exit has risen from approximately six years in 2005 and four years in 1999.

Source: National Venture Capital Association.

Pre-IPO Stage

Selecting the Managing Underwriter

Selecting the managing underwriter or underwriters for your IPO will depend on (1) the size of the offering, (2) the needs of your company, and (3) the experience and specializations of the various investment banks.

Companies will often select more than one underwriter and up to four or five to serve as the managing underwriter (if more than one, then "comanagers"), especially in the case of larger offerings or if specialized or local expertise is needed. In the case of comanagers, one of the firms will serve as the lead underwriter, but it is not unheard of to have colead underwriters. The managing underwriters will form a "syndicate" of other investment banks to share in marketing your shares to investors and to share in some of the risk.

In making your selection, your company may choose to consider several investment banks at once, allowing them to compete for your business in a "beauty contest" or "bake off" in which each bank will present itself. The alternative is to deal with the investment banks one at a time until you find your underwriter. Underwriters will encourage this latter approach but that is primarily because they would rather not compete with others to get your business. Given the narrow market window, the beauty contest approach may be more time efficient and allow you to get your shares to market sooner and provide a better comparison of the underwriters available.

FACTORS TO CONSIDER WHEN SELECTING THE MANAGING UNDERWRITER

There are many investment banks from the big Wall Street firms to the smaller and more regional firms but bigger is not always better, and the decision depends on a number of factors including the following:

- **Experience:** Not only is it important that the managing underwriter have experience with IPOs generally, but does the investment bank have experience with offerings in your industry? With offerings of this size? Relevant expertise in your industry and with offerings of your size is critical for an underwriter.
- **Reputation:** The reputation of a firm will be largely tied to its experience, but your venture capital investors, lawyers, accountants, and commercial bankers are additional resources to advise you as to the reputation of various investment banks.
- **Attention:** Although the Wall Street firms may have the best reputations, for a smaller offering, you may be able to get more attentive service from one of the smaller or more regional firms in which your transaction is relatively more important.
- **Individual bankers:** You should get to know the individual investment bankers who will be assigned to your transaction and ensure that (1) the company will be comfortable in working with those individuals, and (2) those individuals have the necessary experience for your transaction.
- **Aftermarket performance and support:** You will also want to look at the aftermarket (post-IPO) performance for the stock of other companies from the investment bank's previous offerings because this will affect the return earned by the company and its investors. Additionally, you should consider the investment banking services (additional financing, mergers and acquisitions, etc.) that the underwriter will be able to provide post-IPO.

- **Offering price and underwriter fees:** Finally, a company should consider the proposed offering price and underwriter fees of the various investment banks. Keep in mind that price is subject to change throughout the process and should not be determinative in selecting an underwriter. If one investment bank's proposed offering price is significantly higher than all the other proposals, that may be an indicator that the investment bank has not accurately valued your company, which may ultimately make it difficult to attract investors. Underwriters earn their fees by receiving shares at a substantial discount and then selling those shares at the offering price (the "spread"). Typically, the fee amounts to 7% of the offering price, but, for a particularly attractive IPO candidate, the company may be able to negotiate more favorable terms.

Once the underwriters are selected, company counsel will participate in the negotiation of the underwriter agreement, which will cover all aspects of the offering, and then the real work begins.

Organizational Meetings, Registration Statement, and Due Diligence

Once the company has selected a managing underwriter, the next step of the IPO process is to hold an "organizational meeting" with company management, the underwriters, company counsel, and other key participants to set a timetable for the transaction and to designate responsibilities amongst the parties.

If not before, then soon after the organizational meeting, the company should begin preparing the registration statement to be filed with the SEC under Form S-1. The registration statement contains a "prospectus," which is a detailed description of the business, management, and other required information. The prospectus is part selling document, presenting the company in a positive light to investors, but also part disclosure document, identifying the risks in investment. Identifying the risks is critical to allow the underwriters, directors, and officers to limit liability. Given this delicate balance, the company should rely on an experienced lawyer to take the lead role in drafting the registration statement, with active participation from the underwriters and company management.

While the registration statement is being prepared, the company will also engage in the "due diligence" process, which is a legal and business review of a company (organizational documents, board actions, stockholder agreements, financial documents, environmental compliance, contracts, etc.). This process is performed by the company's counsel and by the underwriters and their counsel to (1) ensure the accuracy and completeness of information contained in the registration statement and (2) assist the underwriters in properly valuing the company. Like in a sale of the company, the company's role in the diligence process will be assisting in gathering all of the relevant documents for the lawyers and underwriters to review.

After filing the registration statement, the SEC will review the statement to ensure compliance with the applicable form and then will respond with comments to the company regarding any deficiencies in disclosure. The company then amends the registration statement per the comments and the SEC reviews the amendments, again issuing comments if appropriate. This process continues until the registration statement is declared "effective."

The underwriters and the syndicate will distribute to potential investors a "preliminary prospectus," also known as a "red herring" because of required red print on the cover, warning that it is merely preliminary and is incomplete. It is advisable to wait until at least one round of comments have been received because the printing of the preliminary prospectus can be quite expensive and the company will want to avoid multiple printings if significant changes may be required by the SEC.

BLUE SKY LAWS

Nearly all states have additional securities laws known as "blue sky laws." These laws are designed to provide additional protections to potential investors against fraudulent stock offerings. Check with your attorney to ensure that you are in compliance with all relevant blue sky laws.

Waiting Stage

After the registration statement has been filed, the company will work with the managing underwriters to prepare a presentation for potential investors as part of the "road show": a series of presentations by the management team in various cities before large groups of investors and one-on-one presentations over the course of two to three weeks and arranged by the underwriters. Increasingly, companies may perform part of its road show virtually, through the use of Internet-based presentations.

Typically, the company will wait to embark on its road show until it has responded to SEC comments to the registration statement, because any material changes to the preliminary prospectus would require the company to recirculate the prospectus reflecting the changes at a substantial printing cost and potential delay. While waiting for approval, it is important to limit company publicity (namely public statements of company officials) because this could result in a violation of the securities laws (typically called the "quiet period").

Up until the moment the SEC declares the registration statement effective, the company, on the advice of the underwriters, may decide to delay the offering to a later quarter if market conditions have changed. In extreme cases, the IPO may even be terminated.

Initial Stock Offering

After the company has responded to the rounds of SEC comments and made all required amendments, the SEC will declare the registration statement effective and then the company and underwriters can decide on a final offering price. At this point, the final prospectus can be printed and distributed to investors, and your company is now public. At this stage, the company will undertake efforts (usually with the help of the underwriters) to sell the public stock to potential investors.

The United Kingdom: Listing on the London AIM

Raising money in the public markets has continued to evolve in recent years, with a growing number of U.S. companies looking to foreign markets as a source of funding. There is not a consensus about listing on a foreign market, with experts weighing in on both sides of the debate regarding the benefits and obstacles companies face in listing on a foreign public market. Even still, more U.S. companies are considering the AIM or the TSX as fundraising tools.

For many companies that are currently unable to consider listing on NASDAQ or NYSE, a foreign market such as the AIM may provide another source or method to raise money. Even if your company does not meet the traditional criteria for listing on a public stock exchange, you may still have options in this area. Increasingly, established stock markets are accommodating smaller, fast-growing companies on specially designed submarkets. Two good examples of this trend are the TSX Venture Exchange, associated with the TSX, and the AIM, a submarket of the London Stock Exchange. Increased internationalization in markets and investing has caused stock exchanges such as these to complete globally for the business of fast-growing, entrepreneurial companies. With options in different markets around the world, going public is no longer the exclusive privilege of established, traditional companies. This section discusses some of the benefits and considerations of listing on one such market, the London AIM, and briefly describes the process for listing.

Some experts strongly believe the AIM provides an excellent alternative for fast-growing companies seeking to list on a public market. Its streamlined regulatory scheme was designed with smaller, entrepreneurial companies in mind, and it provides a compelling option for companies that want to raise capital through a public market but that do not meet the requirements for listing on a traditional exchange. At the same time, as a submarket of the London Stock Exchange, AIM offers its companies an opportunity to gain visibility in Europe and around the world. Today, more companies are giving serious consideration to a foreign listing on markets such as the AIM and other alternative markets. Other commentators have raised doubts about raising funds on the AIM, as it may limit future fundraising efforts.

Introduction to the London AIM

The AIM was created as a submarket of the London Stock Exchange in 1995 and has rapidly grown, attracting over 2,000 companies and raising over £20 billion. The primary attraction for companies has been AIM's simplified regulatory environment, which has made going public possible for many smaller, fast-growing businesses that would not have met traditional market-listing criteria. AIM has attracted companies from around the world, particularly in the wake of the passing of the Sarbanes–Oxley Act in the United States in 2002, which added restrictions on listing publicly in the United States.

Benefits of Listing on the AIM

Listing (or "floating," as it is also called on AIM) can provide a number of benefits for your company. The most obvious benefit is rapid access to capital for further growth of your business. Access to capital can be had first when your company lists initially on AIM and later through additional capital raisings.

Additionally, there are a number of secondary benefits to listing on AIM. AIM is a global market whose membership is increasingly international. Listing on AIM provides visibility on the world stage and signifies that your company has a particularly increased presence in Europe. AIM might be a perfect market for a U.S. company that is seeking to go public and is contemplating expanding in Europe in the near future.

Going public on AIM also provides opportunities to repay investors and employees for their support, although there are some limits to this liquidity. Listing allows you to give early-stage investors a chance to realize their investment and also allows you to reward employees with stock option plans or other stock-based incentives.

Considerations, Costs, and Downsides of Listing on the AIM

At the same time, listing on the AIM is a major commitment by your company that entails costs and potential downsides. You should carefully consider what an AIM listing will require of your company and evaluate the risk of an AIM listing before proceeding.

First, you must be prepared for the regulation and closer scrutiny that comes with having your company's shares traded on a public market. Your management must be comfortable with the AIM's requirements for communication, which seek to ensure that the market is appraised of the company's financial status and business prospects and thereby allow investors to make informed decisions on the value of the company's shares. The AIM requires a listed company to promptly notify the market of any development that could impact the company's share price. Additionally, the AIM requires a listed company's financial statements to conform with internationally recognized accounting standards, such as the U.S. or U.K. Generally Accepted Accounting Principles. You and your company must be ready to comply with these requirements.

Second, you should also be prepared for regular trips to London for meetings with analysts and investors. Frequent trips can lessen the time you have to focus on your business and customers in the United States.

Third, it is important to remember that, although part of the AIM's attraction for U.S. companies lies in the avoidance of strict U.S. regulations such as the Sarbanes–Oxley Act, U.S. companies that conduct an IPO on the AIM are still subject to some U.S. securities laws. The SEC considers U.S. securities rules and regulations to have "extraterritorial reach" in many situations. These extraterritorial regulations include restrictions on certain transfers of shares (depending on whether the shares were issued in the AIM IPO or before) and the possibility that a U.S. company will inadvertently become a "reporting company" under U.S. regulations (requiring registration with the SEC and periodic reporting). These restrictions can increase the cost of the AIM listing and may negatively affect the liquidity of your company's shares on the AIM.

Finally, you should be aware that the AIM has been criticized in the past for being too lax with regard to the requirements it makes of its listed companies. Admittedly, some of this criticism has come from executives of rival exchanges. The ongoing success of the AIM has silenced some of these critics, and the AIM itself has announced that it intends to continue strengthening its monitoring of listing companies. Nevertheless, you should consider whether any residual doubts about the regulatory regimen of the AIM would impact your business or ability to raise money through an AIM listing.

The Importance of Selecting Your Nominated Advisor

AIM's simplified regulatory environment makes listing on the market a little different than for other markets. These differences stem from the fact that AIM is highly accessible; the market requires no minimum company size, no particular business history, and no minimum number of public shares. Instead, a company wanting to list on AIM must work with a nominated advisor (abbreviated to "Nomad" in AIM circles) who shepherds the company through the listing process and stays involved as long as the company is listed on AIM. Thus, the first step in the AIM listing process is to select a Nomad and begin working with that person on your company's application.

A Nomad is a professional advisor selected from a roster of business professionals maintained by AIM, and, in addition to helping your company navigate the listing process, the Nomad will engage in a thorough investigation of your company. Based on the Nomad's due diligence, the Nomad, your company, and your company's other legal and accounting advisors will complete an admissions document that is submitted to the market used to list on AIM. This document forms the company's application and contains all the information needed to provide a complete picture of your company and its value.

The flexibility of AIM's listing process lies in the ability of the Nomad to understand your company's particular circumstances. Because the Nomad makes a thorough and ongoing investigation of your company, AIM is able to do away with the one-size-fits-all mandatory requirements that other markets typically apply. This flexibility can be a boon to companies whose structure or situations are unusual but that nevertheless present strong investment potential. A Nomad will be able to work with your company and will be in a position to understand the value that a small, fast-growing company represents, even if that company would not meet more traditional market-listing criteria.

It should be clear that the company-Nomad relationship is crucial to listing on AIM. AIM has no other regulatory or reporting requirements, so a company that is not working with a Nomad is essentially unregulated, which is an untenable situation for a market. Thus, AIM requires a company to work with its Nomad continuously from the initial, prelisting stages and to maintain that relationship the entire time the company is listed. Your relationship with your Nomad will be as important as your relationships with your attorneys, accountants, and other outside specialists.

Processes for Listing on the AIM

There are two processes for listing on AIM available to international companies. The first is through the standard admissions process. This involves forming a relationship with a Nomad and then working with that person and your company's other advisors to complete an admission document for submission to the market. This admission document forms the main part of your company's application, and, although the contents of the document will vary depending on your company, it will provide a total picture of your company.

A second process for listing on AIM is through the AIM Designated Markets Route. This second method allows companies that have been traded on select international markets (such as the NYSE, NASDAQ, or TSE) for at least 18 months to list on the AIM through an expedited process.

Whichever route your company takes, once your company has listed on AIM, it must keep the market appraised of any developments that might affect the share price, and, as noted above, it must make certain that its financial reports conform to international accounting standards.

24

The Entrepreneurial Circle of Life

He's a serial entrepreneur. Somebody stop him before he makes a killing again.

Carl Zetie
Analyst at Forrester Research

High-tech entrepreneurship is a fast-paced lifestyle. The lifespan of a high-technology startup can oftentimes be as short as a few years, with expectations of a three to five year window after venture capital financing for a successful exit event.

Likewise, the founder of a company might not be the best individual to manage operational aspects of the business after development of the initial product. Research suggests that there is a link between certain events and a founder-CEO being replaced, such as when product development has concluded or on the successful completion of a financing round. Professor Noam Wasserman of Harvard University identified that, once a company had developed its initial product successfully, the CEO's job broadened substantially and the rate of succession of a founder-CEO increased immediately. Likewise, Wasserman suggests that, because some investors make their investments contingent on replacing the founder-CEO, this leads to the increase in succession rates after a successful fundraising round.

So what happens when a founder of a company is forced or chooses to step down from a lead role with the organization they had founded?

Not surprisingly, the results are just as varied as the founders themselves. Here are a few examples of the paths former founders take:

- **Technical role:** Some founders, like the Google founders Larry Page and Sergey Brin, will transition into more product focused roles. In the case of the "Google guys," Page now serves as the President of Products for Google Inc., whereas Brin serves as President of Technology, leaving Eric Schmidt as the CEO of the organization.

- **Serial entrepreneurs:** Other former founders will go on to become serial entrepreneurs, founding additional businesses after their exit from another venture. Netscape founders Jim Clark and Marc Andreesen each went on to found very successful companies (Silicon Graphics, myCFO, and Healtheon in the case of Clark, and Opsware and Ning in the case of Andreesen).
- **Angels and VCs:** Some former founders decide to "pay it forward." Jawed Karim, one of the founders of YouTube, created a venture investment group called Youniverity Ventures aiming to focus on early-stage startups. William "Bing" Gordon left Electronic Arts, the video game company he cofounded in the 1980s, to join Kleiner Perkins Caufield & Byers, a venture capital firm.
- **Advisors and mentors:** Still other former founders end off paying it forward in a different approach, serving as advisors and mentors to various startup companies.

In each of these cases, former founders were able to leverage their experiences as founders and lead executives of their successful startups to transition to new roles.

Entrepreneurs should recognize that the business you grow may not always be yours to operate and be prepared to find the most appropriate way to transition. As evidenced by countless success stories in the small world of entrepreneurship and startups, the entrepreneurial circle of life continues going strong.

Today's exit from the company or your role is tomorrow's new opportunity.

Definitions

Accounting: The process of identifying, recording, and summarizing your business' economic transactions (i.e., revenues and expenses) so the information can be used for decision-making purposes. A list of definitions related to accounting-related activities can be found in Chapter 19: Language of Accounting.

Accredited investor: An individual or entity, including a corporation, LLC, partnership or trust, meeting certain qualifications to be considered appropriately sophisticated to make certain investments. An accredited investor is exempted from certain protection under the securities laws, and companies are exempted from certain obligations with the Securities and Exchange Commission. To be deemed to be an accredited investor under Regulation D of the Securities and Exchange Commission (as of 2008), an investor must have (1) an individual income in excess of $200,000 annually, (2) a household income in excess of $300,000 annually, or (3) a net worth in excess of $1 million. Financial institutions (such as venture capital funds, banks, etc.) and affiliates of the company (such as executive officers and company directors) are also deemed to be accredited investors.

Acquisition: The act of an entity obtaining or acquiring a controlling interest or ownership level of another entity or business. This term is also used in the context of a merger, a takeover, or in a phrase: mergers and acquisitions.

Advisory board: A group of individuals who serve as external advisors to the board of directors or management of a company, providing advice, guidance, information, validation, resources, and assistance. Generally, the advisory board is appointed by the company but has no oversight authority or authority to vote on corporate matters. Boards can meet with some regularity or can provide advice on an informal basis on subjects, including strategy, finances, technical aspects, marketing, or operations. Advisory boards may also be referred to as a scientific advisory board, a technical advisory board, or an advisory board of directors.

Angel investors: An individual, entity, or organization investing directly into private startup companies or new ventures. Angel financing is oftentimes used to provide additional fund necessary for a business to grow from a business funded by the founders to a business more likely to need and attract venture capital. Typically, angel funding will range from $50,000 up to $2 million. The term "angel investor" originated from wealthy individuals who would provide financial backing to Broadway plays and musicals.

Articles/ certificate of incorporation: The document that is filed with the state or state agency to serve as the governing rules of the corporation. The articles, certificate, or charter (as the term may be used) provide the primary rules for the management of the corporation.

Blue Sky: Individual state laws and regulations governing the sales and purchases of securities. The laws and regulations govern sales and purchases of securities in the state. Private entities that sell any of the company's securities (stock, notes, etc.) into any state should find an applicable exemption from registration of the securities. (Remember that securities must be exempted under both federal and state laws.)

Board of directors: The group of individuals elected by the shareholders that oversees the conduct and management of the corporation. The board of directors has a fiduciary duty for the oversight of the company's operations on behalf of the shareholders. Referred to as the board of managers for a limited liability company.

Bootstrapping: A method for founding a company without the use of or with limited use of investment capital from outside investors. Bootstrapping involves use of numerous funding sources, including personal credit cards, factoring receivables, home equity mortgages, use of revenue from the business, or other means. Dell Computers was founded and initially operated using bootstrapping methods.

Bridge financing: Financing that is intended to "bridge" the company until a later fundraising. Generally, a bridge financing will be structured as a convertible debt instrument (note) for a short time period, oftentimes between 6 and 12 months, that converts into equity or debt to be issued at a later point, frequently the preferred stock of the company. The traditional structure of a bridge financing provides that the holder of the note will have the right or the obligation to convert her note into preferred stock, usually at a price that is a discount to the price paid by other investors at that time or will provide warrants that may be exercised for additional stock of the company. The discount or warrant coverage oftentimes ranges from 10% to 40% depending on the time period the note is outstanding and the risk of the underlying investment.

Burn rate: A metric used by startup organizations to calculate the rate at which the company will use up its capital (which may come from a variety of sources, including private investors or debt financing). Also referred to as "negative cash flow."

Buy-sell agreement or provision: An agreement or provision in documents governing issuance of shares that defines the terms in which a shareholder must first offer to sell the shares owed to the other shareholders before she is permitted to sell to parties outside the company.

Capitalization table: A document that identifies the holders of stock in the company and the percentages held by each holder. In some cases, a capitalization table will also include debt. Generally, the capitalization table will list various forms of ownership, including common stock, preferred stock, stock options, warrants, convertible debt, senior debt, and subordinated debt.

Closing: The conclusion of a transaction such as a merger, a sale, an acquisition, or a financing round. The closing occurs when all necessary legal documents are signed and capital has been transferred.

Convertible preferred stock: A subset of preferred stock that provides the owner the right to convert the preferred shares into common shares of stock.

Copyright: Protection for published and unpublished literary, scientific, and artistic works that have been fixed in a tangible or material form. The owner of the copyright in an original work of authorship has the right to prevent others from reproducing, distributing, modifying (creating a derivative work), publicly performing, or publicly displaying the work or one that is substantially similar.

Cosale right: The contractual right of an investor to sell some of the investor's stock along with the founder's or majority shareholder's stock if either the founder or majority shareholder elects to sell stock to a third party. Also referred to as a "tag-along right."

Covenant: A binding agreement or promise to do or refrain from doing certain things.

In the context of a typical financing transaction, the management of the company might agree to refrain from incurring any additional debt, which would be defined as a negative covenant, or to provide monthly financials to investors, a positive covenant. In either case, if the management of the company violated the covenant, the investors may be able to require the company to take certain actions such as correcting the mistake or giving up its control of the company.

Deal flow: The number, type, and quality of companies a venture capital fund or private equity firm receives for evaluation and possible investment.

Demand rights: A form of registration right. Provides the investor with the right to require the company to register its shares with the Securities and Exchange Commission and prepare for a public sale of shares.

Due diligence: The investigation stage of the transaction. The purpose of the due diligence process is to investigate details of the potential transaction, including reviewing disclosures of all material events, transactions, documents, and other information that affects the transaction.

Earnings before interest and taxes (EBIT) and earnings before interest, taxes, depreciation, and amortization (EBITDA): Measurements of the financial picture of the company. EBIT measures operating profit of a company, whereas EBITDA measures cash flows. Often used as a methodology in valuing a company based on a comparison with relative values of other private and/or public companies value as a multiple.

Exit strategy: The plan or strategy to provide for an "exit" for owners of the company's stock, since private companies usually have illiquid markets for trading of their stock. Startup companies tend to identify the following alternatives as primary exit strategies: (1) listing of the company's stock on a stock exchange to allow for the sale of stock held by investors and other owners; or (2) sale or merger of the company to another company or individual.

First refusal right: The contractual right of the company or another investor to purchase shares offered for sale by a holder on the same terms as to be sold to another party or individual.

Founder: Any individual that participates in the formation or creation of the company.

Founders' stock: Initial shares of common stock of the company issued to founders, officers, employees, directors, and consultants. The stock is usually priced at a nominal price such as $0.01 or $0.001 per share.

Fully diluted basis: The method for calculating the per-share ratio that assumed (1) all warrants and options are exercised, and (2) all preferred stock is converted into common stock, when determining the total number of shares issued by the company.

Incubator: An organization or company that provides assistance to entrepreneurs and early-stage businesses such as office space, laboratory equipment, computer equipment, mentoring, coaching, and financial or administrative support.

Indemnification: Financial protection for directors, officers, and employees of a company by the company against expenses and liabilities incurred by them in lawsuits alleging a breach of some duty in their service to or on behalf of the corporation.

Independent contractor: One who practices an independent trade, business, or profession in which services are offered to the public. The person contracting for the services must have the right to control or direct only the result of the work and not the means and methods of accomplishing the result. See Internal Revenue Service Publication 1796.

Institutional investor: Professional investors that invest capital on behalf of companies or individuals. Examples include pension plans, insurance companies, and university endowments.

Joint venture: The creation of a new and separate legal entity by two or more businesses joining together to conduct a specific business activity or enterprise in which the intention is for all parties sharing profits and losses. The primary difference with a strategic alliance is that a separate legal entity is created.

Lead investor: The first, the largest, or the most influential investor in any given investment round.

Letter of intent: The document or documents that confirm the intentions of the parties in a transaction or agreement. In the case of a financing event, the letter of intent is prepared by the investor to define the primary terms of the financing for a company. Also referred to as a "term sheet."

Liquidation preference: The contractual right of an investor to receive an agreed-upon priority for receipt of the proceeds from a liquidity event of the company.

Liquidity event: An event or transaction in which holders of shares of a private company will receive cash or stock of another company. The transaction provides holders to more freely transfer their shares of an otherwise illiquid investment vehicle.

Lockup agreement: A contractual agreement among the investors, management, and employees to refrain from sales of shares for a specific time period after an initial public offering, usually from 6 to 12 months. By limiting or avoiding large sales of its stock by these holders, the company is able to use the lockup period to increase interest among potential buyers of its shares.

Management rights: The rights to consult with management on key operational issues, attend board meetings, and review information about the company's financial situation. These rights are usually acquired by venture capital funds at a financing event.

Noncompete/ Noncompetition: A provision or a separate agreement between the company and employees or consultants that such employees agree not to work for competitor companies or form a new competitor company within a certain time period after terminating their employment.

Nondisclosure agreement: An agreement used to protect the privacy of proprietary information, ideas, or technology when disclosing such information to third parties.

No-shop clause: A provision or clause of an acquisition or sale agreement in which the seller agrees not to market or "shop" the company to other potential buyers during a specific time period.

Nonsolicit/ Nonsolicitation: A provision or a separate agreement between the company and employees or consultants that such employees agree not to solicit other employees of the company regarding job opportunities before or after terminating their employment.

Option pool: The number of shares of stock reserved or set aside for future issuance to employees, consultants, and other parties by a private company.

Pay to play: A provision found in a financing agreement in which any investor that does not choose to participate in a future round agrees to have its shares receive substantial dilution when compared with other investors.

Pari passu: Term that describes the equal treatment of two or more parties in an agreement. In the case of a financing event, the investor may agree to have a liquidation preference (on equal terms) as other or previous investors in a financing round.

Patent: Exclusive rights to the invention in exchange for disclosure of the invention. The patent gives its inventor the right to prevent others from making, using, or selling the patented subject matter described in words in the patent's claims.

Postmoney valuation: The valuation of a company that includes the capital provided by the current round of financing. As an example, if an institutional investor invests $10 million into a company valued at $5 million "premoney" (before the investment was made), the company will have a postmoney valuation of $15 million.

Preemptive rights: The contractual rights of shareholders to maintain their equity ownership percentage by purchasing additional shares sold by the company in the future financing rounds or sales of equity of the company.

Premoney valuation: The valuation of a company before the current round of financing. As an example, if an institutional investor invests $10 million into a company valued at $5 million premoney (before the investment was made), the company will have a "postmoney" valuation of $15 million.

Recapitalization/recap: The reorganization of a company's capital structure to reduce its tax burden or to avoid or emerge from bankruptcy, oftentimes by exchanging capital stock for debt instruments.

Redemption rights: The contractual rights of an investor to require the company to repurchase shares.

Registration rights: The contractual rights of an investor regarding the registration of a portion of the startup's shares for sale to the public. These rights may include (1) piggyback rights to provide shareholders the right to have their shares included in a registration, and (2) demand rights to provide the shareholders the option to force management to register the company's shares for a public offering.

Stock option: A contractual right to purchase or sell a share of stock at a specific price within a specific period of time. Generally, startup companies will use stock purchase options for long-term incentive compensation for employees, management, consultants, and directors in high-tech companies.

Strategic alliance: A relationship established between two companies in which they combine efforts for a specific purpose, oftentimes by contract.

Term sheet: The document or documents that confirm the intentions of the parties in a transaction or agreement. In the case of a financing event, the term sheet is prepared by the investor to define the primary terms of the financing for a company. Also referred to as a "letter of intent."

Trademark: A distinctive name, symbol, motto, or design that legally identifies a company or its products and services. The protection for such words, names, symbols, sounds, or colors that distinguish goods and services limits the ability of other parties from using trademarked items for certain uses.

Valuation: An estimate of the worth of a business entity and its assets.

Vesting schedule: A schedule that sets forth the schedule for the release of repurchase rights of the company to shares or stock options reserved for employees, management, directors, or consultants of the company.

Voting rights: The contractual rights of certain holders of preferred and common stock in a company to vote on certain acts affecting the company. Oftentimes, these matters would include dividends payments, issuance of a new class of stock, merger, or liquidation.

Warrant: A security that provides the holder with the right to purchase shares in a company at a predetermined price. Essentially, a warrant functions much like a long-term option.

Additional Resources

MORE INFORMATION

Below are a list of just a few of links to publications, news sites, blogs, networking organizations, fundraising sites, and more. For updated sites and links, please visit the book's website at http://www.myhightechstartup.com.

Business Magazines

- Business 2.0 (http://money.cnn.com/magazines/business2)
- Entrepreneur (http://www.entrepreneur.com)
- Fast Company (http://www.fastcompany.com)
- Wired (http://www.wired.com)

Tech Magazines and News

- C-NET (http://www.news.com)
- CIO (http://www.cio.com)
- ComputerWorld (http://www.computerworld.com)

- First Monday (http://www.uic.edu/htbin/cgiwrap/bin/ojs/index.php/fm/index)
- MIT TechReview (http://www.techreview.com)
- SlashDot (http://www.slashdot.org)
- ZDNet (http://www.zdnet.com/zdnn)

Helpful Startup Websites

- AllBusiness.com (http://www.allbusiness.com/index.jsp)
- All-Biz Network (http://www.all-biz.com)
- Business Owners' Toolkit (http://www.toolkit.cch.com)
- MoreBusiness.com (http://www.morebusiness.com)
- SBA Resources (http://www.sba.gov/hotlist)
- Service Corps of Retired Executives (SCORE) (http://www.score.org)

Startup-Focused Blogs

- GigaOM (http://www.gigaom.com)
- Mashable (http://www.mashable.com)
- OnStartups.com (http://www.onstartups.com)
- ReadWriteWeb (http://www.readwriteweb.com)
- Startup Nation (http://www.startupnation.com)
- TechCrunch (http://www.techcrunch.com)
- VentureBeat (http://www.venturebeat.com)
- VentureWire (http://www.venturewire.com)

Fundraising Resources

- European Private Equity and Venture Capital Association (http://www.evca.com)
- MoneyTree Report (http://www.pwcmoneytree.com)
- National Venture Capital Association (http://www.nvca.org)
- Private Equity HUB (http://www.pehub.com)
- The Funded (http://www.thefunded.com)
- Venture Economics (http://www.ventureeconomics.com)
- vfinance.com (http://www.vfinance.com)

References and Reading List

Adams, Rob. *A Good Hard Kick in the Ass: Basic Training for Entrepreneurs*. Crown Business. 2002.

Bagley, Constance E., and Dauchy, Craig E. *The Entrepreneur's Guide to Business Law*. Second Edition. West Educational Publishing. 2003.

Berkery, Dermot. *Raising Venture Capital for the Serious Entrepreneur*. McGraw-Hill. 2007.

Bolles, Richard Nelson. *What Color Is Your Parachute? 2008: A Practical Manual for Job-Hunters and Career-Changers*. Ten Speed Press. 2007.

Brandt, Steven C. *Entrepreneuring: The Ten Commandments for Building a Growth Company*. Archipelago Publications. 1997.

Collins, Jim. *Good to Great: Why Some Companies Make the Leap . . . and Others Don't*. HarperCollins. 2001.

Collins, Jim, and Porras, Jerry. *Built to Last: Successful Habits of Visionary Companies*. HarperBusiness. 1997.

Davidow, William H. *Marketing High Technology: An Insider's View*. Macmillian, The Free Press. 1986.

Drucker, Peter F. *Innovation and Entrepreneurship*. Collins. 2006.

Gartner, William B., Shaver, Kelly G., Carter, Nancy M., and Reynolds, Paul D. *Handbook of Entrepreneurial Dynamics: The Process of Business Creation*. Sage Publications. 2004.

Gilbert, Jill. *The Entrepreneur's Guide to Patents, Copyrights, Trademarks, Trade Secrets, & Licensing*. Berkley Books. 2004.

Godin, Seth. *Small Is the New Big: And 183 Other Riffs, Rants, and Remarkable Business Ideas*. Portfolio Hardcover. 2006.

Gosden, Freeman F, Jr. *Direct Marketing Success: What Works and Why*. John Wiley &. Sons. 1989.

Heath, Chip, and Heath, Dan. *Made to Stick: Why Some Ideas Survive and Others Die*. Random House. 2007.

Hess, Kenneth L. *Bootstrap: Lessons Learned Building a Successful Company from Scratch*. S-Curve Press. 2001.

Hill, Brian E., and Power, Dee. *Attracting Capital from Angels: How Their Money—And Their Experience—Can Help You Build a Successful Company*. John Wiley & Sons. 2002.

Kaplan, Jerry. *Startup: A Silicon Valley Adventure*. Penguin. 1996.

Kawasaki, Guy. *The Art of the Start: The Time-Tested, Battle-Hardened Guide for Anyone Starting Anything*. Portfolio Hardcover. 2004.

Lerner, Josh, Hardymon, Felda, and Leamon, Ann. *Venture Capital and Private Equity: A Casebook*. Wiley. 2004.

Livingston, Jessica. *Founders at Work: Stories of Startups' Early Days*. Apress. 2007.

MacVicar, Duncan, and Throne, Darwin. *Managing High-Tech Start-Ups*. Butterworth-Heinemann. 1992.

McQuown, Judith H. *Inc. Yourself: How to Profit by Setting up Your Own Corporation*. Tenth Edition. Career Press. 2002.

Mohr, Jakki. *Marketing of High-Technology Products and Innovations*. Prentice Hall. 2001.

Moore, Geoffrey A. *Crossing the Chasm: Marketing and Selling Disruptive Products to Mainstream Customers*. Collins. 2002.

Nesheim, John. *High Tech Start-Up: The Complete Handbook for Creating Successful New High Tech Companies*. Simon & Schuster. 2000.

Pressman, David. *Patent It Yourself*. Twelfth Edition. NOLO Press. 2006.

Rich, Stanley R., and Gumpert, David E. *Business Plans that Win $$$: Lessons from the MIT Enterprise Forum*. Harper & Row 1987.

Roam, Dan. *The Back of the Napkin: Solving Problems and Selling Ideas with Pictures*. Portfolio Hardcover. 2008.

Roberts, Edward B. *Entrepreneurs in High Technology: Lessons from MIT and Beyond*. Oxford University Press. 1991.

Robinson, Robert J., and Van Osnabrugge, Mark. *Angel Investing: Matching Start-Up Funds with Start-Up Companies*. Jossey-Bass. 2000.

Stathis, Michael. *The Startup Company Bible for Entrepreneurs: The Complete Guide for Building Successful Companies and Raising Venture Capital*. AVA Publishing. 2005.

Swanson, James A. and Baird, Michael L. *Engineering Your Start-Up: A Guide for the High-Tech Entrepreneur.* Second Edition. Professional Publications. 2003.

Viardot, Eric. *Successful Marketing Strategy for High-Tech Firms*. Second Edition. Artech House. 1998.

Wilmerding, Alex. *Term Sheets & Valuations: A Line by Line Look at the Intricacies of Venture Capital Term Sheets & Valuations*. Aspatore Books. 2006.

Index

C

D

E

F

G

M

N

O

P

W